IFCoLog Journal of Logics and their Applications

Volume 4, Number 10

November 2017

Disclaimer

Statements of fact and opinion in the articles in IfCoLog Journal of Logics and their Applications are those of the respective authors and contributors and not of the IfCoLog Journal of Logics and their Applications or of College Publications. Neither College Publications nor the IfCoLog Journal of Logics and their Applications make any representation, express or implied, in respect of the accuracy of the material in this journal and cannot accept any legal responsibility or liability for any errors or omissions that may be made. The reader should make his/her own evaluation as to the appropriateness or otherwise of any experimental technique described.

© Individual authors and College Publications 2017
All rights reserved.

ISBN 978-1-84890-268-8
ISSN (E) 2055-3714
ISSN (P) 2055-3706

College Publications
Scientific Director: Dov Gabbay
Managing Director: Jane Spurr

http://www.collegepublications.co.uk

Printed by Lightning Source, Milton Keynes, UK

All rights reserved. No part of this publication may be reproduced, stored in a retrieval system or transmitted in any form, or by any means, electronic, mechanical, photocopying, recording or otherwise without prior permission, in writing, from the publisher.

Editorial Board

Editors-in-Chief
Dov M. Gabbay and Jörg Siekmann

Marcello D'Agostino
Natasha Alechina
Sandra Alves
Arnon Avron
Jan Broersen
Martin Caminada
Balder ten Cate
Agata Ciabttoni
Robin Cooper
Luis Farinas del Cerro
Esther David
Didier Dubois
PM Dung
Amy Felty
David Fernandez Duque
Jan van Eijck

Melvin Fitting
Michael Gabbay
Murdoch Gabbay
Thomas F. Gordon
Wesley H. Holliday
Sara Kalvala
Shalom Lappin
Beishui Liao
David Makinson
George Metcalfe
Claudia Nalon
Valeria de Paiva
Jeff Paris
David Pearce
Brigitte Pientka
Elaine Pimentel

Henri Prade
David Pym
Ruy de Queiroz
Ram Ramanujam
Chrtian Retoré
Ulrike Sattler
Jörg Siekmann
Jane Spurr
Kaile Su
Leon van der Torre
Yde Venema
Rineke Verbrugge
Heinrich Wansing
Jef Wijsen
John Woods
Michael Wooldridge
Anna Zamansky

Scope and Submissions

This journal considers submission in all areas of pure and applied logic, including:

- pure logical systems
- proof theory
- constructive logic
- categorical logic
- modal and temporal logic
- model theory
- recursion theory
- type theory
- nominal theory
- nonclassical logics
- nonmonotonic logic
- numerical and uncertainty reasoning
- logic and AI
- foundations of logic programming
- belief revision
- systems of knowledge and belief
- logics and semantics of programming
- specification and verification
- agent theory
- databases
- dynamic logic
- quantum logic
- algebraic logic
- logic and cognition
- probabilistic logic
- logic and networks
- neuro-logical systems
- complexity
- argumentation theory
- logic and computation
- logic and language
- logic engineering
- knowledge-based systems
- automated reasoning
- knowledge representation
- logic in hardware and VLSI
- natural language
- concurrent computation
- planning

This journal will also consider papers on the application of logic in other subject areas: philosophy, cognitive science, physics etc. provided they have some formal content.

Submissions should be sent to Jane Spurr (jane.spurr@kcl.ac.uk) as a pdf file, preferably compiled in LaTeX using the IFCoLog class file.

CONTENTS

ARTICLES

Preface 3117
   Matthias Baaz

Cubical Type Theory: A Constructive Interpretation of the Univalence
   Axiom 3127
   Cyril Cohen, Thierry Coquand, Simon Huber and Anders Mörtberg

Themes from Gödel: Some Recent Developments 3171
   Peter Koellner

A Glimpse at Polynomials with Quantifiers 3239
   Andrey Bovykin and Michiel De Smet

On Families of Anticommuting Matrices 3263
   Pavel Hrubeš

Worms and Spiders: Reflection Calculi and Ordinal Notation Systems . . 3279
   David Fernández-Duque

Recent Progress in Proof Mining in Nonlinear Analysis 3361
   Ulrich Kohlenbach

The Clique Covering Problem and Other Questions 3411
   Maryanthe Malliaris

Useful Axioms 3431
   Matteo Viale

The Strength of Abstraction with Predicative Comprehension 3467
   Sean Walsh

Perspectives for Proof Unwinding by Programming Languages Techniques 3487
   Danko Ilik

Regular Languages of Infinite Trees and Probability 3509
   Matteo Mio

Orbit Equivalence Relations 3521
   Marcin Sabok

Reasoning about Coalition Structures in Social Environments via Weighted Propositional Logic 3533
   Gianuigi Greco

Preface

Matthias Baaz, Executive Vice President, Kurt Goedel Society
Institute of Discrete Mathematics and Geometry, Vienna University of Technology, Wiedner Hauptstraße 8–10, 1040 Vienna, Austria
baaz@logic.at

Abstract

This volume contains contributions of the thirteen winners of three rounds of the Kurt Goedel Research Prize Fellowship Programs organized by the Kurt Goedel Society and supported by the John Templeton Foundation.

1 Background

The Horizons of Truth, an International Symposium celebrating the 100th birthday of Kurt Goedel, organized by the Kurt Goedel Society, held in 2006, brought together some of the world's most renowned scientists in the fields related to Kurt Goedel's research, particularly in logic. Outstanding lectures have provided deep insights into historical and contemporary scientific developments. A general interpretation of the significance of Kurt Goedel's achievements was offered by Paul Cohen, Georg Kreisel, John Angus MacIntyre, and Avi Wigderson. Advanced research programs in the spirit of Kurt Goedel were formulated by Harvey Friedman and Hugh Woodin. Hilary Putnam and Denys Turner discussed the philosophical and theological implications of Kurt Goedel's outstanding research. The cosmological works were represented by Wolfgang Rindler. The public lecture of the conference provided a unique opportunity for citizens of Vienna to listen to a lecture by Sir Roger Penrose in the Golden Hall of the Viennese City Hall, and this opportunity was widely taken: more than 800 citizens took part. The lecture was followed by a biographical film about Kurt Goedel.

An integral part of the Symposium was the Young Scholars' Competition, in which ten finalists presented their scientific research results in Vienna. Finally, three prize winners were chosen by the Board of Jurors.

The jury consisted of the following members.

- Wolfgang Achtner, Justus Liebig Universitaet and Wolfgang Goethe Universität Frankfurt

- Lev Beklemishev, Russian Academy of Sciences
- Mirna Dzamonja, University of East Anglia
- Solomon Feferman, Stanford University
- Harvey Friedman, Ohio State University
- Petr Hajek, Academy of Sciences of the Czech Republic
- Michael Heller, Pontifical Academy of Theology
- Juliette Kennedy, University of Helsinki
- Daniele Mundici, University of Florence
- Luke Ong, Oxford University Computing Laboratory
- Michel Parigot, University Paris VII
- Jeff Paris, University of Manchester
- Gordon Plotkin, University of Edinburgh
- Jouko Vaananen, University of Helsinki
- Hugh Woodin, University of California, Berkeley
- Jakob Yngvason, University of Vienna

The finalists were:

- Lorenzo Carlucci
- Andrey Bovykin
- Lutz Strassburger
- Laurentiu Leustean
- Mark Van Atten
- Hannes Leitgeb
- Itay Neeman
- Justin Moore

- Eli Ben-Sasson

- Russell O'Connor

The winners were:

- Justin Moore, 1st prize for "The continuum and aleph-2"

- Mark Van Atten, 2nd prize for "Goedel and German Idealism"

- Eli Ben-Sasson, 3rd prize for "Searching for a conditional answer to Goedel's question"

The conference was completed by an exhibition on life and work of Kurt Goedel: Kurt Goedel's Century, carefully assembled by Karl Sigmund, which helped to increase visibility of the conference. The conference and the exhibition were opened on by Dr. Heinz Fischer, the President of the Republic of Austria. The lecture given by Garry Kasparov, chess grandmaster, gave a wider vision to the problems of analyzing the activities of the human mind.

The Symposium has also shown that, Logic was heading toward an inflection point, which could lead to the diminishing returns from the scientific point of view, unless new foundational ideas emerge.

2 Fellowship Competition Rounds

The following criteria of merit for evaluating Fellowship Prize applications were adopted:

1. Intellectual merit, scientific rigor and originality of the submitted paper and work plan. The project should combine visionary thinking with academic excellence.

2. Potential for significant contribution to basic foundational understanding of logic and the likelihood for opening new, fruitful lines of inquiry.

3. Impact of the grant on the project and likelihood that the prize will make this new line of research possible.

4. The probability that the pursuit of this line of research is realistic and feasible for the applicant.

2.1 Round 1 (2008)

The finalists and winners of this round were chosen by the following Board of Jurors:

- Peter Aczel, University of Manchester (GB)
- Lev Beklemishev, Russian Academy of Sciences (RUS)
- John Burgess, Princeton University (USA)
- Harvey Friedman, Ohio State University (USA) CHAIR
- John Harrison, Intel Corporation (USA)
- Wilfried Hodges, Queen Mary University of London (GB)
- Simon Kochen, Princeton University (USA)
- Jan Krajicek, Academy of Sciences of the Czech Republic(CZ)
- Menachem Magidor, Hebrew University (ISRAEL)
- Dave Marker, University of Illinois at Chicago (USA)
- Michel Parigot, Universite Paris 7 (FRANCE)
- Pavel Pudlak, Academy of Sciences of the Czech Republic (CZ)
- Hilary Putnam, Harvard University (USA)
- Jeff Rremmel, University of California at San Diego (USA)
- John Steel, University of California at Berkley (USA)
- Frank Stephan, National University of Singapore (SINGAPORE)
- Albert Visser, University of Utrecht (NL)

Finalists' articles were published in a Special Volume 157 of the Annals of Pure and Applied Logic, Issues 2-3 in 2009.
The finalists of this round were:

- Jeremy Avigad
- Andrey Bovykin
- Vasco Brattka

- Thierry Coquand
- David Fernandez
- Fernando Ferreira
- Andreas Fischer
- Ekaterina Fokina
- Stefan Geschke
- James Hirschorn
- Pavel Hrubes
- Peter Koellner
- Maryanthe Malliaris
- Yuri Matiyasevich
- Kentaro Sato
- Henry Towsner
- Andreas Weiermann

The winners of this round were:

- Andrey Bovykin, post-doc category, US 160,000 prize for the project "Independence results in concrete mathematics"
- Thierry Coquand, senior category, US 120,000 prize for the project "Space of valuations"
- David Fernandez Duque, pre-doc, US 120,000 prize for the project "Non-Deterministic Semantics for Dynamic Topological Logic"
- Pavel Hrubes, pre-doc category, US 120,000 prize for the project "On lengths of proofs in non-classical logics"
- Peter Koellner, post-doc category, US 160,000 prize for the project "On Reflection Principles"

2.2 Round 2 (2010)

Two different juries chose finalists and winners. Both Juries were chaired by Harvey Friedman, Ohio State University, USA.
Jury for determination of finalists consisted of:

- Lenore Blum, MIT, USA
- John Harrison, Intel Corporation, USA
- Kenneth Kunen, University of Wisconsin
- Angus Macintyre, Queen Mary, University of London and Royal Society, UK
- Hiroakira Ono, JAIST Research Center for Integrated Science, Japan
- Pavel Pudlak, Czech Academy of Sciences, Czech Republic
- Michael Rathjen, University of Leeds, UK
- Frank Stephan, National University of Singapore, Singapore
- William Tait, University of Chicago, USA
- Albert Visser, University of Utrecht, The Netherlands
- Andreas Weiermann, Ghent University, Belgium
- Boris Zilber, University of Oxford

Jury for determination of the winners consisted of:

- Lev Beklemishev, Steklov Mathematical Institute of Russian Academy of Sciences and M.V. Lomonossov Moscow State University, Russian Federation
- Dov M. Gabbay, King's College London, UK
- Warren D. Goldfarb, Harvard University, USA
- Howard Jerome Keisler, University of Wisconsin, USA

Finalists' articles were published in a Special Volume of the Annals of Pure and Applied Logic in 2012, Volume 163, Issue 11.
The finalists of this round were:

- Federico Aschieri

Preface

- Giovanni Curi
- Kentaro Fujimoto
- Misha Gavrilovich
- Danko Ilik
- Ulrich Kohlenbach
- Maryanthe Malliaris
- Andre Nies
- Greg Restall
- Alex Simpson
- Lynn Scow
- Matteo Viale
- Sean Walsh
- Christoph Weiss

The winners of this round were:

- Danko Ilik, pre-doc category, EUR 100,000 prize for the project "Towards a new Computational Interpretation of Sub-classical Principles"

- Ulrich Kohlenbach, unrestricted category, EUR 100,000 prize for the project "New Frontiers in Proof Mining"

- Maryanthe Malliaris, post-doc category, EUR 100,000 prize for the project "Comparing the Complexity of Unstable Theories"

- Matteo Viale, post-doc category, EUR 100,000 prize for the project "Three Aspects of Goedel's Program: Supercompactness, Forcing Axioms and Omega Logic"

- Sean Walsh, pre-doc category, EUR 100,000 prize for the project "The Limits of Arithmetical Definability"

2.3 Round 3 (2014)

The third round of the fellowships prize program was held in conjunction with the Vienna Summer of Logic (VSL-14), the largest conference in the history of logic so far uniting three different streams, mathematical logic, logic in computer science and logic in artificial intelligence. In this round, the finalists were consecutively determined by three different juries.

Logical Foundations of Mathematics: Jan Krajíček, Angus Macintyre, and Dana Scott (Chair).

Logical Foundations of Computer Science: Franz Baader, Johann Makowsky, and Wolfgang Thomas (Chair).

Logical Foundations of Artificial Intelligence: Luigia Carlucci Aiello, Georg Gottlob (Chair), and Bernhard Nebel.

The winners were chosen by all three juries together.
The finalists of this round were:
In Logical Foundations of Mathematics category:

- Yong Cheng
- Arno Pauly
- Marcin Sabok
- Sam Sanders

In Logical Foundations of Computer Science category:

- Sicun Gao
- Cameron Hill
- Ori Lahav
- Matteo Mio

In Logical Foundations of Artificial Intelligence category:

- Alessandro Abate
- Vaishak Belle
- Gianluigi Greco

- Sebastian Rudolph

The winners of this round were:

- Gianluigi Greco, logical foundations of artificial intelligence, EUR 100,000 for the project "Collective Behavior in Social Environments: Models and Complexity"

- Matteo Mio, logical foundations of computer science, EUR 100,000 prize for the project "Quantitative Modal Logics"

- Marcin Sabok, logical foundations of mathematics category, EUR 100,000 prize for the project "Classification: in search of groups"

2.4 Summary

Mathematical logic is logic with the rigor of mathematical argumentation. Its conclusions are far-reaching and clear and it is in this field, where the scientific progress of logic as a whole takes place. Mathematical logic is based upon the balance between mathematical knowledge and philosophical incentive. Without philosophical understanding (informal rigor) Gödel would not have been able to find the arguments to prove his famous theorems. Without mathematical knowledge, he would not have been able to communicate them. (Note that a paper on incompleteness had been published by Finsler even before Goedel, but Finsler's arguments were completely insufficient, mathematically).The balance is, however, disturbed nowadays due to the industrialization of scientific work, and industrialization means Taylorism: talented young logicians will focus on small areas of the field, where they will obtain technical superiority after a short time.

The aim of the three fellowships rounds was to restore the balance in two ways:

1. By providing funding and acknowledgement to the individual young scientist and to give him time to breath. It is a paradox of our scientific culture that those, who could be most innovative, namely the young scientists, are forced to be most adaptive, while for the established scientists there exist all degrees of freedom.

2. By propagating the importance of the balance to all scientists in the field of logic.

Cubical Type Theory: A Constructive Interpretation of the Univalence Axiom

Cyril Cohen
Université Côte d'Azur, Inria, Sophia Antipolis, France
cyril.cohen@inria.fr

Thierry Coquand, Simon Huber
Department of Computer Science and Engineering, University of Gothenburg, Gothenburg, Sweden
{thierry.coquand,simon.huber}@cse.gu.se

Anders Mörtberg[*]
School of Mathematics, Institute for Advanced Study, Princeton, NJ, USA
amortberg@math.ias.edu

Abstract

This paper presents a type theory in which it is possible to directly manipulate n-dimensional cubes (points, lines, squares, cubes, etc.) based on an interpretation of dependent type theory in a cubical set model. This enables new ways to reason about identity types, for instance, function extensionality is directly provable in the system. Further, Voevodsky's univalence axiom is provable in this system. We also explain an extension with some higher inductive types like the circle and propositional truncation. Finally we provide semantics for this cubical type theory in a constructive meta-theory.

This article has appeared as [9] licensed under CC BY (http://creativecommons.org/licenses/by/3.0/). The present article improves the treatment of higher inductive types.

[*]This material is based upon work supported by the National Science Foundation under agreement No. DMS-1128155. Any opinions, findings and conclusions or recommendations expressed in this material are those of the author(s) and do not necessarily reflect the views of the National Science Foundation.

1 Introduction

This work is a continuation of the program started in [6, 14] to provide a constructive justification of Voevodsky's univalence axiom [28]. This axiom allows many improvements for the formalization of mathematics in type theory: function extensionality, identification of isomorphic structures, etc. In order to preserve the good computational properties of type theory it is crucial that postulated constants have a computational interpretation. Like in [6, 14, 23] our work is based on a nominal extension of λ-calculus, using *names* to represent formally elements of the unit interval $[0, 1]$. This paper presents two main contributions.

The first one is a refinement of the semantics presented in [6, 14]. We add new operations on names corresponding to the fact that the interval $[0, 1]$ is canonically a de Morgan algebra [3]. This allows us to significantly simplify our semantical justifications. In the previous work, we noticed that it is crucial for the semantics of higher inductive types [27] to have a "diagonal" operation. By adding this operation we can provide a semantical justification of some higher inductive types and we give two examples (the spheres and propositional truncation). Another shortcoming of the previous work was that using path types as equality types did not provide a justification of the computation rule of the Martin-Löf identity type [20] as a judgmental equality. This problem has been solved by Andrew Swan [26], in the framework of [6, 14, 23], who showed that we can define a new type, *equivalent to*, but not judgmentally equal to the path type. This has a simple definition in the present framework.

The second contribution is the design of a type system[1] inspired by this semantics which extends Martin-Löf type theory [21, 20]. We add two new operations on contexts: addition of new names representing dimensions and a restriction operation. Using these we can define a notion of extensibility which generalizes the notion of being connected by a path, and then a Kan composition operation that expresses that being extensible is preserved along paths. We also define a new operation on types which expresses that this notion of extensibility is preserved by equivalences. The axiom of univalence, and composition for the universe, are then both expressible using this new operation.

The paper is organized as follows. The first part, Sections 2 to 7, presents the type system. The second part, Section 8, provides its semantics in cubical sets. Finally, in Section 9, we present two possible extensions: the addition of an identity type, and two examples of higher inductive types.

---

[1] We have implemented a type-checker for this system in HASKELL, which is available at:
https://github.com/mortberg/cubicaltt

2 Basic type theory

In this section we introduce the version of dependent type theory on which the rest of the paper is based. This presentation is standard, but included for completeness. The type theory that we consider has a type of natural numbers, but no universes (we consider the addition of universes in Section 7). It also has β and η-conversion for dependent functions and surjective pairing for dependent pairs.

The syntax of contexts, terms and types is specified by:

$$\Gamma, \Delta ::= () \mid \Gamma, x : A \qquad \text{Contexts}$$

$$\begin{aligned} t, u, A, B ::= & \; x \mid \lambda x : A. \, t \mid t \, u \mid (x : A) \to B & \Pi\text{-types} \\ & \mid (t, u) \mid t.1 \mid t.2 \mid (x : A) \times B & \Sigma\text{-types} \\ & \mid 0 \mid \mathsf{s} \, u \mid \mathsf{natrec} \, t \, u \mid \mathsf{N} & \text{Natural numbers} \end{aligned}$$

We write $A \to B$ for the non-dependent function space and $A \times B$ for the type of non-dependent pairs. Terms and types are considered up to α-equivalence of bound variables. Substitutions, written $\sigma = (x_1/u_1, \ldots, x_n/u_n)$, are defined to act on expressions as usual, i.e., simultaneously replacing x_i by u_i, renaming bound variables whenever necessary. The inference rules of this system are presented in Figure 1 where in the η-rule for Π- and Σ-types we omitted the premises that t and u should have the respective type.

We define $\Delta \vdash \sigma : \Gamma$ by induction on Γ. We have $\Delta \vdash () : ()$ (empty substitution) and $\Delta \vdash (\sigma, x/u) : \Gamma, x : A$ if $\Delta \vdash \sigma : \Gamma$ and $\Delta \vdash u : A\sigma$.

We write J for an arbitrary judgment and, as usual, we consider also *hypothetical* judgments $\Gamma \vdash \mathsf{J}$ in a *context* Γ.

The following lemma will be valid for all extensions of type theory we consider below.

Lemma 1. *Substitution is admissible:*

$$\frac{\Gamma \vdash \mathsf{J} \qquad \Delta \vdash \sigma : \Gamma}{\Delta \vdash \mathsf{J}\sigma}$$

In particular, weakening is admissible, i.e., a judgment valid in a context stays valid in any extension of this context.

3 Path types

As in [6, 23] we assume that we are given a discrete infinite set of names (representing directions) $i, j, k, \ldots$ We define $\mathbb{I}$ to be the free de Morgan algebra [3] on this set of

Well-formed contexts, $\Gamma \vdash$ (The condition $x \notin \mathrm{dom}(\Gamma)$ means that x is not declared in Γ)

$$\frac{}{() \vdash} \qquad \frac{\Gamma \vdash A}{\Gamma, x : A \vdash} \quad (x \notin \mathrm{dom}(\Gamma))$$

Well-formed types, $\Gamma \vdash A$

$$\frac{\Gamma, x : A \vdash B}{\Gamma \vdash (x : A) \to B} \qquad \frac{\Gamma, x : A \vdash B}{\Gamma \vdash (x : A) \times B} \qquad \frac{\Gamma \vdash}{\Gamma \vdash \mathsf{N}}$$

Well-typed terms, $\Gamma \vdash t : A$

$$\frac{\Gamma \vdash t : A \quad \Gamma \vdash A = B}{\Gamma \vdash t : B} \qquad \frac{\Gamma, x : A \vdash t : B}{\Gamma \vdash \lambda x : A.\, t : (x : A) \to B} \qquad \frac{\Gamma \vdash}{\Gamma \vdash x : A} \quad (x : A \in \Gamma)$$

$$\frac{\Gamma \vdash t : (x : A) \to B \quad \Gamma \vdash u : A}{\Gamma \vdash t\, u : B(x/u)} \qquad \frac{\Gamma \vdash t : (x : A) \times B}{\Gamma \vdash t.1 : A} \qquad \frac{\Gamma \vdash t : (x : A) \times B}{\Gamma \vdash t.2 : B(x/t.1)}$$

$$\frac{\Gamma, x : A \vdash B \quad \Gamma \vdash t : A \quad \Gamma \vdash u : B(x/t)}{\Gamma \vdash (t, u) : (x : A) \times B} \qquad \frac{\Gamma \vdash}{\Gamma \vdash 0 : \mathsf{N}} \qquad \frac{\Gamma \vdash n : \mathsf{N}}{\Gamma \vdash \mathsf{s}\, n : \mathsf{N}}$$

$$\frac{\Gamma, x : \mathsf{N} \vdash P \quad \Gamma \vdash a : P(x/0) \quad \Gamma \vdash b : (n : \mathsf{N}) \to P(x/n) \to P(x/\mathsf{s}\, n)}{\Gamma \vdash \mathsf{natrec}\, a\, b : (x : \mathsf{N}) \to P}$$

Type equality, $\Gamma \vdash A = B$ (Congruence and equivalence rules which are omitted)

Term equality, $\Gamma \vdash a = b : A$ (Congruence and equivalence rules are omitted)

$$\frac{\Gamma \vdash t = u : A \quad \Gamma \vdash A = B}{\Gamma \vdash t = u : B} \qquad \frac{\Gamma, x : A \vdash t : B \quad \Gamma \vdash u : A}{\Gamma \vdash (\lambda x : A.\, t)\, u = t(x/u) : B(x/u)} \qquad \frac{\Gamma, x : A \vdash t\, x = u\, x : B}{\Gamma \vdash t = u : (x : A) \to B}$$

$$\frac{\Gamma, x : A \vdash B \quad \Gamma \vdash t : A \quad \Gamma \vdash u : B(x/t)}{\Gamma \vdash (t, u).1 = t : A} \qquad \frac{\Gamma, x : A \vdash B \quad \Gamma \vdash t : A \quad \Gamma \vdash u : B(x/t)}{\Gamma \vdash (t, u).2 = u : B(x/t)}$$

$$\frac{\Gamma, x : A \vdash B \quad \Gamma \vdash t.1 = u.1 : A \quad \Gamma \vdash t.2 = u.2 : B(x/t.1)}{\Gamma \vdash t = u : (x : A) \times B}$$

$$\frac{\Gamma, x : \mathsf{N} \vdash P \quad \Gamma \vdash a : P(x/0) \quad \Gamma \vdash b : (n : \mathsf{N}) \to P(x/n) \to P(x/\mathsf{s}\, n)}{\Gamma \vdash \mathsf{natrec}\, a\, b\, 0 = a : P(x/0)}$$

$$\frac{\Gamma, x : \mathsf{N} \vdash P \quad \Gamma \vdash a : P(x/0) \quad \Gamma \vdash b : (n : \mathsf{N}) \to P(x/n) \to P(x/\mathsf{s}\, n) \quad \Gamma \vdash n : \mathsf{N}}{\Gamma \vdash \mathsf{natrec}\, a\, b\, (\mathsf{s}\, n) = b\, n\, (\mathsf{natrec}\, a\, b\, n) : P(x/\mathsf{s}\, n)}$$

Figure 1: Inference rules of the basic type theory

names. This means that $\mathbb{I}$ is a bounded distributive lattice with top element 1 and bottom element 0 with an involution $1-r$ satisfying:

$$1-0=1 \quad 1-1=0 \quad 1-(r\vee s)=(1-r)\wedge(1-s) \quad 1-(r\wedge s)=(1-r)\vee(1-s)$$

The elements of $\mathbb{I}$ can hence be described by the following grammar:

$$r,s \ ::= \ 0 \ | \ 1 \ | \ i \ | \ 1-r \ | \ r\wedge s \ | \ r\vee s$$

The set $\mathbb{I}$ also has decidable equality, and as a distributive lattice, it can be described as the free distributive lattice generated by symbols i and $1-i$ [3]. As in [6], the elements in $\mathbb{I}$ can be thought as formal representations of elements in $[0,1]$, with $r\wedge s$ representing $min(r,s)$ and $r\vee s$ representing $max(r,s)$. With this in mind it is clear that $(1-r)\wedge r \neq 0$ and $(1-r)\vee r \neq 1$ in general.

Remark. The use of the reverse operation $1-r$ is not essential. Instead of requiring composition operations (see Section 4.3) extending from 0 to 1 we would then also have to require one from 1 to 0.

Remark. We could instead also use a so-called Kleene algebra [16], i.e., a de Morgan algebra satisfying in addition $r\wedge(1-r) \leqslant s\vee(1-s)$. The free Kleene algebra on the set of names can be described as above but by additionally imposing the equations $i\wedge(1-i) \leqslant j\vee(1-j)$ on the generators; this still has a decidable equality. Note that $[0,1]$ with the operations described above is a Kleene algebra. With this added condition, $r=s$ if, and only if, their interpretations in $[0,1]$ are equal. A consequence of using a Kleene algebra instead would be that more terms would be judgmentally equal in the type theory.

3.1 Syntax and inference rules

Contexts can now be extended with name declarations:

$$\Gamma,\Delta \ ::= \ \ldots \ | \ \Gamma, i : \mathbb{I}$$

together with the context rule:

$$\frac{\Gamma \vdash}{\Gamma, i : \mathbb{I} \vdash} \ (i \notin \mathrm{dom}(\Gamma))$$

A judgment of the form $\Gamma \vdash r : \mathbb{I}$ means that $\Gamma \vdash$ and r in $\mathbb{I}$ depends only on the names declared in Γ. The judgment $\Gamma \vdash r = s : \mathbb{I}$ means that r and s are equal as elements of $\mathbb{I}$, $\Gamma \vdash r : \mathbb{I}$, and $\Gamma \vdash s : \mathbb{I}$. Note, that judgmental equality for $\mathbb{I}$ will be re-defined once we introduce restricted contexts in Section 4.

The extension to the syntax of basic dependent type theory is:

$$t, u, A, B ::= \ldots \mid \mathsf{Path}\ A\ t\ u \mid \langle i \rangle\ t \mid t\ r \qquad \text{Path types}$$

Path abstraction, $\langle i \rangle\ t$, binds the name i in t, and path application, $t\ r$, applies a term t to an element $r : \mathbb{I}$. This is similar to the notion of name-abstraction in nominal sets [22].

The substitution operation now has to be extended to substitutions of the form (i/r). There are special substitutions of the form $(i/0)$ and $(i/1)$ corresponding to taking faces of an n-dimensional cube, we write these simply as $(i0)$ and $(i1)$.

The inference rules for path types are presented in Figure 2 where again in the η-rule we omitted that t and u should be appropriately typed.

$$\frac{\Gamma \vdash A \quad \Gamma \vdash t : A \quad \Gamma \vdash u : A}{\Gamma \vdash \mathsf{Path}\ A\ t\ u} \qquad \frac{\Gamma \vdash A \quad \Gamma, i : \mathbb{I} \vdash t : A}{\Gamma \vdash \langle i \rangle\ t : \mathsf{Path}\ A\ t(i0)\ t(i1)}$$

$$\frac{\Gamma \vdash t : \mathsf{Path}\ A\ u_0\ u_1 \quad \Gamma \vdash r : \mathbb{I}}{\Gamma \vdash t\ r : A} \qquad \frac{\Gamma \vdash A \quad \Gamma, i : \mathbb{I} \vdash t : A \quad \Gamma \vdash r : \mathbb{I}}{\Gamma \vdash (\langle i \rangle\ t)\ r = t(i/r) : A}$$

$$\frac{\Gamma, i : \mathbb{I} \vdash t\ i = u\ i : A}{\Gamma \vdash t = u : \mathsf{Path}\ A\ u_0\ u_1} \qquad \frac{\Gamma \vdash t : \mathsf{Path}\ A\ u_0\ u_1}{\Gamma \vdash t\ 0 = u_0 : A} \qquad \frac{\Gamma \vdash t : \mathsf{Path}\ A\ u_0\ u_1}{\Gamma \vdash t\ 1 = u_1 : A}$$

Figure 2: Inference rules for path types

We define $1_a : \mathsf{Path}\ A\ a\ a$ as $1_a = \langle i \rangle\ a$, which corresponds to a proof of reflexivity.

The intuition is that a type in a context with n names corresponds to an n-dimensional cube:

$(\,)\vdash A$	$\bullet\ A$
$i : \mathbb{I} \vdash A$	$A(i0) \xrightarrow{A} A(i1)$
$i : \mathbb{I}, j : \mathbb{I} \vdash A$	$\begin{array}{c} A(i0)(j1) \xrightarrow{A(j1)} A(i1)(j1) \\ A(i0) \uparrow \qquad\qquad \uparrow A(i1) \\ A(i0)(j0) \xrightarrow{A(j0)} A(i1)(j0) \end{array}$
$\vdots$	$\vdots$

Note that $A(i0)(j0) = A(j0)(i0)$. The substitution (i/j) corresponds to renaming a dimension, while $(i/1-i)$ corresponds to the inversion of a path. If we have $i : \mathbb{I} \vdash p$ with $p(i0) = a$ and $p(i1) = b$ then it can be seen as a line

$$a \xrightarrow{p} b$$

in direction i, then:

$$b \xrightarrow{p(i/1-i)} a$$

The substitutions $(i/i \wedge j)$ and $(i/i \vee j)$ correspond to special kinds of degeneracies called *connections* [7]. The connections $p(i/i \wedge j)$ and $p(i/i \vee j)$ can be drawn as the squares:

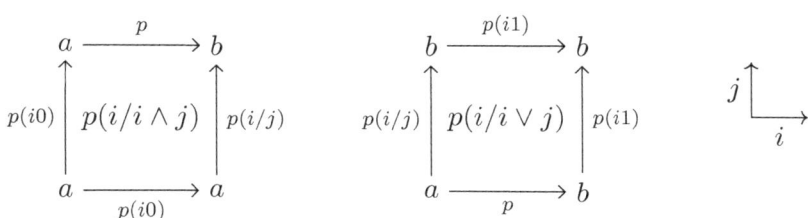

where, for instance, the right-hand side of the left square is computed as

$$p(i/i \wedge j)(i1) \quad = \quad p(i/1 \wedge j) \quad = \quad p(i/j)$$

and the bottom and left-hand sides are degenerate.

3.2 Examples

Representing equalities using path types allows novel definitions of many standard operations on identity types that are usually proved by identity elimination. For instance, the fact that the images of two equal elements are equal can be defined as:

$$\frac{\Gamma \vdash a : A \quad \Gamma \vdash b : A \quad \Gamma \vdash f : A \to B \quad \Gamma \vdash p : \mathsf{Path}\ A\ a\ b}{\Gamma \vdash \langle i \rangle\ f\ (p\ i) : \mathsf{Path}\ B\ (f\ a)\ (f\ b)}$$

This operation satisfies some judgmental equalities that do not hold judgmentally when the identity type is defined as an inductive family (see Section 7.2 of [6] for details).

We can also define new operations, for instance, function extensionality for path types can be proved as:

$$\frac{\Gamma \vdash g : (x : A) \to B \qquad \Gamma \vdash p : (x : A) \to \mathsf{Path}\ B\ (f\ x)\ (g\ x)}{\Gamma \vdash \langle i \rangle\ \lambda x : A.\ p\ x\ i : \mathsf{Path}\ ((x : A) \to B)\ f\ g}$$

with $\Gamma \vdash f : (x : A) \to B$.

To see that this is correct we check that the term has the correct faces, for instance:

$$(\langle i \rangle\ \lambda x : A.\ p\ x\ i)\ 0\ =\ \lambda x : A.\ p\ x\ 0\ =\ \lambda x : A.\ f\ x\ =\ f$$

We can also justify the fact that singletons are contractable, that is, that any element in $(x : A) \times (\mathsf{Path}\ A\ a\ x)$ is equal to $(a, 1_a)$:

$$\frac{\Gamma \vdash p : \mathsf{Path}\ A\ a\ b}{\Gamma \vdash \langle i \rangle\ (p\ i, \langle j \rangle\ p\ (i \wedge j)) : \mathsf{Path}\ ((x : A) \times (\mathsf{Path}\ A\ a\ x))\ (a, 1_a)\ (b, p)}$$

As in the previous work [6, 14] we need to add *composition operations*, defined by induction on the type, in order to justify the elimination principle for paths.

4 Systems, composition, and transport

In this section we define the operation of context *restriction* which will allow us to describe new geometrical shapes corresponding to "sub-polyhedra" of a cube. Using this we can define the composition operation. From this operation we will also be able to define the transport operation and the elimination principle for Path types.

4.1 The face lattice

The *face lattice*, $\mathbb{F}$, is the distributive lattice generated by symbols $(i = 0)$ and $(i = 1)$ with the relation $(i = 0) \wedge (i = 1) = 0_\mathbb{F}$. The elements of the face lattice, called *face formulas*, can be described by the grammar

$$\varphi, \psi\ ::=\ 0_\mathbb{F}\ |\ 1_\mathbb{F}\ |\ (i = 0)\ |\ (i = 1)\ |\ \varphi \wedge \psi\ |\ \varphi \vee \psi$$

There is a canonical lattice map $\mathbb{I} \to \mathbb{F}$ sending i to $(i = 1)$ and $1 - i$ to $(i = 0)$. We write $(r = 1)$ for the image of $r : \mathbb{I}$ in $\mathbb{F}$ and we write $(r = 0)$ for $(1 - r = 1)$. We have $(r = 1) \wedge (r = 0) = 0_\mathbb{F}$ and we define the lattice map $\mathbb{F} \to \mathbb{F}$, $\psi \mapsto \psi(i/r)$ sending $(i = 1)$ to $(r = 1)$ and $(i = 0)$ to $(r = 0)$.

Any element of $\mathbb{F}$ is the join of the irreducible elements below it. An irreducible element of this lattice is a *face*, i.e., a conjunction of elements of the form $(i = 0)$ and

($j = 1$). This provides a disjunctive normal form for face formulas, and it follows from this that the equality on $\mathbb{F}$ is decidable.

Geometrically, the face formulas describe "sub-polyhedra" of a cube. For instance, the element $(i = 0) \vee (j = 1)$ can be seen as the union of two faces of the square in directions j and i. If I is a finite set of names, we define the *boundary* of I as the element ∂_I of $\mathbb{F}$ which is the disjunction of all $(i = 0) \vee (i = 1)$ for i in I. It is the greatest element depending at most on elements in I which is $< 1_{\mathbb{F}}$.

We write $\Gamma \vdash \psi : \mathbb{F}$ to mean that ψ is a face formula using only the names declared in Γ. We introduce then the new *restriction* operation on contexts:

$$\Gamma, \Delta ::= \ldots \mid \Gamma, \varphi$$

together with the rule:

$$\frac{\Gamma \vdash \varphi : \mathbb{F}}{\Gamma, \varphi \vdash}$$

This allows us to describe new geometrical shapes: as we have seen above, a type in a context $\Gamma = i : \mathbb{I}, j : \mathbb{I}$ can be thought of as a square, and a type in the restricted context Γ, φ will then represent a compatible union of faces of this square. This can be illustrated by:

$i : \mathbb{I}, (i=0) \vee (i=1) \vdash A$	$A(i0) \bullet \quad A(i1) \bullet$
$i : \mathbb{I}, j : \mathbb{I}, (i=0) \vee (j=1) \vdash A$	$A(i0)(j1) \xrightarrow{A(j1)} A(i1)(j1)$ $A(i0) \uparrow$ $A(i0)(j0)$
$i : \mathbb{I}, j : \mathbb{I}, (i=0) \vee (i=1) \vee (j=0) \vdash A$	$A(i0)(j1) \quad\quad A(i1)(j1)$ $A(i0) \uparrow \quad\quad\quad \uparrow A(i1)$ $A(i0)(j0) \xrightarrow{A(j0)} A(i1)(j0)$

There is a canonical map from the lattice $\mathbb{F}$ to the congruence lattice of $\mathbb{I}$, which is distributive [3], sending $(i = 1)$ to the congruence identifying i with 1 (and $1 - i$ with 0) and sending $(i = 0)$ to the congruence identifying i with 0 (and $1 - i$ with 1). In this way, any element ψ of $\mathbb{F}$ defines a congruence $r = s \pmod{\psi}$ on $\mathbb{I}$.

This congruence can be described as a substitution if ψ is irreducible; for instance, if ψ is $(i = 0) \wedge (j = 1)$ then $r = s$ (mod. ψ) is equivalent to $r(i0)(j1) = s(i0)(j1)$. The congruence associated to $\psi = \varphi_0 \vee \varphi_1$ is the meet of the congruences associated to φ_0 and φ_1 respectively, so that we have, e.g., $i = 1 - j$ (mod. ψ) if $\varphi_0 = (i = 0) \wedge (j = 1)$ and $\varphi_1 = (i = 1) \wedge (j = 0)$.

To any context Γ we can associate recursively a congruence on $\mathbb{I}$, the congruence on Γ, ψ being the join of the congruence defined by Γ and the congruence defined by ψ. The congruence defined by () is equality in $\mathbb{I}$, and an extension $x : A$ or $i : \mathbb{I}$ does not change the congruence. The judgment $\Gamma \vdash r = s : \mathbb{I}$ then means that $r = s$ (mod. Γ), $\Gamma \vdash r : \mathbb{I}$, and $\Gamma \vdash s : \mathbb{I}$.

In the case where Γ does not use the restriction operation, this judgment means $r = s$ in $\mathbb{I}$. If i is declared in Γ, then $\Gamma, (i = 0) \vdash r = s : \mathbb{I}$ is equivalent to $\Gamma \vdash r(i0) = s(i0) : \mathbb{I}$. Similarly any context Γ defines a congruence on $\mathbb{F}$ with $\Gamma, \psi \vdash \varphi_0 = \varphi_1 : \mathbb{F}$ being equivalent to $\Gamma \vdash \psi \wedge \varphi_0 = \psi \wedge \varphi_1 : \mathbb{F}$.

As explained above, the elements of $\mathbb{I}$ can be seen as formal representations of elements in the interval $[0, 1]$. The elements of $\mathbb{F}$ can then be seen as formulas on elements of $[0, 1]$. We have a simple form of *quantifier elimination* on $\mathbb{F}$: given a name i, we define $\forall i : \mathbb{F} \to \mathbb{F}$ as the lattice morphism sending $(i = 0)$ and $(i = 1)$ to $0_\mathbb{F}$, and being the identity on all the other generators. If ψ is independent of i, we have $\psi \leqslant \varphi$ if, and only if, $\psi \leqslant \forall i.\varphi$. For example, if φ is $(i = 0) \vee ((i = 1) \wedge (j = 0)) \vee (j = 1)$, then $\forall i.\varphi$ is $(j = 1)$. This operation will play a crucial role in Section 6.2 for the definition of composition of glueing.

Since $\mathbb{F}$ is not a Boolean algebra, we don't have in general $\varphi = (\varphi \wedge (i = 0)) \vee (\varphi \wedge (i = 1))$, but we always have the following decomposition:

Lemma 2. *For any element φ of $\mathbb{F}$ and any name i we have*

$$\varphi = (\forall i.\varphi) \vee (\varphi \wedge (i = 0)) \vee (\varphi \wedge (i = 1))$$

We also have $\varphi \wedge (i = 0) \leqslant \varphi(i0)$ and $\varphi \wedge (i = 1) \leqslant \varphi(i1)$.

4.2 Syntax and inference rules for systems

Systems allow to introduce "sub-polyhedra" as compatible unions of cubes. The extension to the syntax of dependent type theory with path types is:

$$
\begin{array}{lll}
t, u, A, B & ::= & \ldots \\
& | & [\,\varphi_1\ t_1, \ldots, \varphi_n\ t_n\,] \qquad \text{Systems}
\end{array}
$$

We allow $n = 0$ and get the empty system [ ]. As explained above, a context now corresponds in general to the union of sub-faces of a cube. In Figure 3 we provide operations for combining compatible systems of types and elements, the side condition for these rules is that $\Gamma \vdash \varphi_1 \vee \cdots \vee \varphi_n = 1_\mathbb{F} : \mathbb{F}$. This condition requires Γ to be sufficiently restricted: for example $\Delta, (i = 0) \vee (i = 1) \vdash (i = 0) \vee (i = 1) = 1_\mathbb{F}$. The first rule introduces systems of types, each defined on one φ_l and requiring the types to agree whenever they overlap; the second rule is the analogous rule for terms. The last two rules make sure that systems have the correct faces. The third inference rule says that that any judgment which is valid locally at each φ_l is valid; note that in particular $n = 0$ is allowed (then the side condition becomes $\Gamma \vdash 0_\mathbb{F} = 1_\mathbb{F} : \mathbb{F}$).

$$\frac{\Gamma, \varphi_1 \vdash A_1 \quad \cdots \quad \Gamma, \varphi_n \vdash A_n \quad \Gamma, \varphi_i \wedge \varphi_j \vdash A_i = A_j \quad (1 \leqslant i, j \leqslant n)}{\Gamma \vdash [\,\varphi_1\ A_1, \ldots, \varphi_n\ A_n\,]}$$

$$\frac{\Gamma \vdash A \quad \Gamma, \varphi_1 \vdash t_1 : A \quad \cdots \quad \Gamma, \varphi_n \vdash t_n : A \quad \Gamma, \varphi_i \wedge \varphi_j \vdash t_i = t_j : A \quad (1 \leqslant i, j \leqslant n)}{\Gamma \vdash [\,\varphi_1\ t_1, \ldots, \varphi_n\ t_n\,] : A}$$

$$\frac{\Gamma, \varphi_1 \vdash \mathsf{J} \quad \cdots \quad \Gamma, \varphi_n \vdash \mathsf{J}}{\Gamma \vdash \mathsf{J}} \qquad \frac{\Gamma \vdash [\,\varphi_1\ A_1, \ldots, \varphi_n\ A_n\,] \quad \Gamma \vdash \varphi_i = 1_\mathbb{F} : \mathbb{F}}{\Gamma \vdash [\,\varphi_1\ A_1, \ldots, \varphi_n\ A_n\,] = A_i}$$

$$\frac{\Gamma \vdash [\,\varphi_1\ t_1, \ldots, \varphi_n\ t_n\,] : A \quad \Gamma \vdash \varphi_i = 1_\mathbb{F} : \mathbb{F}}{\Gamma \vdash [\,\varphi_1\ t_1, \ldots, \varphi_n\ t_n\,] = t_i : A}$$

Figure 3: Inference rules for systems with side condition $\Gamma \vdash \varphi_1 \vee \cdots \vee \varphi_n = 1_\mathbb{F} : \mathbb{F}$

Note that when $n = 0$ the second of the above rules should be read as: if $\Gamma \vdash 0_\mathbb{F} = 1_\mathbb{F} : \mathbb{F}$ and $\Gamma \vdash A$, then $\Gamma \vdash [\,] : A$.

We extend the definition of the substitution judgment by $\Delta \vdash \sigma : \Gamma, \varphi$ if $\Delta \vdash \sigma : \Gamma$, $\Gamma \vdash \varphi : \mathbb{F}$, and $\Delta \vdash \varphi\sigma = 1_\mathbb{F} : \mathbb{F}$.

If $\Gamma, \varphi \vdash u : A$, then $\Gamma \vdash a : A[\varphi \mapsto u]$ is an abbreviation for $\Gamma \vdash a : A$ and $\Gamma, \varphi \vdash a = u : A$. In this case, we see this element a as a witness that the partial element u, defined on the "extent" φ (using the terminology from [11]), is *extensible*. More generally, we write $\Gamma \vdash a : A[\varphi_1 \mapsto u_1, \ldots, \varphi_k \mapsto u_k]$ for $\Gamma \vdash a : A$ and $\Gamma, \varphi_l \vdash a = u_l : A$ for $l = 1, \ldots, k$.

For instance, if $\Gamma, i : \mathbb{I} \vdash A$ and $\Gamma, i : \mathbb{I}, \varphi \vdash u : A$ where $\varphi = (i = 0) \vee (i = 1)$ then the element u is determined by two elements $\Gamma \vdash a_0 : A(i0)$ and $\Gamma \vdash a_1 : A(i1)$ and an element $\Gamma, i : \mathbb{I} \vdash a : A[(i = 0) \mapsto a_0, (i = 1) \mapsto a_1]$ gives a path connecting a_0 and a_1.

Lemma 3. *The following rules are admissible:*[2]

$$\frac{\Gamma \vdash \varphi \leqslant \psi : \mathbb{F} \quad \Gamma, \psi \vdash J}{\Gamma, \varphi \vdash J} \qquad \frac{\Gamma, 1_\mathbb{F} \vdash J}{\Gamma \vdash J} \qquad \frac{\Gamma, \varphi, \psi \vdash J}{\Gamma, \varphi \wedge \psi \vdash J}$$

Furthermore, if φ is independent of i, the following rules are admissible

$$\frac{\Gamma, i : \mathbb{I}, \varphi \vdash J}{\Gamma, \varphi, i : \mathbb{I} \vdash J}$$

and it follows that we have in general:

$$\frac{\Gamma, i : \mathbb{I}, \varphi \vdash J}{\Gamma, \forall i.\varphi, i : \mathbb{I} \vdash J}$$

4.3 Composition operation

The syntax of compositions is given by:

$$t, u, A, B ::= \ldots$$
$$\mid \mathsf{comp}^i \ A \ [\varphi \mapsto u] \ a_0 \qquad \text{Compositions}$$

where u is a system on the extent φ.

The composition operation expresses that being extensible is preserved along paths: if a partial path is extensible at 0, then it is extensible at 1.

$$\frac{\Gamma \vdash \varphi : \mathbb{F} \quad \Gamma, i : \mathbb{I} \vdash A \quad \Gamma, \varphi, i : \mathbb{I} \vdash u : A \quad \Gamma \vdash a_0 : A(i0)[\varphi \mapsto u(i0)]}{\Gamma \vdash \mathsf{comp}^i \ A \ [\varphi \mapsto u] \ a_0 : A(i1)[\varphi \mapsto u(i1)]}$$

Note that comp^i binds i in A and u and that we have in particular the following equality judgments for systems:

$$\Gamma \vdash \mathsf{comp}^i \ A \ [1_\mathbb{F} \mapsto u] \ a_0 = u(i1) : A(i1)$$

If we have a substitution $\Delta \vdash \sigma : \Gamma$, then

$$(\mathsf{comp}^i \ A \ [\varphi \mapsto u] \ a_0)\sigma = \mathsf{comp}^j \ A(\sigma, i/j) \ [\varphi\sigma \mapsto u(\sigma, i/j)] \ a_0\sigma$$

where j is fresh for Δ, which corresponds semantically to the *uniformity* [6, 14] of the composition operation.

We use the abbreviation $[\varphi_1 \mapsto u_1, \ldots, \varphi_n \mapsto u_n]$ for $[\bigvee_l \varphi_l \mapsto [\varphi_1 \ u_1, \ldots, \varphi_n \ u_n]]$ and in particular we write $[\,]$ for $[0_\mathbb{F} \mapsto [\,]]$.

---

[2] The inference rules with double line are each a pair of rules, because they can be used in both directions.

Example 4. With composition we can justify transitivity of path types:

$$\frac{\Gamma \vdash p : \mathsf{Path}\ A\ a\ b \qquad \Gamma \vdash q : \mathsf{Path}\ A\ b\ c}{\Gamma \vdash \langle i \rangle\ \mathsf{comp}^j\ A\ [(i=0) \mapsto a, (i=1) \mapsto q\ j]\ (p\ i) : \mathsf{Path}\ A\ a\ c}$$

This composition can be visualized as the dashed arrow in the square:

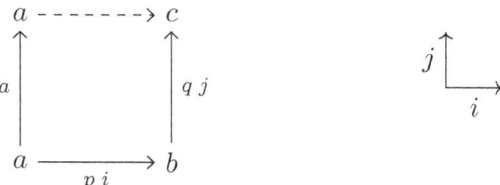

4.4 Kan filling operation

As we have connections we also get Kan filling operations from compositions:

$$\Gamma, i : \mathbb{I} \vdash \mathsf{fill}^i\ A\ [\varphi \mapsto u]\ a_0 = \mathsf{comp}^j\ A(i/i \wedge j)\ [\varphi \mapsto u(i/i \wedge j), (i=0) \mapsto a_0]\ a_0 : A$$

where j is fresh for Γ. The element $\Gamma, i : \mathbb{I} \vdash v = \mathsf{fill}^i\ A\ [\varphi \mapsto u]\ a_0 : A$ satisfies:

$$\Gamma \vdash v(i0) = a_0 : A(i0)$$
$$\Gamma \vdash v(i1) = \mathsf{comp}^i\ A\ [\varphi \mapsto u]\ a_0 : A(i1)$$
$$\Gamma, \varphi, i : \mathbb{I} \vdash v = u : A$$

This means that we can not only compute the lid of an open box but also its filling. If φ is the boundary formula on the names declared in Γ, we recover the Kan operation for cubical sets [17].

4.5 Equality judgments for composition

The equality judgments for $\mathsf{comp}^i\ C\ [\varphi \mapsto u]\ a_0$ are defined by cases on the type C which depends on i, i.e., $\Gamma, i : \mathbb{I} \vdash C$. The right hand side of the definitions are all equal to $u(i1)$ on the extent φ by the typing rule for compositions. There are four cases to consider:

Product types, $C = (x : A) \to B$

Given $\Gamma, \varphi, i : \mathbb{I} \vdash u : C$ and $\Gamma \vdash \lambda_0 : C(i0)[\varphi \mapsto u(i0)]$ the composition will be of type $C(i1)$. For $\Gamma \vdash u_1 : A(i1)$, we first let:

$$w = \mathsf{fill}^i\ A(i/1-i)\ []\ u_1 \qquad \text{(in context } \Gamma, i : \mathbb{I} \text{ and of type } A(i/1-i))$$
$$v = w(i/1-i) \qquad \text{(in context } \Gamma, i : \mathbb{I} \text{ and of type } A)$$

Using this we define the equality judgment:

$$\Gamma \vdash (\mathsf{comp}^i\ C\ [\varphi \mapsto \mu]\ \lambda_0)\ u_1 = \mathsf{comp}^i\ B(x/v)\ [\varphi \mapsto \mu\ v]\ (\lambda_0\ v(i0)) : B(x/v)(i1)$$

Sum types, $C = (x : A) \times B$

Given $\Gamma, \varphi, i : \mathbb{I} \vdash w : C$ and $\Gamma \vdash w_0 : C(i0)[\varphi \mapsto w(i0)]$ we let:

$a = \mathsf{fill}^i\ A\ [\varphi \mapsto w.1]\ w_0.1$ (in context $\Gamma, i : \mathbb{I}$ and of type A)

$c_1 = \mathsf{comp}^i\ A\ [\varphi \mapsto w.1]\ w_0.1$ (in context Γ and of type $A(i1)$)

$c_2 = \mathsf{comp}^i\ B(x/a)\ [\varphi \mapsto w.2]\ w_0.2$ (in context Γ and of type $B(x/a)(i1)$)

From which we define:

$$\Gamma \vdash \mathsf{comp}^i\ C\ [\varphi \mapsto w]\ w_0 = (c_1, c_2) : C(i1)$$

Natural numbers, $C = \mathsf{N}$

In this we define $\mathsf{comp}^i\ C\ [\varphi \mapsto n]\ n_0$ by recursion:

$$\Gamma \vdash \mathsf{comp}^i\ C\ [\varphi \mapsto 0]\ 0 = 0 : C$$
$$\Gamma \vdash \mathsf{comp}^i\ C\ [\varphi \mapsto \mathsf{s}\ n]\ (\mathsf{s}\ n_0) = \mathsf{s}\ (\mathsf{comp}^i\ C\ [\varphi \mapsto n]\ n_0) : C$$

Path types, $C = \mathsf{Path}\ A\ u\ v$

Given $\Gamma, \varphi, i : \mathbb{I} \vdash p : C$ and $\Gamma \vdash p_0 : C(i0)[\varphi \mapsto p(i0)]$ we define:

$$\Gamma \vdash \mathsf{comp}^i\ C\ [\varphi \mapsto p]\ p_0 =$$
$$\langle j \rangle\ \mathsf{comp}^i\ A\ [\varphi \mapsto p\ j, (j = 0) \mapsto u, (j = 1) \mapsto v]\ (p_0\ j) : C(i1)$$

4.6 Transport

Composition for $\varphi = 0_\mathbb{F}$ corresponds to transport:

$$\Gamma \vdash \mathsf{transp}^i\ A\ a = \mathsf{comp}^i\ A\ []\ a : A(i1)$$

Together with the fact that singletons are contractible, from Section 3.2, we get the elimination principle for Path types in the same manner as explained for identity types in Section 7.2 of [6].

5 Derived notions and operations

This section defines various notions and operations that will be used for defining compositions for the **glue** operation in the next section. This operation will then be used to define the composition operation for the universe and to prove the univalence axiom.

5.1 Contractible types

We define $\text{isContr } A = (x : A) \times ((y : A) \to \text{Path } A\ x\ y)$. A proof of $\text{isContr } A$ witnesses the fact that A is *contractible*.

Given $\Gamma \vdash p : \text{isContr } A$ and $\Gamma, \varphi \vdash u : A$ we define the operation[3]

$$\Gamma \vdash \text{contr } p\ [\varphi \mapsto u] = \text{comp}^i\ A\ [\varphi \mapsto p.2\ u\ i]\ p.1 : A[\varphi \mapsto u]$$

Conversely, we can state the following characterization of contractible types:

Lemma 5. *Let $\Gamma \vdash A$ and assume that we have one operation*

$$\frac{\Gamma, \varphi \vdash u : A}{\Gamma \vdash \text{contr } [\varphi \mapsto u] : A[\varphi \mapsto u]}$$

then we can find an element in $\text{isContr } A$.

Proof. We define $x = \text{contr } [] : A$ and prove that any element $y : A$ is path equal to x. For this, we introduce a fresh name $i : \mathbb{I}$ and define $\varphi = (i = 0) \vee (i = 1)$ and $u = [(i = 0) \mapsto x, (i = 1) \mapsto y]$. Using this we obtain $\Gamma, i : \mathbb{I} \vdash v = \text{contr } [\varphi \mapsto u] : A[\varphi \mapsto u]$. In this way, we get a path $\langle i \rangle \text{contr } [\varphi \mapsto u]$ connecting x and y. □

5.2 The equiv operation

We define $\text{isEquiv } T\ A\ f = (y : A) \to \text{isContr } ((x : T) \times \text{Path } A\ y\ (f\ x))$ and $\text{Equiv } T\ A = (f : T \to A) \times \text{isEquiv } T\ A\ f$. If $w : \text{Equiv } T\ A$ and $t : T$, we may write $w\ t$ for $w.1\ t$.

Lemma 6. *If $\Gamma \vdash w : \text{Equiv } T\ A$, we have an operation*

$$\frac{\Gamma, \varphi \vdash t : T \qquad \Gamma \vdash a : A[\varphi \mapsto w\ t]}{\Gamma \vdash \text{equiv } w\ [\varphi \mapsto t]\ a : ((x : T) \times \text{Path } A\ a\ (w\ x))[\varphi \mapsto (t, \langle j \rangle a)]}$$

Proof. We define $\text{equiv } w\ [\varphi \mapsto t]\ a = \text{contr } (w.2\ a)\ [\varphi \mapsto (t, \langle j \rangle a)]$ using the **contr** operation defined above. □

---
[3]This expresses that the restriction map $\Gamma, \varphi \to \Gamma$ has the left lifting property w.r.t. any "trivial fibration", i.e., contractible extensions $\Gamma, x : A \to \Gamma$. The restriction maps $\Gamma, \varphi \to \Gamma$ thus represent "cofibrations" while the maps $\Gamma, x : A \to \Gamma$ represent "fibrations".

6 Glueing

In this section, we introduce the glueing operation. This operation expresses that to be "extensible" is invariant by equivalence. From this operation, we can define a composition operation for universes, and prove the univalence axiom.

6.1 Syntax and inference rules for glueing

We introduce the *glueing* construction at type and term level by:

$$t, u, A, B ::= \ldots$$
$$| \ \mathsf{Glue}\ [\varphi \mapsto (T, w)]\ A \qquad \text{Glue type}$$
$$| \ \mathsf{glue}\ [\varphi \mapsto t]\ u \qquad \text{Glue term}$$
$$| \ \mathsf{unglue}\ [\varphi \mapsto w]\ u \qquad \text{Unglue term}$$

We may write simply $\mathsf{unglue}\ b$ for $\mathsf{unglue}\ [\varphi \mapsto w]\ b$. The inference rules for these are presented in Figure 4.

$$\frac{\Gamma \vdash A \quad \Gamma, \varphi \vdash T \quad \Gamma, \varphi \vdash w : \mathsf{Equiv}\ T\ A}{\Gamma \vdash \mathsf{Glue}\ [\varphi \mapsto (T, w)]\ A} \qquad \frac{\Gamma \vdash b : \mathsf{Glue}\ [\varphi \mapsto (T, w)]\ A}{\Gamma \vdash \mathsf{unglue}\ b : A[\varphi \mapsto w\ b]}$$

$$\frac{\Gamma, \varphi \vdash w : \mathsf{Equiv}\ T\ A \quad \Gamma, \varphi \vdash t : T \quad \Gamma \vdash a : A[\varphi \mapsto w\ t]}{\Gamma \vdash \mathsf{glue}\ [\varphi \mapsto t]\ a : \mathsf{Glue}\ [\varphi \mapsto (T, w)]\ A}$$

$$\frac{\Gamma \vdash T \quad \Gamma \vdash w : \mathsf{Equiv}\ T\ A}{\Gamma \vdash \mathsf{Glue}\ [1_\mathbb{F} \mapsto (T, w)]\ A = T} \qquad \frac{\Gamma \vdash t : T \quad \Gamma \vdash w : \mathsf{Equiv}\ T\ A}{\Gamma \vdash \mathsf{glue}\ [1_\mathbb{F} \mapsto t]\ (f\ t) = t : T}$$

$$\frac{\Gamma \vdash b : \mathsf{Glue}\ [\varphi \mapsto (T, w)]\ A}{\Gamma \vdash b = \mathsf{glue}\ [\varphi \mapsto b]\ (\mathsf{unglue}\ b) : \mathsf{Glue}\ [\varphi \mapsto (T, w)]\ A}$$

$$\frac{\Gamma, \varphi \vdash w : \mathsf{Equiv}\ T\ A \quad \Gamma, \varphi \vdash t : T \quad \Gamma \vdash a : A[\varphi \mapsto w\ t]}{\Gamma \vdash \mathsf{unglue}\ (\mathsf{glue}\ [\varphi \mapsto t]\ a) = a : A}$$

Figure 4: Inference rules for glueing

It follows from these rules that if $\Gamma \vdash b : \mathsf{Glue}\ [\varphi \mapsto (T, w)]\ A$, then $\Gamma, \varphi \vdash b : T$. In the case $\varphi = (i = 0) \vee (i = 1)$ the glueing operation can be illustrated as the dashed line in:

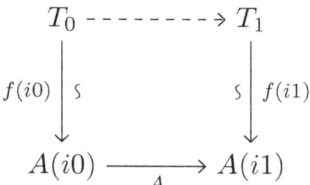

This illustrates why the operation is called glue: it *glues* together along a partial equivalence the partial type T and the total type A to a total type that extends T.

Remark. In general $\mathsf{Glue}\ [\varphi \mapsto (T, w)]\ A$ can be illustrated as:

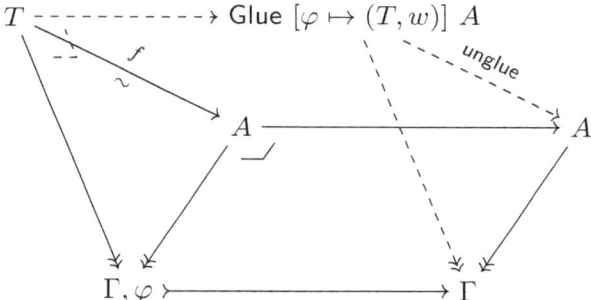

This diagram suggests that a construction similar to Glue also appears in the simplicial set model. Indeed, the proof of Theorem 3.4.1 in [18] contains a similar diagram where $\overline{E}_1$ corresponds to $\mathsf{Glue}\ [\varphi \mapsto (T, w)]\ A$.

Example 7. Using glueing we can construct a path from an equivalence $\Gamma \vdash w : \mathsf{Equiv}\ A\ B$ by defining

$$\Gamma, i : \mathbb{I} \vdash E = \mathsf{Glue}\ [(i = 0) \mapsto (A, w), (i = 1) \mapsto (B, \mathsf{id}_B)]\ B$$

so that $E(i0) = A$ and $E(i1) = B$, where $\mathsf{id}_B : \mathsf{Equiv}\ B\ B$ is defined as:

$\mathsf{id}_B = (\lambda x : B.\ x, \lambda x : B.\ ((x, 1_x), \lambda u : (y : B) \times \mathsf{Path}\ B\ x\ y.\ \langle i \rangle\ (u.2\ i, \langle j \rangle\ u.2\ (i \wedge j))))$

In Section 7 we introduce a universe of types U and we will be able to define a function of type $(A\ B : \mathsf{U}) \to \mathsf{Equiv}\ A\ B \to \mathsf{Path}\ \mathsf{U}\ A\ B$ by:

$$\lambda A\ B : \mathsf{U}.\ \lambda w : \mathsf{Equiv}\ A\ B.\ \langle i \rangle\ \mathsf{Glue}\ [(i = 0) \mapsto (A, w), (i = 1) \mapsto (B, \mathsf{id}_B)]\ B$$

6.2 Composition for glueing

We assume $\Gamma, i : \mathbb{I} \vdash B = \mathsf{Glue}\ [\varphi \mapsto (T, w)]\ A$, and define the composition in B. In order to do so, assume

$$\Gamma, \psi, i : \mathbb{I} \vdash b : B \qquad\qquad \Gamma \vdash b_0 : B(i0)[\psi \mapsto b(i0)]$$

and define:

$a = \mathsf{unglue}\ b$ (in context $\Gamma, \psi, i : \mathbb{I}$ and of type $A[\varphi \mapsto w\ b]$)

$a_0 = \mathsf{unglue}\ b_0$ (in context Γ and of type $A(i0)[\varphi(i0) \mapsto w(i0)\ b_0, \psi \mapsto a(i0)]$)

The following provides the algorithm for composition $\mathsf{comp}^i\ B\ [\psi \mapsto b]\ b_0 = b_1$ of type $B(i1)[\psi \mapsto b(i1)]$.

$$
\begin{array}{llll}
\delta & = & \forall i.\varphi & \Gamma \\
\tilde{t} & = & \mathsf{fill}^i\ T\ [\psi \mapsto b]\ b_0 & \Gamma, \delta, i : \mathbb{I} \\
a'_1 & = & \mathsf{comp}^i\ A\ [\delta \mapsto w\ \tilde{t}, \psi \mapsto a]\ a_0 & \Gamma \\
(t_1, \alpha) & = & \mathsf{equiv}\ w(i1)\ [\delta \mapsto \tilde{t}(i1), \psi \mapsto b(i1))]\ a'_1 & \Gamma, \varphi(i1) \\
a_1 & = & \mathsf{comp}^j\ A(i1)\ [\varphi(i1) \mapsto \alpha\ j, \psi \mapsto a(i1)]\ a'_1 & \Gamma \\
b_1 & = & \mathsf{glue}\ [\varphi(i1) \mapsto t_1]\ a_1 & \Gamma \\
\end{array}
$$

We can check that whenever $\Gamma, i : \mathbb{I} \vdash \varphi = 1_\mathbb{F} : \mathbb{F}$ the definition of b_1 coincides with $\mathsf{comp}^i\ T\ [\psi \mapsto b]\ b_0$, which is consistent with the fact that $B = T$ in this case.

In the next section we will use the glue operation to define the composition for the universe and to prove the univalence axiom.

7 Universe and the univalence axiom

As in [21], we now introduce a universe U à la Russell by reflecting all typing rules and

$$\frac{\Gamma \vdash}{\Gamma \vdash \mathsf{U}} \qquad\qquad \frac{\Gamma \vdash A : \mathsf{U}}{\Gamma \vdash A}$$

In particular, we have $\Gamma \vdash \mathsf{Glue}\ [\varphi \mapsto (T, w)]\ A : \mathsf{U}$ whenever $\Gamma \vdash A : \mathsf{U}$, $\Gamma, \varphi \vdash T : \mathsf{U}$, and $\Gamma, \varphi \vdash w : \mathsf{Equiv}\ T\ A$.

3144

7.1 Composition for the universe

In order to describe the composition operation for the universe we first have to explain how to construct an equivalence from a line in the universe. Given $\Gamma \vdash A$, $\Gamma \vdash B$, and $\Gamma, i : \mathbb{I} \vdash E$, such that $E(i0) = A$ and $E(i1) = B$, we will construct $\mathsf{equiv}^i\ E : \mathsf{Equiv}\ A\ B$. In order to do this we first define

$f = \lambda x : A.\, \mathsf{transp}^i\ E\ x$ (in context Γ and of type $A \to B$)

$g = \lambda y : B.(\mathsf{transp}^i\ E(i/1-i)\ y)(i/1-i)$ (in context Γ and of type $B \to A$)

$u = \lambda x : A.\mathsf{fill}^i\ E\ []\ x$ (in context $\Gamma, i : \mathbb{I}$ and of type $A \to E$)

$v = \lambda y : B.(\mathsf{fill}^i\ E(i/1-i)\ []\ y)(i/1-i)$ (in context $\Gamma, i : \mathbb{I}$ and of type $B \to E$)

such that:

$$u(i0) = \lambda x : A.x \qquad u(i1) = f \qquad v(i0) = g \qquad v(i1) = \lambda y : B.y$$

We will now prove that f is an equivalence. Given $y : B$ we see that $(x : A) \times \mathsf{Path}\ B\ y\ (f\ x)$ is inhabited as it contains the element $(g\ y, \langle j \rangle\ \theta_0(i1))$ where

$$\theta_0 = \mathsf{fill}^i\ E\ [(j = 0) \mapsto v\ y, (j = 1) \mapsto u\ (g\ y)]\ (g\ y).$$

Next, given an element (x, β) of $(x : A) \times \mathsf{Path}\ B\ y\ (f\ x)$ we will construct a path from $(g\ y, \langle j \rangle\ \theta_0(i1))$ to (x, β). Let

$$\theta_1 = (\mathsf{fill}^i\ E(i/1-i)\ [(j = 0) \mapsto (v\ y)(i/1-i), (j = 1) \mapsto (u\ x)(i/1-i)]\ (\beta\ j))(i/1-i)$$

and $\omega = \theta_1(i0)$ so $\Gamma, i : \mathbb{I}, j : \mathbb{I} \vdash \theta_1 : E$, $\omega(j0) = g\ y$, and $\omega(j1) = x$. And further with δ defined as

$$\mathsf{comp}^i\ E[(k = 0) \mapsto \theta_0, (k = 1) \mapsto \theta_1, (j = 0) \mapsto v\ y, (j = 1) \mapsto u\ \omega(j/k)]\ \omega(j/j \wedge k)$$

we obtain

$$\langle k \rangle\ (\omega(j/k), \langle j \rangle\ \delta) : \mathsf{Path}\ ((x : A) \times \mathsf{Path}\ B\ y\ (f\ x))\ (g\ y, \langle j \rangle\ \theta_0(i1))\ (x, \beta)$$

as desired. This concludes the proof that f is an equivalence and thus also the construction of $\mathsf{equiv}^i\ E : \mathsf{Equiv}\ A\ B$.

Using this we can now define the composition for the universe:

$$\Gamma \vdash \mathsf{comp}^i\ \mathsf{U}\ [\varphi \mapsto E]\ A_0 = \mathsf{Glue}\ [\varphi \mapsto (E(i1), \mathsf{equiv}^i\ E(i/1-i))]\ A_0 : \mathsf{U}$$

Remark. Given $\Gamma, i : \mathbb{I} \vdash E$ we can also get an equivalence in $\mathsf{Equiv}\ A\ B$ (where $A = E(i0)$ and $B = E(i1)$) with a less direct description by

$$\Gamma \vdash \mathsf{transp}^i\ (\mathsf{Equiv}\ A\ E)\ \mathsf{id}_A : \mathsf{Equiv}\ A\ B$$

where id_A is the identity equivalence as given in Example 7.

7.2 The univalence axiom

Given $B = \text{Glue } [\varphi \mapsto (T, w)] \, A$ the map $\text{unglue} : B \to A$ extends w, in the sense that $\Gamma, \varphi \vdash \text{unglue } b = w \, b : A$ if $\Gamma \vdash b : B$.

Theorem 8. *The map* $\text{unglue} : B \to A$ *is an equivalence.*

Proof. By Lemma 6 it suffices to construct

$$\tilde{b} : B[\psi \mapsto b] \qquad \tilde{\alpha} : \text{Path } A \, u \, (\text{unglue } \tilde{b})[\psi \mapsto \alpha]$$

given $\Gamma, \psi \vdash b : B$ and $\Gamma \vdash u : A$ and $\Gamma, \psi \vdash \alpha : \text{Path } A \, u \, (\text{unglue } b)$.

Since $\Gamma, \varphi \vdash w : T \to A$ is an equivalence and

$$\Gamma, \varphi, \psi \vdash b : T \qquad \Gamma, \varphi, \psi \vdash \alpha : \text{Path } A \, u \, (f \, b)$$

we get, using Lemma 6

$$\Gamma, \varphi \vdash t : T[\psi \mapsto b] \qquad \Gamma, \varphi \vdash \beta : \text{Path } A \, u \, (w \, t) \, [\psi \mapsto \alpha]$$

We then define $\tilde{a} = \text{comp}^i \, A \, [\varphi \mapsto \beta \, i, \psi \mapsto \alpha \, i] \, u$, and using this we conclude by letting $\tilde{b} = \text{glue } [\varphi \mapsto t] \, \tilde{a}$ and $\tilde{\alpha} = \text{fill}^i \, A \, [\varphi \mapsto \beta \, i, \psi \mapsto \alpha \, i] \, u$. □

Corollary 9. *For any type* $A : \mathsf{U}$ *the type* $C = (X : \mathsf{U}) \times \text{Equiv } X \, A$ *is contractible.*[4]

Proof. It is enough by Lemma 5 to show that any partial element $\varphi \vdash (T, w) : C$ is path equal to the restriction of a total element. The map unglue extends w and is an equivalence by the previous theorem. Since any two elements of the type $\text{isEquiv } X \, A \, w.1$ are path equal, this shows that any partial element of type C is path equal to the restriction of a total element. We can then conclude by Theorem 8. □

Corollary 10 (Univalence axiom). *For any term*

$$t : (A \, B : \mathsf{U}) \to \text{Path } \mathsf{U} \, A \, B \to \text{Equiv } A \, B$$

the map $t \, A \, B : \text{Path } \mathsf{U} \, A \, B \to \text{Equiv } A \, B$ *is an equivalence.*

Proof. Both $(X : \mathsf{U}) \times \text{Path } \mathsf{U} \, A \, X$ and $(X : \mathsf{U}) \times \text{Equiv } A \, X$ are contractible. Hence the result follows from Theorem 4.7.7 in [27]. □

Two alternative proofs of univalence can be found in Appendix A.

---

[4]This formulation of the univalence axiom can be found in the message of Martín Escardó in: https://groups.google.com/forum/#!msg/homotopytypetheory/HfCB_b-PNEU/Ibb48LvUMeUJ
This is also used in the (classical) proofs of the univalence axiom, see Theorem 3.4.1 of [18] and Proposition 2.18 of [8], where an operation similar to the glueing operation appears implicitly.

8 Semantics

In this section we will explain the semantics of the type theory under consideration in cubical sets. We will first review how cubical sets, as a presheaf category, yield a model of basic type theory, and then explain the additional so-called composition structure we have to require to interpret the full cubical type theory.

8.1 The category of cubes and cubical sets

Consider the monad dM on the category of sets associating to each set the free de Morgan algebra on that set. The *category of cubes* $\mathcal{C}$ is the small category whose objects are finite subsets $I, J, K, \ldots$ of a fixed, discrete, and countably infinite set, called *names*, and a morphism $\mathrm{Hom}(J, I)$ is a map $I \to \mathsf{dM}(J)$. Identities and compositions are inherited from the Kleisli category of dM, i.e., the identity on I is given by the unit $I \to \mathsf{dM}(I)$, and composition $fg \in \mathrm{Hom}(K, I)$ of $g \in \mathrm{Hom}(K, J)$ and $f \in \mathrm{Hom}(J, I)$ is given by $\mu_K \circ \mathsf{dM}(g) \circ f$ where $\mu_K \colon \mathsf{dM}(\mathsf{dM}(K)) \to \mathsf{dM}(K)$ denotes multiplication of dM. We will use f, g, h for morphisms in $\mathcal{C}$ and simply write $f \colon J \to I$ for $f \in \mathrm{Hom}(J, I)$. We will often write unions with commas and omit curly braces around finite sets of names, e.g., writing I, i, j for $I \cup \{i, j\}$ and $I - i$ for $I - \{i\}$ etc.

If i is in I and b is $0_\mathbb{I}$ or $1_\mathbb{I}$, we have maps (ib) in $\mathrm{Hom}(I - i, I)$ whose underlying map sends $j \neq i$ to itself and i to b. A *face map* is a composition of such maps. A *strict map* $\mathrm{Hom}(J, I)$ is a map $I \to \mathsf{dM}(J)$ which never takes the value $0_\mathbb{I}$ or $1_\mathbb{I}$. Any f can be uniquely written as a composition $f = gh$ where g is a face map and h is strict.

Definition 11. A *cubical set* is a presheaf on $\mathcal{C}$.

Thus, a cubical set Γ is given by sets $\Gamma(I)$ for each $I \in \mathcal{C}$ and maps (called restrictions) $\Gamma(f) \colon \Gamma(I) \to \Gamma(J)$ for each $f \colon J \to I$. If we write $\Gamma(f)(\rho) = \rho f$ for $\rho \in \Gamma(I)$ (leaving the Γ implicit), these maps should satisfy $\rho \, \mathrm{id}_I = \rho$ and $(\rho f)g = \rho(fg)$ for $f \colon J \to I$ and $g \colon K \to J$.

Let us discuss some important examples of cubical sets. Using the canonical de Morgan algebra structure of the unit interval, $[0, 1]$, we can define a functor

$$\mathcal{C} \to \mathbf{Top}, \quad I \mapsto [0, 1]^I. \tag{1}$$

If u is in $[0, 1]^I$ we can think of u as an environment giving values in $[0, 1]$ to each $i \in I$, so that iu is in $[0, 1]$ if $i \in I$. Since $[0, 1]$ is a de Morgan algebra, this extends uniquely to ru for $r \in \mathsf{dM}(I)$. So any $f \colon J \to I$ in $\mathcal{C}$ induces $f \colon [0, 1]^J \to [0, 1]^I$ by $i(fu) = (if)u$.

To any topological space X we can associate its *singular cubical set* $\mathrm{S}(X)$ by taking $\mathrm{S}(X)(I)$ to be the set of continuous functions $[0,1]^I \to X$.

For a finite set of names I we get the formal cube $\mathbf{y}\,I$ where $\mathbf{y}\colon \mathcal{C} \to [\mathcal{C}^{\mathrm{op}}, \mathbf{Set}]$ denotes the Yoneda embedding. Note that since **Top** is cocomplete the functor in (1) extends to a cocontinuous functor assigning to each cubical set its *geometric realization* as a topological space, in such a way that $\mathbf{y}\,I$ has $[0,1]^I$ as its geometric realization.

The formal interval $\mathbb{I}$ induces a cubical set given by $\mathbb{I}(I) = \mathsf{dM}(I)$. The face lattice $\mathbb{F}$ induces a cubical set by taking as $\mathbb{F}(I)$ to be those $\varphi \in \mathbb{F}$ which only use symbols in I. The restrictions along $f\colon J \to I$ are in both cases simply *substituting* the symbols $i \in I$ by $f(i) \in \mathsf{dM}(J)$.

As any presheaf category, cubical sets have a subobject classifier Ω where $\Omega(I)$ is the set of sieves on I (i.e., subfunctors of $\mathbf{y}\,I$). Consider the natural transformation $(\cdot = 1)\colon \mathbb{I} \to \Omega$ where for $r \in \mathbb{I}(I)$, $(r = 1)$ is the sieve on I of all $f\colon J \to I$ such that $rf = 1_\mathbb{I}$. The image of $(\cdot = 1)$ is $\mathbb{F} \to \Omega$, assigning to each φ the sieve of all f with $\varphi f = 1_\mathbb{F}$.

8.2 Presheaf semantics

The category of cubical sets (with morphisms being natural transformations) induce– as does any presheaf category–a category with families (CwF) [10] where the category of contexts and substitutions is the category of cubical sets. We will review the basic constructions but omit verification of the required equations (see, e.g., [13, 14, 6] for more details).

Basic presheaf semantics

As already mentioned the category of (semantic) contexts and substitutions is given by cubical sets and their maps. In this section we will use Γ, Δ to denote cubical sets and (semantic) substitutions by $\sigma\colon \Delta \to \Gamma$, overloading previous use of the corresponding meta-variables to emphasize their intended role.

Given a cubical set Γ, the types A in context Γ, written $A \in \mathrm{Ty}(\Gamma)$, are given by sets $A\rho$ for each $I \in \mathcal{C}$ and $\rho \in \Gamma(I)$ together with restriction maps $A\rho \to A(\rho f)$, $u \mapsto uf$ for $f\colon J \to I$ satisfying $u\,\mathrm{id}_I = u$ and $(uf)g = u(fg) \in A(\rho f g)$ if $g\colon K \to J$. Equivalently, $A \in \mathrm{Ty}(\Gamma)$ are the presheaves on the category of elements of Γ. For a type $A \in \mathrm{Ty}(\Gamma)$ its terms $a \in \mathrm{Ter}(\Gamma; A)$ are given by families of elements $a\rho \in A\rho$ for each $I \in \mathcal{C}$ and $\rho \in \Gamma(I)$ such that $(a\rho)f = a(\rho f)$ for $f\colon J \to I$. Note that our notation leaves a lot implicit; e.g., we should have written $A(I, \rho)$ for $A\rho$; $A(I, \rho, f)$ for the restriction map $A\rho \to A(\rho f)$; and $a(I, \rho)$ for $a\rho$.

For $A \in \text{Ty}(\Gamma)$ and $\sigma \colon \Delta \to \Gamma$ we define $A\sigma \in \text{Ty}(\Delta)$ by $(A\sigma)\rho = A(\sigma\rho)$ and the induced restrictions. If we also have $a \in \text{Ter}(\Gamma; A)$, we define $a\sigma \in \text{Ter}(\Delta; A\sigma)$ by $(a\sigma)\rho = a(\sigma\rho)$. For a type $A \in \text{Ty}(\Gamma)$ we define the cubical set $\Gamma.A$ by $(\Gamma.A)(I)$ being the set of all (ρ, u) with $\rho \in \Gamma(I)$ and $u \in A\rho$; restrictions are given by $(\rho, u)f = (\rho f, uf)$. The first projection yields a map $\mathsf{p} \colon \Gamma.A \to \Gamma$ and the second projection a term $\mathsf{q} \in \text{Ter}(\Gamma.A; A\mathsf{p})$. Given $\sigma \colon \Delta \to \Gamma$, $A \in \text{Ty}(\Gamma)$, and $a \in \text{Ter}(\Delta; A\sigma)$ we define $(\sigma, a) \colon \Delta \to \Gamma.A$ by $(\sigma, a)\rho = (\sigma\rho, a\rho)$. For $u \in \text{Ter}(\Gamma; A)$ we define $[u] = (\text{id}_\Gamma, u) \colon \Gamma \to \Gamma.A$.

The basic type formers are interpreted as follows. For $A \in \text{Ty}(\Gamma)$ and $B \in \text{Ty}(\Gamma.A)$ define $\Sigma_\Gamma(A, B) \in \text{Ty}(\Gamma)$ by letting $\Sigma_\Gamma(A, B)\rho$ contain all pairs (u, v) where $u \in A\rho$ and $v \in B(\rho, v)$; restrictions are defined as $(u, v)f = (uf, vf)$. Given $w \in \text{Ter}(\Gamma; \Sigma(A, B))$ we get $w.1 \in \text{Ter}(\Gamma; A)$ and $w.2 \in \text{Ter}(\Gamma; B[w.1])$ by $(w.1)\rho = \mathsf{p}(w\rho)$ and $(w.2)\rho = \mathsf{q}(w\rho)$ where $\mathsf{p}(u, v) = u$ and $\mathsf{q}(u, v) = v$ are the set-theoretic projections.

Given $A \in \text{Ty}(\Gamma)$ and $B \in \text{Ty}(\Gamma.A)$ the dependent function space $\Pi_\Gamma(A, B) \in \text{Ty}(\Gamma)$ is defined by letting $\Pi_\Gamma(A, B)\rho$ for $\rho \in \Gamma(I)$ contain all families $w = (w_f \mid J \in \mathcal{C}, f \colon J \to I)$ where

$$w_f \in \prod_{u \in A(\rho f)} B(\rho f, u) \quad \text{such that} \quad (w_f u)g = w_{fg}(ug) \quad \text{for } u \in A(\rho f),\ g \colon K \to J.$$

The restriction by $f \colon J \to I$ of such a w is defined by $(wf)_g = w_{fg}$. Given $v \in \text{Ter}(\Gamma.A; B)$ we have $\lambda_{\Gamma; A} v \in \text{Ter}(\Gamma; \Pi(A, B))$ given by $((\lambda v)\rho)_f u = v(\rho f, u)$. Application $\mathsf{app}(w, u) \in \text{Ter}(\Gamma; B[u])$ of $w \in \text{Ter}(\Gamma; \Pi(A, B))$ to $u \in \text{Ter}(\Gamma; A)$ is defined by

$$\mathsf{app}(w, u)\rho = (w\rho)_{\text{id}_I}(u\rho) \in (B[u])\rho. \tag{2}$$

Basic data types can be interpreted as discrete presheaves, i.e., $\mathsf{N} \in \text{Ty}(\Gamma)$ is given by $\mathsf{N}\rho = \mathbb{N}$; the constants are interpreted by the lifts of the corresponding set-theoretic operations on $\mathbb{N}$. This concludes the outline of the basic CwF structure on cubical sets.

Remark. Following Aczel [1] we will make use of that our semantic entities are actual sets in the ambient set theory. This will allow us to interpret syntax in Section 8.3 with fewer type annotations than are usually needed for general categorical semantics of type theory (see [25]). E.g., the definition of application $\mathsf{app}(w, u)\rho$ as defined in (2) is independent of Γ, A and B, since set-theoretic application is a (class) operation on all sets. Likewise, we don't need annotations for first and second projections. But note that we will need the type A for λ-abstraction for $(\lambda_{\Gamma; A} v)\rho$ to be a set by the replacement axiom.

Semantic path types

Note that we can consider any cubical set X as $X' \in \mathrm{Ty}(\Gamma)$ by setting $X'\rho = X(I)$ for $\rho \in \Gamma(I)$. We will usually simply write X for X'. In particular, for a cubical set Γ we can form the cubical set $\Gamma.\mathbb{I}$.

For $A \in \mathrm{Ty}(\Gamma)$ and $u, v \in \mathrm{Ter}(\Gamma; A)$ the semantic path type $\mathtt{Path}_A^\Gamma(u, v) \in \mathrm{Ty}(\Gamma)$ is given by: for $\rho \in \Gamma(I)$, $\mathtt{Path}_A(u, v)\rho$ consists of equivalence classes $\langle i \rangle\, w$ where $i \notin I$, $w \in A(\rho s_i)$ such that $w(i0) = u\rho$ and $w(i1) = v\rho$; two such elements $\langle i \rangle\, w$ and $\langle j \rangle\, w'$ are equal iff $w(i/j) = w'$. Here $s_i \colon I, i \to I$ is induced by the inclusion $I \subseteq I, i$ and (i/j) setting i to j. We define $(\langle i \rangle\, w)f = \langle j \rangle\, w(f, i/j)$ for $f \colon J \to I$ and $j \notin J$. For $r \in \mathbb{I}(I)$ we set $(\langle i \rangle\, w) r = w(i/r)$. Both operations, name abstraction and application, lift to terms, i.e., if $w \in \mathrm{Ter}(\Gamma.\mathbb{I}; A)$, then $\langle\,\rangle w \in \mathrm{Ter}(\Gamma; \mathtt{Path}_A(w[0], w[1]))$ given by $(\langle\,\rangle w)\rho = \langle i \rangle\, w(\rho s_i)$ for a fresh i; also if $u \in \mathrm{Ter}(\Gamma; \mathtt{Path}_A(a, b))$ and $r \in \mathrm{Ter}(\Gamma; \mathbb{I})$, then $u\, r \in \mathrm{Ter}(\Gamma; A)$ defined as $(u\, r)\rho = (u\rho)\,(r\rho)$.

Composition structure

For $\varphi \in \mathrm{Ter}(\Gamma; \mathbb{F})$ we define the cubical set Γ, φ by taking $\rho \in (\Gamma, \varphi)(I)$ iff $\rho \in \Gamma(I)$ and $\varphi\rho = 1_\mathbb{F} \in \mathbb{F}$; the restrictions are those induced by Γ. In particular, we have $\Gamma, 1 = \Gamma$ and $\Gamma, 0$ is the empty cubical set. (Here, $0 \in \mathrm{Ter}(\Gamma; \mathbb{F})$ is $0\rho = 0_\mathbb{F}$ and similarly for $1_\mathbb{F}$.) Any $\sigma \colon \Delta \to \Gamma$ gives rise to a morphism $\Delta, \varphi\sigma \to \Gamma, \varphi$ which we also will denote by σ.

If $A \in \mathrm{Ty}(\Gamma)$ and $\varphi \in \mathrm{Ter}(\Gamma; \mathbb{F})$, we define a *partial element of* $A \in \mathrm{Ty}(\Gamma)$ *of extent* φ to be an element of $\mathrm{Ter}(\Gamma, \varphi; A\iota_\varphi)$ where $\iota_\varphi \colon \Gamma, \varphi \hookrightarrow \Gamma$ is the inclusion. So, such a partial element u is given by a family of elements $u\rho \in A\rho$ for each $\rho \in \Gamma(I)$ such that $\varphi\rho = 1$, satisfying $(u\rho)f = u(\rho f)$ whenever $f \colon J \to I$. Each $u \in \mathrm{Ter}(\Gamma; A)$ gives rise to the partial element $u\iota \in \mathrm{Ter}(\Gamma, \varphi; A\iota)$; a partial element is *extensible* if it is induced by such an element of $\mathrm{Ter}(\Gamma; A)$.

For the next definition note that if $A \in \mathrm{Ty}(\Gamma)$, then $\rho \in \Gamma(I)$ corresponds to $\rho \colon \mathbf{y}I \to \Gamma$ and thus $A\rho \in \mathrm{Ty}(\mathbf{y}I)$; also, any $\varphi \in \mathbb{F}(I)$ corresponds to $\varphi \in \mathrm{Ter}(\mathbf{y}I; \mathbb{F})$.

Definition 12. A *composition structure* for $A \in \mathrm{Ty}(\Gamma)$ is given by the following operations. For each I, $i \notin I$, $\rho \in \Gamma(I, i)$, $\varphi \in \mathbb{F}(I)$, u a partial element of $A\rho$ of extent φ, and $a_0 \in A\rho(i0)$ with $a_0 f = u_{(i0)f}$ for all $f \colon J \to I$ with $\varphi f = 1_\mathbb{F}$ (i.e., $a_0 \iota_\varphi = u(i0)$ if a_0 is considered as element of $\mathrm{Ter}(\mathbf{y}I; A\rho(i0))$), we require

$$\mathtt{comp}(I, i, \rho, \varphi, u, a_0) \in A\rho(i1)$$

such that for any $f\colon J\to I$ and $j\notin J$,
$$(\mathsf{comp}(I,i,\rho,\varphi,u,a_0))f = \mathsf{comp}(J,j,\rho(f,i=j),\varphi f,u(f,i=j),a_0 f),$$
and $\mathsf{comp}(I,i,\rho,1_\mathbb{F},u,a_0) = u_{(i1)}$.

A type $A\in\mathrm{Ty}(\Gamma)$ together with a composition structure comp on A is called a *fibrant type*, written $(A,\mathsf{comp})\in\mathrm{FTy}(\Gamma)$. We will usually simply write $A\in\mathrm{FTy}(\Gamma)$ and comp_A for its composition structure. But observe that $A\in\mathrm{Ty}(\Gamma)$ can have different composition structures. Call a cubical set Γ *fibrant* if it is a fibrant type when Γ considered as type $\Gamma\in\mathrm{Ty}(\top)$ is fibrant where $\top$ is a terminal cubical set. A prime example of a fibrant cubical set is the singular cubical set of a topological space (see Appendix B).

Theorem 13. *The CwF on cubical sets supporting dependent products, dependent sums, and natural numbers described above can be extended to fibrant types.*

Proof. For example, if $A\in\mathrm{FTy}(\Gamma)$ and $\sigma\colon\Delta\to\Gamma$, we set
$$\mathsf{comp}_{A\sigma}(I,i,\rho,\varphi,u,a_0) = \mathsf{comp}_A(I,i,\sigma\rho,\varphi,u,a_0)$$
as the composition structure on $A\sigma$ in $\mathrm{FTy}(\Delta)$. Type formers are treated analogously to their syntactic counterpart given in Section 4. Note that one also has to check that all equations between types are also preserved by their associated composition structures. $\square$

Note that we can also, like in the syntax, define a composition structure on $\mathtt{Path}_A(u,v)$ given that A has one.

Semantic glueing

Next we will give a semantic counterpart to the **Glue** construction. To define the semantic glueing as an element of $\mathrm{Ty}(\Gamma)$ it is not necessary that the given types have composition structures or that the functions are equivalences; this is only needed later to give the composition structure. Assume $\varphi\in\mathrm{Ter}(\Gamma;\mathbb{F})$, $T\in\mathrm{Ty}(\Gamma,\varphi)$, $A\in\mathrm{Ty}(\Gamma)$, and $w\in\mathrm{Ter}(\Gamma,\varphi;T\to A\iota)$ (where $A\to B$ is $\Pi(A,B\mathsf{p})$).

Definition 14. The *semantic glueing* $\mathtt{Glue}_\Gamma(\varphi,T,A,w)\in\mathrm{Ty}(\Gamma)$ is defined as follows. For $\rho\in\Gamma(I)$, we let $u\in\mathtt{Glue}(\varphi,T,A,w)\rho$ iff either

- $u\in T\rho$ and $\varphi\rho = 1_\mathbb{F}$; or

- $u = \mathtt{glue}(\varphi\rho, t, a)$ and $\varphi\rho \neq 1_{\mathbb{F}}$, with $t \in \mathrm{Ter}(\mathbf{y}\, I, \varphi\rho; T\rho)$ and $a \in \mathrm{Ter}(\mathbf{y}\, I; A\rho)$ such that $\mathtt{app}(w\rho, t) = a\iota \in \mathrm{Ter}(\mathbf{y}\, I, \varphi\rho; A\rho\iota)$.

For $f\colon J \to I$ we define the restriction uf of $u \in \mathtt{Glue}(\varphi, T, A, w)$ to be given by the restriction of $T\rho$ in the first case; in the second case, i.e., if $\varphi\rho \neq 1_{\mathbb{F}}$, we let $uf = \mathtt{glue}(\varphi\rho, t, a)f = t_f \in T\rho f$ in case $\varphi\rho f = 1_{\mathbb{F}}$, and otherwise $uf = \mathtt{glue}(\varphi\rho f, tf, af)$.

Here $\mathtt{glue}$ was defined as a constructor; we extend $\mathtt{glue}$ to any $t \in \mathrm{Ter}(\mathbf{y}\, I; T\rho)$, $a \in \mathrm{Ter}(\mathbf{y}\, I; A\rho)$ such that $\mathtt{app}(w\rho, t) = a$ (so if $\varphi\rho = 1_{\mathbb{F}}$) by $\mathtt{glue}(1_{\mathbb{F}}, t, a) = t_{\mathrm{id}_I}$. This way any element of $\mathtt{Glue}(\varphi, T, A, w)\rho$ is of the form $\mathtt{glue}(\varphi\rho, t, a)$ for suitable t and a, and restriction is given by $(\mathtt{glue}(\varphi\rho, t, a))f = \mathtt{glue}(\varphi\rho f, tf, af)$. Note that we get

$$\mathtt{Glue}_\Gamma(1_{\mathbb{F}}, T, A, w) = T \text{ and } (\mathtt{Glue}_\Gamma(\varphi, T, A, w))\sigma = \mathtt{Glue}_\Delta(\varphi\sigma, T\sigma, A\sigma, w\sigma) \qquad (3)$$

for $\sigma \colon \Delta \to \Gamma$. We define $\mathtt{unglue}(\varphi, w) \in \mathrm{Ter}(\Gamma.\mathtt{Glue}(\varphi, T, A, w); A\mathsf{p})$ by

$$\mathtt{unglue}(\varphi, w)(\rho, t) = \mathtt{app}(w\rho, t)_{\mathrm{id}_I} \in A\rho \qquad \text{whenever } \varphi\rho = 1_{\mathbb{F}}, \text{ and}$$
$$\mathtt{unglue}(\varphi, w)(\rho, \mathtt{glue}(\varphi, t, a)) = a \qquad \text{otherwise,}$$

where $\rho \in \Gamma(I)$.

Definition 15. For $A, B \in \mathrm{Ty}(\Gamma)$ and $w \in \mathrm{Ter}(\Gamma; A \to B)$ an *equivalence structure* for w is given by the following operations such that for each

- $\rho \in \Gamma(I)$,
- $\varphi \in \mathbb{F}(I)$,
- $b \in B\rho$, and
- partial elements a of $A\rho$ and ω of $\mathtt{Path}_B(\mathtt{app}(w\rho, a), b\iota)\rho$ with extent φ,

we are given $\mathsf{e}_0(\rho, \varphi, b, a, \omega) \in A\rho$, and a path $\mathsf{e}_1(\rho, \varphi, b, a, \omega)$ between

$$\mathtt{app}(w\rho, \mathsf{e}_0(\rho, \varphi, b, a, \omega))$$

and b such that $\mathsf{e}_0(\rho, \varphi, b, a, \omega)\iota = a$, $\mathsf{e}_1(\rho, \varphi, b, a, \omega)\iota = \omega$ (where $\iota\colon \mathbf{y}\, I, \varphi \to \mathbf{y}\, I$) and for any $f\colon J \to I$ and $\nu = 0, 1$:

$$(\mathsf{e}_\nu(\rho, \varphi, b, a, \omega))f = \mathsf{e}_\nu(\rho f, \varphi f, bf, af, \omega f).$$

Following the argument in the syntax we can use the equivalence structure to explain a composition for $\mathtt{Glue}$.

Theorem 16. *If $A \in \mathrm{FTy}(\Gamma)$, $T \in \mathrm{FTy}(\Gamma, \varphi)$, and we have an equivalence structure for w, then we have a composition structure for $\mathtt{Glue}(\varphi, T, A, w)$ such that the equations (3) also hold for the respective composition structures.*

Semantic universes

Assuming a Grothendieck universe of small sets in our ambient set theory, we can define $A \in \text{Ty}_0(\Gamma)$ iff all $A\rho$ are small for $\rho \in \Gamma(I)$; and $A \in \text{FTy}_0(\Gamma)$ iff $A \in \text{Ty}_0(\Gamma)$ when forgetting the composition structure of A.

Definition 17. The semantic universe U is the cubical set defined by $\mathsf{U}(I) = \text{FTy}_0(\mathbf{y}\,I)$; restriction along $f\colon J \to I$ is simply substitution along $\mathbf{y}\,f$.

We can consider U as an element of $\text{Ty}(\Gamma)$. As such we can, as in the syntactic counterpart, define a composition structure on U using semantic glueing, so that $\mathsf{U} \in \text{FTy}(\Gamma)$. Note that semantic glueing preserves smallness.

For $T \in \text{Ter}(\Gamma;\mathsf{U})$ we can define decoding $\mathsf{El}\,T \in \text{FTy}_0(\Gamma)$ by $(\mathsf{El}\,T)\rho = (T\rho)\,\text{id}_I$ and likewise for the composition structure. For $A \in \text{FTy}_0(\Gamma)$ we get its code $\ulcorner A \urcorner \in \text{Ter}(\Gamma;\mathsf{U})$ by setting $\ulcorner A \urcorner \rho \in \text{FTy}_0(\mathbf{y}\,I)$ to be given by the sets $(\ulcorner A \urcorner \rho)f = A(\rho f)$ and likewise for restrictions and composition structure. These operations satisfy $\mathsf{El}\,\ulcorner A \urcorner = A$ and $\ulcorner \mathsf{El}\,T \urcorner = T$.

8.3 Interpretation of the syntax

Following [25] we define a partial interpretation function from raw syntax to the CwF with fibrant types given in the previous section.

To interpret the universe rules à la Russell we assume two Grothendieck universes in the underlying set theory, say *tiny* and *small* sets. So that any tiny set is small, and the set of tiny sets is small. For a cubical set X we define $\text{FTy}_0(X)$ and $\text{FTy}_1(X)$ as in the previous section, now referring to tiny and small sets, respectively. We get semantic universes $\mathsf{U}_i(I) = \text{FTy}_i(\mathbf{y}\,I)$ for $i = 0,1$; we identify those with their lifts to types. As noted above, these lifts carry a composition structure, and thus are fibrant. We also have $\mathsf{U}_0 \subseteq \mathsf{U}_1$ and thus $\text{Ter}(X;\mathsf{U}_0) \subseteq \text{Ter}(X;\mathsf{U}_1)$. Note that coding and decoding are, as set-theoretic operations, the same for both universes. We get that $\ulcorner \mathsf{U}_0 \urcorner \in \text{Ter}(X;\mathsf{U}_1)$ which will serve as the interpretation of U.

In what follows, we define a partial interpretation function of raw syntax: $[\![\Gamma]\!]$, $[\![\Gamma;t]\!]$, and $[\![\Delta;\sigma]\!]$ by recursion on the raw syntax. Since we want to interpret a universe à la Russell we cannot assume terms and types to have different syntactic categories. The definition is given below and should be read such that the interpretation is defined whenever all interpretations on the right-hand sides are defined *and* make sense; so, e.g., for $[\![\Gamma]\!].\mathsf{El}\,[\![\Gamma;A]\!]$ below, we require that $[\![\Gamma]\!]$ is defined and a cubical set, $[\![\Gamma;A]\!]$ is defined, and $\mathsf{El}\,[\![\Gamma;A]\!] \in \text{FTy}([\![\Gamma]\!])$. The interpretation for raw contexts is given by:

$$[\![()]\!] = \top \qquad [\![\Gamma, x : A]\!] = [\![\Gamma]\!].\mathsf{El}\,[\![\Gamma; A]\!] \quad \text{if } x \notin \mathrm{dom}(\Gamma)$$
$$[\![\Gamma, \varphi]\!] = [\![\Gamma]\!], [\![\Gamma; \varphi]\!] \qquad [\![\Gamma, i : \mathbb{I}]\!] = [\![\Gamma]\!].\mathbb{I} \qquad \text{if } i \notin \mathrm{dom}(\Gamma)$$

where $\top$ is a terminal cubical set and in the last equation $\mathbb{I}$ is considered as an element of $\mathrm{Ty}([\![\Gamma]\!])$. When defining $[\![\Gamma; t]\!]$ we require that $[\![\Gamma]\!]$ is defined and a cubical set; then $[\![\Gamma; t]\!]$ is a (partial) family of sets $[\![\Gamma; t]\!](I, \rho)$ for $I \in \mathcal{C}$ and $\rho \in [\![\Gamma]\!](I)$ (leaving I implicit in the definition). We define:

$$[\![\Gamma; \mathsf{U}]\!] = \ulcorner \mathsf{U}_0 \urcorner \in \mathrm{Ter}([\![\Gamma]\!]; \mathsf{U}_1)$$
$$[\![\Gamma; \mathsf{N}]\!] = \ulcorner \mathsf{N} \urcorner \in \mathrm{Ter}([\![\Gamma]\!]; \mathsf{U}_0)$$
$$[\![\Gamma; (x : A) \to B]\!] = \ulcorner \Pi_{[\![\Gamma]\!]}(\mathsf{El}\,[\![\Gamma; A]\!], \mathsf{El}\,[\![\Gamma, x : A; B]\!]) \urcorner$$
$$[\![\Gamma; (x : A) \times B]\!] = \ulcorner \Sigma_{[\![\Gamma]\!]}(\mathsf{El}\,[\![\Gamma; A]\!], \mathsf{El}\,[\![\Gamma, x : A; B]\!]) \urcorner$$
$$[\![\Gamma; \mathsf{Path}\, A\, a\, b]\!] = \ulcorner \mathsf{Path}^{[\![\Gamma]\!]}_{\mathsf{El}\,[\![\Gamma; A]\!]}([\![\Gamma; a]\!], [\![\Gamma; b]\!]) \urcorner$$
$$[\![\Gamma; \mathsf{Glue}\,[\varphi \mapsto (T, w)]\, A]\!] = \ulcorner \mathsf{Glue}_{[\![\Gamma]\!]}([\![\Gamma; \varphi]\!], \mathsf{El}\,[\![\Gamma, \varphi; T]\!], \mathsf{El}\,[\![\Gamma; A]\!], [\![\Gamma, \varphi; w]\!]) \urcorner$$
$$[\![\Gamma; \lambda x : A.t]\!] = \lambda_{[\![\Gamma]\!]; \mathsf{El}\,[\![\Gamma; A]\!]}([\![\Gamma, x : A; t]\!])$$
$$[\![\Gamma; t\, u]\!] = \mathsf{app}([\![\Gamma; t]\!], [\![\Gamma; u]\!])$$
$$[\![\Gamma; \langle i \rangle\, t]\!] = \langle \rangle_{[\![\Gamma]\!]} [\![\Gamma, i : \mathbb{I}; t]\!]$$
$$[\![\Gamma; t\, r]\!] = [\![\Gamma; t]\!][\![\Gamma; r]\!]$$

where for path application, juxtaposition on the right-hand side is semantic path application. In the case of a bound variable, we assume that x (respectively i) is a *chosen* variable fresh for Γ; if this is not possible the expression is undefined. Moreover, all type formers should be read as those on fibrant types, i.e., also defining the composition structure. In the case of Glue, it is understood that the function part, i.e., the fourth argument of Glue in Definition 14 is $\mathsf{p} \circ [\![\Gamma, \varphi; w]\!]$ and the required (by Theorem 16) equivalence structure is to be extracted from $\mathsf{q} \circ [\![\Gamma, \varphi; w]\!]$ as in Section 5.2. In virtue of the remark in Section 8.2 we don't need type annotations to interpret applications. Note that coding and decoding tacitly refer to $[\![\Gamma]\!]$ as well. For the rest of the raw terms we also assume we are given $\rho \in [\![\Gamma]\!](I)$. Variables are interpreted by:

$$[\![\Gamma, x : A; x]\!]\rho = \mathsf{q}(\rho) \qquad [\![\Gamma, x : A; y]\!]\rho = [\![\Gamma; y]\!](\mathsf{p}(\rho)) \qquad [\![\Gamma, \varphi; y]\!]\rho = [\![\Gamma; y]\!]\rho$$

These should also be read to include the case when x or y are name variables; if x is a name variable, we require A to be $\mathbb{I}$. The interpretations of $[\![\Gamma; r]\!]\rho$ where r is not a name and $[\![\Gamma; \varphi]\!]\rho$ follow inductively as elements of $\mathbb{I}$ and $\mathbb{F}$, respectively.

Constants for dependent sums are interpreted by:

$$\llbracket \Gamma; (t,u) \rrbracket \rho = (\llbracket \Gamma; t \rrbracket \rho, \llbracket \Gamma; u \rrbracket \rho) \quad \llbracket \Gamma; t.1 \rrbracket \rho = \mathsf{p}(\llbracket \Gamma; t \rrbracket \rho) \quad \llbracket \Gamma; t.2 \rrbracket \rho = \mathsf{q}(\llbracket \Gamma; t \rrbracket \rho)$$

Likewise, constants for N will be interpreted by their semantic analogues (omitted). The interpretations for the constants related to glueing are

$$\llbracket \Gamma; \mathsf{glue}\,[\varphi \mapsto t]\,u \rrbracket \rho = \mathtt{glue}(\llbracket \Gamma; \varphi \rrbracket \rho, \llbracket \Gamma, \varphi; t \rrbracket \hat{\rho}, \llbracket \Gamma; u \rrbracket \rho)$$
$$\llbracket \Gamma; \mathsf{unglue}\,[\varphi \mapsto w]\,u \rrbracket \rho = \mathtt{unglue}(\llbracket \Gamma; \varphi \rrbracket, \mathsf{p} \circ \llbracket \Gamma; w \rrbracket)(\rho, \llbracket \Gamma; u \rrbracket \rho)$$

where $\llbracket \Gamma, \varphi; t \rrbracket \hat{\rho}$ is the family assigning $\llbracket \Gamma, \varphi; t \rrbracket(\rho f)$ to $J \in \mathcal{C}$ and $f \colon J \to I$ (and ρf refers to the restriction given by $\llbracket \Gamma \rrbracket$ which is assumed to be a cubical set). Partial elements are interpreted by

$$\llbracket \Gamma; [\,\varphi_1\ u_1, \ldots, \varphi_n\ u_n\,] \rrbracket \rho = \llbracket \Gamma, \varphi_i; u_i \rrbracket \rho \quad \text{if } \llbracket \Gamma; \varphi_i \rrbracket \rho = 1_{\mathbb{F}},$$

where for this to be defined we additionally assume that all $\llbracket \Gamma, \varphi_i; u_i \rrbracket$ are defined and $\llbracket \Gamma, \varphi_i; u_i \rrbracket \rho' = \llbracket \Gamma, \varphi_j; u_j \rrbracket \rho'$ for each $\rho' \in \llbracket \Gamma \rrbracket(I)$ with $\llbracket \Gamma; \varphi_i \wedge \varphi_j \rrbracket \rho' = 1_{\mathbb{F}}$.

Finally, the interpretation of composition is given by

$$\llbracket \Gamma; \mathsf{comp}^i\,A\,[\varphi \mapsto u]\,a_0 \rrbracket \rho = \mathsf{comp}_{\mathsf{El}\,\llbracket \Gamma, i : \mathbb{I}; A \rrbracket}(I, j, \rho', \llbracket \Gamma; \varphi \rrbracket \rho, \llbracket \Gamma, \varphi, i \colon \mathbb{I}; u \rrbracket \rho', \llbracket \Gamma; a_0 \rrbracket \rho)$$

if $i \notin \mathrm{dom}(\Gamma)$, and where j is fresh and $\rho' = (\rho \mathsf{s}_j, i = j)$ with $\mathsf{s}_j \colon I, j \to I$ induced from the inclusion $I \subseteq I, j$.

The interpretation of substitutions $\llbracket \Delta; \sigma \rrbracket$ is a (partial) family of sets $\llbracket \Delta; \sigma \rrbracket(I, \rho)$ for $I \in \mathcal{C}$ and $\rho \in \llbracket \Delta \rrbracket(I)$. We set

$$\llbracket \Delta; () \rrbracket \rho = *, \qquad \llbracket \Delta; (\sigma, x/t) \rrbracket \rho = (\llbracket \Delta; \sigma \rrbracket \rho, \llbracket \Delta; t \rrbracket \rho) \quad \text{if } x \notin \mathrm{dom}(\sigma),$$

where $*$ is the unique element of $\top(I)$. This concludes the definition of the interpretation of syntax.

In the following α stands for either a raw term or raw substitution. In the latter case, $\alpha \sigma$ denotes composition of substitutions.

Lemma 18. *Let Γ' be like Γ but with some φ's inserted, and assume both $\llbracket \Gamma \rrbracket$ and $\llbracket \Gamma' \rrbracket$ are defined; then:*

1. *$\llbracket \Gamma' \rrbracket$ is a sub-cubical set of $\llbracket \Gamma \rrbracket$;*

2. *if $\llbracket \Gamma; \alpha \rrbracket$ is defined, then so is $\llbracket \Gamma'; \alpha \rrbracket$ and they agree on $\llbracket \Gamma' \rrbracket$.*

Lemma 19 (Weakening). *Let $\llbracket \Gamma \rrbracket$ be defined.*

1. If $[\![\Gamma, x : A, \Delta]\!]$ is defined, then so is $[\![\Gamma, x : A, \Delta; x]\!]$ which is moreover the projection to the x-part.[5]

2. If $[\![\Gamma, \Delta]\!]$ is defined, then so is $[\![\Gamma, \Delta; \mathrm{id}_\Gamma]\!]$ which is moreover the projection to the Γ-part.

3. If $[\![\Gamma, \Delta]\!]$, $[\![\Gamma; \alpha]\!]$ are defined and the variables in Δ are fresh for α, then $[\![\Gamma, \Delta; \alpha]\!]$ is defined and for $\rho \in [\![\Gamma, \Delta]\!](I)$:

$$[\![\Gamma; \alpha]\!]([\![\Gamma, \Delta; \mathrm{id}_\Gamma]\!]\rho) = [\![\Gamma, \Delta; \alpha]\!]\rho$$

Lemma 20 (Substitution). *For $[\![\Gamma]\!]$, $[\![\Delta]\!]$, $[\![\Delta; \sigma]\!]$, and $[\![\Gamma; \alpha]\!]$ defined with $\mathrm{dom}(\Gamma) = \mathrm{dom}(\sigma)$ (as lists), also $[\![\Delta; \alpha\sigma]\!]$ is defined and for $\rho \in [\![\Delta]\!](I)$:*

$$[\![\Gamma; \alpha]\!]([\![\Delta; \sigma]\!]\rho) = [\![\Delta; \alpha\sigma]\!]\rho$$

Lemma 21. *If $[\![\Gamma]\!]$ is defined and a cubical set, and $[\![\Gamma; \alpha]\!]$ is defined, then*

$$([\![\Gamma; \alpha]\!]\rho)f = [\![\Gamma; \alpha]\!](\rho f)$$

To state the next theorem let us set $[\![\Gamma; \mathbb{I}]\!] = \ulcorner \mathbb{I} \urcorner$ and $[\![\Gamma; \mathbb{F}]\!] = \ulcorner \mathbb{F} \urcorner$ as elements of $\mathrm{Ty}_0([\![\Gamma]\!])$.

Theorem 22 (Soundness). *We have the following implications, and all occurrences of $[\![-]\!]$ in the conclusions are defined. In (3) and (5) we allow A to be $\mathbb{I}$ or $\mathbb{F}$.*

1. *if $\Gamma \vdash$, then $[\![\Gamma]\!]$ is a cubical set;*

2. *if $\Gamma \vdash A$, then $[\![\Gamma; A]\!] \in \mathrm{Ter}([\![\Gamma]\!]; \mathsf{U}_1)$;*

3. *if $\Gamma \vdash t : A$, then $[\![\Gamma; t]\!] \in \mathrm{Ter}([\![\Gamma]\!]; \mathsf{El}\,[\![\Gamma; A]\!])$;*

4. *if $\Gamma \vdash A = B$, then $[\![\Gamma; A]\!] = [\![\Gamma; B]\!]$;*

5. *if $\Gamma \vdash a = b : A$, then $[\![\Gamma; a]\!] = [\![\Gamma; b]\!]$;*

6. *if $\Gamma \vdash \sigma : \Delta$, then $[\![\Gamma; \sigma]\!]$ restricts to a natural transformation $[\![\Gamma]\!] \to [\![\Delta]\!]$.*

9 Extensions: identity types and higher inductive types

In this section we consider possible extensions to cubical type theory. The first is an identity type defined using path types whose elimination principle holds as a judgmental equality. The second are two examples of higher inductive types.

---

[5]E.g., if Γ is $y : B, z : C$, the projection to the x-part maps $(b, (c, (a, \delta)))$ to a, and the projection to the Γ-part maps $(b, (c, \delta))$ to (b, c).

9.1 Identity types

We can use the path type to represent equalities. Using the composition operation, we can indeed build a substitution function $P(a) \to P(b)$ from any path between a and b. However, since we don't have in general the judgmental equality $\mathsf{transp}^i\ A\ a_0 = a_0$ if A is independent of i (which is an equality that we cannot expect geometrically in general, as shown in Appendix B), this substitution function does not need to be the constant function when the path is constant. This means that, as in the previous model [6, 14], we don't get an interpretation of Martin-Löf identity type [20] with the standard judgmental equalities.

However, we can define another type which *does* give an interpretation of this identity type following an idea of Andrew Swan.

Identity types

The basic idea of $\mathsf{Id}\ A\ a_0\ a_1$ is to define it in terms of $\mathsf{Path}\ A\ a_0\ a_1$ but also mark the paths where they are known to be constant. Formally, the formation and introduction rules are

$$\frac{\Gamma \vdash A \quad \Gamma \vdash a_0 : A \quad \Gamma \vdash a_1 : A}{\Gamma \vdash \mathsf{Id}\ A\ a_0\ a_1} \qquad \frac{\Gamma \vdash \omega : \mathsf{Path}\ A\ a_0\ a_1[\varphi \mapsto \langle i \rangle\ a_0]}{\Gamma \vdash (\omega, \varphi) : \mathsf{Id}\ A\ a_0\ a_1}$$

and we can define $\mathsf{r}\ a = (1_a, 1_\mathbb{F}) : \mathsf{Id}\ A\ a\ a$ for $a : A$. The elimination rule, given $\Gamma \vdash a : A$, is

$$\frac{\Gamma, x : A, \alpha : \mathsf{Id}\ A\ a\ x \vdash C \quad \Gamma \vdash d : C(x/a, \alpha/\mathsf{r}\ a) \quad \Gamma \vdash b : A \quad \Gamma \vdash \beta : \mathsf{Id}\ A\ a\ b}{\Gamma \vdash \mathsf{J}_{x,\alpha.C}\ d\ b\ \beta : C(x/b, \alpha/\beta)}$$

together with the following judgmental equality in case β is of the form (ω, φ)

$$\mathsf{J}\ d\ b\ \beta = \mathsf{comp}^i\ C(x/\omega\ i, \alpha/\beta^*(i))\ [\varphi \mapsto d]\ d$$

where $\Gamma, i : \mathbb{I} \vdash \beta^*(i) : \mathsf{Id}\ A\ a\ (\omega\ i)$ is given by

$$\beta^*(i) = (\langle j \rangle\ \omega\ (i \wedge j), \varphi \vee (i = 0)).$$

Note that with this definition we get $\mathsf{J}\ d\ a\ (\mathsf{r}\ a) = d$ as desired.

The composition operation for Id is explained as follows. Given $\Gamma, i : \mathbb{I} \vdash \mathsf{Id}\ A\ a_0\ a_1$, $\Gamma, \varphi, i : \mathbb{I} \vdash (\omega, \psi) : \mathsf{Id}\ A\ a_0\ a_1$, and $\Gamma \vdash (\omega_0, \psi_0) : (\mathsf{Id}\ A\ a_0\ a_1)(i0)[\varphi \mapsto (\omega(i0), \psi(i0))]$ we have the judgmental equality

$$\mathsf{comp}^i\ (\mathsf{Id}\ A\ a_0\ a_1)[\varphi \mapsto (\omega, \psi)](\omega_0, \psi_0) = (\mathsf{comp}^i\ (\mathsf{Path}\ A\ a_0\ a_1)[\varphi \mapsto \omega]\omega_0, \varphi \wedge \psi(i1))$$

3157

It can then be shown that the types Id $A\ a\ b$ and Path $A\ a\ b$ are (Path)-equivalent. In particular, a type is (Path)-contractible if, and only if, it is (Id)-contractible. The univalence axiom, proved in Section 7.2 for the Path-type, hence holds as well for the Id-type.[6]

Cofibration-trivial fibration factorization

The same idea can be used to factorize an arbitrary map of (not necessary fibrant) cubical sets $f : A \to B$ into a cofibration followed by a trivial fibration. We define a "trivial fibration" to be a first projection from a total space of a contractible family of types and a "cofibration" to be a map that has the left lifting property against any trivial fibration. For this we define, for $b : B$, the type $T_f(b)$ to be the type of elements $[\varphi \mapsto a]$ with $\varphi \vdash a : A$ and $\varphi \vdash f\ a = b : B$.

Theorem 23. *The type $T_f(b)$ is contractible and the map*

$$A \to (b : B) \times T_f(b), \qquad a \longmapsto (f\ a, [1_\mathbb{F} \mapsto a])$$

is a cofibration.

The definition of the identity type can be seen as a special case of this, if we take the B the type of paths in A and for f the constant path function.

9.2 Higher inductive types

In this section we consider the extension of cubical type theory with two different higher inductive types: spheres and propositional truncation. The presentation in this section is syntactical, but it can be directly translated into semantic definitions.

Extension to dependent path types

In order to formulate the elimination rules for higher inductive types, we need to extend the path type to *dependent path type*, which is described by the following rules. If $i : \mathbb{I} \vdash A$ and $\vdash a_0 : A(i0)$, $a_1 : A(i1)$, then $\vdash$ Path$^i\ A\ a_0\ a_1$. The introduction rule is that $\vdash \langle i \rangle\ t :$ Path$^i\ A\ t(i0)\ t(i1)$ if $i : \mathbb{I} \vdash t : A$. The elimination rule is $\vdash p\ r : A(i/r)$ if $\vdash p :$ Path$^i\ A\ a_0\ a_1$ with equalities $p\ 0 = a_0 : A(i0)$ and $p\ 1 = a_1 : A(i1)$.

---

[6]This has been formally verified using the HASKELL implementation:
https://github.com/mortberg/cubicaltt/blob/v1.0/examples/idtypes.ctt

Spheres, syntactical presentation

We define the circle S^1 by the rules

$$\frac{\Gamma \vdash}{\Gamma \vdash S^1} \qquad \frac{\Gamma \vdash}{\Gamma \vdash \mathsf{base} : S^1} \qquad \frac{\Gamma \vdash r : \mathbb{I}}{\Gamma \vdash \mathsf{loop}\ r : S^1}$$

with the equalities $\mathsf{loop}\ 0 = \mathsf{loop}\ 1 = \mathsf{base}$.

Since we want to represent the "free" type with one base point and a loop, we add composition as a *constructor* operation hcomp^i (which binnds i in u)

$$\frac{\Gamma, \varphi, i : \mathbb{I} \vdash u : S^1 \qquad \Gamma \vdash u_0 : S^1[\varphi \mapsto u(i0)]}{\Gamma \vdash \mathsf{hcomp}^i\ [\varphi \mapsto u]\ u_0 : S^1[\varphi \mapsto u(i1)]}$$

Given a dependent type $x : S^1 \vdash A$ and $a : A(x/\mathsf{base})$ and $l : \mathsf{Path}^i\ A(x/\mathsf{loop}\ i)\ a\ a$ we can define a function $g : \Pi(x : S^1)A$ by the equations[7]

$$g\ \mathsf{base} = a \qquad g\ (\mathsf{loop}\ r) = l\ r$$

This definition is non ambiguous since $l\ 0 = l\ 1 = a$ and we get *judgmental* computation rules. Finally

$$g\ (\mathsf{hcomp}^i\ [\varphi \mapsto u]\ u_0) = \mathsf{comp}^i\ A(x/v)\ [\varphi \mapsto g\ u]\ (g\ u_0)$$

where $v = \mathsf{fill}^i\ S^1\ [\varphi \mapsto u]\ u_0 = \mathsf{hcomp}^j\ [\varphi \mapsto u(i/i \wedge j), (i=0) \mapsto u_0]\ u_0$.

We have a similar definition for S^n taking as constructors base and $\mathsf{loop}\ r_1\ \ldots\ r_n$.

Spheres, semantical presentation

We suppose to have a fresh name function on the set of names, with $\mathsf{fresh}(I)$ being a name not in I, and we write $I^+ = I, \mathsf{fresh}(I)$. We can define in a functorial way $f^+ : J^+ \to I^+$ extending $f : J \to I$ by sending $\mathsf{fresh}(I)$ to $\mathsf{fresh}(J)$. We also have for natural transformations the projection $p : I^+ \to I$ and the map $e_0 : I \to I^+$ (resp. $e_1 : I \to I^+$) sending $\mathsf{fresh}(I)$ to 0 (resp. 1).

We define first a family of sets $X(I)$ which is an "upper approximation" of the circle, together with maps $X(I) \to X(J)$, $v \mapsto vf$ for $f : J \to I$. An element of $X(I)$ is of the form base or $\mathsf{loop}\ r$ with $r \neq 0, 1$ in $\mathbb{I}(I)$ or of the form $\mathsf{hcomp}\ [\psi \mapsto u]\ u_0$ with $\psi \neq 1$ in $\mathbb{F}(I)$ and u_0 in $X(I)$ and u a family of elements u_f in $X(J^+)$ for

---

[7] For the equation $g\ (\mathsf{loop}\ r) = l\ r$, it may be that l and r are dependent on the same name i, and this could not work without a diagonal operation on names.

$f : J \to I$ such that $\psi f = 1$. In this way an element of $X(I)$ can be seen as a well-founded tree. We can define uf in $X(J)$ for $f : J \to I$ by induction on u. We take base $f = $ base and (loop $r)f = $ loop (rf) if $rf \neq 0, 1$ and (loop $r)f = $ base if rf is 0 or 1. Finally (hcomp $[\psi \mapsto u]\ u_0)f$ is $u_f e_1$ if $\psi f = 1$ and hcomp $[\psi f \mapsto uf^+]\ (u_0 f)$ if $\psi f \neq 1$ where uf^+ is the family $(uf^+)_g = u_{fg}$ for $g : K \to J$.

We then define the subset $\mathsf{S}^1(I) \subseteq X(I)$ by taking the elements base and loop r and hcomp $[\psi \mapsto u]\ u_0$ such that u_0 in $\mathsf{S}^1(I)$ and each u_f in $\mathsf{S}^1(J^+)$ and $u_0 f = u_f e_0$ and $u_f g^+ = u_{fg}$ for $f : J \to I$ and $g : K \to J$. This defines a cubical set S^1, such that $\mathsf{S}^1(I)$ is a subset of $X(I)$ for each I.

Propositional truncation, syntactical presentation

We define the propositional truncation $\|A\|$ of a type A by the rules:

$$\frac{\Gamma \vdash A}{\Gamma \vdash \|A\|} \qquad \frac{\Gamma \vdash a : A}{\Gamma \vdash \mathsf{inc}\ a : \|A\|} \qquad \frac{\Gamma \vdash u_0 : \|A\| \quad \Gamma \vdash u_1 : \|A\| \quad \Gamma \vdash r : \mathbb{I}}{\Gamma \vdash \mathsf{squash}\ u_0\ u_1\ r : \|A\|}$$

with the equalities $\mathsf{squash}\ u_0\ u_1\ 0 = u_0$ and $\mathsf{squash}\ u_0\ u_1\ 1 = u_1$.

As before, we add composition as a constructor, but only in the form[8]

$$\frac{\Gamma \vdash A \quad \Gamma, \varphi, i : \mathbb{I} \vdash u : \|A\| \quad \Gamma \vdash u_0 : \|A\|\,[\varphi \mapsto u(i0)]}{\Gamma \vdash \mathsf{hcomp}^i\ [\varphi \mapsto u]\ u_0 : \|A\|\,[\varphi \mapsto u(i1)]}$$

This only provides a composition operation $\mathsf{comp}^i\ \|A\|\ [\varphi \mapsto u]\ u_0$ in the case where A is independent of i, and we have to explain how to define the general case.

Given $x : \|A\| \vdash B$ and

$$q : \Pi(x_0 : \|A\|)(y_0 : B(x_0))(x_1 : \|A\|)(y_1 : B(x_1))\mathsf{Path}^i\ B(\mathsf{squash}\ x_0\ x_1\ i)\ y_0\ y_1$$

and $f : \Pi(x : A)B(\mathsf{inc}\ x)$ we define $g : \Pi(x : \|A\|)B$ by the equations

$$\begin{aligned}g\ (\mathsf{inc}\ a) &= f\ a \\ g\ (\mathsf{squash}\ u_0\ u_1\ r) &= q\ u_0\ (g\ u_0)\ u_1\ (g\ u_1)\ r \\ g\ (\mathsf{hcomp}^i\ [\varphi \mapsto u]\ u_0) &= \mathsf{comp}^i\ B(v)\ [\varphi \mapsto g\ u]\ (g\ u_0)\end{aligned}$$

where $v = \mathsf{hcomp}^j\ [\varphi \mapsto u(i/i \wedge j), (i = 0) \mapsto u_0]\ u_0$.

---

[8] The restriction of A being independent of i on the constructor is essential for the justification of the elimination rule, as explained in the Comments at the end.

We still have to define the general composition operation. We first define an operation of *flattening an open box*

$$\frac{\Gamma, i : \mathbb{I} \vdash A \quad \Gamma \vdash r : \mathbb{I} \quad \Gamma \vdash u : \|A(i/r)\|}{\Gamma \vdash \mathsf{forward}_{i.A}\ r\ u : \|A(i/1)\|\ [(r=1) \mapsto u]}$$

by the equations

$$\begin{aligned}
\mathsf{forward}\ r\ (\mathsf{inc}\ a) &= \mathsf{inc}\ (\mathsf{comp}^i\ A(i \vee r)\ [(r=1) \mapsto a]\ a) \\
\mathsf{forward}\ r\ (\mathsf{squash}\ u_0\ u_1\ s) &= \mathsf{squash}\ (\mathsf{forward}\ r\ u_0)\ (\mathsf{forward}\ r\ u_1)\ s \\
\mathsf{forward}\ r\ (\mathsf{hcomp}^j\ [\varphi \mapsto u]\ u_0) &= \mathsf{hcomp}^j\ [\varphi \mapsto \mathsf{forward}\ r\ u]\ (\mathsf{forward}\ r\ u_0)
\end{aligned}$$

Using this operation, we can define a general composition operation[9]

$$\frac{\Gamma, i : \mathbb{I} \vdash A \quad \Gamma, \varphi, i : \mathbb{I} \vdash u : \|A\| \quad \Gamma \vdash u_0 : \|A(i0)\|\ [\varphi \mapsto u(i0)]}{\Gamma \vdash \mathsf{comp}^i\ \|A\|\ [\varphi \mapsto u]\ u_0 : \|A(i1)\|\ [\varphi \mapsto u(i1)]}$$

by $\Gamma \vdash \mathsf{comp}^i\ \|A\|\ [\varphi \mapsto u]\ u_0 = \mathsf{hcomp}^i\ [\varphi \mapsto \mathsf{forward}\ i\ u]\ (\mathsf{forward}\ 0\ u_0) : \|A(i1)\|$.

Propositional truncation, semantical presentation

Given A in $\mathrm{FTy}(\Gamma)$ we define $\|A\|$ in $\mathrm{FTy}(\Gamma)$. For this, we define first an "upper approximation" given by a family of sets $X\rho$ for ρ in $\Gamma(I)$ and maps $X\rho \to X\rho f$ for $f : J \to I$. An element of $X\rho$ is of the form $\mathsf{inc}\ a$ with a in $A\rho$ or $\mathsf{squash}\ u_0\ u_1\ r$ with $r \neq 0, 1$ in $\mathbb{I}(I)$ and u_0 in $X\rho$ and u_1 in $X\rho$ or of the form $\mathsf{hcomp}\ [\psi \mapsto u]\ u_0$ with $\psi \neq 1$ in $\mathbb{F}(I)$ and u_0 in $X\rho$ and u a family of elements u_f in $X(\rho f p)$ for $f : J \to I$ such that $\psi f = 1$. Each element in $X\rho$ can be seen as a well-founded tree.

We can then define uf in $X\rho f$ for u in $X\rho$ and $f : J \to I$ by induction on u.

Then $\|A\|\rho$ is defined to be the subset of $X\rho$ of elements $\mathsf{inc}\ a$ or $\mathsf{squash}\ u_0\ u_1\ r$ with u_0 and u_1 in $\|A\|\rho$ and $\mathsf{hcomp}\ [\psi \mapsto u]\ u_0$ with u_0 in $\|A\|\rho$ and $u_f e_0 = u_0 f$ and each u_f in $\|A\|\ (\rho f p)$ and $u_f g^+ = u_{fg}$ for $g : J \to K$.

It is then possible to define a composition structure for $\|A\|$ if we have a composition structure for A exactly as it is done syntactically.

Comments

Universes

This operation is stable under substitution: if $\sigma : \Delta \to \Gamma$ and A is in $\mathrm{FTy}(\Gamma)$ then $\|A\|\sigma = \|A\sigma\|$. Also each $\|A\|\rho$ is small if each $A\rho$ is a small set. This means that

---

[9] The open box is given by $\varphi \mapsto u$ and u_0 and it is flattened in the $\|A(i/1)\|$ type by the $\mathsf{forward}$ operation.

the univalent universe that we have defined previously is stable by propositional truncation.

We expect that the same method of defining a composition by "flattening an open box" can be used to define other higher inductive types (suspension, push-out, ...). It avoids coherence issues, and an application is that the addition of higher inductive types and univalence to type theory does not raise its proof-theoretic power. Indeed, all we do can be modelled in Aczel's system $\text{CZFu}_{<\omega}$, which is interpretable in type theory with universes.

Flattening open boxes

One key step is the restriction of the constructor to the form

$$\frac{\Gamma \vdash T \quad \Gamma, \varphi, i : \mathbb{I} \vdash u : \|T\| \quad \Gamma \vdash u_0 : \|T\| [\varphi \mapsto u(i0)]}{\Gamma \vdash \mathsf{hcomp}^i \, [\varphi \mapsto u] \, u_0 : \|T\| [\varphi \mapsto u(i1)]}$$

instead of representing directly composition as a constructor (which is what we tried first to implement)

$$\frac{\Gamma, i : \mathbb{I} \vdash T \quad \Gamma, \varphi, i : \mathbb{I} \vdash u : \|T\| \quad \Gamma \vdash u_0 : \|T(i/0)\| [\varphi \mapsto u(i0)]}{\Gamma \vdash \mathsf{comp}^i \, [\varphi \mapsto u] \, u_0 : \|T(i/1)\| [\varphi \mapsto u(i1)]}$$

Indeed, with this later choice, it does not seem possible to define even a non-dependent function $g : \|A\| \to B$ given $f : A \to B$ and $q : \Pi(x\ y : B)B$. We can define g (inc a) = $f\ a$ and g (squash $u_0\ u_1\ r$) = $q\ (g\ u_0)\ (g\ u_1)\ r$ but it is not clear how to define g ($\mathsf{hcomp}^i\ [\varphi \mapsto u]\ u_0$) since we only know at this point that we have some path $i : \mathbb{I} \vdash T$ such that $A = T(i/1)$ and $u_0 : T(i/0)$ and there is no way to apply an induction for defining g ($\mathsf{comp}^i\ [\varphi \mapsto u]\ u_0$).

Inductive definition

We have used a generalized inductive definition in the definition of $\mathsf{S}^1(I)$. Actually, it is possible to see each element of $\mathsf{S}^1(I)$ as a finite object, since a partial element u of extent ψ, which is a family u_f in $\mathsf{S}^1(J)$ for each $f : J \to I$ such that $\psi f = 1$, is actually completely determined by the finite set of elements u_f where f is a face map (J is a subset of I and $f(i)$ can only take the value i or 0 or 1).

10 Related and future work

Cubical ideas have proved useful to reason about equality in homotopy type theory [19]. In cubical type theory these techniques could be simplified as there are new judgmental equalities and better notations for manipulating higher dimensional cubes. Indeed some simple experiments using the HASKELL implementation have shown that we can simplify some constructions in synthetic homotopy theory.[10]

Other approaches to extending intensional type theory with extensionality principles can be found in [2, 24]. These approaches have close connections to techniques for internalizing parametricity in type theory [5]. Further, nominal extensions to λ-calculus and semantical ideas related to the ones presented in this paper have recently also proved useful for justifying type theory with internalized parametricity [4].

The paper [12] provides a general framework for analyzing the uniformity condition, which applies to simplicial and cubical sets.

Large parts of the semantics presented in this paper have been formally verified in NuPrl by Mark Bickford[11], in particular, the definition of Kan filling in terms of composition as in Section 4.4 and composition for glueing as given in Section 6.2 and composition for the universe as in Section 7.1.

Following the usual reducibility method, we expect it to be possible to adapt our presheaf semantics to a proof of normalization and decidability of type checking. A first step in this direction is the proof of canonicity in [15]. We end the paper with a list of open problems and conjectures:

1. Extend the semantics of identity types to the semantics of inductive families.

2. Give a general syntax and semantics of higher inductive types.

3. Extend the system with resizing rules and show normalization.

4. Is there a model where **Path** and **Id** coincide?

Acknowledgements This work originates from discussions between the four authors around an implementation of a type system corresponding to the model described in [6]. This implementation indicated a problem with the representation of higher inductive types, e.g., the elimination rule for the circle, and suggested the need of extending this cubical model with a diagonal operation. The general framework (uniformity condition, connections, semantics of spheres and propositional truncation) is due to the second author. In particular, the glueing operation with its composition was introduced as a generalization of the operation described in [6] transforming an equivalence into a path, and with the condition $A = \mathsf{Glue}\,[]\,A$. In a first attempt, we tried to force "regularity", i.e., the equation $\mathsf{transp}\ i\ A\ a_0 = a_0$

---

[10]For details see: https://github.com/mortberg/cubicaltt/tree/master/examples/
[11]For details see: http://www.nuprl.org/wip/Mathematics/cubical!type!theory/

if A is independent of i (which seemed to be necessary in order to get filling from compositions, and which implies Path = Id). There was a problem however for getting regularity for the universe, that was discovered by Dan Licata (from discussions with Carlo Angiuli and Bob Harper). Thanks to this discovery, it was realized that regularity is actually not needed for the model to work. In particular, the second author adapted the definition of filling from composition as in Section 4.4, the third author noticed that we can remove the condition $A = $ Glue [] A, and together with the last author, they derived the univalence axiom from the glueing operation as presented in the appendix. This was surprising since glueing was introduced a priori only as a way to transform equivalences into paths, but was later explained by a remark of Dan Licata (also presented in the appendix: we get univalence as soon as the transport map associated to this path is path equal to the given equivalence). The second author introduced then the restriction operation Γ, φ on contexts, which, as noticed by Christian Sattler, can be seen as an explicit syntax for the notion of cofibration, and designed the other proof of univalence in Section 7.2 from discussions between Nicola Gambino, Peter LeFanu Lumsdaine and the third author. Not having regularity, the type of paths is not the same as the Id type but, as explained in Section 9.1, we can recover the usual identity type from the path type, following an idea of Andrew Swan. This version incorporates also an important simplification for the composition of propositional truncation, due to the third author.

The authors would like to thank the referees and Martín Escardó, Georges Gonthier, Dan Grayson, Peter Hancock, Dan Licata, Peter LeFanu Lumsdaine, Christian Sattler, Andrew Swan, Vladimir Voevodsky for many interesting discussions and remarks.

References

[1] P. Aczel. On Relating Type Theories and Set Theories. In T. Altenkirch, B. Reus, and W. Naraschewski, editors, *Types for Proofs and Programs*, volume 1657 of *Lecture Notes in Computer Science*, pages 1–18. Springer Verlag, Berlin, Heidelberg, New York, 1999.

[2] T. Altenkirch. Extensional Equality in Intensional Type Theory. In *14th Annual IEEE Symposium on Logic in Computer Science*, pages 412–420, 1999.

[3] R. Balbes and P. Dwinger. *Distributive Lattices*. Abstract Space Publishing, 2011.

[4] J.-P. Bernardy, T. Coquand, and G. Moulin. A Presheaf Model of Parametric Type Theory. *Electronic Notes in Theoretical Computer Science*, 319:67–82, 2015.

[5] J.-P. Bernardy and G. Moulin. Type-theory in Color. *SIGPLAN Not.*, 48(9):61–72, September 2013.

[6] M. Bezem, T. Coquand, and S. Huber. A Model of Type Theory in Cubical Sets. In R. Matthes and A. Schubert, editors, *19th International Conference on Types for Proofs and Programs (TYPES 2013)*, volume 26 of *Leibniz International Proceedings in Informatics (LIPIcs)*, pages 107–128. Schloss Dagstuhl–Leibniz-Zentrum fuer Informatik, 2014.

[7] R. Brown, P.J. Higgins, and R. Sivera. *Nonabelian Algebraic Topology: Filtered Spaces, Crossed Complexes, Cubical Homotopy Groupoids*, volume 15 of *EMS tracts in mathematics*. European Mathematical Society, 2011.

[8] D.-C. Cisinski. Univalent universes for elegant models of homotopy types. *arXiv:1406.0058*, May 2014. Preprint.

[9] C. Cohen, T. Coquand, S. Huber, and A. Mörtberg. Cubical Type Theory: A Constructive Interpretation of the Univalence Axiom. In T. Uustalu, editor, *21st International Conference on Types for Proofs and Programs (TYPES 2015)*, volume 69 of *Leibniz International Proceedings in Informatics (LIPIcs)*. Schloss Dagstuhl–Leibniz-Zentrum fuer Informatik, 2018. doi:10.4230/LIPIcs.TYPES.2015.5.

[10] P. Dybjer. Internal Type Theory. In *Lecture Notes in Computer Science*, pages 120–134. Springer Verlag, Berlin, Heidelberg, New York, 1996.

[11] M. Fourman and D. Scott. Sheaves and logic. In M. Fourman, C. Mulvey, and D. Scott, editors, *Applications of Sheaves*, volume 753 of *Lecture Notes in Mathematics*, pages 302–401. Springer Berlin Heidelberg, 1979.

[12] N. Gambino and C. Sattler. Uniform Fibrations and the Frobenius Condition. *arXiv:1510.00669*, October 2015. Preprint.

[13] M. Hofmann. Syntax and semantics of dependent types. In A.M. Pitts and P. Dybjer, editors, *Semantics and logics of computation*, volume 14 of *Publ. Newton Inst.*, pages 79–130. Cambridge University Press, Cambridge, 1997.

[14] S. Huber. *A Model of Type Theory in Cubical Sets*. Licentiate thesis, University of Gothenburg, May 2015.

[15] S. Huber. Canonicity for cubical type theory. *arXiv:1607.04156v1*, July 2016. Preprint.

[16] J. A. Kalman. Lattices with involution. *Transactions of the American Mathematical Society*, 87:485–491, 1958.

[17] D. M. Kan. Abstract homotopy. I. *Proceedings of the National Academy of Sciences of the United States of America*, 41(12):1092–1096, 1955.

[18] C. Kapulkin and P. LeFanu Lumsdaine. The Simplicial Model of Univalent Foundations (after Voevodsky). *arXiv:1211.2851v4*, November 2012. Preprint.

[19] D. R. Licata and G. Brunerie. A Cubical Approach to Synthetic Homotopy Theory. In *30th Annual ACM/IEEE Symposium on Logic in Computer Science*, pages 92–103, 2015.

[20] P. Martin-Löf. An intuitionistic theory of types: predicative part. In *Logic Colloquium '73 (Bristol, 1973)*, pages 73–118. Studies in Logic and the Foundations of Mathematics, Vol. 80. North-Holland, Amsterdam, 1975.

[21] P. Martin-Löf. An intuitionistic theory of types. In G. Sambin and J. M. Smith, editors,

Twenty-five years of constructive type theory (Venice, 1995), volume 36 of *Oxford Logic Guides*, pages 127–172. Oxford University Press, 1998.

[22] A. M. Pitts. *Nominal Sets: Names and Symmetry in Computer Science*. Cambridge University Press, New York, NY, USA, 2013.

[23] A. M. Pitts. Nominal Presentation of Cubical Sets Models of Type Theory. In H. Herbelin, P. Letouzey, and M. Sozeau, editors, *20th International Conference on Types for Proofs and Programs (TYPES 2014)*, volume 39 of *Leibniz International Proceedings in Informatics (LIPIcs)*, pages 202–220. Schloss Dagstuhl – Leibniz-Zentrum fuer Informatik, 2015.

[24] Andrew Polonsky. Extensionality of lambda-*. In H. Herbelin, P. Letouzey, and M. Sozeau, editors, *20th International Conference on Types for Proofs and Programs (TYPES 2014)*, volume 39 of *Leibniz International Proceedings in Informatics (LIPIcs)*, pages 221–250. Schloss Dagstuhl–Leibniz-Zentrum fuer Informatik, 2015.

[25] T. Streicher. *Semantics of Type Theory*. Progress in Theoretical Computer Science. Birkhäuser Basel, 1991.

[26] A. Swan. An Algebraic Weak Factorisation System on 01-Substitution Sets: A Constructive Proof. *arXiv:1409.1829*, September 2014. Preprint.

[27] The Univalent Foundations Program. *Homotopy Type Theory: Univalent Foundations of Mathematics*. http://homotopytypetheory.org/book, Institute for Advanced Study, 2013.

[28] V. Voevodsky. The equivalence axiom and univalent models of type theory. (Talk at CMU on February 4, 2010). *arXiv:1402.5556*, February 2014. Preprint.

A Univalence from glueing

We also give two alternative proofs of the univalence axiom for **Path** only involving the glue construction.[12] The first is a direct proof of the standard formulation of the univalence axiom while the second goes through an alternative formulation as in Corollary 9.[13]

Lemma 24. *For $\Gamma \vdash A : \mathsf{U}$, $\Gamma \vdash B : \mathsf{U}$, and an equivalence $\Gamma \vdash w : \mathsf{Equiv}\ A\ B$ we have the following constructions:*

1. $\Gamma \vdash \mathsf{eqToPath}\, w : \mathsf{Path}\, \mathsf{U}\, A\, B$;

2. $\Gamma \vdash \mathsf{Path}\, (A \to B)\, (\mathsf{transp}^i(\mathsf{eqToPath}\, w\, i)))\, w.1$ *is inhabited; and*

---

[12] The proofs of the univalence axiom have all been formally verified inside the system using the HASKELL implementation. We note that the proof of Theorem 8 can be given such that it extends $w.2$ and hence in Corollary 9 we do not need the fact that $\mathsf{isEquiv}\, X\, A\, w.1$ is a proposition. For details see: https://github.com/mortberg/cubicaltt/blob/v1.0/examples/univalence.ctt

[13] The second of these proofs is inspired by a proof by Dan Licata from: https://groups.google.com/d/msg/homotopytypetheory/j2KBIvDw53s/YTDK4DONFQAJ

3. *if $w = \mathsf{equiv}^i(P\,i)$ for $\Gamma \vdash P : \mathsf{Path}\,\mathsf{U}\,A\,B$, then the following type is inhabited:*

$$\Gamma \vdash \mathsf{Path}\,(\mathsf{Path}\,\mathsf{U}\,A\,B)\,(\mathsf{eqToPath}\,(\mathsf{equiv}^i(P\,i)))\,P$$

Proof. For (1) we define

$$\mathsf{eqToPath}\,w = \langle i \rangle\,\mathsf{Glue}\,[(i=0) \mapsto (A,w), (i=1) \mapsto (B, \mathsf{equiv}^k B)]\,B. \qquad (4)$$

Note that here $\mathsf{equiv}^k B$ is an equivalence between B and B (see Section 7.1). For (2) we have to closely look at how the composition was defined for Glue. By unfolding the definition, we see that the left-hand side of the equality is equal $w.1$ composed with multiple transports in a constant type; using filling and functional extensionality, these transports can be shown to be equal to the identity; for details see the formal proof.

The term for (3) is given by:

$$\begin{aligned}
\langle j \rangle\,\langle i \rangle\,\mathsf{Glue}\,[&(i=0) \mapsto (A, \mathsf{equiv}^k(P\,k)), \\
&(i=1) \mapsto (B, \mathsf{equiv}^k B), \\
&(j=1) \mapsto (P\,i, \mathsf{equiv}^k(P(i \vee k)))] \\
&B
\end{aligned}$$

$\square$

Corollary 25 (Univalence axiom). *For the canonical map*

$$\mathsf{pathToEq} : (A\,B : \mathsf{U}) \to \mathsf{Path}\,\mathsf{U}\,A\,B \to \mathsf{Equiv}\,A\,B$$

we have that $\mathsf{pathToEq}\,A\,B$ is an equivalence for all $A : \mathsf{U}$ and $B : \mathsf{U}$.

Proof 1. Let us first show that the canonical map $\mathsf{pathToEq}$ is path equal to:

$$\mathsf{equiv} = \lambda A\,B : \mathsf{U}.\,\lambda P : \mathsf{Path}\,\mathsf{U}\,A\,B.\,\mathsf{equiv}^i(P\,i)$$

By function extensionality, it suffices to check this pointwise. Using path-induction, we may assume that P is reflexivity. In this case $\mathsf{pathToEq}\,A\,A\,1_A$ is the identity equivalence by definition. Because being an equivalence is a proposition, it thus suffices that the first component of $\mathsf{equiv}^i A$ is propositionally equal to the identity. By definition, this first component is given by transport (now in the constant type A) which is easily seen to be the identity using filling (see Section 4.4).

Thus it suffices to prove that $\mathsf{equiv}\,A\,B$ is an equivalence. To do so it is enough to give an inverse (see Theorems 4.2.3 and 4.2.6 of [27]). But $\mathsf{eqToPath}$ is a left inverse by Lemma 24 (3), and a right inverse by Lemma 24 (2) using that being an equivalence is a proposition. $\square$

Proof 2. Points (1) and (2) of Lemma 24 imply that Equiv $A\,B$ is a retract of Path $\mathsf{U}\,A\,B$. Hence $(X:\mathsf{U}) \times$ Equiv $A\,X$ is a retract of $(X:\mathsf{U}) \times$ Path $\mathsf{U}\,A\,X$. But $(X:\mathsf{U}) \times$ Path $\mathsf{U}\,A\,X$ is contractible, so $(X:\mathsf{U}) \times$ Equiv $A\,X$ is also contractible as a retract of a contractible type. As discussed in Section 7.2 this is an alternative formulation of the univalence axiom and the rest of this proof follows as there. □

Note that the first proof uses all three of the points of Lemma 24 while the second proof only uses the first two. As the second proof only uses the first two points it is possible to prove it if point (1) is defined as in Example 7 leading to a slightly simpler proof of point (2).

B Singular cubical sets

Recall the functor $\mathcal{C} \to \mathbf{Top}, I \mapsto [0,1]^I$ given at (1) in Section 8.1. In particular, the face maps $(ib): I - i \to I$ (for $b = 0_\mathbb{I}$ or $1_\mathbb{I}$) induce the maps $(ib): [0,1]^{I-i} \to [0,1]^I$ by $i(ib)u = b$ and $j(ib)u = ju$ if $j \neq i$ is in I. If ψ is in $\mathbb{F}(I)$ and u in $[0,1]^I$, then ψu is a truth value.

We assume given a family of idempotent functions $r_I : [0,1]^I \times [0,1] \to [0,1]^I \times [0,1]$ such that

1. $r_I(u,z) = (u,z)$ iff $\partial_I u = 1$ or $z = 0$, and

2. for any *strict* f in $\mathsf{Hom}(I,J)$ we have $r_J(f \times \mathrm{id})r_I = r_J(f \times \mathrm{id})$.

Such a family can for instance be defined as in the following picture ("retraction from above center"). If the center has coordinate $(1/2, 2)$, then $r_I(u,z) = r_I(u',z')$ is equivalent to $(2-z')(-1+2u) = (2-z)(-1+2u')$.

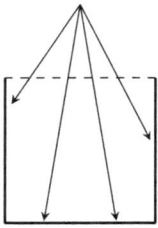

Property (1) holds for the retraction defined by this picture. The property (2) can be reformulated as $r_I(u,z) = r_I(u',z') \to r_J(fu,z) = r_J(fu',z')$. It also holds in this case, since $r_I(u,z) = r_I(u',z')$ is then equivalent to $(2-z')(-1+2u) = (2-z)(-1+2u')$, which implies $(2-z')(-1+2fu) = (2-z)(-1+2fu')$ if f is strict.

Using this family, we can define for each ψ in $\mathbb{F}(I)$ an idempotent function

$$r_\psi : [0,1]^I \times [0,1] \to [0,1]^I \times [0,1]$$

having for fixed-points the element (u,z) such that $\psi u = 1$ or $z = 0$. This function r_ψ is completely characterized by the following properties:

1. $r_\psi = \mathrm{id}$ if $\psi = 1$

2. $r_\psi = r_\psi r_I$ if $\psi \neq 1$

3. $r_\psi(u,z) = (u,z)$ if $z = 0$

4. $r_\psi((ib) \times \mathrm{id}) = ((ib) \times \mathrm{id}) r_{\psi(ib)}$

These properties imply for instance $r_{\partial_I}(u,z) = (u,z)$ if $\partial_I u = 1$ or $z = 0$ and so they imply $r_{\partial_I} = r_I$. They also imply that $r_\psi(u,z) = (u,z)$ if $\psi u = 1$.

From these properties, we can prove the uniformity of the family of functions r_ψ.

Theorem 26. *If f is in $\mathsf{Hom}(I,J)$ and ψ is in $\mathbb{F}(J)$, then $r_\psi(f \times \mathrm{id}) = (f \times \mathrm{id})r_{\psi f}$.*

This is proved by induction on the number of element of I (the result being clear if I is empty).

A particular case is $r_J(f \times \mathrm{id}) = (f \times \mathrm{id})r_{\partial_J f}$. Note that, in general, $\partial_J f$ is not ∂_I.

A direct consequence of the previous theorem is the following.

Corollary 27. *The singular cubical set associated to a topological space has a composition structure.*

Themes from Gödel: Some Recent Developments

Peter Koellner
Harvard University, USA.
koellner@fas.harvard.edu

I have found it extremely inspiring to read Gödel. He always seems to ask the right questions, the big questions, and yet he is able to bring mathematical results to bear on them in a fruitful way. This dual yearning—for the heavens and for the earth—seems to me to combine the philosophical and mathematical spirit in an ideal way.

What I would like to do in this paper is describe some recent progress along various fronts. I will in effect be giving a high-level overview of a series of papers that collectively tell a story, one that touches on several themes from Gödel and presents us with an emerging foundational picture that I think he would have found intriguing.[1]

The themes involve, in one way or another, the question of whether there are "absolutely undecidable" sentences in mathematics. In 1931 Gödel was quick to point out that although his incompleteness theorems provided us with sentences that are relatively undecidable they do not provide us with statements that are "absolutely undecidable":

> [These statements are] not at all absolutely undecidable; rather, one can always pass to "higher" systems in which the sentence in question is decidable. (Some sentences, of course, nevertheless remain undecidable.) In particular, for example, it turns out that analysis is a system higher in this sense than number theory, and the axiom system of set theory is higher still than analysis.[2]

---

[1] The main papers I shall be overviewing are: Bagaria, Koellner, and Woodin [1], Koellner [32], [33], [34], [35], [36], and Koellner and Woodin [37]. The reader is directed to these papers for proofs of the main results and for additional discussion.

[2] Gödel [19], p. 35.

What he had in mind is this: We know that if PA is consistent then it is incomplete and, in particular, that it misses the Π_1^0-truth Con(PA). Let PA_2 be the natural axiomatization of *second*-order arithmetic. It turns out that Con(PA) is provable in PA_2; so, in ascending from PA to PA_2 we capture the Π_1^0-truth that was missed by PA. Of course, the second incompleteness theorem applies to PA_2 as well, and so, assuming that PA_2 is consistent, it misses the Π_1^0-truth Con(PA_2). But now if we let PA_3 be the natural axiomatization of *third*-order arithmetic we find that it proves Con(PA_2) and so captures the Π_1^0-truth that was missed by PA_2. This pattern continues up through the orders of arithmetic, up through the hierarchy of set-theoretic systems, and ultimately up through the large cardinal hierarchy. At each stage a missing Π_1^0-truth is captured and a new one is revealed and that new missing Π_1^0-truth is captured at the next stage.

Thus, although, for a given system, the incompleteness theorems provide us with statements that are undecidable *relative* to that system, they do not *on the face of it* provide us with statements that are "absolutely undecidable" (in the sense of not being decided by *any* system of axioms that is justified). This is not to deny that there might be some more subtle argument based on the incompleteness theorems which does indeed show that there are "absolutely undecidable" sentences. Indeed, as we shall see below, others have made rather strong claims for the consequences of the incompleteness theorems. In any case, the first plausible candidates of "absolutely undecidable" statements came not through the incompleteness theorems but rather through the much more radical forms of independence that arise in set theory.

This came about through the dual results of Gödel and Cohen which showed that CH is independent of ZFC. In 1938 Gödel proved one direction by inventing the method of *inner models*. He constructed the inner model L (the constructible universe) and showed that it satisfies ZFC, $V=L$, and CH, thereby demonstrating that ZFC cannot refute either $V=L$ or CH. In 1963 Cohen established the other direction by inventing the method of *outer models* (also known as the method of *forcing*). He used this method to construct models which satisfied ZFC, $V \neq L$, and $\neg$CH, thereby demonstrating that ZFC cannot prove either $V=L$ or CH. Together these results show that assuming ZFC is consistent then $V=L$ and CH are independent of ZFC.

In the wake of his result on CH Gödel entertained the idea that $V=L$ and CH are in fact "absolutely undecidable" and, for a brief period, he said things that have led some to think that he even thought that the powerset operation (when applied to infinite sets) was "indefinite" and hence that the statement CH does not even have a well-defined meaning.[3] He was aware that one could sharpen the powerset operation

---

[3] For the former see Koellner [29], §1.3. The main quote that has led some (e.g. Martin Davis) to think that Gödel also thought that the powerset operation was "indefinite" is the following, which appeared in 1938 in his note announcing his results: "The proposition [$V=L$] added as a new axiom

by asserting $V = L$. But for philosophical reasons he found the axiom $V = L$ to be untenable.

Later he was to change his views on the question of whether the primitive notions of set theory (like the powerset operation) are "indefinite" and whether certain statements of set theory (like $V = L$ and CH) are "absolutely undecidable." He swung in the other direction and maintained that "if the meanings of the primitives of set theory" (as he explained them) "are accepted as sound" (which he certainly believed at the time he wrote this passage) then "it follows that the set-theoretical concepts and theorems describe some well-determined reality, in which Cantor's conjecture must be either true or false."[4] He concluded that "its undecidability from the axioms being assumed today can only mean that these axioms do not contain a complete description of that reality."[5] What we needed, then, were new axioms, axioms that are both justified and sufficiently strong to overcome significant instances of incompleteness.

In 1946 he proposed new axioms that he thought were up to the task. Recall, from our discussion above, that the instances of independence involved in the incompleteness theorems can be captured by passing to "higher" systems, that is, by ascending up through the layers of higher-order arithmetic, the layers of set theory, and ultimately through the layers of the large cardinal hierarchy. Gödel thought that large cardinal axioms might also settle the statements like $V = L$ and CH. In fact he had much higher hopes. He thought that large cardinal axioms might provide us with a notion of "absolute provability," one that would settle *all* statements of set theory and thereby erase independence. He was aware, of course, that in order for this to be the case one would have to treat the large cardinal axioms as completely open ended, the reason being that any recursive delimitation would lead to a system that succumbed to the incompleteness theorems. He noted that "[i]t is certainly impossible to give a combinatorial and decidable characterization of what an axiom of infinity is."[6] But he thought that "there might exist, e.g., a characterization of the following sort: An axiom of infinity is a proposition which has a certain (decidable) formal structure and which in addition is true."[7] On the basis of this he entertained the possibility of a *generalized completeness theorem*:

> It is not impossible that for such a concept of demonstrability some completeness theorem would hold which would say that every proposition

---

seems to give a natural completion of the axioms of set theory, in so far as it determines the vague notion of an arbitrary infinite set in a definite way" (Gödel [20], p. 27).

[4] Gödel [23], p. 260.
[5] Gödel [23], p. 260.
[6] The expressions 'axiom of infinite' and 'large cardinal axiom' are synonymous.
[7] Gödel [21], p. 151.

expressible in set theory is decidable from the present axioms plus some true assertion about the largeness of the universe of all sets.[8]

Thus, as a notion of "absolute provability" he proposed "provable from ZFC+LCA," where here "LCA" stands for the non-precisely specifiable open-ended scheme of "true large cardinal axioms."

Subsequently, there were two major developments in set theory that had important implications for this proposal. In 1961 Scott showed that large cardinal axioms—in particular the axiom asserting that there is a measurable cardinal—imply $V \neq L$. This bolstered Gödel's view that although $V = L$ provided us with a sharpening of the primitives of set theory it could not be accepted as a new axiom. And, by showing that large cardinal axioms have substantive consequences concerning the universe of sets, it bolstered the hope that large cardinal axioms might indeed provide us with a notion of "absolute provability." This hope, however, was to be shaken in 1967 by result of Levy and Solovay used Cohen's method of forcing to show that measurable cardinals cannot settle CH. And the hope was dashed by the subsequent generalizations of this result which collectively showed that none of the traditional large cardinal axioms (or anything resembling them in certain basic ways) can settle CH. Thus, it seemed that Gödel's proposed notion of "absolute provability" was a failure.

As we shall see below, very recent results may in fact show that Gödel's proposal was on the right track. The trick is to combine *both* of Gödel's ideas, that is, to combine a version of $V = L$ with large cardinal axioms. The trouble with $V = L$ is that it is incompatible (via Scott's result) with large cardinal axioms. But in recent work in inner model theory, Woodin has isolated a new axiom, the axiom $V =$ Ultimate-L, which *is* compatible with all of the traditional large cardinal axioms. This leads to a new proposed notion of "absolute provability," namely, the notion of "provable from ZFC + $V =$ Ultimate-L + LCA." But I am getting ahead of myself.

⁓

The above discussion of Gödel's ideas on "absolute provability" gives a quick overview of some of the themes from Gödel that will guide our discussion. As we proceed additional themes will enter.

In §1 I will approach the question of whether "there are absolutely undecidable statements" from an abstract and general point of view. We saw above that the incompleteness theorems do not *on the face of it* provide us with sentences that are absolutely undecidable, but we left open the question of whether they might tell us

---

[8]Gödel [21], p. 151.

something about absolute undecidability. Gödel, in fact, thought that they did tell us something. But he thought that it was tied up with another issue. He though that the incompleteness theorems implied that *either* "the mind cannot be mechanized" *or* "there are absolutely undecidable statements." In fact, he went so far as to maintain that this disjunction is a "mathematically established fact."[9] Others, who have gone further, have argued that the incompleteness theorems actually imply the first disjunct. I will discuss and assess these arguments. This will involve both a philosophical critique and a mathematical critique. The mathematical critique will consist in showing that when suitably formalized, although the disjunction is provable, neither disjunct can be proved or refuted. On the basis of this I will conclude with my own disjunctive conclusion: Either the statement that "there are absolutely undecidable statements" is indefinite (as the philosophical critique maintains) or it is definite and the above results (from the mathematical critique) provide evidence that it itself is about as good an example of an "absolutely undecidable" proposition that one might find.

In §2 I will turn to the question of whether certain primitive notions of set theory (like the powerset operation) and statements of set theory (like $V=L$ and CH) are "indefinite" and, again, I will consider an approach that proceeds at a high level of abstraction and generality. We saw above that Gödel came to maintain that the primitives of set theory are definite and that the statements of set theory have a determinate truth-value. In recent years, the foremost critic of this view has been Feferman. He presented five central arguments for the claim that the powerset operation is "inherently unclear" and "indefinite," and that statements like CH are "indefinite." I will examine his arguments and argue that in the end the entire case rests on the brute claim that the concepts of set theory are "not clear enough to secure definiteness." My response to this final point will be that the concept of "being clear enough to secure definiteness" is about as clear a case of an "inherently unclear" and "indefinite" concept as one might find, and, as such, it can bear little weight in an argument to the effect that the powerset operation and CH—items which, on the face of it at least, would appear to be clearer than the of "being clear enough to secure definiteness"—are "inherently unclear" and "indefinite."

Having thus seen little progress in approaching the questions of absolute undecidability and indefiniteness at this high level of abstraction and generality, in the remainder of the paper I will bring things down to earth, and focus on approaches that involve more engagement with developments in mathematics. For the remainder of the paper I will adopt the default view that the concepts and statements of set theory are to be taken at face value (in particular, that they are definite) and I

---

[9]Gödel [22], p. 310.

will focus on the question of absolute undecidability.

In §3 I will turn to the case that has been made for axioms that settle CH. In 1964, Gödel suggested that "an axiom in some sense opposite to [$V=L$]," one that asserted "some maximum property of the system of all sets" would be one that would "harmonize with the concept of set" and may very well enable one to derive the negation of CH.[10] In 1970 he proposed such axioms and argued that they led to the probable conclusion that the continuum has size $\aleph_2$.[11] This idea of "maximality" has found its most sophisticated development in the modern field of forcing axioms. These axioms provide us with the strongest current case we have for axioms settling CH and, as Gödel had hoped, they imply that the continuum has size $\aleph_2$. I will begin by reviewing the case for forcing axioms, discussing both the approach that leads to MM and the approach that leads to (∗). I will then argue that the case ultimately rests on an oversight. This will involve outlining a new perspective on forcing axioms, *the envelope perspective*, a perspective developed in joint work with Hugh Woodin. I will argue that this perspective undermines the current case for forcing axioms.

Having thus concluded that at present we do not have a strong case for or against CH, in the remainder of the paper I will stand back and examine the mathematical landscape from a broader vantage point. But the approach will continue to remain engaged with developments in mathematics. In fact, the entire discussion will be guided by a dichotomy theorem, one that points to two radically different possible futures. We will have thus passed from a philosophical disjunction (in §1) to a mathematical dichotomy (in §4). Just as the philosophical disjunction guided our discussion (by raising the question of which disjunct held), so too the mathematical dichotomy will guide our discussion (by raising the question of which side of the dichotomy holds). The difference will be that in the mathematical case the terms involved will be perfectly precise. We will be dealing with clear and distinct propositions and we will be in a position where we know that exactly one side of the dichotomy must hold.

In §4 I will lay the groundwork for the remainder of the paper by discussing the dichotomy theorem. The dichotomy involves another theme from Gödel, one that also originates in the 1946 paper in which discussed the concept of "absolute provability." In that paper he also discussed the concept of "absolute definability." He proposed "ordinal definability" as a concept of "absolute definability," and he thought that this would lead to simpler proof of the relative consistency of AC with ZFC than the one provided by his inner model L. It was subsequently shown (by

---

[10] Gödel [23], pp. 262–3, fn. 23.
[11] Gödel [24].

several people) that (as Gödel may have already known) the class HOD of *hereditarily ordinal definable* sets is an inner model of ZFC and so provides an alternative proof of the relative consistency of AC with ZFC. The dichotomy theorem that I shall be discussing involves HOD and is due to Woodin. It is known as the HOD *Dichotomy Theorem*. Roughly speaking, it asserts that (assuming large cardinal axioms) either HOD is "close" to V or HOD is "far" from V. The question is: Which side of the HOD Dichotomy holds?

There is a program aimed at establishing the first side of HOD Dichotomy—namely, the program of *inner model theory*—and a program aimed at establishing the second side of the HOD Dichotomy—namely, the program of *large cardinals beyond choice*.

In §5 I will discuss the program of inner model theory and the recent advances that have been made in it. This will culminate in Woodin's recent investigations into Ultimate-L, which, as the name indicates, is an ultimate version of Gödel's constructible universe L. The key difference between the axiom $V=L$ and the axiom $V=$ Ultimate-L is that the latter is conjectured to be compatible with *all* traditional large cardinal axioms. There is a certain conjecture concerning this axiom—the Ultimate-L Conjecture—which, if true implies that (assuming large cardinal axioms) we are, provably from large cardinal hypotheses, in the first half of the HOD Dichotomy, where HOD is "close" to V. This is the future in which *pattern* prevails.

In §6 I will discuss the program of large cardinals beyond choice. These large cardinal axioms are incredibly strong. And, as the name suggests, they imply that AC fails. If they are consistent then the Ultimate-L Conjecture must fail and so we will have lost our best current reason for thinking that the first side of the HOD Dichotomy holds. And, if certain consequences of these large cardinals hold—consequences that do not directly conflict with AC—then (assuming large cardinals) we will be in the second half of the HOD Dichotomy, where HOD is "far" from V. This is the future in which *chaos* prevails.

Finally, in §7, the various threads we have considered will merge and we will see that they are interwoven in an intricate fashion. In particular, we will see that the prospect of an absolute notion of provability and an absolute notion of definability turns on which of the two futures transpires, that is, whether pattern or chaos will prevail.[12]

---

[12] ACKNOWLEDGEMENTS: I would like to thank an anonymous referee and especially Gabriel Goldberg for very helpful comments.

1 Gödel's Disjunction

Let us start by examining the concept of absolute provability from a very general point of view.

The incompleteness theorems show that for any formal system meeting certain minimal conditions there are statements that are not provable *relative to the system in question*. But they do not, on the face of it at least, provide us with statements that are not probable relative to *all systems that are justified*, that is, they do not provide us with statements that are "absolutely unprovable." Nevertheless, Gödel thought that the incompleteness theorems enabled us to conclude *something* about the scope of absolute provability.

He thought it was connected with a very different question, namely, the question of whether "the mind can be mechanized" (understood here in the rather specific sense that "the mathematical outputs of the idealized finite human mind coincide with the mathematical outputs of an idealized finite machine (that is, a Turing machine))." He argued that his incompleteness theorems implied that *either* "the mind cannot be mechanized" (in the sense just described) *or* "there are absolutely undecidable sentences." This disjunction is known as *Gödel's Disjunction*.

Now, it is rarely the case in philosophy that claims are actually established beyond a shadow of a doubt, and this is especially true when those claims concern such large matters as the relationship between mechanism, mind, and mathematical truth. Nevertheless, Gödel—who was generally quite cautious in the claims he made—went so far as to call the disjunction a "mathematically established fact."[13] If it is indeed a "mathematically established fact" then it is something to be reckoned with.

He was convinced that the first disjunct was true and the second disjunct was false; that is, he was convinced that the mind could not be mechanized and that human reason was sufficiently powerful to capture all mathematical truths. But although he was convinced of these stronger claims he did not believe that he was in a position to establish either. He did, however, think that *one day* we would be in a position to establish more. What was missing, as he saw it, was an adequate resolution of the paradoxes involving self-applicable concepts, like the concept of truth. And he maintained that

> [i]f one could clear up the intensional paradoxes somehow, one would get a clear proof that mind is not machine.[14]

---

[13] Gödel [22], p. 310.
[14] This statement is reported in Wang [61], p. 187.

But he did not think that we had yet arrived at an adequate resolution of the paradoxes. And, lacking such a resolution, he felt that the most that he could claim to have established was the disjunctive conclusion.

Others, who have discussed these matters since Gödel, have claimed more. Famously, Lucas and Penrose have argued that the incompleteness theorems actually imply the first disjunct.[15]

In a series of papers I examined both the argument for the disjunction and the various arguments for each disjunct. But the approach that I took is quite different from the approach that is customary in the literature. One difficulty with the discussion in the literature is that the background assumptions on the underlying concepts—most notably, the concepts of "idealized finite machine" ($\sim$ "relative provability"), "idealized finite mind" ($\sim$ "absolute provability"), and "truth"—are seldom fully articulated and, as a consequence, it is difficult to assess the cogency of the arguments. One of my goals was to sharpen the discussion by making the background assumptions on the fundamental concepts explicit. Once this is done we are able to pull the entire discussion into a framework where we can establish definitive results of the form: "If the principles governing the fundamental concepts are such-and-such, then there is no hope of proving or refuting the first disjunct."

What I would like to do here is give a high-level overview of that discussion and the results obtained.[16]

1.1 Gödel

The disjunction concerns the concepts of "relative provability," "absolute provability," and "truth," and, in a variant formulation, the related concepts of an "idealized finite machine" and an "idealized human mind." We shall formulate our discussion in terms of the first three concepts.[17]

---

[15]There are really two different generations of arguments for the first disjunct. The first generation began with Nagel and Newman in 1956 ([44], [45]), continued with Lucas in 1961 ([40]), and culminated in a book length account by Penrose in 1989 ([46]). The second generation of arguments—one argument really—appeared in another book-length account by Penrose in 1994 ([47]). Penrose has continued to defend his argument—for example, in his address at the Gödel Centenary in 2006 and the subsequent published version of 2011 ([49]).

[16]For more on the philosophical subtleties, the strength of the claims, and the proofs of the various theorems the reader should consult Koellner [32], [35], and [36].

[17]There is no loss of generality in doing this, for it is assumed in the discussion that notions are understood in a way such that (1) the concept of what is relatively provable with respect to a formal system is co-extensive with the concept of what can be produced by an idealized finite machine (a Turing machine) and (2) the concept of what is absolutely provable is co-extensive with the concept of what can be produced by an idealized human mind.

Let us use 'F' for the set of sentences that are provable relative to a given formal system, 'K' for the set of sentences that are absolutely provable, and 'T' for the set of sentences that are true. A formal system F is said to be *correct* if $F \subseteq T$. And K is understood in such a way that $K \subseteq T$.

Gödel argues for three main claims on the basis of the incompleteness theorems:

Claim 1. For any F, $F \subseteq T \rightarrow F \subsetneq T$.

("For any correct formal system F there are truths that cannot be proved by F.")

Claim 2. For any F, $K(F \subseteq T) \rightarrow F \subsetneq K$.

("For any formal system F, *if* it is absolutely provable that F is correct, then K outstrips F.")

Claim 3. *Either* $\neg \exists F (F = K)$ *or* $\exists \varphi (\varphi \in T \wedge \varphi \notin K \wedge \neg \varphi \notin K)$.

("Either the mind cannot be mechanized or there are absolutely undecidable statements.")

The last of these claims is the disjunction, and, as we noted above, Gödel thought that he could actually prove it. He did not, however, think that he was in a position to prove the first disjunct. In fact, concerning his arguments he explicitly said:

> [I]t is not precluded that there should exist a finite rule producing all of its evident axioms. However, if such a rule exists ... we could never know with mathematical certainty that all the propositions it produces are correct.[18]

In other words, the arguments do not preclude the existence of a "master system" F^* such that $F^* = K$; they just show, by (2) above, that *if* there is such a system, then we can't have $K(F^* \subseteq T)$.

In order to assess the above claims we need to spell out the background assumptions on F, K and T, and thereby place the above informal discussion in a precise, formal setting, where one can establish definitive results.

It is straightforward to sharpen the notions of F and T: The informal notion of being "provable relative to a given formal system" is rendered mathematically precise in terms of the notion of being "provable relative to a recursive set of axioms" (this

---

[18] Gödel [22], p. 309.

latter notion being provably co-extensive with the notion of "what can be produced by a Turing machine"). So, in the case of F we have a *substantive* analysis of the notion, along with a precise way of quantifying over formal systems F_e, with e ranging over the indexes of the systems. The case of T is a little more delicate since in this case it would be difficult to give a substantive analysis of the notion in more fundamental terms. Nevertheless, we can hope to give a *structural* analysis, in the sense of articulating the fundamental principles that govern the notion. In our present setting, where we are dealing with the *typed* notion of truth, we have a perfectly adequate analysis, namely, that of Tarski.

The case of K is more difficult. The trouble is that, in contrast to the case of truth, there is little agreement even on what principles are supposed to govern the notion "absolute provability" (or the notion of what can be produced by an "idealized human mind") since there is little agreement on how absolute absolute provability is supposed to be (or how ideal the idealized human mind is supposed to be). What I shall do is follow a charitable course and consider a very strong notion of K. The reason for doing this is as follows: I am engaging with the opponent who wishes to show that K outstrips any F ("the mind cannot be mechanized") and T does not outstrip K ("there are no absolutely undecidable sentences"). The point is that the more I grant my opponent concerning K, the easier his task, and, correspondingly, the stronger any negative result that I might establish to the effect that even if we grant such a strong notion of K one cannot show that it outstrips any F or that it coincides with T. The strength of a criticism is proportional to the degree to which it is charitable.

The result of sharpening F, K, and T in the above manner results in the system EA_T, due to Reinhardt and Shapiro. I won't pause to spell out the details. The main point is that in this systems each of Gödel's three main claims can be formalized and proved:

Theorem 1.1. EA_T proves Claims 1, 2, and 3.

In this sense, Gödel was correct in maintaining that the disjunction is a "mathematically established fact."

Now that we know that disjunction holds (from the vantage point of the sharpening given by EA_T) the following question arises with added significance:

Question. Which disjunct holds?

Our main interest (in this paper) is in the second disjunct (that is the statement that there are absolutely undecidable sentences), but to orient the discussion it will be helpful to focus on the arguments that have been given for the first disjunct.

1.2 First Generation

The first generation of arguments for the first disjunct are really just versions of Gödel's argument for his second claim, namely that *if* the correctness of F is absolutely provable then $F \subsetneq K$. In the words of Penrose: "Human mathematicians are not using a knowably sound algorithm in order to ascertain mathematical truth" ([47], p. 76). As we saw, this conclusion, when suitably formalized, is provable in EA_T.

However, we also saw that in addition to his three postive claims, Gödel also made a negative claim to the effect that the argument did *not* preclude the possibility of "master system" F^* such that $F^* = K$; they just show, by (2) above, that *if* there is such a system, then we can't have $K(F^* \subseteq T)$. In short, the argument does not yield the first disjunct; rather it provides us with a *conditional* statement, and to arrive at the consequent of the conditional one needs to discharge the antecedent, that is, one needs the additional premise that the soundness of the system is absolutely provable.

The question, then, is whether for *any* F one can determine (in the sense of absolutely prove or refute) whether or not F is correct. This would involve, in the very least, being able to determine whether or not F is consistent. But this is no small task. For example, let S be the system $\text{PA} + R$ where 'R' stands for the famous open problem known as the Riemann Hypothesis. A result of Kreisel shows that R can be formulated as a Π_1^0-sentence. If $\text{PA} + R$ is consistent then R must be true.[19] And if $\text{PA} + R$ is inconsistent then R must be false. So to know whether or not $\text{PA} + R$ is consistent is to know whether or not R is true. But R is a major outstanding problem in mathematics, so outstanding that the Clay Institute has offered one million dollars for its resolution. No one at present knows the answer to R and it is no small task to determine the answer. It follows that no one at present knows whether or not $\text{PA} + R$ is consistent.

Now one might question this example by pointing out that although Lucas and Penrose do not know whether $\text{PA} + R$ is consistent, they might plausibly maintain that the answer is indeed within the reach of what is absolutely provable. But the choice of R was merely representative and the point is much stronger: To know of *every* system F whether or not F is consistent is to have an oracle for Π_1^0-truth, and that is not something we can claim to have at the *start* of an argument for the first disjunct since it trivially contains the conclusion of the argument.

These considerations show that this *particular* argument for the first disjunct

---

[19]This is because R is Π_1^0 and PA proves all true Σ_1^0-statements.

fails. But perhaps there is another argument ...

⌒

... In fact, we can say something much stronger. There is *no* argument for the first disjunct in EA_T.

To describe the limitative results and the subtle issues involved it is useful to distinguish (following Reinhardt) three grades of the mechanistic thesis. To begin with let us use the notation 'F_e' to range over formal systems (recursively enumerable sets) as e ranges over the natural numbers. The three mechanistic theses are:

(1) (WMT) $\exists e \, (K = F_e)$

(2) (SMT) $K \, \exists e \, (K = F_e)$

(3) (SSMT) $\exists e \, K(K = F_e)$

The first thesis is the *weak mechanistic thesis*.[20] It asserts that there is a Turing machine which coincides with the idealized human mind (in the sense that the two have the same outputs). This is simply the first disjunct of Gödel's Disjunction. The second thesis is the *strong mechanistic thesis*. It asserts that the idealized human mind knows that there is a Turing machine which coincides with the idealized human mind. The third thesis is the *super strong mechanistic thesis*. It asserts that there is a particular Turing machine such that the idealized human mind knows that *that* particular machine coincides with the idealized human mind.

The first result that is of relevance to this discussion is the following:

Theorem 1.2 (Reinhardt [52]). 'EA_T + SSMT' is inconsistent.

In other words, in the context of EA_T, it is true that there cannot be a Turing machine such that the idealized human mind knows that it coincides with *that* machine.

This fact does not vindicate the proponents of the first disjunct. For they are claiming that ¬WMT holds, that is, they are claiming that there is *no* Turing machine that produces exactly the same outputs as the idealized human mind. But this stronger conclusion cannot be obtained in EA_T, as the following result demonstrates:

Theorem 1.3 (Reinhardt [53]). 'EA_T + WMT' is consistent.

---

[20] Here I will express matters in terms of the variant formulation.

In other words, from the point of view of EA_T it is entirely possible that the idealized human mind is in fact a Turing machine. It just can't know which one.[21] This shows that there is *no* argument for the first disjunct in EA_T and since EA_T would seem to embody all of the assumptions held by the proponents of the first disjunct it shows that there is a fundamental obstacle.

In fact, there is an even stronger conclusion. Reinhardt conjectured that even SMT is consistent with EA_T and Carlson proved this conjecture, via a sophisticated construction:

Theorem 1.4 (Carlson [4]). 'EA_T + SMT' is consistent.

In other words, from the point of view of EA_T it is entirely possible that the idealized human mind *knows* that it is a Turing machine. It just can't know which one.

These results show that if one is to have a hope of establishing the first disjunct, one is either going to have to invoke stronger assumptions or shift to an entirely new framework.

The results show that there is *no* argument for the first disjunct in EA_T. And since EA_T would seem to embody all of the principles that our opponent would be willing to make[22] this shows that not only does their *particular* argument for the first disjunct fail, but that there can be *no argument* for the first disjunct within the framework that they are working.

I hope that this puts to rest the first generation of arguments for the first disjunct and that all participants in the dispute can agree on this.

1.3 Second Generation

The above results show that if one is to have a hope of establishing the first disjunct, one is either going to have to invoke stronger assumptions or shift to an entirely new framework. Now, there are independent reasons for shifting to a new framework since in the above framework the notion of truth is *typed* whereas the notion of truth that we employ in everyday life appears to be *type-free*. Moreover, Gödel had hoped that once we had an adequate resolution of the paradoxes—most notably an adequate type-free theory of truth—we would be in a position to establish the first disjunct. We now have many type-free theories of truth. And it turns out that to formalize Penrose's new argument one must employ a type-free theory of truth. So perhaps Penrose has fulfilled Gödel's hope.

---

[21] This result gives precise mathematical substance to the possibility raised by Gödel (see §1.1 above) and later raised by Benacerraf in [3].

[22] I say that it would *seem to cover* and not that *it does cover* since the assumptions on F, T, and K are seldom explicitly articulated in the arguments that are given for the first disjunct.

Penrose gave his new argument in [47] and, after much discussion, he distilled the essence of the argument as follows:

> Though I don't know that I necessarily am F, I conclude that if I were, then the system F would have to be sound and, more to the point, F' would have to be sound, where F' is F supplemented by the further assertion "I am F". I perceive that it follows from the assumption that I am F that the Gödel statement $G(F')$ would have to be true and, furthermore, that it would not be a consequence of F'. But I have just perceived that "if I happened to be F, then $G(F')$ would have to be true", and perceptions of this nature would be precisely what F is supposed to achieve. Since I am therefore capable of perceiving something beyond the powers of F, I deduce that I cannot be F after all.[23]

This argument is something of a mind-bender. In order to get a grip on it let us go though it line by line and articulate the principles employed at each point.

I will use the concise notation that we have already employed, involving F_e, K, and T. It will be useful at the outset to isolate three principles that are employed in the course of the argument:

$$K \subseteq T \qquad (K \to T)$$

$$\varphi \to T(\varphi) \qquad (T\text{-In})$$

$$\frac{\varphi}{K(\varphi)} \qquad (K\text{-Intro})$$

Here is the argument: Let e be an arbitrary natural number. We wish to show that $K \neq F_e$.

(1) $K = F_e \to F_e \subseteq T$

[This follows by the rule '$K \to T$'.]

(2) $K = F_e \to F_{e^+} \subseteq T$

[Here n^+ is the index of the machine obtained by adding the sentence '$K = F_e$' to F_e. This line follows from (1) by the rule 'T-In'.]

(3) $K = F_e \to G(F_{e^+})$

[Here '$G(F_{e^+})$' is the Gödel sentence for F_{e^+}. This line follows from (2) by reflection and the nature of the Gödel sentence.]

---

[23]Penrose [48], §3.2.

(4) $K = F_e \to F_{e+} \nvdash G(F_{e+})$

[This line follows from (2) by the incompleteness theorem.]

(5) $K(K = F_e \to G(F_{e+}))$

[This line follows from (3) by the rule 'K-Intro'.]

(6) $K = F_e \to (F_e \nvdash (K = F_e \to G(F_{e+})))$

[This line follows from (4).]

(7) $K = F_e \to K \neq F_e$

[This line follows from (5) and (6) since under the assumption that $K = F_e$ lines (5) and (6) show that K and F_e disagree on "$K = F_e \to G(F_{e+})$".]

(8) $K \neq F_e$

[This line follows from (7) by logic.]

Notice that in (2) 'T' is applied to 'K' and in the rule 'K-Intro' 'K' is applied to 'T'. So 'T' is applied to 'T' and 'K' is applied to 'K'. For this reason the argument cannot be formalized in EA_T. We need a type-free analogue of EA_T.

The first thing we need to do is select a type-free theory of truth from the large collection of type-free theories of truth that now exist. A particularly nice type-free theory of truth is Feferman's recent system DT. In order to fix ideas I will select this system.[24]

The base system of DT is PA. The language is extended by adding a type-free truth predicate 'T' and the base system is extended by allowing 'T' to figure in induction and adding axioms of *determinateness* and *truth*. Here determinateness is symbolized as '$D(x)$', a defined symbol, which is short for '$T(x) \vee T(\neg x)$'. I won't pause to give the details. The important point for our purposes is that this system has the attractive feature that it enables us to prove the Tarski biconditionals, *provided one conditions on determinateness*:

Theorem 1.5 (Feferman [12]). *For all φ in the language of DT,*

$$\text{DT} \vdash D(\ulcorner\varphi\urcorner) \to (T(\ulcorner\varphi\urcorner) \leftrightarrow \varphi).$$

---

[24] In Koellner [36] I give reasons for this choice and I give reasons for thinking that the limitative results would persist if we proceeded with one of the other existing type-free theories of truth.

It follows from this that in DT certain defective sentences (like the liar λ) can be seen to be defective and actually proved to be indeterminate.

The next step is to combine DT with a theory of absolute provability. It turns out that if one does this in the naive way (treating K as an operator) then inconsistency ensues. However, if one treats K as a predicate then the resulting system is consistent:

Theorem 1.6. DTK is consistent.

This is perhaps surprising since it enables us to circumvent the limitative results of Gödel, Montague, and Thomason, which showed that under certain general conditions inconsistency arises if one treats K as a predicate. The guiding philosophical idea here is that the paradoxes of type-free K are to be solved by tethering the theory of K to the theory of T and then importing the solution to the paradoxes given by the theory of truth over to the theory of absolute provability.[25]

We are now in a position to sharpen our discussion and examine the Penrose argument in a precise setting. To begin with, it turns out that in this new setting on can prove Gödel's three claims.

Theorem 1.7. DTK proves Claims 1, 2, and 3.

In particular, from the vantage point of DTK the disjunction holds. And so once again the following question arises with added significance:

Question. Which disjunct holds?

Let us begin by concentrating on the *particular* argument that Penrose gives for the first disjunct.

Recall that the argument uses the principles $K \to T$, T-In, and K-Intro. Now, in DTK we have $K \to T$, but we do not have T-In and K-Intro. In fact, if one adds T-In and K-Intro then the resulting system is inconsistent. This is not surprising since when one enters the type-free setting and encounters indeterminate sentences these rules are only plausible *provided one conditions on determinateness*. In DTK we have these restricted versions. They are labeled 'DT-In' and 'DK-Intro'.

Now, in the Penrose argument, the rules 'T-In' and 'K-Intro' were applied to the statements '$K = F_e \to F_{e+} \subseteq T$' and '$K(K = F_e \to G(F_{e+}))$' (in steps (2) and (5), respectively). This application is legitimate only if these statements are determinate. It turns out, however, these statements are in fact provably indeterminate.

---

[25]This strategy can be applied in other realms where type-free notions lead to paradox. If we tie all of the type-free notions to truth then all of our beasts will have one head. This idea was independently developed by Stern [55], where he applies the idea to the paradoxes of modality.

Theorem 1.8. (1) $\mathrm{DTK} \vdash \neg D(\ulcorner K = F_e \to F_{e^+} \subseteq T\urcorner)$

(2) $\mathrm{DTK} \vdash \neg D(\ulcorner K(K = F_e \to G(F_{e^+}))\urcorner)$

Thus, from the vantage point of DTK, Penrose's argument is invalid.

The above discussion shows that Penrose's *particular* argument for the first disjunct fails in the context of DTK. But perhaps there is another argument ...

... In fact, we can say something much stronger. There is *no* argument for the first disjunct in DTK.

To describe the limitative results we will distinguish various versions of the disjunction and its disjuncts, where we specify which fragment of the language the quantifiers range over. The *full* versions are the versions where there is no restriction on the language. For convenience let us use 'GD' for the disjunction, '¬WMT' for the first disjunct, and 'AU' for the second disjunct.

The *restricted* versions—which we will indicate by writing 'GD_L', 'WMT_L' and 'AU_L'—are obtained by restricting the quantifiers to a sublanguage L. The case of interest is where L is some determinate sublanguage of the language of DTK. For specificity we shall focus on the case where L is the language of PA.

We noted that the disjunction, GD, is provable in DTK. And we noted that Penrose's argument for the first disjunct fails in DTK since it encounters indeterminate sentences. This last point raises the concern that perhaps the very disjunction is indeterminate. In fact, in the fully general case, the disjunction and each of its disjuncts is provably indeterminate:

Theorem 1.9. $\mathrm{DTK} \vdash \neg D(\mathrm{GD}) \land \neg D(\neg \mathrm{WMT}) \land \neg D(\mathrm{AU})$.

It might seem counter-intuitive that GD is provable in DTK while at the same time being provably indeterminate. But in the type-free context this can happen with certain indeterminate sentences. For example, it turns out that the liar sentence λ is also provable in DTK while at the same time being provably indeterminate. The point is that in the type-free context the provability of a statement is not sufficient to have confidence in a statement; one must, in addition to having a proof, also have assurance that the statement in question is determinate.

There is a particular irony here. For to have a hope of proving the first disjunct in its full generality we have had to switch to a system in which truth is type-free. But when we make this shift we have to be on the lookout for sentences which, like the liar sentence, are indeterminate. The irony is that *the very statement we set out to prove*—namely, the first disjunct—is provably indeterminate.

This motivates the turn to the restricted versions of the disjunction and its disjuncts. In this case it turns out that the statements in question are indeed determinate.

Theorem 1.10. $\text{DTK} \vdash D(\text{GD}_{\text{PA}}) \wedge D(\neg \text{WMT}_{\text{PA}}) \wedge D(\text{AU}_{\text{PA}})$.

So in this case there is at least a prospect of getting started. Moreover, once again, we have that the disjunction is provable:

Theorem 1.11. $\text{DTK} \vdash \text{GD}_{\text{PA}}$.

And so, given that the disjunction is both provable and provably determinate we have reason to accept it, and the question arises as to which disjunct holds.

In turns out, however, neither disjunct is provable or refutable:

Theorem 1.12. Assume that DTK is correct for arithmetical statements. Then DTK can neither prove nor refute either $\neg\text{WMT}_{\text{PA}}$ or AU_{PA}.

In other words, in the restricted case the disjunction and its disjuncts are indeed determinate, but while the disjunction itself is provable neither disjunct is provable or refutable in DTK.[26]

I hope that this puts to rest the second generation of arguments for the first disjunct. And equally I hope it puts to rest the prospect of proving or refuting the statement that "there are absolutely undecidable propositions" on an approach where one proceeds with general principles governing K.

1.4 Conclusion

Throughout the discussion thus far I have been following the charitable course in presuming that the questions at hand—namely, the questions of whether "the mind can be mechanized" and whether "there are absolutely undecidable sentences"—are definite. But in fact, for reasons that I give at the end of [36], I think that there are problems with the notion of "an idealized human mind" and with treating the notion of "absolute provability" as if it were a fixed notion. The main problem with the first notion—that of "an idealized human mind"—is that unless we have an adequate theory of the mind (which we certainly do not) it is hard to make sense of the lines along which we are to idealize. For that reason the entire discussion might as well concern "the angelic mind." The main problem with the second notion—that

---

[26]In Koellner [32] and [36] I go on to show that this independence result persists even if one adjoins stronger principles governing absolute provability, and I give reasons for thinking that the situation will not improve if one alters the underlying type-free theory of truth.

of "absolute provability"—is that evidence and justification in mathematics come in grades. For that reason there appears to me to be a misstep in dealing with the notion of absolute provability as if it were a fixed, absolute notion.

But I have not wanted to rest my critical case on these philosophical misgivings; instead I have wanted to grant my opponent the notions in question and place the weight on limitative results that the opponent must accept. In any case, I think that we can draw a disjunctive conclusion: Either the statement that "there are absolutely undecidable statements" is indefinite (as the philosophical critique maintains) or it is definite and the above results provide evidence that it itself is about as good an example of an "absolutely undecidable" proposition that one might find.

For these reasons I don't think that at this level of abstraction and generality we can gain much insight into the question of whether "there are absolutely undecidable statements."

2 Indeterminateness

I want now to turn to the related question of whether the primitive notions and statements of set theory are "definite" and, for the time being, I want to continue to consider approaches that proceed at a high level of abstraction and generality.

Recall that after a brief period of entertaining the idea that the powerset operation is "indefinite," and statements like $V=L$ and CH are indefinite, Gödel came to the view that all of the primitive notions of set theory, along with all of the statements of set theory, are "definite." In most recent years, Feferman has been the most forceful critic of this view.

Feferman maintains that while the statements of first-order number theory are "completely clear" and "completely definite," many of the statements of analysis and set theory are "inherently vague" and "indefinite." He gives five main arguments:

(1) Feferman maintains that CH has effectively ceased to be regarded as definite by the mathematical community and that this fact provides "considerable circumstantial evidence to support the view that CH is not definite."[27]

(2) Feferman thinks that the concept of an arbitrary subset (of a given infinite domain (like the natural numbers)) is inherently vague in the sense that (a) it is vague and (b) it cannot be sharpened without violating what it is supposed to be about.

---

[27]Feferman [13], p. 1.

(3) Feferman argues that given the alleged lack of clarity of the concept of an arbitrary subset (of a given infinite domain) the only recourse to establishing the definiteness of statements of set theory (or even analysis) is an untenable form of platonism, one which faces certain insurmountable philosophical problems.

(4) Feferman's reasons for thinking that CH is indefinite are partly based on the metamathematical results in set theory; in particular the results showing that "CH is independent of all remotely plausible axioms extending ZFC, including all large cardinal axioms that have been proposed so far."[28]

(5) Feferman takes the formal results on indefiniteness—in particular the result of Rathjen showing that CH is indefinite relative to the semi-constructive system SCS^+—as providing evidence that CH is indefinite.

In [33] and [34] I examined these arguments and argued that (1) has no force and that when the dust settles (3), (4), and (5) all reduce to (2) and that in the end the entire case rests on the brute intuition that the concept of subsets of natural numbers—along with the richer concepts of set theory—is not "clear enough to secure definiteness." My response to this final, remaining argument is that the concept of being "clear enough to secure definiteness" is about as clear a case of an unclear and indefinite concept as one might find and, as such, and as such it can bear little weight in making a case against the definiteness of analysis and set theory.

What I would like to do here is illustrate some of the key steps components of this critique by criticizing (2) and showing that (5) reduces to (2).[29]

2.1 Conceptual Clarity

Feferman is an avowed anti-platonist. In place of platonism he espouses what he calls *conceptual structuralism*, an "ontologically non-realist philosophy of mathematics"[30] according to which "the basic objects of mathematical thought exist only as socially shared mental conceptions."[31] For present purposes I will only need one component of this view, namely, the part which maintains that "there are differences in clarity or definiteness between basic conceptions."[32]

This component of the view enables him to take an asymmetrical stance toward number theory and set theory. With regard to number theory he maintains that the

---

[28] Feferman [11], p. 127.
[29] The reader is referred to Koellner [33] and [34] for the full critique.
[30] Feferman [14], pp. 74.
[31] Feferman [15], p. 235.
[32] Feferman [15], p. 236.

concept of the structure of natural numbers is "completely clear"; in fact, it is *so clear* that (i) (on the epistemic dimension) the standard axioms of PA (with open-ended induction) are "evident on our conception"[33] and (ii) (on the semantical, or metaphysical dimension) it is "completely definite"[34] (in the sense that each statement of number theory has a definite truth value regardless of whether or not we can determine that truth value on the basis of evident principles.) In contrast, he thinks that the distinctive notions and statements of analysis and set theory are "inherently unclear" and "indefinite." Indeed he thinks that the concept of an arbitrary subset (of an infinite domain) is "inherently unclear" (or "inherently vague") in the sense that it "cannot be sharpened."[35]

In saying that the concept of an arbitrary subset (of an infinite domain) is "inherently unclear" (or "inherently vague") Feferman means that the concept cannot be sharpened *without violating what that notion is supposed to be about*:

> Moreover, I would argue that it is *inherently vague,* in the sense that there is no reasonable way the notion can be sharpened without violating what the notion is supposed to be about. For example, the assumption that all subsets of the reals are in L or even $L(\mathbb{R})$ would be such a sharpening, since that violates the idea of "arbitrariness." In the other direction, it is hard to see how there could be any non-circular sharpening of the form that there [are] as many such sets as possible. It is from such considerations that I have been led to the view that the statement CH is inherently vague and that it is meaningless to speak of its truth value.[36]

This point is repeated in his most recent writings on the topic[37] and we are told there that this is the "*main reason* that has led [him] to the view that CH is not definite."[38]

I agree that "[w]hat we are dealing with here are *questions of relative conceptual clarity and foundational status*"[39] and I am willing to grant that the concept of the natural numbers is clearer than the concept of an arbitrary subset (of a given infinite

---

[33] Feferman [8], p. 70.
[34] Feferman [15], p. 240.
[35] Feferman [9], p. 405.
[36] Feferman [9], pp. 410–411.
[37] See, for example, Feferman [11], p. 130 and Feferman [13], p. 2 and p. 21.
[38] Feferman [13], p. 2, my emphasis.
[39] Feferman [10], p. 619.

domain, say, the natural numbers). The question, however, is whether our conception of the former is "sufficiently clear to secure definiteness" while our conception of the latter is not.

I do not strictly speaking agree with the claim that there can be no way to sharpen the concept of an arbitrary subset (of a given infinite domain) without *violating what it is supposed to be about*, since this makes it look as though there is something implicit in the concept that *implies* that V cannot be any fine-structural inner model. But I do agree with the related claim that there is no way to sharpen the concept of an arbitrary subset (of a given infinite domain) that *merely unfolds that concept*. That is, I agree that it is not *analytic* of our concept of arbitrary subset of say the natural numbers that all such subsets appear in L, or any other fine-structural inner model. The reason is that there is nothing on the face of the concept of arbitrary subsets of natural numbers that involves mention of these models (the definitions of which are quite technical) and it would be far-fetched to maintain that a "deeper analysis" of the concept would imply, say, that all arbitrary subsets of natural numbers are in L, or any of the other fine-structural inner models.

But although I agree that the essential nature of the concept of an arbitrary subset (of a given infinite domain) cannot be clarified in more fundamental terms (there can be "no non-circular sharpening") I don't see this *alone* as sufficient to imply indefiniteness. It is rather a sign that we are dealing here with a *primitive* concept. For example, consider the concept of natural numbers. The essential nature of this concept cannot be clarified in more fundamental terms. All attempts to give a more primitive explication lead back to the same concept in a different guise. For example, one might try to explain the domain of natural numbers as the domain obtained by starting with 0 and applying the successor operation a finite number of times. But here the conception of natural numbers appears, hidden, in a different guise, in the reference to *finite number of times*. The hallmark of a primitive concept—indeed the defining characteristic of being a primitive concept—is that such a concept cannot be defined or explained in more fundamental terms (there can be "no non-circular sharpening"). So in *this respect alone* our conception of an arbitrary subset (of a given infinite domain, like the domain of natural numbers) is on a par with our conception of natural numbers.

The key difference, then, between our conception of the natural numbers and our conception of an arbitrary subset (of a given infinite domain, like the domain of the natural numbers)) is that the former is *clearer than* the latter. That is something with which I am happy to agree. But the key question at hand is whether the former, but not the latter, is "clear enough to secure definiteness." This is where I think the case falters. Feferman is making essential use of the concept of "being sufficiently clear to secure definiteness." I would like to say that the concept of "being sufficiently clear to secure definiteness" is not sufficiently clear to secure definiteness. It is about as good an example of an inherently unclear and indefinite concept as one might find. In this regard it bears more kinship to the concept of a "feasible number" than to the concept (taken at face value) of an arbitrary subset of a given infinite domain (like that of the natural numbers), and as such I don't think it can bear much weight in a case against the definiteness of the latter concept. To find an argument to that effect we will have to dig even deeper.

In [33] and [34] I go on to examine Feferman's arguments (3), (4), and (5) and I argue that in each case the argument ultimately reduces to (2). What I would like to do here is illustrate how this comes about in the case of (5).

2.2 Formal Results on Indefiniteness

In the spirit of getting exact about informal claims, Feferman "proposed [a] logical framework for distinguishing definite from indefinite concepts."[40] The basic idea is that classical logic is the logic appropriate for definite concepts, while intuitionistic logic is the logic appropriate for indefinite concepts. Thus, for example, if one wishes to articulate and investigate the view of someone who, like Feferman, maintains that the concept of natural numbers is definite, while the concept of subsets of natural numbers is not definite, then an appropriate system would be a system of *semi-constructive* set theory involving two logics: classical logic for the number-theoretic component and intuitionistic logic for the remainder. A statement φ is then defined to be (formally) *definite* with regard to such a system S if $S \vdash \varphi \vee \neg \varphi$. So, in the case under consideration, it will be immediate that statements of number theory are definite (this having been built in from the start). The question then arises: *Which other statements can be shown to be definite relative to such a system S?*

Feferman introduced two semi-constructive systems of set theory, which will here be labeled 'SCS' and 'SCS$^+$'. The first system is the one described above. The second system aims to capture the view of a descriptive set theorist who maintains that the concept of subsets of natural numbers is definite, but makes no explicit claims about the definiteness of the concept of subsets of the set of subsets of natural numbers, or concepts involving further iterations of the powerset operation. It involves classical logic for second-order number theory and intuitionistic logic beyond.

Feferman conjectured that CH is indefinite relative to SCS$^+$ and this subsequently proved by Rathjen.[41] Feferman takes this result "as further evidence in support" of the claim that CH is indefinite.[42] The question is whether this result

---

[40]Feferman [13], p. 2.
[41]Rathjen [50].
[42]Feferman [16], p. 23.

does indeed provide further evidence in support of the claim that CH is indefinite.

∽

Let's examine definiteness in SCS^+ and see how well it articulates the views of the descriptive set theorists. We have that all statements of second-order number theory are definite, this having been built in from the start. So far so good—this is in accord with the views of the descriptive set theorists. Let us now march our way up the hierarchy of definability. It turns out that statements about Borel sets and projective sets come out as definite; for example, the statements that all Borel sets and all projective sets are Lebesgue measurable, have the perfect set property, and are determined, all come out as definite in SCS^+. Thus far we are doing well—all of this is in accord with the views of the descriptive set theorists.

But things change when we get to $L(\mathbb{R})$. Let $AC^{L(\mathbb{R})}$ be the statement that "there is a well-ordering of $\mathbb{R}$ in $L(\mathbb{R})$". It turns out that $AC^{L(\mathbb{R})}$ is indefinite from the point of view of SCS^+, assuming the consistency of large cardinals at the level of $AD^{L(\mathbb{R})}$.

Theorem 2.1 (K. and Woodin). Assume that "ZFC + There are ω-many Woodin cardinals" is consistent. Then $SCS^+ \nvdash AC^{L(\mathbb{R})} \vee \neg AC^{L(\mathbb{R})}$.

Likewise $AD^{L(\mathbb{R})}$ comes out as indefinite in SCS^+. So, in this regard, $AD^{L(\mathbb{R})}$ is just like CH. But in contrast to CH it does not concern arbitrary subsets of reals. Rather, it only concerns definable ones, and it is a statement that the descriptive set theorists, at least modern descriptive set theorists, regard as definite.

So the system SCS^+ gives a mixed verdict on statements of descriptive set theory, when measured with regard to the community of modern descriptive set theorists. In some cases (like PD) it is in alignment with that community, and in other cases (like $AD^{L(\mathbb{R})}$) it is out of alignment with that community. Furthermore, if SCS^+ were indeed an articulation of Feferman's considered view then his focus should be on $AD^{L(\mathbb{R})}$ and not CH. The difference is that large cardinal axioms *do* settle $AD^{L(\mathbb{R})}$ (cf. Feferman's argument (4)) and, indeed, there has been a sophisticated case in favour of $AD^{L(\mathbb{R})}$. So if the focus were put on $AD^{L(\mathbb{R})}$ one would have to engage with that case.

But when it comes to Feferman's view even SCS^+ itself is a bit of a red herring. For recall that Feferman maintains something much stronger than that CH is indefinite. He maintains that the concept of the subsets of natural numbers is indefinite. In particular, he maintains not only that $AD^{L(\mathbb{R})}$ is indefinite, but also that PD and much more is indefinite. So, of the two systems that he has introduced, SCS is much closer than SCS$^+$ to articulating his own position with regard to matters of definiteness. Let us then investigate this system and see what comes out as indefinite within it.

It turns out that any statement which is equivalent to ATR$_0$ over RCA$_0$ comes out as indefinite.[43] For example:

Theorem 2.2 (K. and Woodin). SCS $\nvdash \varphi \vee \neg \varphi$ where φ is any of the following statements:

(a) Perfect Set Theorem: Every uncountable closed (or analytic) set has a perfect subset.

(b) The Ulm theory for countable reduced Abelian p-groups.

(c) Δ_0 Determinacy.

Despite this result, Feferman has made it clear to me in conversation that he thinks that open determinacy is in fact definite. Feferman has told me that to fully capture his view concerning the extent of what is definite one would have to supplement SCS by adding structure beyond that of the natural numbers, with the aim of ensuring that open determinacy and its kin come out as definite. I think that this is the right course for him to take. It is certainly the right course for *us* to take, since I don't think there can be reasonable doubt that open determinacy is definite.

What the above two case studies illustrate is that the present approach is not a foundational approach that is going to give us insight into what's definite and what's not; rather, the systems are designed to articulate the pre-theoretic views of a particular community. In the first case, where the community is that of the descriptive set theorists, PD comes out as definite. Good. But AD$^{L(\mathbb{R})}$ comes out as indefinite. Bad. So we have to revise the system to bring it into alignment with the target community. In the second case, where the community is Feferman and those who agree with him on matters of definiteness, the Riemann Hypothesis comes out as definite. Good. Open determinacy comes out as indefinite. Bad. So we have to revise the system to get the right outputs and thereby bring it into alignment with the target community.

The point I wish to make is that once again we have not been given an independent argument for the claim that the concept of subsets of natural numbers is inherently unclear and indefinite. Rather the entire enterprise is being guided by brute, pre-theoretic intuitions as to what is completely clear and definite and what is inherently unclear and indefinite. So, we are back to (2).

---

[43]This result also follows from a more general result independently proved by Rathjen.

2.3 Conclusion

I have argued that when the dust settles the entire case rests on the brute intuition that the concept of arbitrary subset (of a given infinite domain, like that of the natural numbers) is not "clear enough to secure definiteness." My response to this final point is that the concept of "being clear enough to secure definiteness" is about as clear a case of an inherently unclear and indefinite concept as one might find. For these reasons it cannot carry much weight in a foundational enterprise, especially one aimed at arguing that the concept of an arbitrary subset of a given infinite domain, like that of the natural numbers—a conception that on the face of it would appear to be *clearer* than the concept of "being clear enough to secure definiteness"—is not a definite concept.

3 The Continuum Hypothesis

In the above discussion the questions of absolute undecidability and indefiniteness were addressed at a very high level of abstraction and generality. And we saw—in the approaches we considered—that very little could be concluded about our questions at this level of abstraction and generality. This is perhaps not so surprising since the discussion involved such philosophical concepts as "absolute undecidability" and "indefiniteness" (as well as the concepts of "an idealized human mind" and being "clear enough to secure indefiniteness") and these concepts are not sharp enough for us to draw definite, substantive conclusions by reasoning with them alone. One can try to sharpen them—as, for example, we have done in terms of DTK in the first case, and in terms of SCS^+ and SCS in the second case—but, as we have seen, doing so tends to lead to results that are in conflict with the central claims that the proponents wish to maintain. Of course, one can go back to the drawing board and provide a new sharpening ... But I am not optimistic about this approach. I think that if we are going to make a substantial advance then we must broaden our horizon and engage more seriously with developments in mathematics. That is what I would like to do in the remainder of this paper.

The focus will be on the question of absolute undecidability and, for the purposes of the remaining discussion, I will adopt the default stance according to which the concepts and statements of set theory are to be taken at face value and regarded as definite.[44]

---

[44] Despite this stance the question of indefiniteness will not be entirely left behind. For to the extent that the question of indefiniteness turns on the question of absolute undecidability I will be implicitly address the former as well.

In early work I argued that we now have new axioms that are justified and which settle the central statements of second-order arithmetic and the central statements concerning $L(\mathbb{R})$ and somewhat beyond. It would take us too far afield to review this case. I will just say that it is a case that has developed in conjunction with a mathematical development over a 50 year period, and that as the mathematic development panned out in such a way that the philosophical case grew stronger at each step. It is on the basis of this case that I would maintain that "absolute undecidability" does not arise at the level of $L(\mathbb{R})$.[45]

Unfortunately the case for these axioms falls short of yielding axioms that have much of an impact on the level where CH appears, namely, the level of Σ_1^2-statements. In particular, it does not yield axioms that settle CH. It is for this reason that I would like to focus on this level and, more specifically, the case of CH.

The principles that I shall be discussing are in the spirit of a suggestion made by Gödel in 1964:

> [F]rom an axiom in some sense opposite to [$V = L$], the negation of Cantor's conjecture could perhaps be derived. I am thinking of an axiom which (similar to Hilbert's completeness axiom in geometry) would state some maximum property of the system of all sets, whereas [$V = L$] states a minimum property. Note that only a maximum property would seem to harmonize with the concept of set.[46]

Later, in 1970, he proposed specific principles of this kind and he argued that they led to the probably conclusion that $\mathfrak{c} = \aleph_2$.[47]

At around the same time, in 1970, the first forcing axiom was born—*Martin's Axiom* ($\text{MA}_{\aleph_1}$).[48] This opened up a new subfield of set theory—the study of forcing axioms—which eventually generated axioms that led to the current case against CH. These axioms are "maximality" principles of the kind entertained by Gödel, and they provide us with the strongest current case for axioms that settle the size of the continuum. Curiously, they imply that $\mathfrak{c} = \aleph_2$, something that Gödel would have found congenial.[49] In this section I would like to describe the case for forcing axioms and then present a new perspective (arrived at jointly with Woodin), called the *envelope perspective*. I will argue that this new perspective deflates the case

---

[45] See Koellner [29] and [31] for the details of this case.
[46] Gödel [23], pp. 262–3, fn. 23.
[47] Gödel [24].
[48] Martin and Solovay [41].
[49] See Koellner [30] for a general discussion of CH and for further details concerning the case for CH that I shall be reviewing below.

for forcing axioms and that if one adopts it then forcing axioms have an entirely different significance.[50]

3.1 Forcing Axioms

There are really two approaches to forcing axioms. First, there is the approach based on generalizations of the Baire Category Theorem, which leads to a hierarchy of forcing axioms, ranging from MA (Martin's axiom), to PFA (proper forcing axiom) to MM (Martin's Maximum). Second, there is the approached based on $\mathbb{P}_{\max}$, which leads to the axiom (*).

3.1.1 Generalizations of the Baire Category Theorem

The first approach to forcing axioms is through generalizations of the Baire Category Theorem, which, for present purposes, we will take to be the statement $BC_{\aleph_0}$ which asserts that for every poset $\mathbb{P}$, every family of countably many dense sets has a generic filter.[51] This theorem is at the heart of many fundamental results concerning $H(\omega_1)$, and so, in order to settle some of the independent propositions concerning $H(\omega_2)$, it is natural to try to extend $BC_{\aleph_0}$ to the level of $\aleph_1$.

The most natural way to do this is to replace '$\aleph_0$' with '$\aleph_1$' and consider the principle $BC_{\aleph_1}$ asserting that for every poset $\mathbb{P}$, every family of no more than $\aleph_1$-many dense sets has a generic filter. The trouble is that this principle is inconsistent.

So we have to restrict the principle in some way. A natural way to do this is to consider not *all* posets but rather various collections Γ of posets.

Definition 3.1. For a cardinal κ and a class Γ of posets, let $BC_\kappa(\Gamma)$ be the statement that for every poset $\mathbb{P} \in \Gamma$, and for every family D of no more than κ-many dense subsets of $\mathbb{P}$, there is a filter meeting every set in D.

The first axiom along these lines was *Martin's Axiom* ($MA_{\aleph_1}$). This is just the principle $BC_{\aleph_1}(ccc)$ where one restricts to posets that have the countable chain condition. To arrive at stronger principles one allows broader classes of posets. For example, if one considers the class of *proper* posets then the principle $BC_{\aleph_1}(proper)$ is known at the *proper forcing axiom* (PFA), a principle much stronger that $MA_{\aleph_1}$.[52]

---

[50] For further details see Koellner and Woodin [37].
[51] See Fremlin [18] and Todorcevic [59] for further details on this approach, as well as for references concerning the central notions and main results.
[52] In what follows I will focus on the case where $\kappa = \aleph_1$.

In 1988 Foreman, Magidor, and Shelah isolated what is provably the strongest forcing axioms in this hierarchy. It is the principle $BC_{\aleph_1}(\text{stat})$ where one restricts to posets that are stationary set preserving. It is known as *Martin's Maximum* (MM).[53]

It is quite straightforward to show that $BC_{\mathfrak{c}}(\text{ccc})$ is false, and so $MA_{\aleph_1}$ immediately implies that CH fails. In fact, the general idea behind all of these principles is to assert that $\aleph_1$ "resembles" $\aleph_0$ and, in so doing, they quite *explicitly* render CH false.

As a reminder of this we will employ the notation of Baire category numbers. Let $\mathfrak{m}$ be the least cardinal κ such that $BC_\kappa(\text{ccc})$ fails, and let $\mathfrak{mm}$ be the least cardinal κ such that $BC_\kappa(\text{stat})$ fails. In this notation, $\mathfrak{m} > \aleph_1$ is a restatement of $MA_{\aleph_1}$, and $\mathfrak{mm} > \aleph_1$ is a restatement of MM. We have

$$\aleph_1 \leqslant \mathfrak{mm} \leqslant \mathfrak{m} \leqslant \mathfrak{c}.$$

Although it is immediate that $\mathfrak{mm} > \aleph_1$ implies that $\mathfrak{c} > \aleph_1$, a remarkable fact is that $\mathfrak{mm} > \aleph_1$ actually *fixes* the size of the continuum.

Theorem 3.2 (Foreman-Magidor-Shelah, [17]). *Assume $\mathfrak{mm} > \aleph_1$. Then $\mathfrak{c} = \aleph_2$.*

3.1.2 $\mathbb{P}_{\max}$

The second approach to justifying forcing axioms is motivated by the attempt to find a model of $\neg CH$ which is canonical in the sense that its theory cannot be altered by set forcing in the presence of large cardinals.[54] The background motivation is this: First, we know that in the presence of large cardinal axioms the theory of $L(\mathbb{R})$ is invariant under set forcing (a result of Woodin). The importance of this is that it demonstrates that our main independence techniques cannot be used to establish the independence of statements about $L(\mathbb{R})$ in the presence of large cardinals. It follows that if $\mathbb{P}$ is a definable, homogeneous partial order in $L(\mathbb{R})$ then the generic extension $L(\mathbb{R})^\mathbb{P}$ inherits the generic absoluteness of $L(\mathbb{R})$. Woodin discovered that there is a partial order, $\mathbb{P}_{\max}$, that has this feature and which has additional remarkable properties—in particular, it does not add reals and the model $L(\mathbb{R})^{\mathbb{P}_{\max}}$ satisfies $ZFC + \neg CH$. Thus, $L(\mathbb{R})^{\mathbb{P}_{\max}}$ is a model of $\neg CH$ that is canonical in the desired sense and has the same reals as V.

This model also has another remarkable property—it is "maximal" (or "saturated") with respect to sentences that are of a certain complexity and which can be shown to be consistent via set forcing; in other words, if these sentences *can hold* (in

---
[53] Foreman-Magidor-Shelah [17].
[54] See Woodin [62] for further details on this approach, as well as for references concerning the central notions and main results.

the sense that they hold in a set generic extension) then they *do hold* in the model. The class of sentences in question involves the structure $\langle H(\omega_2), \in, I_{NS}, A^G \rangle \models \varphi$, where I_{NS} is the non-stationary ideal on ω_1 and A^G is the interpretation of (the canonical representation of) a set of reals A in $L(\mathbb{R})$. The details will not be important for our purposes and the reader is asked to just think of $H(\omega_2)$ along with some "extra structure" and not worry about the details concerning the "extra structure."

We are now in a position to state the main results:

Theorem 3.3 (Woodin [62]). Assume that there is a proper class of Woodin cardinals. There is a partial order $\mathbb{P}_{max}$ such that the following hold:

(1) $\mathbb{P}_{max} \subseteq H(\omega_1)$ and $\mathbb{P}_{max}$ is definable in $H(\omega_1)$.

(2) $\mathbb{P}_{max}$ is ω-closed and homogenous.

(3) $L(\mathbb{R})^{\mathbb{P}_{max}} \models \text{ZFC}$.

(4) Suppose φ is a Π_2-sentence in the language for the structure

$$\langle H(\omega_2), \in, I_{NS}, A : A \in L(\mathbb{R}) \cap P(\mathbb{R}) \rangle$$

and

$$\langle H(\omega_2), \in, I_{NS}, A : A \in L(\mathbb{R}) \cap P(\mathbb{R}) \rangle \models \varphi.$$

Then

$$L(\mathbb{R})^{\mathbb{P}_{max}} \models \text{``} \langle H(\omega_2), \in, I_{NS}, A : A \in L(\mathbb{R}) \cap P(\mathbb{R}) \rangle \models \varphi\text{''}.$$

Theorem 3.4 (Woodin [62]). Assume that there is a proper class of Woodin cardinals and that $V[G]$ is a set-generic extension of V. Then there exists an elementary embedding

$$j : L(\mathbb{R}) \to L(\mathbb{R})^{V[G]}.$$

Corollary 3.5 (Woodin [62]). Assume that there is a proper class of Woodin cardinals. Suppose φ is a a Π_2-sentence in the language for the structure

$$\langle H(\omega_2), \in, I_{NS}, A : A \in P(\mathbb{R}) \cap L(\mathbb{R}) \rangle$$

and there is a set-generic extension $V[G]$ of V such that

$$\langle H(\omega_2), \in, I_{NS}, A : A \in P(\mathbb{R}) \cap L(\mathbb{R}) \rangle^{V[G]} \models \varphi.$$

Then

$$L(\mathbb{R})^{\mathbb{P}_{max}} \models \text{``}\langle H(\omega_2), \in, I_{NS}, A : A \in P(\mathbb{R}) \cap L(\mathbb{R}) \rangle \models \varphi\text{''}.$$

To summarize: Assuming that there is a proper class of Woodin cardinals we have: The partial order $\mathbb{P}_{\max}$ forces AC over a model of AD. The resulting model $L(\mathbb{R})^{\mathbb{P}_{\max}}$ is a definable, homogeneous extension of $L(\mathbb{R})$ and, as such, it is *canonical* in that like $L(\mathbb{R})$ its theory is invariant under set forcing. The resulting model $L(\mathbb{R})^{\mathbb{P}_{\max}}$ is also *maximal* in the sense that it satisfies all Π_2-sentences (about the relevant structure) that can possibly hold (in the sense that they can be shown to be consistent by set forcing over V).

One would like to get a handle on the theory of this structure by axiomatizing it. There is, in fact, an axiom that does this:

Definition 3.6 (Woodin [62]). Axiom $(*)$: $\mathrm{AD}^{L(\mathbb{R})}$ holds and $L(P(\omega_1))$ is a $\mathbb{P}_{\max}$-generic extension of $L(\mathbb{R})$.

In parallel to the case with $\mathfrak{mm} > \aleph_1$, this axiom settles CH; moreover, is gives the same size of the continuum:

Theorem 3.7 (Woodin [62]). Assume $(*)$. Then $\mathfrak{c} = \aleph_2$.

More generally, by the maximality theorem, this axiom captures the main consequences that $\mathfrak{mm} > \aleph_1$ has for the structure of $H(\omega_2)$.

3.2 The Case for $\mathfrak{mm} > \aleph_1$ and the Case for $(*)$

Now that we have described the two approaches to forcing axioms—that based on the Baire Category Theorem (an approach that leads to $\mathfrak{mm} > \aleph_1$) and that based on finding a canonical model for the failure of CH (an approach that leads to $(*)$)—let us now examine the cases that have been put forward for accepting $\mathfrak{mm} > \aleph_1$ and $(*)$ as new axioms. Notice that since $\mathfrak{mm} > \aleph_1$ and $(*)$ each imply the failure of CH—and, in particular, that $\mathfrak{c} = \aleph_2$—any case for these axioms is going to have great relevance for Gödel's program for new axioms, the main point of which was to settle CH.

Gödel distinguished between *intrinsic* and *extrinsic* justifications for new axioms. Some set theorists have argued that forcing axioms are intrinsically justified on the basis of the component of "maximality" that is claimed to be inherent in the concept of set. In [37] we discuss these justifications and find them lacking. The most promising justifications are extrinsic, and that will be my focus here.

There are two kinds of extrinsic justifications—those based on *global* consequences for the universe of sets and those based on *local* consequences for the universe of sets (more specifically, consequences for the structure of $H(\omega_2)$). In [37] we give reasons for thinking that the extrinsic justifications based on global consequences (like the Singular Cardinal Hypothesis) are rather weak. The most promising extrinsic case is based on local consequences (concerning the structure theory of $H(\omega_2)$), and so that is what I would like to concentrate on here.

3.2.1 The Case for $\mathfrak{mm} > \aleph_1$

The extrinsic case for $\mathfrak{mm} > \aleph_1$ is largely based on the structural consequences that it has for $H(\omega_2)$. The consequences of $\mathfrak{mm} > \aleph_1$ for the structure $H(\omega_2)$ are radically different than the consequences of CH for this structure. The general argument maintains that $\mathfrak{mm} > \aleph_1$ gives the "right" consequences, while CH gives the "wrong" ones. In order to describe the case it will be useful to have some specific examples of the contrasting structure theories at hand.

In Subsections A, B, and C below I will discuss three central examples. This will involve introducing several mathematical notions and results that may be new to the reader. For the reader anxious to see the main take-away points, I suggest skipping ahead to the discussion that continues after Subsection C, and then referring back to the details as needed. [55]

A. Uncountable Separable Linear Orderings

To motivate our discussion let us first recall some classic results concerning the class $\text{LO}_{\aleph_0}$ of *countable linear orderings*.

The structure theory of $\text{LO}_{\aleph_0}$ turns out to be quite simple. A central place is occupied by the countable dense linear orderings without endpoints.

Theorem 3.8 (Cantor). *All countable dense linear orderings without endpoints are isomorphic.*

This theorem implies that $\text{LO}_{\aleph_0}$ has a *universal element*; that is, there is an element $L' \in \text{LO}_{\aleph_0}$ such that for every element $L \in \text{LO}_{\aleph_0}$, we have $L \leqslant L'$ (which is to say that an isomorphic copy of L sits inside L'). The universal element is in fact $\mathbb{Q}$. In addition, Ramsey's Theorem implies that $\{\omega, \omega^*\}$ is a *basis* for $\text{LO}_{\aleph_0}$, where ω^* is ω with the reverse ordering; that is, it implies that for all $L \in \text{LO}_{\aleph_0}$ we have $\omega \leqslant L$ or $\omega^* \leqslant L$. Thus, $\text{LO}_{\aleph_0}$ has a very simple structure theory.

Let us now turn to the class of uncountable linear orderings, $\text{LO}_{>\aleph_0}$. To begin with, let us first consider those uncountable linear orderings that are separable, $\text{SLO}_{>\aleph_0}$.

Definition 3.9. A linear ordering $(L, \leqslant_L)$ is *separable* if it contains a countable dense subset, that is, if there is a countable set $D \subseteq L$ such that for all $x, y \in L$ if $x <_L y$ then there exists $z \in D$ such that $x \leqslant_L z \leqslant y$.

---

[55] My account will follow the account in Todorcevic [58].

It is natural to ask whether one can generalize Cantor's theorem to the next level, that is, to the level of $\aleph_1$. To describe what we have in mind here it will be helpful to introduce two more definitions:

Definition 3.10. For a cardinal κ a linear ordering L is κ-*dense* if for all non-trivial intervals of L (including the entire ordering) have cardinality κ.

Definition 3.11. For a cardinal κ, $\mathrm{BA}(\kappa)$ is the statement that all separable κ-dense linear orderings are isomorphic.

For example, it follows from Cantor's theorem that $\mathrm{BA}(\aleph_0)$ is true. The question is whether $\mathrm{BA}(\kappa)$ can hold for $\kappa > \aleph_0$.

Theorem 3.12 (Dushnik-Miller [7]). $\mathrm{BA}(\mathfrak{c})$ fails. In fact, there are $2^{2^{\aleph_0}}$-many non-isomorphic suborders of $\mathbb{R}$. In particular, if CH holds then $\mathrm{BA}(\aleph_1)$ fails and, in fact, there are $2^{\aleph_1}$-many pairwise non-isomorphic $\aleph_1$-dense suborders of $\mathbb{R}$.

Theorem 3.13 (Baumgartner [2]). Assume $\mathfrak{mm} > \aleph_1$. Then $\mathrm{BA}(\aleph_1)$ holds.

It is not too hard to show that as a consequence we have the following: Assume $\mathfrak{mm} > \aleph_1$. Let B be an arbitrary subset of $\mathbb{R}$ of cardinality $\aleph_1$. Then $\{B\}$ is a one-element basis for $\mathrm{SLO}_{>\aleph_0}$.

B. UNCOUNTABLE LINEAR ORDERINGS

Let us now broaden our perspective and look at the class of *all* uncountable linear orderings, $\mathrm{LO}_{>\aleph_0}$. The question is whether $\mathrm{LO}_{>\aleph_0}$ has a finite basis. Of course, by the Dushnik-Miller result from above we have:

Theorem 3.14 (Dushnik-Miller [7]). Assume CH. Then $\mathrm{LO}_{>\aleph_0}$ does not have a basis of size less than $2^{\aleph_1}$.

On the other hand, Baumgartner's result raises the hope that under $\mathfrak{mm} > \aleph_1$ the class $\mathrm{LO}_{>\aleph_0}$ admits a finite basis. It remains to deal with the following class of orderings:

Definition 3.15. A linear ordering L is an *Aronszajn line* if it is uncountable and if it contains no uncountable subordering which is well-ordered, reverse well-ordered, or separable.

The name comes from the connection with Aronszajn trees: The lexicographic ordering of an Aronszajn tree is an Aronszajn line and any partition tree of an Aronszajn line is an Aronszajn tree.[56]

---

[56] See Todorcevic [56].

By the results of the previous section we know that assuming $\mathfrak{mm} > \aleph_1$ the elements of $\mathrm{LO}_{>\aleph_0}$ which are *not* Aronszajn lines have a three element basis, namely, $\{\omega_1, \omega_1^*, B\}$ where B is an arbitrary subset of $\mathbb{R}$ of cardinality $\aleph_1$ and ω_1^* is ω_1 with the reverse ordering. It remains to find a basis for the Aronszajn lines.

Definition 3.16. A *Countryman line* is an uncountable linear ordering C such that $C \times C$ can be decomposed into countably many chains.

Shelah [54] showed that Countryman lines exist and he conjectured $\mathfrak{mm} > \aleph_1$ implies that every Aronszajn line must contain a Countryman line.[57]

Theorem 3.17 (Moore, [42]). Assume $\mathfrak{mm} > \aleph_1$. Then every Aronszajn line contains a Countryman line.

Corollary 3.18. Assume $\mathfrak{mm} > \aleph_1$. Then $\mathrm{LO}_{>\aleph_0}$ has a five-element basis, namely, $\{\omega_1, \omega_1^*, B, C, C^*\}$, where B is any subset of $\mathbb{R}$ of cardinality $\aleph_1$, C is any Countryman line, and C^* is C with the reverse ordering.

C. Directed Sets and Cofinal Types

Recall that a *partial ordering* is a pair $(D, \leqslant_D)$ where D is a set and $\leqslant$ is a relation on D which is reflexive, antisymmetric, and transtive. A partial ordering $(D, \leqslant_D)$ is *directed* if for any $x, y \in D$ there exists $z \in D$ such that $x \leqslant_D z$ and $y \leqslant_D z$. A subset $X \subseteq D$ of a partial ordering $(D, \leqslant_D)$ is *cofinal in D* if for every $x \in D$ there exists $y \in X$ such that $x \leqslant_D y$.

Suppose that $(D, \leqslant_D)$ and $(E, \leqslant_E)$ are directed partial orderings. Tukey [60] introduced the following central notion: A function $f : E \to D$ is *cofinal* if for each $X \subseteq E$ which is cofinal in E, the pointwise image $f``X$ is cofinal in D. If there exists such a map then we say that $(D, \leqslant_D)$ is *Tukey reducible* (or *cofinally finer*) than $(E, \leqslant_E)$ and we write $D \leqslant_T E$.

There is an equivalent formulation of Tukey reducibility that is quite useful. A subset $X \subseteq D$ of a partially ordered set $(D, \leqslant_D)$ is *unbounded in D* if there is no single element $x \in D$ which bounds every member of X; that is, for each $x \in D$ there exists $y \in X$ such that $y \not\leqslant_D x$. A function $f : D \to E$ is called a *Tukey map* (or an *unbounded map*) if for each $X \subseteq D$ which is unbounded in D, the pointwise image $f``X$ is unbounded in E. It is not too hard to show the following: Suppose that $(D, \leqslant_D)$ and $(E, \leqslant_E)$ are directed partial orderings. Then $D \leqslant_T E$ iff there is a Tukey map from D into E. Let us write $D \equiv_T E$ if $D \leqslant_T E$ and $E \leqslant_T D$.

---

[57]Shelah actually conjectured that PFA implies this. But for unity of exposition I will continue to frame the discussion with respect to the stronger axiom $\mathfrak{mm} > \aleph_1$.

Tukey [60] showed that for two directed partial orderings $(D, \leqslant_D)$ and $(E, \leqslant_E)$ we have $D \equiv_T E$ iff there is a partially ordered set $(C, \leqslant_C)$ into which both D and E can be embedded as cofinal subsets. For this reason the equivalence classes of directed partial orderings are often called *cofinal types*.

For a given cardinal κ, let $\mathscr{D}_\kappa$ denote the set of all cofinal types of directed sets of size $\leqslant \kappa$. We are interested in determining $\mathscr{D}_\kappa$ for various κ. It is easy to see that $\mathscr{D}_{\aleph_0} = \{1, \omega\}$. So the first nontrivial problem is to determine $\mathscr{D}_{\aleph_1}$.

A canonical class of examples of directed sets are sets of the form $[\kappa]^{<\lambda}$ ordered under $\subseteq$. Notice that by the above characterization in terms of Tukey maps we have that $[\kappa]^{<\omega}$ is cofinally finer than any directed set of size $\leqslant \kappa$. In particular, $[\omega_1]^{<\omega}$ is cofinally finer than any set in $\mathscr{D}_{\aleph_1}$. Tukey [60] showed that $1, \omega, \omega_1, \omega \times \omega_1$, and $[\omega_1]^{<\omega}$ (each with their natural ordering) are distinct elements of $\mathscr{D}_{\aleph_1}$. The question arises: Are there any other elements in $\mathscr{D}_{\aleph_1}$?

Theorem 3.19 (Todorcevic [57]). $\mathscr{D}_\mathfrak{c}$ has at least $2^{\aleph_1}$-many elements. In particular, if CH holds, then $\mathscr{D}_{\aleph_1}$ has $2^{\aleph_1}$-many elements.

Theorem 3.20 (Todorcevic [57]). Assume $\mathfrak{mm} > \aleph_1$. Then $\mathscr{D}_{\aleph_1}$ has just the five canonical elements: $1, \omega, \omega_1, \omega \times \omega_1, [\omega_1]^{<\omega}$.

In each of the above three cases the structure theory of the domain in question is drastically different under the two hypotheses CH and $\mathfrak{mm} > \aleph_1$. Under the assumption CH, the three classes under consideration have a large number of elements and do not admit a tidy structure theory: $\text{SLO}_{>\aleph_0}$ does not have a basis of size less than $2^{\aleph_1}$, $\text{LO}_{>\aleph_0}$ does not have a basis of size less than $2^{\aleph_1}$, and $\mathscr{D}_{\aleph_1}$ has $2^{\aleph_1}$ many elements. In contrast, under the assumption $\mathfrak{mm} > \aleph_1$, the three classes under consideration have few elements and admit of a tidy structure theory: $\text{SLO}_{>\aleph_0}$ has a one-element basis, $\text{LO}_{>\aleph_0}$ has a five-element basis, and $\mathscr{D}_{\aleph_1}$ has five-elements. In slogan form: "Under CH we have chaos, under $\mathfrak{mm} > \aleph_1$ we have pattern."

The extrinsic case for $\mathfrak{mm} > \aleph_1$ is intended to mirror the extrinsic case for $\text{AD}^{L(\mathbb{R})}$, a case that is taken to be a paradigm of extrinsic justification in set theory. For purposes of comparison, let us briefly review that case.

(1) $\text{AD}^{L(\mathbb{R})}$ has "intrinsically plausible" consequences; for example, it implies that sets of reals in $L(\mathbb{R})$ are Lebesgue measurable, have the property of Baire, and have the perfect set property; and it implies that Σ_1^2-uniformization holds in $L(\mathbb{R})$.

(2) $\text{AD}^{L(\mathbb{R})}$ in turn is implied by its intrinsically plausible consequences.

(3) $\mathrm{AD}^{L(\mathbb{R})}$ is implied by large cardinal axioms.

(4) Large cardinals provide a set-generically invariant theory of $L(\mathbb{R})$ and $\mathrm{AD}^{L(\mathbb{R})}$ is at the heart of this theory.

(5) $\mathrm{AD}^{L(\mathbb{R})}$ is equivalent to the existence of inner models at the level of the large cardinals that imply it.

(6) $\mathrm{AD}^{L(\mathbb{R})}$ is implied any sufficiently strong natural principle, including principles which are incompatible with each other; in this regard it lies within the "overlapping consensus."

These six points illustrate the systematic manner in which evidence for $\mathrm{AD}^{L(\mathbb{R})}$ has been accrued.[58]

In the above case for $\mathfrak{mm} > \aleph_1$ we don't have anything like (2)–(6). The case is rather modeled on (1), where one attempts to justify a principle in terms of its intrinsically plausible consequences.

But are the consequences of $\mathfrak{mm} > \aleph_1$ really intrinsically plausible? It is true that the assumption of $\mathfrak{mm} > \aleph_1$ provides us with a tidy structure theory for $H(\omega_2)$ and, in contrast, the assumption of CH provides us with a great many counterexamples to that structure theory. But notice that in the case for $\mathrm{AD}^{L(\mathbb{R})}$ the structure theory in question pertains not to *all* of $P(\mathbb{R})$ but only to a restricted fragment, namely, $P(\mathbb{R}) \cap L(\mathbb{R})$; and the argument supports not *full* AD, but only for a restricted fragment, namely, $\mathrm{AD}^{L(\mathbb{R})}$, the fragment that applies to $P(\mathbb{R}) \cap L(\mathbb{R})$. It is intrinsically plausible that the sets in the restricted fragment $P(\mathbb{R}) \cap L(\mathbb{R})$ admit the regularity properties and enjoy a tidy structure theory, but it is *not* intrinsically plausible that *all* of $P(\mathbb{R})$ admit the regularity properties and enjoy a tidy structure theory. Similarly, in the case of $\mathfrak{mm} > \aleph_1$ it would seem that what is intrinsically plausible is that a *restricted fragment* of $H(\omega_2)$—say, a fragment of the form $H(\omega_2) \cap M$ for some inner model M—satisfies the tidy structure theory; and that it is not intrinsically plausible that *all* of $H(\omega_2)$ satisfies the tidy structure theory. If this is right then the case for $\mathfrak{mm} > \aleph_1$ misses its mark. What one is really getting is a case for a restricted fragment of $\mathfrak{mm} > \aleph_1$, one that applies to a restricted fragment of $H(\omega_2)$.[59]

So far this reply is largely schematic. It shows that something is amiss in the above extrinsic case for $\mathfrak{mm} > \aleph_1$. To strengthen our reply we must fill in some details. We must say something about the restricted fragment $H(\omega_2) \cap M$ that we

---

[58] See Koellner [29] and [31] for details.

[59] Bear in mind that here I am only considering *local* consequences of $\mathfrak{mm} > \aleph_1$ and I am only considering extrinsic justifications modeled on (1) above.

have in mind and something about the restricted fragment of $\mathfrak{mm} > \aleph_1$ that we have in mind. For this we must turn to the case for $(*)$.

3.2.2 The Case for $(*)$

The extrinsic case for $(*)$ is based on generic absoluteness. To describe this it will be useful to reformulate generic absoluteness in terms of Ω-logic.

Definition 3.21 (Woodin [62]). Suppose that T is a countable theory in the language of set theory and φ is a sentence. Then

$$T \models_\Omega \varphi$$

if for all complete Boolean algebras $\mathbb{B}$ and for all ordinals α,

$$\text{if } V_\alpha^\mathbb{B} \models T \text{ then } V_\alpha^\mathbb{B} \models \varphi.$$

We say that a statement φ is Ω-*satisfiable* if there exists an ordinal α and a complete Boolean algebra $\mathbb{B}$ such that $V_\alpha^\mathbb{B} \models \varphi$, and we say that φ is Ω-*valid* if $\varnothing \models_\Omega \varphi$.

Theorem 3.22 (Woodin [62]). *Assume ZFC and that there is a proper class of Woodin cardinals. Suppose that T is a countable theory in the language of set theory and φ is a sentence. Then for all complete Boolean algebras $\mathbb{B}$,*

$$T \models_\Omega \varphi \text{ iff } V^\mathbb{B} \models \text{``}T \models_\Omega \varphi\text{.''}$$

Thus, this logic is robust in the sense that the question of what implies what is generically invariant under large cardinal assumptions.

Corresponding to the semantic relation $\models_\Omega$ there is a quasi-syntactic proof relation $\vdash_\Omega$. The "proofs" are certain "robust" sets of reals (universally Baire sets of reals) and the test structures are models that are "closed" under these proofs. The precise notions of "closure" and "proof" are somewhat technical and so we will pass over them in silence.

Like the semantic relation, this quasi-syntactic proof relation is generically invariant under large cardinal assumptions:

Theorem 3.23 (Woodin [62]). *Assume ZFC and that there is a proper class of Woodin cardinals. Suppose T is a countable theory in the language of set theory, φ is a sentence, and $\mathbb{B}$ is a complete Boolean algebra. Then*

$$T \vdash_\Omega \varphi \text{ iff } V^\mathbb{B} \models \text{``}T \vdash_\Omega \varphi\text{''}.$$

Thus, we have a semantic consequence relation and a quasi-syntactic proof relation, both of which are generically invariant under the assumption of large cardinal axioms. It is natural to ask whether the soundness and completeness theorems hold for these relations. The soundness theorem is known to hold:

Theorem 3.24 (Woodin [62]). Assume ZFC. Suppose T is a countable theory in the language of set theory and φ is a sentence. If $T \vdash_\Omega \varphi$ then $T \models_\Omega \varphi$.

It is open whether the corresponding completeness theorem holds. Woodin's Ω Conjecture is simply the assertion that it does:

Conjecture 3.25 (Ω **Conjecture**). Assume ZFC and that there is a proper class of Woodin cardinals. Then for each sentence φ,

$$\varnothing \models_\Omega \varphi \text{ iff } \varnothing \vdash_\Omega \varphi.$$

We will need a strong form of this conjecture which we shall call the Strong Ω Conjecture. It is somewhat technical and so we will pass over it in silence.[60]

Recall that one key virtue of large cardinal axioms is that they "effectively settle" the theory of second-order arithmetic (and, in fact, the theory of $L(\mathbb{R})$ and more) in the sense that in the presence of large cardinals one cannot use the method of set forcing to establish independence with respect to statements about $L(\mathbb{R})$. This notion of invariance under set forcing played a key role in the case for $\mathrm{AD}^{L(\mathbb{R})}$. We can now rephrase this notion in terms of Ω-logic.

Definition 3.26. A theory T is Ω-*complete* for a collection of sentences Γ if for each $\varphi \in \Gamma$, $T \models_\Omega \varphi$ or $T \models_\Omega \neg\varphi$.

The invariance of the theory of $L(\mathbb{R})$ under set forcing can now be rephrased as follows:

Theorem 3.27 (Woodin). Assume ZFC and that there is a proper class of Woodin cardinals. Then ZFC is Ω-complete for the collection of sentences of the form "$L(\mathbb{R}) \models \varphi$".

Unfortunately, it follows from the aforementioned generalizations of the Levy-Solovay result that traditional large cardinal axioms do not alone constitute an Ω-complete theory at the level of Σ_1^2 since one can always use a "small" (and hence large cardinal preserving) forcing to alter the truth-value of CH. Nevertheless, if one supplements large cardinal axioms then Ω-complete theories can be obtained. This is the centerpiece of Woodin's case against CH.

---

[60]See Koellner [30] for the details.

Theorem 3.28 (Woodin [62]). Assume that there is a proper class of Woodin cardinals and that the Strong Ω Conjecture holds.

(1) There is an axiom A such that

    (i) ZFC $+ A$ is Ω-satisfiable and

    (ii) ZFC $+ A$ is Ω-complete for the structure $H(\omega_2)$.

(2) Any such axiom A has the feature that

$$\text{ZFC} + A \models_\Omega \text{`}H(\omega_2) \models \neg\text{CH'}.$$

Let us rephrase this as follows: For each A satisfying (1), let

$$T_A = \{\varphi \mid \text{ZFC} + A \models_\Omega \text{`}H(\omega_2) \models \varphi\text{'}\}.$$

The theorem says that if there is a proper class of Woodin cardinals and the Strong Ω Conjecture holds, then there are (non-trivial) Ω-complete theories T_A of $H(\omega_2)$ and all such theories contain $\neg$CH.

It is natural to ask whether there is greater agreement among the Ω-complete theories T_A. Ideally, there would be just one. Unfortunately, there isn't.

Theorem 3.29 (K. and Woodin [38]). Assume that there is a proper class of Woodin cardinals. Suppose that A is an axiom such that

  (i) ZFC $+ A$ is Ω-satisfiable and

  (ii) ZFC $+ A$ is Ω-complete for the structure $H(\omega_2)$.

Then there is an axiom B such that

  (i') ZFC $+ B$ is Ω-satisfiable and

  (ii') ZFC $+ B$ is Ω-complete for the structure $H(\omega_2)$

and $T_A \neq T_B$.

How then shall one select from among these theories? Woodin's work in this area goes a good deal beyond Theorem 3.28. In addition to isolating an axiom that satisfies (1) of Theorem 3.28 (assuming Ω-satisfiability), he isolates an axiom—namely, the axiom $(*)$ mentioned earlier—which has additional important features. This axiom can be phrased in terms of (the provability notion of) Ω-logic:

Theorem 3.30 (Woodin [62]). Assume ZFC and that there is a proper class of Woodin cardinals. Then the following are equivalent:

(1) $(*)$.

(2) For each Π_2-sentence φ in the language for the structure

$$\langle H(\omega_2), \in, I_{\text{NS}}, A : A \in P(\mathbb{R}) \cap L(\mathbb{R})\rangle$$

if

$$\text{ZFC} + \text{``}\langle H(\omega_2), \in, I_{\text{NS}}, A : A \in P(\mathbb{R}) \cap L(\mathbb{R})\rangle \models \varphi\text{''}$$

is Ω-consistent, then

$$\langle H(\omega_2), \in, I_{\text{NS}}, A : A \in P(\mathbb{R}) \cap L(\mathbb{R})\rangle \models \varphi.$$

It follows that of the various theories T_A involved in Theorem 3.28, assuming the Strong Ω Conjecture, there is one that stands out: The theory $T_{(*)}$ given by $(*)$. This theory maximizes the Π_2-theory of the structure $\langle H(\omega_2), \in, I_{\text{NS}}, A : A \in P(\mathbb{R}) \cap L(\mathbb{R})\rangle$. The continuum hypothesis fails in this theory. Moreover, in the maximal theory $T_{(*)}$ given by $(*)$ the size of the continuum is $\aleph_2$.

To summarize: Assuming the Strong Ω Conjecture, there is a "good" theory of $H(\omega_2)$ and all such theories imply that CH fails. Moreover, (again, assuming the Strong Ω Conjecture) there is a maximal such theory and in that theory $2^{\aleph_0} = \aleph_2$.

The key open question in this area is whether $(*)$ is ZFC + Ω-satisfiable. It is known that the answer is 'yes' if the Strong Ω Conjecture holds, but this conjecture is open. Nevertheless, even without the Strong Ω Conjecture there is a nice parallel between the case of $L(\mathbb{R})$ (which is equivalent to $L(H(\omega_1))$) and $L(H(\omega_2))$: Assume that there is a proper class of Woodin cardinals. Then, in the first case we have that for every formula φ, either

$$\text{ZFC} \vdash_\Omega \text{``}L(\mathbb{R}) \models \varphi\text{''}$$

or

$$\text{ZFC} \vdash_\Omega \text{``}L(\mathbb{R}) \models \neg\varphi\text{''}.$$

Similarly, in the second case, for every formula φ, either

$$\text{ZFC} + (*) \vdash_\Omega \text{``}L(H(\omega_2)) \models \varphi\text{''}$$

or

$$\text{ZFC} + (*) \vdash_\Omega \text{``}L(H(\omega_2)) \models \neg\varphi\text{''}.$$

This, in this sense, $(*)$ lifts the "effective completeness" that we have for $L(\mathbb{R})$ up to the level of $L(H(\omega_2))$.

3.3 The Envelope Perspective

Let us now return to our discussion of the case for $\mathfrak{mm} > \aleph_1$. Recall that in that discussion we had reached the following juncture:

The case extrinsic case for $\mathfrak{mm} > \aleph_1$ that we are considering is supposed to be modeled on (1) in the case for $\mathrm{AD}^{L(\mathbb{R})}$, where a principle is justified in terms of its intrinsically plausible consequences. In the case of $\mathrm{AD}^{L(\mathbb{R})}$, the extrinsic case argues that (i) $\mathrm{AD}^{L(\mathbb{R})}$ implies that the restricted collection $P(\mathbb{R}) \cap L(\mathbb{R})$ satisfies a tidy structure theory—where all sets of reals are Lebesgue measurable, have the other regularity properties, and satisfy various structural principles—which is intrinsically plausible (when it comes to definable sets of reals), and (ii) that this constitutes evidence for $\mathrm{AD}^{L(\mathbb{R})}$. Notice that the same case cannot be made for full AD since this broader axiom implies that the full collection $P(\mathbb{R})$ has this tidy structure theory and it is *not* intrinsically plausible that the full collection satisfies this tidy structure theory (for example, it is not intrinsically plausible that all sets of reals are Lebesgue measurable, have the other regularity properties, and satisfy the structural principles given by determinacy).

Now the parallel extrinsic case for $\mathfrak{mm} > \aleph_1$ we are considering argues that (i) $\mathfrak{mm} > \aleph_1$ implies that the collection $H(\omega_2)$ satisfies a tidy structure theory which is intrinsically plausible, and (ii) that this constitutes evidence for $\mathfrak{mm} > \aleph_1$. It is not clear, however, that it is intrinsically plausible that the full collection $H(\omega_2)$ satisfies the tidy structure theory in question (like that discussed in Subsections A, B, and C); rather, what is intrinsically plausible is that for some inner model M the restricted collection $H(\omega_2) \cap M$ satisfies the tidy structure theory. And so, if one is to truly parallel the case for $\mathrm{AD}^{L(\mathbb{R})}$ then we would get an argument not for full $\mathfrak{mm} > \aleph_1$ but rather some restricted fragment $\mathfrak{mm} > \aleph_1$ (restricted in the way that $\mathrm{AD}^{L(\mathbb{R})}$ is a restricted fragment of AD) that applies only to $H(\omega_2) \cap M$, for some suitable M.

We could leave the matter there and simply pose the following challenge to the advocate of $\mathfrak{mm} > \aleph_1$:

> Provide an argument for why we should regard the consequences that $\mathfrak{mm} > \aleph_1$ has for all of $H(\omega_2)$ (as opposed to some restricted fragment of the form $H(\omega_2) \cap M$) as intrinsically plausible.

But we would like to go further and outline a perspective—the *envelope perspective*—which follows the parallel with $\mathrm{AD}^{L(\mathbb{R})}$ more closely and provides us with both a natural restricted fragment $H(\omega_2) \cap M$ and a natural restricted fragment of $\mathfrak{mm} > \aleph_1$.

The model M that we have in mind is simply the $\mathbb{P}_{\max}$-extension of $L(\mathbb{R})$ and the restricted fragment of $\mathfrak{mm} > \aleph_1$ is simply the axiom $(*)$.

The envelope perspective on forcing axioms is a perspective from the point of view where CH holds and large cardinals exist. So, for the purposes of this discussion, assume that the ambient universe satisfies CH and has large cardinals at the level of a proper class of Woodin cardinals. (In the next section we will introduce a candidate for an ultimate inner model in which CH holds, namely, Ultimate-L.)

Within our background universe (think of Ultimate-L) we have the canonical inner model $L(\mathbb{R})$. It satisfies AD but does not satisfy AC. It is a "paradise for analysts" in that in it all sets of reals are Lebesgue measurable, have the property of Baire, have the perfect set property and admit a tidy structure theory. It is also canonical in the sense that its theory is generically invariant. Within this model there is a definable, homogenous partial order $\mathbb{P}_{\max}$. Let $G \subseteq \mathbb{P}_{\max}$ be $L(\mathbb{R})$-generic. The generic G is equivalent to a set $A \subseteq \omega_1$ and we shall think of the $\mathbb{P}_{\max}$-extensions of $L(\mathbb{R})$ as having the form $L(\mathbb{R})[A]$ where $A \subseteq \omega_1$ is the associated set. The point is that since $\mathbb{P}_{\max}$ is ω-closed and since we are assuming CH and since we have large cardinals (in particular, $\mathbb{R}^\#$) we *can actually build the generics*. So, inside our ambient universe we have the inner model $L(\mathbb{R})$ and we also have various inner models $L(\mathbb{R})[A_0]$, $L(\mathbb{R})[A_1]$, etc. extending it like spokes emanating out from the hub of a wheel. These are the $\mathbb{P}_{\max}$-extensions. For any two such extensions $L(\mathbb{R})[A_0]$ and $L(\mathbb{R})[A_1]$ we have $(H(\omega_2) \cap L(\mathbb{R})[A_0]) \cap (H(\omega_2) \cap L(\mathbb{R})[A_1]) = H(\omega_2) \cap L(\mathbb{R})$. They are elementarily equivalent (with arbitrary parameters in $L(\mathbb{R})$).

Let us focus on one such $\mathbb{P}_{\max}$-extension $L(\mathbb{R})[A]$. It actually exists as an inner model and it is canonical in that its theory is generically invariant. The $H(\omega_2)$ of $L(\mathbb{R})[A]$ is not the full $H(\omega_2)$ of V, but is rather a restricted fragment—it is an *envelope* of the full $H(\omega_2)$ of V. The structural consequences of $\mathfrak{mm} > \aleph_1$ that apply to $H(\omega_2)$ hold in this envelope since the model $L(\mathbb{R})[A]$ has the Π_2-maximality property discussed above.

From this perspective the structural consequences of $\mathfrak{mm} > \aleph_1$ that pertain to $H(\omega_2)$ do not concern the full $H(\omega_2)$; rather they just concern the $H(\omega_2)$ of the $\mathbb{P}_{\max}$-extension $L(\mathbb{R})[A]$. The structure theory that emerges is simply the "definable trace" of the structure theory that holds in $L(\mathbb{R})$ under $\mathrm{AD}^{L(\mathbb{R})}$, as it is "imported" through the definable, homogenous forcing to the extension $L(\mathbb{R})[A]$.

In hindsight, from the present perspective—where we are assuming CH and large cardinal axioms—this is what one should have expected all along. The structural consequences of $\mathfrak{mm} > \aleph_1$ that pertain to $H(\omega_2)$ are captured by $(*)$ and so hold in the $\mathbb{P}_{\max}$-extensions $L(\mathbb{R})[A]$. In the present setting these $\mathbb{P}_{\max}$-extensions actually exist. Their theories are canonical since they are the "definable trace" of the theory of $L(\mathbb{R})$ and so inherit the canonicity of $L(\mathbb{R})$. Consequently, one would expect these models to have a nice structure theory. In particular, it is entirely plausible that all $\aleph_1$-dense sets of reals are isomorphic in such a model, just as it is entirely plausible that all sets of reals are Lebesgue measurable in $L(\mathbb{R})$. The model $L(\mathbb{R})$ is a paradise for analysts. But in the real world there are non-Lebesgue measurable sets. Likewise, the models $L(\mathbb{R})[A]$ are paradises for combinatorial set theorists. But in the real world (on this perspective) there are $\aleph_1$-dense sets of reals that are non-isomorphic.

I think that this perspective largely deflates the force of the extrinsic case (of type (1)) that has been made for forcing axioms. It deflates the case against CH since from the perspective of CH it makes perfect sense to see all of the structural results as holding in inner models of the form $L(\mathbb{R})[A]$, just as from the perspective of AC it makes sense to see all of the structural results of AD as holding in $L(\mathbb{R})$.

I want to stress that I am not claiming that CH should be regarded as parallel to AC in terms of its intrinsic plausibility. In fact, I would say that it has next to no intrinsic plausibility. I am also not saying that we should accept CH. I am not making a case for CH at all. But I am saying that the fact that CH implies counter-examples to a tidy structure theory should not *alone* be a mark against it, any more than the fact alone that AC implies counter-examples to a tidy structure theory should be regarded as a mark against it; and that this is especially so given that we have been given no independent reasons for thinking that the structure theory in question applies to anything but a restricted fragment.

The main points I have been making are these:

(1) It should be noted that the above case for $\mathfrak{mm} > \aleph_1$ is not really parallel to the case for $\mathrm{AD}^{L(\mathbb{R})}$ since it concerns full $H(\omega_2)$ and not a restricted fragment.

(2) It is unclear that the consequences that $\mathfrak{mm} > \aleph_1$ has for full $H(\omega_2)$ are intrinsically plausible. The advocate of this kind of justification for $\mathfrak{mm} > \aleph_1$ needs to say more about why it isn't just plausible that the consequences hold for some restricted fragment.

(3) If the arguments have force then they should have force to a person who is open to the idea that CH holds.

(4) But from the point of view of CH, the envelope perspective provides us with a case that truly parallels the case for $\mathrm{AD}^{L(\mathbb{R})}$ (cf. (1) above) and provides us with a very natural restricted fragment of $H(\omega_2)$ and a restricted fragment of $\mathfrak{mm} > \aleph_1$ (cf. (2) above), namely, $H(\omega_2) \cap L(\mathbb{R})[A]$ (where the latter is a $\mathbb{P}_{\max}$ extension) and $(*)$.

(5) From this perspective everything fits together perfectly: The structure theory that emerges for the restricted fragment of $H(\omega_2)$ is simply the "definable trace" of $\mathrm{AD}^{L(\mathbb{R})}$ as it is imported by the definable, homogenous forcing from $L(\mathbb{R})$ to the $\mathbb{P}_{\max}$-extension $L(\mathbb{R})[A]$.

There is much more to be said about the envelope perspective. Here I will confine myself to three additional points. First, the method applies not just to $L(\mathbb{R})$ but rather applies to much richer models of determinacy. Second, there are $\mathbb{P}_{\max}$ variations. For example, there is the $\mathbb{Q}_{\max}$ variation. This forcing produces canonical extensions $L(\mathbb{R})[B_0]$, $L(\mathbb{R})[B_1]$, etc. which also emanate from $L(\mathbb{R})$ like spokes. Like the $\mathbb{P}_{\max}$ extensions these models are all canonical, they all look alike, and they have a canonical structure theory. But they look different than the $\mathbb{P}_{\max}$ models. From the envelope perspective this makes perfect sense. What is happening is that we are constructing (via CH) inner models that are canonical extensions of $L(\mathbb{R})$. The structure theory of $L(\mathbb{R})$ gets imported to the structure theory of the canonical extension. But it gets imported one way via $\mathbb{P}_{\max}$ and another way via $\mathbb{Q}_{\max}$. All of these models exist alongside one another. We don't have to choose. We let a thousand flowers bloom. Finally, in this last respect the envelope perspective is quite different than the perspective of $\mathfrak{mm} > \aleph_1$. For $\mathfrak{mm} > \aleph_1$ implies that any two models containing the reals also contain $P(\omega_1)$. So from the point of view of $\mathfrak{mm} > \aleph_1$ the variations do not exist. There is only one flower in the garden. The advocate of $\mathfrak{mm} > \aleph_1$ would have to say something about why there is only this one flower—why the structure theories given in the $\mathbb{P}_{\max}$ variations are not intrinsically plausible and why we must jettison them in favour of one the one true structure theory that stands above the rest.

4 The HOD Dichotomy

The reader might feel at this point that the results discussed in this paper so far have been largely negative. In the first two sections I considered approaches that proceeded at a high level of abstraction and generality and concluded that we could learn little about the questions of absolute undecidability and indefiniteness at that level. That negative conclusion motived the turn to deeper engagement with developments in mathematics. In the previous section we did just that, engaging with the strongest case that has been made for axioms that settle CH. But again, our conclusion was negative, since we concluded that those arguments—even though they engage with developments in mathematics—turn on an oversight and are ultimately undermined by other developments in mathematics. As it stands I think it is fair to say that at the present time we do not have a strong case for or against CH.

So perhaps it is time to stand back and survey the landscape of possibilities. That is what I would like to do now. In this section I will present a dichotomy theorem. Recall, that in the first section we considered a dichotomy theorem—namely, Gödel's Disjunction—and this dichotomy oriented our discussion by presenting us with two alternatives. The trouble was that the dichotomy involved unclear philosophical notions that were hard to pin down and this impeded our progress. Indeed, in retrospect it is unclear that the dichotomy was even clear enough to truly orient our discussion, for it is unclear what was even meant by each alternative. In contrast, the dichotomy that I will now present is a purely mathematical dichotomy, one involving only notions that are clear and exact. It is a dichotomy that can truly guide us. It points to two possible, radically different futures.

⁓

The dichotomy theorem involves another theme from Gödel, one that also originates in the 1946 paper in which he introduced the notion of "absolute provability" and proposed being "provable from "large cardinal axioms" " as an absolute notion of provability. In that paper he also introduced the notion of "absolute definability" and he proposed "being ordinal definable" as an absolute notion of definability.

The motivating idea in each case was Turing's analysis of computability. Gödel thought that the importance of Turing's analysis was "largely due to the fact that with this concept one has for the first time succeeded in giving an absolute definition of an interesting epistemological notion," one that is not relative to a given language or formal system, one where by "a kind of miracle it is not necessary to distinguish orders, and the diagonal procedure does not lead outside the defined notion."[61] He wanted to do the same for provability and definability. In the first case he wanted to avoid diagonalization through the incompleteness theorems and he did this by employing the vague and open-ended notion of a "large cardinal axiom." In the second case he wanted to avoid diagonalization through Richard's paradox (which he explicitly mentions) and Berry's paradox (which he does not).

A natural way to arrive at such a candidate for an absolute notion of definability is to notice first that any notion of definability—call it 'D-definable'—according to which there are ordinal numbers that are not D-definable is susceptible to transcendence through diagonalization: For consider (as in Berry's paradox) "the least ordinal which is not D-definable." We have, in doing so, just given a higher-level definition of an ordinal that is not D-definable; and so, by reflecting on the notion, we have been led to a notion of definability that transcends D-definability. It follows

---

[61]Gödel [21], p. 150.

that any candidate for an absolute notion of definability must have the feature that according to it all ordinals are definable. So it is natural to take the minimal step and simply consider the notion of being definable from an ordinal. This is the notion of *ordinal definability* that Gödel proposed as an absolute notion of definability.

The question is whether it is indeed an absolute notion. Gödel claimed that "[i]t can be proved that it has the required closure property" by which he meant that if you try to transcend the notion by introducing a truth predicate you will find that you do not get a richer notion.[62] This is true, and the way to prove it is via the reflection principle, something that Gödel seems to have known. He also said that he thinks that the notion of ordinal definability "will lead to another, and probably simpler, proof for the consistency of the axiom of choice."[63] He was right about this as well. It seems likely that he had a proof in mind. In any case, the proof was subsequently given by others.[64] It involves considering HOD, the class of sets which are hereditarily ordinal definable.

The inner models L and HOD are, in many respects, at opposite ends of the inner model spectrum. L is the most slender of inner models, while, in some sense, HOD is the broadest; L is defined locally, while HOD is defined globally; and L cannot accomodate modest large cardinals, while HOD can accomodate all traditional large cardinals.

This last point is relevant to our present discussion. For although the axiom $V = L$ has the virtue that it settles many propositions of set theory—and for that reason might be considered as the basis of a proposed notion of absolute provability—it has the drawback that it is incompatible with large cardinals, and so is incompatible with Gödel's proposed notion of absolute provability. In contrast, the axiom $V = \text{HOD}$ is *not* in conflict with traditional large cardinals. This leads to the prospect of a *merger* between Gödel's proposed notion of absolute provability and his proposed notion of absolute definability. For perhaps there is an ultimate version of L that satisfies $V = \text{HOD}$. This prospect of such a model will guide our discussion throughout the rest of the paper. But first, to the dichotomy.

∽

The dichotomy theorem that I have in mind concerns HOD. But it is motivated by the following remarkable dichotomy theorem—the *L Dichotomy Theorem*:

Theorem 4.1 (Jensen). Exactly one of the following hold.

---

[62] Gödel [21], pp. 151-2.
[63] Gödel [21], pp. 151-2.
[64] See Myhill and Scott [43].

(1) For every singular cardinal γ, γ is singular in L and $(\gamma^+)^L = \gamma^+$.

(2) Every uncountable cardinal is an inaccessible cardinal in L

The first alternative is one in which L is "close" to V, in that it correctly computes much of the cardinal structure of V. The second alternative is on in which L is "far" from V, in that it radically fails to capture the cardinal structure of V, thinking, for example, that ω_1 is an inaccessible cardinal. [65]

Woodin proved a similar dichotomy theorem for HOD—the HOD *Dichotomy Theorem*—a weak version of which is the following:

Theorem 4.2 (Woodin). Suppose that κ is an extendible cardinal. Then exactly one of the following hold.

(1) For every singular cardinal $\gamma > \kappa$, γ is singular in HOD and $(\gamma^+)^{\text{HOD}} = \gamma^+$.

(2) Every regular cardinal $\gamma \geqslant \kappa$ is a measurable cardinal in HOD.

Again, in the first alternative HOD is "close" to V and in the second alternative HOD is "far" from V.[66]

There is an important foundational difference between these the L Dichotomy Theorem and the HOD Dichotomy Theorem. In the case of the L Dichotomy, granting modest large cardinals, we know which side of the dichotomy holds; in particular, if $0^\#$ exists then the second side of the dichotomy must hold. But in the case of the HOD Dichotomy, *no* traditional large cardinal axiom can force us into the second side of the dichotomy (since every traditional large cardinal axiom is compatible with $V=\text{HOD}$). So, perhaps the first side of the HOD Dichotomy is true. Or, perhaps there are new large cardinals (a higher analogue of $0^\#$) which force us into the second side of the HOD Dichotomy.

There are thus, at present, two possible futures (assuming an extendible cardinal). The first is the future in which the first side of the HOD Dichotomy holds and HOD is "close" to V. The second is the future in which the second side of the HOD Dichotomy holds and HOD is "far" from V. The question is: Which side of the HOD Dichotomy holds?

There is a program aimed at establishing the first future—the first half of the HOD Dichotomy. This is the program of *inner model theory*. Recent work of Woodin has shown that if inner model theory reaches the level of one supercompact cardinal then it "goes all the way." This suggests that there might be an "Ultimate-L." It is not presently known how to construct such a model but a lot of machinery has been

---

[65] The proof of this theorem can be found in a number of texts; for example, Devlin [6].

[66] See Woodin [63] and [66] for the proof and for discussion.

developed and there is an axiom ("$V =$ Ultimate-L") and an associated conjecture (the "Ultimate-L Conjecture") concerning the existence of such an inner model. The point, for our present purposes, is that if the Ultimate-L conjecture holds then (assuming that there is an extendible cardinal with a huge cardinal above it) the first side of the HOD Dichotomy must hold. In this future, HOD is "close" to V and there are no large cardinals that transcend HOD; in other words, there is no analogue of $0^\#$ and HOD is indeed compatible with all large cardinals (not just the traditional ones).

There is a very different program, one aimed at a radically different future, one in which the second half of the HOD Dichotomy holds. This is the program of *large cardinals beyond choice*.

In the next two sections I will discuss each of these programs, and in the final section I will return to the HOD Dichotomy and discuss the two possible futures that lie before us.

5 Inner Model Theory

In this section I would like to discuss the first program.[67] To do this it will be useful to say more about the HOD Dichotomy.

5.1 The HOD Dichotomy

The official version of the HOD Dichotomy Theorem involves the notion of a cardinal being ω-strongly measurable in HOD.

Definition 5.1 (Woodin [63]). Let γ be an uncountable regular cardinal. Let $S_\omega^\gamma = \{\alpha < \gamma : \mathrm{cof}(\alpha) = \omega\}$. Then γ is ω-*strongly measurable in* HOD if there exists $\kappa < \gamma$ such that

(1) $(2^\kappa)^{\mathrm{HOD}} < \gamma$ and

(2) There is no partition $\langle S_\alpha : \alpha < \kappa \rangle$ of S_ω^γ into stationary sets such that $\langle S_\alpha : \alpha < \kappa \rangle \in \mathrm{HOD}$.

Lemma 5.2 (Woodin [63]). *Assume that γ is ω-strongly measurable in* HOD. *Then*

$$\mathrm{HOD} \models \gamma \text{ is a measurable cardinal.}$$

---

[67]For further discussion and for the proofs of the results discussed in this section see Woodin [63], [65], and [66].

Theorem 5.3 (Woodin [63]). (The HOD Dichotomy Theorem) Suppose that κ is an extendible cardinal. Then exactly one of the following hold.

(1) For every singular cardinal $\gamma > \kappa$, γ is singular in HOD and $(\gamma^+)^{\text{HOD}} = \gamma^+$.

(2) Every regular cardinal $\gamma \geqslant \kappa$ is ω-strongly measurable in HOD.

Definition 5.4 (Woodin [66]). The HOD *Hypothesis* is the statement that there exists a proper class of regular cardinals γ which are not ω-strongly measurable in HOD.

There is a series of conjectures to the effect that the HOD Hypothesis is provable from ZFC, possibly supplemented with large cardinals.

Definition 5.5 (Woodin [66]). The *Weak HOD Conjecture* is the conjecture that

ZFC + "There is an extendible cardinal with a huge cardinal above"

proves the HOD Hypothesis. The *HOD Conjecture* is this conjecture with 'supercompact' in place of 'extendible cardinal with a huge cardinal above'. The *Strong HOD Conjecture* is the conjecture that ZFC alone proves the HOD Hypothesis.

Notice that these conjectures are number-theoretic conjectures; in fact, they are Σ_1^0-statements, asserting that a certain statement is provable in a certain system. The point is that these conjectures are not going to run up against the rock of undecidability, as happened in the case of CH. Notice also that these conjectures become more plausible as one strengthens the large cardinal assumption. In what follows we shall focus on the Weak HOD Conjecture, the most plausible of the three conjectures.

If the Weak HOD Conjecture is true then, assuming that there is an extendible cardinal with a huge cardinal above, we must be in the first half of the HOD Dichotomy, where HOD is "close" to V.

It is natural to ask why one might make such a conjecture. It is really quite a surprising conjecture. For it posits that (in the presence of an extendible cardinal with a huge cardinal above) there are arbitrarily large regular cardinals γ such that for every $\kappa < \gamma$ such that $(2^\kappa)^{\text{HOD}} < \gamma$ there is a partition $\langle S_\alpha : \alpha < \kappa \rangle \in \text{HOD}$ of S_ω^γ into sets which are stationary in V. We know, by Solovay's theorem on stationary splitting that there are always such partitions in V, but there is little reason to expect that such splittings can exist in HOD, that is, that the splitting can be done "definably." In fact, when I was a graduate student this conjecture was known by a different name. It was called "the silly conjecture."

The evidence for the conjecture comes from inner model theory.

5.2 Weak Extender Models

Inner model theory began, of course, with Gödel's L, but it entered the large cardinal hierarchy with Solovay's $L[U]$. Since then the holy grail of inner model theory has been an inner model for a supercompact cardinal.

Prior to the actual construction of such a model it is hard to know what it will look like. But one can isolate a very basic feature that such a model should have.

Definition 5.6 (Woodin [63]). A transtive class $N \models$ ZFC is a *weak extender model for the supercompactness of* κ if for every $\lambda > \kappa$ there exists a κ-complete normal fine measure U on $P_\kappa(\lambda)$ such that

(1) $N \cap P_\kappa(\lambda) \in U$ and

(2) $U \cap N \in N$.

Each of these conditions is motivated by the case of $L[U]$ and the other inner models. They are what one would expect in the case of an inner model for the supercompactness of κ.

In general, when one constructs an inner model for a given large cardinal, the existence of stronger large cardinals implies that the model is "far" from V. Remarkably, in the case of a supercompact the above conditions ensure that the model is "close" to V regardless of which other large cardinals live in V.

Theorem 5.7 (Woodin [63]). Suppose that N is a weak extender model for the supercompactness of κ. Then for every singular cardinal $\gamma > \kappa$, γ is singular in N and $(\gamma^+)^N = \gamma^+$.

In particular, if HOD is a weak extender model for the supercompactness of κ then we must be in the first half of the HOD Dichotomy.

This result suggests that a weak extender model for the supercompactness of κ also captures the large cardinal structure of V. This turns out to be the case. The key result is the following theorem, the Universality Theorem.

Theorem 5.8 (Woodin [63]). Suppose that N is a weak extender model for the supercompactness of κ. Suppose E is an N-extender of length η with critical point $\kappa_E \geqslant \kappa$. Let

$$j_E : N \to M_E$$

be the ultrapower embedding given by E. Then the following are equivalent.

(1) For each $A \subseteq [\eta]^{<\omega}$, $j_E(A) \cap [\eta]^{<\omega} \in N$.

(2) $E \in N$.

This theorem lies at the heart of a cluster of results which show that weak extender models for the supercompactness of κ capture the large cardinal structure of V. For example:

Theorem 5.9 (Woodin [63]). Suppose that N is a weak extender model for the supercompactness of κ. Suppose that for each $n < \omega$ there is a proper class of n-huge cardinals. Then, in N, for each $n < \omega$, there is a proper class of n-huge cardinals.

It is worthwhile pausing to underscore the unexpected nature of these developments. As mentioned earlier, in general in inner model theory when one targets a given large cardinal, the resulting model cannot accomodate stronger large cardinals; in fact, the existence of stronger large cardinals typically implies that the model is "far" from V (in parallel to the manner in which the existence of $0^{\#}$ implies that L is "far" from V). But the case of a weak extender model for the supercompactness of κ is completely different. Here one is just targeting one supercompact cardinal and the result is a model that is not only "close" to V with regard to its computation of cardinal structure (above κ) but is also "close" to V in that it inherits all of the traditional large cardinals existing in V. This includes large cardinals (like n-huge cardinals) which are far beyond the level of supercompactness. In short, in the case of supercompacts there is an "overflow" and the model "goes all the way". This suggests that the problem of inner model theory is reduced to the problem of finding an inner model for a supercompact cardinal. The question is whether there is such an "L-like" model, a model with a fine-structure, an ultimate version of L, one that cannot be transcended by large cardinals, one that is "close" to V. But before turning to that let us return to the connection with the HOD Dichotomy Theorem.

5.3 The HOD Dichotomy and Weak Extender Models

The relevance of weak extender models to the HOD Dichotomy is contained in the following theorem:

Theorem 5.10 (Woodin [63]). Suppose that κ is an extendible cardinal. Then the following are equivalent.

(1) The HOD Hypothesis holds.

(2) There is a regular cardinal $\gamma \geqslant \kappa$ which is not ω-strongly measurable in HOD.

(3) No regular cardinal $\gamma \geq \kappa$ is ω-strongly measurable in HOD.

(4) There is a cardinal $\gamma \geqslant \kappa$ such that $(\gamma^+)^{\mathrm{HOD}} = \gamma^+$.

(5) HOD is a weak extender model for the supercompactness of κ.

(6) There is a weak extender model N for the supercompactness of κ such that $N \subseteq \mathrm{HOD}$.

It is this last equivalence which leads to the expectation that the HOD Hypothesis actually holds. For, assuming large cardinals, it is natural to expect that there is a weak extender model for the supercompactness of κ and, given the course of inner model theory, it is natural to expect that such a model lives (as a proper class) in HOD. But what would such a model look like?

5.4 The Ultimate-L Conjecture

Here we will have to invoke some notions that are beyond the scope of this paper, but in the interest of providing the reader with a broader picture we will give a brief account.[68]

In the case of the canonical inner models M that have been built to date the constructions are quite complex and one is not always in a position to even state the axiom "$V = M$". Moreover, in the cases where one can state the axiom "$V = M$" one has to first construct the model. But curiously in the case of the candidate for the ultimate inner model one can state the axiom prior to the actual construction, and in this case the axiom is comparatively simple. The definition is motivated by the discovery that the HODs of determinacy models turn out to be canonical (strategic) inner models, and the definition involves the reflection of the Σ_2-truth of V into such models.

Definition 5.11 (Woodin [66]). "$V = $ Ultimate-L" is the conjunction of the following two statements:

(1) There is a proper class of Woodin cardinals.

(2) For each Σ_2-sentence φ, if φ holds in V, then there exists a universally Baire set $A \subseteq \mathbb{R}$ such that
$$\mathrm{HOD}^{L(A,\mathbb{R})} \models \varphi.$$

---

[68] For further details see Woodin [66].

The question, of course, is whether there are models of "$V=$Ultimate-L" that have a supercompact cardinal. But the virtue of knowing the final axiom prior to the actual construction is that one can start to mine its consequences before the construction is completed. In the case of Ultimate-L the consequences are profound.

Theorem 5.12 (Woodin [66]). Assume $V=$Ultimate-L. Then

(1) CH holds.

(2) $V=$HOD.

(3) V is the minimum universe of the Generic Multiverse.[69]

The Ultimate-L Conjecture is the conjecture that such a model exists.

Definition 5.13 (Woodin [66]). The *Ultimate-L Conjecture* is the conjecture that

ZFC + "There is an extendible cardinal κ with a huge cardinal above κ"

proves: There exists a weak extender model N for the supercompactness of κ such that

(1) N is weakly Σ_2-definable and $N \subseteq $ HOD.[70]

(2) $N \models $ "$V=$Ultimate-L".[71]

Notice that like the HOD Conjecture this conjecture is a Σ_1^0-statement and as such it will not run up against the rock of undecidability.

5.5 The First Future

The first future is the future in which the first side of the HOD Dichotomy holds, where HOD is "close" to V.

The program aimed at realizing this future is the program of inner model theory described above, in particular, the program to build Ultimate-L. The Ultimate-L Conjecture implies the Weak HOD Conjecture (by the equivalence of (1) and (6) in Theorem 5.10) and so (assuming an extendible cardinal with a huge cardinal above)

---

[69]See Woodin [64] for a definition of 'Generic Multiverse'.
[70]See Woodin [66] for a definition of 'weakly Σ_2-definable'.
[71]Earlier published versions of the Ultimate-L Conjecture are stated with a weaker large cardinal assumption, namely, that of an extendible cardinal. But the present strategy for proving the conjecture appears to need a huge cardinal above the extendible. I have accordingly strengthened the hypothesis.

in this future we are on the first side of the HOD Dichotomy, where HOD is "close" to V.

But the Ultimate-L Conjecture implies much more. It implies that there is a canonical inner model—Ultimate-L—which is contained in HOD and which is itself "close" to V. Ultimate-L is quite different from L in that it cannot be transcended by the traditional large cardinals; that is, it "absorbs" all of the traditional large cardinals. Even though the model has not yet been constructed we already know—through the axiom $V=$ Ultimate-L—that it satisfies CH, that it thinks $V=$ HOD, and that it is the minimum universe its Generic Multiverse. Once the model has been actually constructed (assuming that it can be constructed) we will know much more. Ultimate-L, like L, will admit a complete analysis, and, given that the model is "close" to V this analysis will give us great insight into V itself. Indeed, some set theorists would maintain that if such a model exists then it is a candidate for V itself. In any case, whether or not one accepts $V=$ Ultimate-L, given that in the first future Ultimate-L is "close" to V, in the first future we will have a detailed fine-structural insight into V. The first future is the future where *pattern* prevails.

6 Large Cardinals Beyond Choice

Let us now turn to the program of *large cardinals beyond choice*, a program that points toward the second side of the HOD Dichotomy, where HOD is "far" from V. This program was initiated in joint work with Bagaria and Woodin. The results that I would like to report on here are from our joint paper [1].

Recall that a natural template for formulating large cardinal axioms is to assert that there is a non-trivial elementary embedding $j : V \to M$, where M is a transitive class. The critical point, $\mathrm{crit}(j)$, of the embedding is the first ordinal moved by the embedding and is generally the large cardinal associated with the embedding. It follows immediately that for any such embedding, if κ is the critical point, then M resembles V to the extent that $(V_{\kappa+1})^M = V_{\kappa+1}$. It is this resemblance which is responsible for the strong reflection properties that hold at κ. For example, it readily implies that there are many inaccessible cardinals below κ. To obtain embeddings with greater strength one demands that M resembles V to a higher degree. For example, if one demands that $(V_{\kappa+2})^M = V_{\kappa+2}$ then it follows that there are many measurable cardinals below κ. In the limit, it is natural to consider, as Reinhardt did in his dissertation,[72] the "ultimate axiom," where one demands full resemblance, by positing a non-trivial elementary embedding $j : V \to V$. Let us call a cardinal κ a *Reinhardt cardinal* if there is a non-trivial elementary embedding $j : V \to V$ with

---
[72]See Reinhardt [51].

critical point κ. Kunen famously showed that Reinhardt cardinals are inconsistent in the context of ZFC.[73] It has remained a longstanding open question whether Reinhardt cardinals are inconsistent in the context of ZF alone.

In [1] we investigated the hierarchy of such "choiceless" large cardinal axioms, a hierarchy that starts with a Reinhardt cardinal and passes upward through strong forms of Reinhardt cardinals and then onward through Berkeley cardinals and strong forms of Berkeley cardinals. These large cardinals are, of course, inconsistent with AC. However, each of the "choiceless" large cardinals in this hierarchy has a "HOD-analogue" which is consistent with AC (if it is consistent, of course). As we shall see below, the relevance of this to the HOD Dichotomy is this: If the choiceless large cardinals are consistent then the Ultimate-L Conjecture must fail and so we will have lost our main reason for believing that the first future must hold. Moreover, if the HOD-analogues of the choiceless large cardinals exist then there is indeed a higher analogue of $0^\#$ and the second future *must* hold.

6.1 Very Large Cardinals

We are interested in the relative strength of the large cardinals beyond choice. Here it is helpful to distinguish three grades of reflection.

Definition 6.1. Suppose that Φ_1 and Φ_2 are large cardinal notions. We say that Φ_1 *reflects* Φ_2 if for all κ such that $\Phi_1(\kappa)$ there exists $\bar\kappa < \kappa$ such that $\Phi_2(\bar\kappa)$. We say that Φ_1 *rank-reflects* Φ_2 if for all κ such that $\Phi_1(\kappa)$ there are $\bar\kappa < \gamma \leqslant \kappa$ such that $\langle V_\gamma, V_{\gamma+1}\rangle \models \text{ZF}_2 + \Phi_2(\bar\kappa)$. Finally, we say that Φ_1 *strongly rank-reflects* Φ_2 if for all κ such that $\Phi_1(\kappa)$ there are $\bar\kappa < \gamma < \kappa$ such that $\langle V_\gamma, V_{\gamma+1}\rangle \models \text{ZF}_2 + \Phi_2(\bar\kappa)$.

The first choiceless large cardinal was introduced by Reinhardt.

Definition 6.2. A cardinal κ is *Reinhardt* if there exists a non-trivial elementary embedding $j : V \to V$ such that $\text{CRT}(j) = \kappa$.

There is a natural way to strengthen this notion: one simply follows the template involved in defining strong cardinals, by demanding that for each ordinal λ there is an embedding that sends κ above λ.

Definition 6.3. A cardinal κ is *super Reinhardt* if for all ordinals λ there exists a non-trivial elementary embedding $j : V \to V$ such that $\text{CRT}(j) = \kappa$ and $j(\kappa) > \lambda$.

This notion can be strengthened in turn by following the template employed in defining Woodin cardinals.

---

[73]See Kunen [39] for the original proof. See Kanamori [28] for several alternative proofs as well as for additional background on the traditional large cardinal hierarchy.

Definition 6.4. Let A be a proper class. A cardinal κ is *A-super Reinhardt* if for all ordinals λ there exists a non-trivial elementary embedding $j : V \to V$ such that $\mathrm{CRT}(j) = \kappa$, $j(\kappa) > \lambda$, and $j(A) = A$, where $j(A) = \bigcup_{\alpha \in \mathrm{On}} j(A \cap V_\alpha)$. A cardinal κ is *totally Reinhardt* if for each $A \in V_{\kappa+1}$,

$$\langle V_\kappa, V_{\kappa+1} \rangle \models \mathrm{ZF}_2 + \text{"There is an } A\text{-super Reinhardt cardinal."}$$

There is another series of large cardinal notions that has a someone different flavour. These are the Berkeley cardinals.

Definition 6.5. For a transitive set M, let $\mathscr{E}(M)$ be the set of all non-trivial elementary embeddings $j : M \to M$.

Definition 6.6. A cardinal δ is a *Berkeley cardinal* if for every transitive set M such that $\delta \in M$, and for every ordinal $\eta < \delta$, there exists $j \in \mathscr{E}(M)$ with $\eta < \mathrm{CRT}(j) < \delta$.

In other words, a Berkeley cardinal is so large that it "shatters" any transitive set that contains it. Notice that if δ is a Berkeley cardinal then for any $\lambda > \delta$ there is an non-trivial elementary embedding from $V_\lambda \to V_\lambda$ with critical point less than δ. This notion can be further strengthened as follows:

Definition 6.7. A cardinal δ is a *club Berkeley cardinal* if δ is regular and for all clubs $C \subseteq \delta$ and for all transitive M with $\delta \in M$ there exists $j \in \mathscr{E}(M)$ with $\mathrm{CRT}(j) \in C$.

Definition 6.8. A cardinal δ is a *limit club Berkeley cardinal* if δ is a club Berkeley cardinal which is a limit of Berkeley cardinals.

Let us now turn to the question of the relative strengths of these large cardinal notions. These large cardinals are all known to be stronger than the large cardinals in the traditional large cardinal hierarchy, as summarized in the diagram at the end of Kanamori's book [28] (with the exception of the one he has at the very top, namely, '0 = 1'). Moreover, they form a hierarchy of their own. Here is a sample result.

Theorem 6.9. Suppose that κ is a super Reinhardt cardinal. Then there exists $\gamma < \kappa$ such that

$$\langle V_\gamma, V_{\gamma+1} \rangle \models \mathrm{ZF}_2 + \text{"There is a Reinhardt cardinal."}$$

Thus, super Reinhardt cardinals strongly rank-reflect Reinhardt cardinals. And since totally Reinhardt cardinals trivially rank-reflect Reinhardt cardinals we have a proper hierarchy in terms of strength.

Instead of stating the other results let us summarize what is known in a diagram, one that can be inserted at the top of the diagram at the end of Kanamori's book [28], above I_0 and below $0=1$.[74]

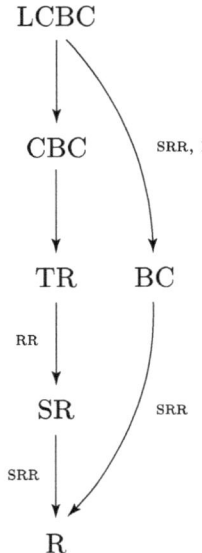

Here 'LCBC' stands for 'limit club Berkeley cardinal', 'CBC' stands for 'club Berkeley cardinal', 'BC' stands for 'Berkeley cardinal', 'TR' stands for 'totally Reinhardt', 'SR' stands for 'super Reinhardt', and 'R' stands for 'Reinhardt'. The arrows are to be interpreted as follows. And on the arrows 'SRR' stands for 'strongly rank reflects', 'RR' stands for 'rank reflects', and an unlabeled arrow between X and Y means that 'X is a Y'.

6.2 The Axiom of Choice

The above large cardinals imply, of course, that AC fails. It turns out that there is a connection between the degree to which AC fails and the cofinality of the least Berkeley cardinal.

---

[74]This follows from a result of Goldberg [27], who proved that (assuming DC) if there is a Reinhardt cardinal κ then there is a forcing extension $V[G]$ such that $V[G]_\kappa \models \text{ZFC} + I_0$.

Theorem 6.10. Suppose that δ_0 is the least Berkeley cardinal. Let $\gamma = \text{cof}(\delta_0)$. Then γ-DC fails.

This raises an interesting question: What *is* the cofinality of the least Berkeley cardinal?

It turns out that the answer to this question is independent! In early work we showed that if there is a club Berkeley cardinal then there are forcing extensions where (in a rank initial segment) the least Berkeley cardinal has countable cofinality and there are forcing extensions where (in a rank initial segment) the least Berkeley cardinal has uncountable cofinality. The forcing construction involved Prikry forcing in the choiceless setting and the proof was rather involved. Subsequently, Raffaella Cutolo (a student of mine and Woodin) found much simpler proofs of even sharper results.

Theorem 6.11 (Cutolo [5]). Assume ZF + DC + BC. Then there is a forcing extension $V[G]$ such that
$$V[G] \models \text{``cof}(\gamma_0) = \omega_1\text{''}$$
where γ_0 is the least Berkeley cardinal as computed in $V[G]$.

Theorem 6.12 (Cutolo [5]). Assume ZF + BC. Then there is a forcing extension $V[G]$ such that
$$V[G] \models \text{``cof}(\delta_0) = \omega\text{.''}$$

This independence result is quite surprising. For in general, when it comes to large cardinal axioms formulated in terms of elementary embeddings, such basic questions are usually readily settled. Moreover, this result represents something much more general. In the context of the choiceless large cardinal hierarchy there are basic questions that are beyond the reach of current technology and which raise the prospect of questions that are truly absolutely undecidable.[75] This is something we shall return to in the final section.

6.3 The Second Future

The second future is the future in which the second side of the HOD Dichotomy holds, where HOD is "far" from V.

The reasons for thinking that this future might transpire come from the program for large cardinals beyond choice described above. But here the situation is a bit more subtle than in the case of the first future. The subtleties turn on the difference between (i) assuming that large cardinals beyond choice are *consistent* and (ii)

---

[75] See Bagaria, Koellner, and Woodin [1] for further examples and further discussion.

assuming that certain consequences of large cardinals beyond choice actually *hold* (something we shall explain in more detail below.)

Let's start with consistency. The key result is the following (which is a theorem of ZF):

Theorem 6.13 (Bagaria, K., and Woodin [1]). Suppose that the Weak HOD Conjecture holds. Then there cannot be an non-trivial elementary embedding $j : V \to V$ and an ω-huge cardinal above $\kappa_\omega(j)$.

(Here $\kappa_\omega(j)$ is defined as follows: Let $\kappa_0 = \text{crit}(j)$ and, for $n < \omega$, let $\kappa_{n+1} = j(\kappa_n)$. Then $\kappa_\omega(j) = \sup_{n<\omega} \kappa_n$.)

Corollary 6.14 (ZF). Suppose that the Weak HOD Conjecture holds. Then there cannot be a super Reinhardt cardinal and there cannot be a Berkeley cardinal.

In short, if large cardinals beyond choice are *consistent* then the Weak HOD Conjecture fails and hence that the Ultimate-L Conjecture fails, and so we will have lost our only reason for thinking that the first future holds. Moreover, in this case we also have an anti-inner model theorem.

It is important to note that the result is not just about a single inner model, namely, Ultimate-L. It is much more general. Inner model theory proceeds in a general setting, by assuming a large cardinal hypothesis and then showing that one can build a canonical inner model for that large cardinal. But the failure of the Weak HOD Conjecture would imply that even if one assumes that there is an extendible cardinal κ with a huge cardinal above it (or any stronger assumption on κ) then one cannot show that there is a weak extender model N for the supercompactness of κ such that $N \subseteq \text{HOD}$. The canonical inner models to date are all definable. The notion of definability embodied in HOD is the most general notion of definability. It is hard to see what meaning there could be to inner model theory if it cannot be executed even at this most general level of definability. For this reason, in this scenario, where we merely assume that large cardinals beyond choice are *consistent* there can be no detailed, fine-structural insight into V. The second future is the future in which *chaos* prevails.

The consistency of large cardinals beyond choice does not strictly speaking imply that we are on the second side of the HOD Dichotomy but it opens the way for a higher analogue of $0^\#$, which *does* imply that we are on the second side of the HOD Dichotomy.

For a given inner model N of ZFC one can consider relativized versions of the choiceless large cardinals in the context of ZFC. For example, a cardinal κ is N-*Reinhardt* if there exists a non-trivial elementary embedding $j : N \to N$ with

$\operatorname{crit}(j) = \kappa$; a cardinal κ is N-*super Reinhardt* if for all γ there exists a non-trivial elementary embedding $j : N \to N$ with $\operatorname{crit}(j) = \kappa$ and $j(\kappa) > \gamma$; a cardinal δ is an N-*Berkeley cardinal* if for all transitive sets $M \in N$ such that $\delta \in M$, and for every ordinal $\eta < \delta$, there exists $j \in \mathscr{E}(M)$ with $\eta < \operatorname{crit}(j) < \delta$; and so on.

In the case where N is L all of the notions collapse—they are all equivalent to the existence of $0^\#$; in particular, the existence of an L-Berkeley cardinal is equivalent to the existence of $0^\#$. Let us focus on the other extreme, the case where N is the inner model HOD. These are the "HOD-analogues" of the choiceless large cardinals. Notice that these large cardinal notions are formulated in the context of ZFC. It is, of course, entirely possible that they are inconsistent. But if they are consistent then we would have a new hierarchy of large cardinal principles in ZFC.

The relevance of this to our present discussion is that these HOD-analogues would provide us with a higher analogue of $0^\#$. Recall that the existence of $0^\#$ (equivalently, an L Berkeley cardinal) implies that we are on the second side of the L Dichotomy. A HOD-Berkeley cardinal provides us with a higher analogue of $0^\#$ (equivalently, an L-Berkeley cardinal) in this sense: Assuming that there is an extendible cardinal, then the existence of a HOD-Berkeley cardinal implies that we must be on the second side of the HOD Dichotomy, where HOD is "far" from V.[76]

To summarize: If the large cardinals beyond choice are consistent then inner model theory (as we know it) fails and there can be no fine-structural understanding of V. And if the HOD-analogues of large cardinals beyond choice are accepted (in the context of ZFC and traditional large cardinals) then we must be on the second side of the HOD Dichotomy, where HOD is "far" from V. In either case, this is the future in which *chaos* prevails.

7 Themes from Gödel

We thus have a purely mathematical dichotomy, one that is firmly established and presents us with two, radically different possible futures. And we have a program aimed at realizing each possible future. We do not yet know which future will hold, but either way it is going to be interesting.

There is much more to be said about these two programs and the differences between the two possible futures that lie before us. What I would like to do in the remainder of this paper is discuss the different implications each future has for the questions of absolute undecidability and absolute definability.

---

[76] It also wipes out inner model theory in the sense described above.

7.1 The First Future

Suppose that the Ultimate-L Conjecture holds.[77] Then, as we saw above, the Weak HOD Conjecture holds and so, assuming large cardinals, we are in the first side of the HOD Dichotomy where HOD is "close" to V. Moreover, in this scenario we would have a detailed, fine-structural understanding of V.

For the rest of this subsection let us assume that the Ultimate-L Conjecture holds and let us assume large cardinal axioms as the level of an an extendible cardinal with a huge cardinal above. This puts us in the first side of the HOD Dichotomy.

7.1.1 Absolute Provability

As a candidate for "absolute provability" Gödel proposed "provable from ZFC + LCA," where here "LCA" is non-precisely specifiable open-ended scheme standing for "true large cardinal axioms." We saw that this proposal was undermined by results of Levy-Solovay and others. For while it has the virtue of being able to handle "vertical independence" (the kind of independence involved, e.g. with consistency statements), by the aforementioned results it cannot adequately handle "horizontal independence" (the kind of independence established by forcing). Now, the axiom $V=L$ has complementary virtues and shortcomings, for while it has the virtue of being able to handle "horizontal independence" it is incompatible with the large cardinal axioms needed to handle "vertical independence". It would be ideal if we had a version of $V=L$ which had the same virtue but lacked the shortcoming.

The axiom $V=$ Ultimate-L is such an axiom. Like $V=L$ it handles "horizontal independence" but at the same time it is compatible with *all* traditional large cardinals. So it provides us with an amended proposal: As a candidate for "absolute provability" we can now consider "provable from ZFC $+$ $V=$ Ultimate-$L+$LCA." The first component—$V=$ Ultimate-L—would erase "horizontal independence" while the second component—LCA, the non-recursive series of "all true large cardinal axioms"—would erase "vertical independence."[78] The task of discovering new axioms that overcome incompleteness would now be reduced to the task of discovering stronger and stronger large cardinal axioms, as Gödel had hoped. In this future we would have the kind of generalized completness theorem that Gödel envisaged. Incompleteness would no longer be the serious problem that it is today.

---

[77] Recall that the Ultimate-L Conjecture is a Σ^0_1-statement.

[78] One might raise the concern that perhaps the consistency hierarchy goes much further than the traditional large cardinal hierarchy, on upward through the large cardinals beyond choice. But notice that on the present scenario the large cardinals beyond choice are provably inconsistent.

7.1.2 Absolute Definability

As a candidate for "absolute definability" Gödel proposed "ordinal definability." He showed that the notion is absolute in some sense (namely that it is immune to a kind of diagonalization). But there are other senses in which a notion can be considered absolute. For example, another sense in which ordinal definability could be absolute is if there can be no non-trivial elementary embedding from HOD into itself.

It is not known that such an embedding is ruled out under or present background assumptions. But it is a reasonable conjecture that they do rule it out. And something very closely related is known. Let T be the Σ_2-theory of V with ordinal parameters and consider the structure (HOD, T).[79] A result of Woodin shows that (under our present assumptions) there can be no non-trivial elementary embedding from (HOD, T) into itself.[80] In this sense then, we will have garnered further evidence that ordinal definability does indeed provide us with an absolute notion of definability.

7.2 The Second Future

Suppose that large cardinals beyond choice—say, at the level of a Berkeley cardinal—are consistent. Then the Weak HOD Conjecture fails, the Ultimate-L Conjecture fails, and, as discussed above, the prospect of inner model theory as we know it is doomed.

7.2.1 Absolute Provability

In the present scenario we will have lost our best current approach to finding amended version of Gödel's proposal that provides us with an adequate notion of "absolute provability." Moreover, with the failure of inner model theory we will have lost the hope of a fine-structural insight into V. We will be presented with a proliferation of candidates for sentences that are truly "absolutely undecidable." In particular, many of the questions concerning large cardinals beyond choice are completely beyond current technology and would themselves be candidates for absolutely undecidable statements.[81]

---

[79] This is essentially the structure consisting of HOD where we add to HOD a predicate that allows it to identify itself.

[80] See Woodin [63], p. 306.

[81] See the final section of Bagaria, Koellner, and Woodin [1] for examples and for further discussion.

7.2.2 Absolute Definability

Assume now not only that large cardinals beyond choice are consistent but also that their HOD analogues hold—that is, assume the axioms asserting the existence of HOD Reinhardt cardinals, HOD Berkeley cardinals, etc.[82] For the reasons just discussed this would show that in a certain sense the notion of ordinal definability is *not* a legitimate candiate for absolute definability.

7.3 Conclusion

In the end all of the themes have become intertwined.

If, on the one hand, there is an ultimate version of Gödel's L then we will have a deeper understanding of the large cardinal hierarchy, support for his proposed notion of absolute definability, and, most importantly, a true candidate for an absolute notion of provability, one which would put us in a position of effectively erasing independence.

If, on the other hand, the large cardinals beyond choice are consistent and the traditional large cardinal hierarchy extends to include HOD Reinhardt cardinals, HOD Berkeley cardinals and so forth, then this will shake the support for Gödel's proposed notion of absolute definability and, more importantly, it will shatter our current hopes for a fine-structural understanding of V and unleash a whole host of candidates for absolutely undecidable statements.

I think that Gödel would have found these developments intriguing. He would want to know whether pattern or chaos prevailed. Fortunately, that is something that we will likely one day know, perhaps very soon.

References

[1] Joan Bagaria, Peter Koellner, and W. Hugh Woodin. Large cardinals beyond choice. Forthcoming, 2018.

[2] James Baumgartner. All $\aleph_1$-dense sets of reals can be isomorphic. *Fundamenta Mathematicae*, 79(2):101–106, 1973.

[3] Paul Benacerraf. God, the devil, and Gödel. *The Monist*, 51(1):9–32, January 1967.

[4] Timothy J. Carlson. Knowledge, machines, and the consistency of Reinhardt's strong mechanistic thesis. *Annals of Pure and Applied Logic*, 105(1–3):51–81, 2005.

[5] Raffaella Cutolo. *Berkeley cardinals and the search for V*. PhD thesis, University of Naples "Federico II", 2017.

---

[82] Notice that we are continuing to work in the context of AC. We are not assuming the axioms asserting the existence of large cardinals beyond choice. To do wo would be an even more radical option. But it is not one that I will consider in this paper.

[6] Keith J. Devlin. *Constructibility*. Perspectives in Mathematical Logic. Springer-Verla, 1984.

[7] B. Dushnik and E. W. Miller. Concerning similarity transformations of linearly ordered sets. *Bulletin of the American Mathematical Society*, 46:322–326, 1940.

[8] Solomon Feferman. A more perspicuous formal system for predicativity. In K. Lorenz, editor, *Konstruktionen versus Positionen*, volume I, pages 69–93. Walter de Gruyter, Berlin, 1979.

[9] Solomon Feferman. Why the programs for new axioms need to be questioned. *Bulletin of Symbolic Logic*, 6(4):401–413, 2000.

[10] Solomon Feferman. Predicativity. In Stewart Shapiro, editor, *The Oxford Handbook of Philosophy of Mathematics and Logic*, pages 590–624. Oxford University Press, 2005.

[11] Solomon Feferman. The 5 questions. In V. F. Hendricks and H. Leitgeb, editors, *Philosophy of Mathematics: 5 Questions*, pages 115–135. Automatic Press/VIP, 2007.

[12] Solomon Feferman. Axioms for determinateness and truth. *Review of Symbolic Logic*, 1(2):204–217, 2008.

[13] Solomon Feferman. Is the continuum hypothesis a definite mathematical problem? For *Exploring the Frontiers of Incompleteness*, Harvard University, 2010–2012, 2011.

[14] Solomon Feferman. Logic, mathematics, and conceptual structuralism. In P. Rush, editor, *The Metaphysics of Logic*. Cambridge University Press, 2014.

[15] Solomon Feferman. Parsons and I: Sympathies and differences. *Journal of Philosophy*, CXIII(5/6):234–246, 2016.

[16] Solomon Feferman. The continuum hypothesis is neither a definite mathematical problem nor a definite logical problem. In Peter Koellner, editor, *Exploring the Frontiers of Incompleteness*. 2018. To appear.

[17] Matthew Foreman, Menachem Magidor, and Saharon Shelah. Martin's Maximum, saturated ideals, and non-regular ultrafilters. Part I. *Annals of Mathematics*, 127:1–47, 1988.

[18] D. H. Fremlin. *Consequences of Martin's Axiom*, volume 84 of *Cambridge Tracts in Mathematics*. Cambridge University Press, 1984.

[19] Kurt Gödel. On undecidable sentences. In *[26]*, pages 31–35. Oxford University Press, *1931?

[20] Kurt Gödel. The consistency of the axiom of choice and of the generalized continuum hypothesis. In *[25]*, pages 26–27. Oxford University Press, 1938.

[21] Kurt Gödel. Remarks before the Princeton bicentennial conference on problems in mathematics. In *[25]*, pages 150–153. Oxford University Press, 1946.

[22] Kurt Gödel. Some basic theorems on the foundations of mathematics and their implications. In *[26]*, pages 304–323. Oxford University Press, *1951.

[23] Kurt Gödel. What is Cantor's continuum problem? In *[25]*, pages 254–270. Oxford University Press, 1964.

[24] Kurt Gödel. Some considerations leading to the probably conclusion that the true power of the continuum is $\aleph_2$. In *[26]*, pages 420–422. Oxford University Press, 1970.

[25] Kurt Gödel. *Collected Works, Volume II: Publications 1938–1974*. Oxford University Press, New York and Oxford, 1990.

[26] Kurt Gödel. *Collected Works, Volume III: Unpublished Essays and Lectures*. Oxford University Press, New York and Oxford, 1995.

[27] Gabriel Goldberg. On the consistency strength of Reinhardt cardinals. Unpublished.

[28] Akihiro Kanamori. *The Higher Infinite: Large Cardinals in Set Theory from their Beginnings*. Springer Monographs in Mathematics. Springer, Berlin, second edition, 2003.

[29] Peter Koellner. On the question of absolute undecidability. *Philosophia Mathematica*, 14(2):153–188, 2006. Revised and reprinted with a new postscript in *Kurt Gödel: Essays for his Centennial*, edited by Solomon Feferman, Charles Parsons, and Stephen G. Simpson. Lecture Notes in Logic, 33. Association of Symbolic Logic, 2009.

[30] Peter Koellner. The Continuum Hypothesis. In Edward N. Zalta, editor, *Stanford Encyclopedia of Philosophy*. Stanford University, 2013. URL = ⟨http://plato.stanford.edu/entries/continuum-hypothesis⟩.

[31] Peter Koellner. Large cardinals and determinacy. In Edward N. Zalta, editor, *Stanford Encyclopedia of Philosophy*. Stanford University, 2013. URL = ⟨http://plato.stanford.edu/entries/large-cardinals-determinacy⟩.

[32] Peter Koellner. Gödel's disjunction. In Leon Horsten and Philip Welch, editors, *Gödel's Disjunction: The Scope and Limits of Mathematical Knowledge*, pages 148–188. Oxford university Press, 2016.

[33] Peter Koellner. Infinity up on trial: Reply to Feferman. *Journal of Philosophy*, CXIII(5/6):247–60, 2016.

[34] Peter Koellner. Feferman on set theory: Infinity up on trial. To appear in Gerhard Jaeger and Wilfried Sieg, editors, *Feferman on Foundations—Logic, Mathematics, Philosophy*, Springer, 2018.

[35] Peter Koellner. On the question of whether the mind can be mechanized. Part 1: From Gödel to Penrose. To appear in the *Journal of Philosophy*, 2018.

[36] Peter Koellner. On the question of whether the mind can be mechanized. Part 2: Penrose's new argument. To appear in the *Journal of Philosophy*, 2018.

[37] Peter Koellner and W. Hugh Woodin. The envelope perspective on forcing axioms. In preparation.

[38] Peter Koellner and W. Hugh Woodin. Incompatible Ω-complete theories. *The Journal of Symbolic Logic*, 74(4), December 2009.

[39] Kenneth Kunen. Elementary embeddings and infinitary combinatorics. *Journal of Symbolic Logic*, 36:407–413, 1971.

[40] John Randolph Lucas. Minds, machines, and Gödel. *Philosophy*, 36:112–137, 1961.

[41] Donald A. Martin and Robert M. Solovay. Internal Cohen extensions. *Annals of Mathematical Logic*, 2(2):143–178, October 1970.

[42] Justin Moore. A five element basis for the uncountable linear orders. *Annals of Mathematics*, 163(2):669–688, 2006.

[43] John Myhill and Dana S. Scott. Ordinal definability. In Dana S. Scott, editor, *Axiomatic Set Theory*, volume XIII of *Proceedings of Symposia in Pure Mathematics*, pages 271–278. American Mathematical Society, 1971.

[44] Ernest R. Nagel and James R. Newman. Gödel's proof. *Scientific American*, 3:1668–1695, 1956.

[45] Ernest R. Nagel and James R. Newman. *Gödel's Proof.* New York University Press, 1958. Revised edition, 2001, edited with a new forward by Douglas R. Hofstadter.

[46] Roger Penrose. *The Emperor's New Mind: Concerning Computeres, Minds, and the Laws of Physics.* Oxford University Press, 1989.

[47] Roger Penrose. *Shadows of the Mind: A Search for the Missing Science of Consciousness.* Oxford University Press, 1994.

[48] Roger Penrose. Beyond the doubting of a shadow. *Psyche*, 2:89–129, 1995.

[49] Roger Penrose. Gödel, the mind, and the laws of physics. In Matthias Baaz, Christos H. Papadimitriou, Hilary W. Putnam, Dana S. Scott, and Charles L. Harper, editors, *Kurt Gödel and the Foundations of Mathematics: Horizons of Truth*, chapter 16, pages 339–358. Cambridge University Press, 2011.

[50] Michael Rathjen. Indefiniteness in semi-intuitionistic set theories: On a conjecture of Feferman. *The Journal of Symbolic Logic*, 81(2):742–754, June 2016.

[51] William Reinhardt. *Topics in the Metamathematics of Set Theory*. PhD thesis, University of California, Berkeley, 1967.

[52] William N. Reinhardt. Absolute versions of incompleteness theorems. *Noûs*, 19(3):317–346, September 1985.

[53] William N. Reinhardt. The consistency of a variant of Church's thesis with an axiomatic theory of an epistemic notion. In *Special Volume for the Proceedings of the 5th Latin American Symposium on Mathematical Logic, 1981*, volume 19 of *Revista Colombiana de Matemáticas*, pages 177–200, 1985.

[54] Saharon Shelah. Decomposing uncountable squares into countably many chains. *Journal of Combinatorial Theory, Series A*, 21(1):110–114, 1976.

[55] Johannes Stern. *Toward Predicate Approaches to Modality*. PhD thesis, University of Geneva, 2012.

[56] Stevo Todorcevic. Trees and linearly ordered sets. In Kenneth Kunen and Jerry E. Vaughan, editors, *Handbook of Set-Theoretic Topology*, pages 253–293. North-Holland, 1984.

[57] Stevo Todorcevic. Directed sets and cofinal types. *Transactions of the American Mathematical Society*, 290(2):711–723, August 1985.

[58] Stevo Todorcevic. The powerset of ω_1 and the continuum problem. Paper written for the EFI Project at Harvard University, 2013.

[59] Stevo Todorcevic. *Notes on Forcing Axioms*, volume 26 of *Lecture Notes Series*. World Scientific, 2014.

[60] John W. Tukey. *Convergence and Uniformity in Topology*, volume 2 of *Annals of Mathematics Studies*. Princeton University Press, 1940.

[61] Hao Wang. *A Logical Journey: From Gödel to Philosophy*. MIT Press, 1996.
[62] W. Hugh Woodin. *The Axiom of Determinacy, Forcing Axioms, and the Nonstationary Ideal*, volume 1 of *de Gruyter Series in Logic and its Applications*. de Gruyter, Berlin, 1999.
[63] W. Hugh Woodin. Suitable extender models I. *Journal of Mathematical Logic*, 10(1-2):101–339, June 2010.
[64] W. Hugh Woodin. The Continuum Hypothesis, the generic-multiverse of sets, and the Ω-conjecture. In Juliette Kennedy and Roman Kossak, editors, *Set Theory, Arithmetic, and Foundations of Mathematics: Theorems, Philosophies*, volume 36 of *Lecture Notes in Logic*. Cambridge University Press, 2011.
[65] W. Hugh Woodin. Suitable extender models II. *Journal of Mathematical Logic*, 11(2):115–436, 2011.
[66] W. Hugh Woodin. In search of Ultimate-L: The 19th Midrasha Mathematicae lectures. *Bulletin of Symbolic Logic*, 23(1), 2017.

A GLIMPSE AT POLYNOMIALS WITH QUANTIFIERS

ANDREY BOVYKIN
Institute of Mathematics, Universidade Federal da Bahia, Brazil
andrey.bovykin@gmail.com

MICHIEL DE SMET
University of Gent, Belgium
michiel.de.smet@gmail.com

Abstract

A prefixed polynomial equation, or a polynomial expression with a quantifier prefix, is an equation of the form $P(x_1, x_2, \ldots, x_n) = 0$, where P is a polynomial with integer coefficients whose variables $x_1, x_2, \ldots, x_n$ range over natural numbers, that is preceded by some quantifiers over all of its variables $x_1, x_2, \ldots, x_n$. Here is a typical, seemingly random, example of such an expression, Φ:

$$\forall\ m\ e\ \exists\ N\ \forall\ a\ b\ \exists\ c\ d\ A\ X\ \forall\ x\ y\ \exists\ BCF\ \exists\ hijk\ell npqrst$$

$$x \cdot \big(y+B-x\big) \cdot \big(A+m+B-y\big) \cdot \Big((A+h-d)^2 + ((d+1)\cdot i + A - c)^2 + (B+n-dx)^2 +$$

$$+ ((dx+1)\cdot j + B - c)^2 + (C+r-dy)^2 + ((dy+1)\cdot k + C - c)^2 + (B+s+1-C)^2 +$$

$$+ (C+t-N)^2 + (F+p-b(B+C^2))^2 + (a-\ell b(B+C^2) - F - \ell)^2 + (X-F+eq)^2\Big) = 0$$

In this note we initiate the study of the collection of all possible such expressions ('the Atlas'), equipped with the equivalence relation of "being EFA-provably equivalent" on its members. The Atlas is partially ordered by EFA-implication. Here is the first abstract picture of the Atlas to have in mind:

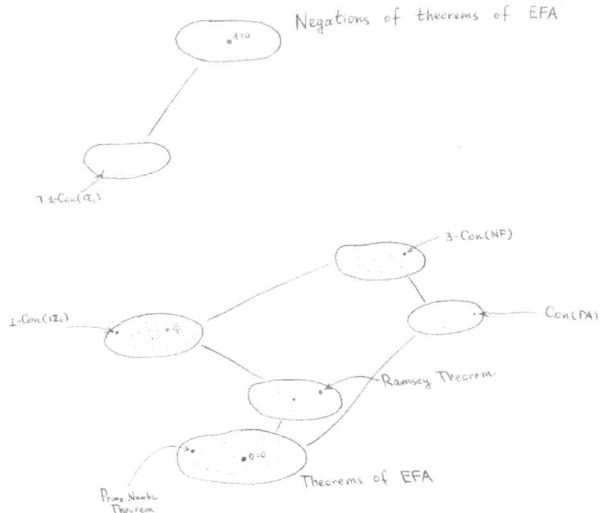

Notice that the set of all prefixed polynomial equations is arithmetically complete, that is, every first-order arithmetical formula is EFA-equivalent to some prefixed polynomial expression. In this sense, the Atlas is just another way of talking about first-order arithmetical statements. Gödel's Incompleteness theorems guarantee existence of many distinct equivalence classes. Our first task is to find examples.

We start off with examples of distinct equivalence classes of metamathematical interest: the 1-consistency of $I\Sigma_1$ (the expression Φ above), the 1-consistency of $I\Sigma_2$, the 1-consistency of full Peano Arithmetic, the highly unprovable Finite Kruskal's Theorem.

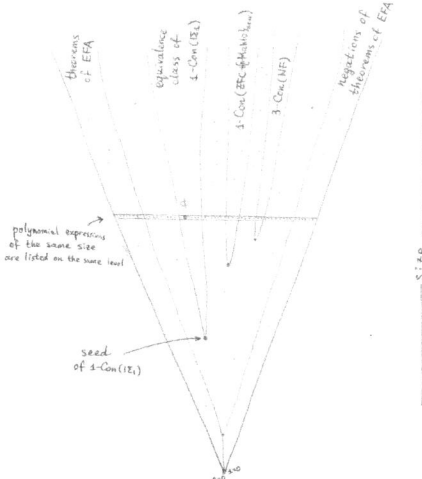

Then we give an example of the phase transition phenomenon. We produce a polynomial expression that has two free variables m and n such that whenever $\frac{m}{n}$ is smaller or equal to Weiermann's constant $w \approx 0.63957768999472013311\ldots$, the expression is EFA-provable, otherwise it is unprovable in the theory ATR_0 (so we witness a phase transition between EFA-provability and predicative unprovability). Then we give a crude example of a polynomial equation with quantifiers that is equivalent to the famous Graph Minor Theorem and, hence, is of the unknown high strength beyond that of Π_1^1-CA_0. A 'seed' is a prefixed polynomial equation that is of minimal length in its EFA-equivalence class. We discuss seeds and notice that the seed of the 1-consistency of $I\Sigma_1$ is smaller than 131. We discuss the role of the Atlas and its possible future partial implementation as a metamathematically-sensitive database of mathematical knowledge. The purpose of this note is to give definitions, to arrange the set-up, give non-trivial examples, introduce the right unprovability-sensitive notions, and ask some first questions about the Atlas. The current note omits all proofs. All proofs can be found in the grand manuscript [4]. Eventually, a bigger article, stemming from [4] will appear. Some of the material from this project became a chapter in the second author's doctoral thesis [6] at the University of Gent, Belgium.

The authors genuinely and cordially thank the John Templeton Foundation for its interest and support for Unprovability research. The first author thanks the John Templeton Foundation for its financial support.

1 Introduction

Definition 1.
A prefixed polynomial equation, or a polynomial expression with a quantifier-prefix, or, occasionally, simply a polynomial expression, is an expression of the form

$$Q_1 x_1\, Q_2 x_2\, \ldots\, Q_n x_n\, P(x_1, x_2, \ldots, x_n) = 0,$$

where P is a polynomial with integer coefficients whose variables $x_1, x_2, \ldots, x_n$ range over natural numbers (including zero), that is preceded by a block of various existential or universal quantifiers $Q_1, Q_2, \ldots, Q_n$ over its variables $x_1, x_2, \ldots, x_n$. It is important to interpret subtraction correctly: our statement is evaluated over natural number inputs, and outputs are also natural numbers.

Definition 2. Atlas
The set of all possible polynomial expressions with quantifiers will be called the Atlas. We shall sometimes speak of the Atlas as a 'template' in the sense that it is the set of all substitution instances of concrete polynomial expressions P and quantifier-blocks into our formula above.

Definition 3. Equivalence

We say that two prefixed polynomial equations P and Q are equivalent or EFA-equivalent, or EFA-provably equivalent if there is a proof of $P \longleftrightarrow Q$ in the theory EFA, the Exponential Function Arithmetic, or its variant $I\Delta_0 + \exp$. This is an equivalence relation on the Atlas.

All familiar statements in number theory can be converted into polynomial expressions. Some very easily, and some others using encoding techniques. For example, here is the infinitude of primes:

$$\forall n \; \exists p \; \exists k \; \forall i \; \forall j \; \exists \ell$$

$$\left(n+k+1-p\right)^2 + \left((i+2)(j+2)+\ell+1-p\right) \cdot \left(p+\ell+1-(i+2)(j+2)\right) = 0.$$

And here is the Goldbach Conjecture:

$$\forall n \; \exists p \; \exists q \; \forall i \; \forall j \; \exists k \; \exists \ell$$

$$\left((i+2)(j+2)+k+1-p\right) \cdot \left(p+k+1-(i+2)(j+2)\right) + \left((i+2)(j+2)+\ell+1-q\right) \cdot$$

$$\cdot \left(q+\ell+1-(i+2)(j+2)\right) + \left(2(n+2)-(p+q)\right)^2 = 0.$$

An even shorter open problem is the Diophantine statement from [16]:

$$\exists x \; \exists y \; \exists z \quad x^3 - y^3 - z^3 - 33 = 0.$$

The important measure of any mathematical statement is its 'true' quantifier complexity. For example, the Riemann Hypothesis has the 'true' complexity Π_1^0, i.e., is provably equivalent to a Π_1^0 statement. The Goldbach Conjecture also has quantifier complexity Π_1^0. It is easy to see that in the expression above, only the first quantifier is truly free: all others can be bounded and eliminated in the standard manner described in, say, [13] or [17]. This transformation would normally explode the size of the polynomial. This is a recurring theme in this kind of projects: the eternal and unavoidable trade-off between the size of the polynomial and the quantifier complexity. Low quantifier complexity is very expensive.

Definition 4. Order (or strength)

For two polynomial expressions P and Q, we set $P < Q$ if EFA proves $Q \to P$ but does not prove $P \to Q$. Then we say that Q is strictly provably stronger than P. Clearly, we can also introduce the rest of the Boolean algebra structure on the EFA-equivalence classes of the Atlas.

Definition 5. Length (or size)
Let us fix the following method of counting the length of a polynomial expression with quantifiers. We ignore the quantifier-prefix and the final " $= 0$" and count only the length of the polynomial expression, as follows: every occurrence of a variable or multiplication or addition operations contributes 1 to the total size, a coefficient n contributes $(n-1)$ to the total size, $+n$ or $-n$ contribute n, and the power n contributes $(n-1)$ to the total size.

This is the first moment in this note that we made a fairly arbitrary decision. We could of course instead count coefficients and powers as contributing $\log_2(n)$ or $\log_{10}(n)$ to the size, or in any other monotone way. However, the way we fixed is already very natural. Note that we are counting the length of an expression, not of the polynomial function itself. Clearly, every expression can be re-written into infinitely-many equivalent expressions of various lengths.

Definition 6. Seeds
A prefixed polynomial expression that is shortest in its EFA-equivalence class is called a seed of its equivalence class. There may occasionally exist several seeds of the same equivalence class of the same size. The notion of a seed is of course sensitive to the choice of the language of the Atlas and the method of counting size.

The Atlas, as we defined it, is not really a new object. Some people in the XXth century, given a theory T in any language, called the set of equivalence classes of formulas under T-provability, the 'Lindenbaum-Tarski algebra' of T. However, as far as we know, the arithmetical Atlas defined here has never been systematically studied.

Three possible polynomial templates: over $\mathbb{N}$, $\mathbb{Z}$, $\mathbb{Q}$

There are at least three natural interesting templates involving polynomials with quantifiers which we could use in this project:

1. to let variables range over natural numbers (non-negative integers);

2. to let variables range over all integers;

3. to let variables range over all rationals.

All three set-ups should be equally interesting mathematically. Of course polynomials with quantifiers mean completely different things when the variables are allowed to range over these different sets. In building examples and proving various theorems, we can expect advantages and disadvantages of each of these templates and trade-offs between them. We chose to concentrate on template (1) for now because it is more familiar to logicians, although the other two are also good and important and should wait for their turn to be explored. By Julia Robinson's theorem on the extraction of $\mathbb{Z}$ from $\mathbb{Q}$, the template (3) above is also arithmetically complete (although adding a bunch of quantifiers on the way: the original Robinson's formula is Π_4^0, namely $\forall \exists \forall \exists$, the recent improvement by Poonen [16] is Π_2^0, namely $\forall\forall\exists\exists\exists\exists\exists\exists\exists$).

As for non-polynomial templates, the natural next step is to allow exponentiation and other familiar functions into the equations. Losing the purity of pure polynomials but continuing on the route of enlarging our permitted language, it is natural to introduce more symbols: concrete functions, relations, sets, cardinality and a range of familiar combinatorial symbols.

Diophantine equations

Diophantine equations are a particular subclass of instances of our template, that, in which all quantifiers in front of the polynomial are existential:

$$\exists x_1 \, \exists x_2 \ldots \exists x_n \, P(x_1, x_2, \ldots, x_n) = 0.$$

Solvability of Diophantine equations (over $\mathbb{Z}$) by a single algorithm was the question of Hilbert's Tenth Problem, spectacularly resolved in the second half of the XXth century, first for exponential-polynomial equations by Martin Davis, Hilary Putnam and Julia Robinson, and finally, young Yuri Matiyasevich eliminating the need for the exponent. The story is now well-known and we refer the reader elsewhere for the mathematical and personal accounts of the events.

What is important for us in this note:

1. there is no algorithm that resolves Diophantine equations or even picks out the resolvable ones (the ones that have solutions);

2. every Σ_1^0 formula is EFA-equivalent to solvability of a certain Diophantine equation [7];

3. every Π_1^0 formula is EFA-equivalent to non-solvability of a certain Diophantine equation.

In particular, Π_1^0 statements, like Con(ZFC), that belong to non-trivial equivalence classes of the Atlas, are equivalent to non-solvability of their respective Diophantine equations.

In our proofs, we used some tricks from the study of Hilbert's Tenth Problem. However, not being restricted by one block of quantifiers pays off very well and we end up with very short metamathematically-interesting expressions.

On the choice of our background theory EFA

We chose EFA as our base theory in this project. This may seem as the second arbitrary decision we took (apart from the definition of the size of a polynomial). Indeed, we could theoretically choose another theory: $I\Delta_0$, $I\Sigma_1$, ATR_0 or Z_2. Each would factorise the Atlas differently. So why did we choose EFA?

We could choose a weaker theory, like $I\Delta_0$ or $I\Delta_0 + \Omega$ but then the Atlas would dissolve into a cloud of myriads of unreasonably-small equivalence classes, where two naturally equivalent formulas (truly easily equivalent) would be made to belong to different equivalence classes not for a mathematical reason but just because the base theory lacks the coding power to conduct an innocent equivalence proof.

We could choose a stronger theory, say ZFC+ "there is a huge cardinal". But this would make the Atlas blind to all metamathematical distinctions we want to emphasise. We want "0 = 0", the Finite Ramsey Theorem, the Paris-Harrington Principle and Con(ZFC+"there is a measurable cardinal") belong to distinct equivalence classes, because they represent different phenomena, and we care about these differences.

The choice of the base theory is the choice of metamathematical sensitivity. Instead of EFA, we could of course choose $I\Sigma_1$ (or its variants PRA, RCA_0, WKL_0), PA (ACA_0) or even a very strong theory KP+ Inf if model-theoretic equivalences involving L are used in proofs.

More on EFA and its strength can be found in Jeremy Avigad's article [1].

Arithmetisation of mathematics

Early pioneers of computability theory (Kurt Gödel, Alfred Tarski, Alan Turing, Alonzo Church with disciples, and all early intuitionists and constructivists) made an important discovery. On the one hand, they understood that the arithmetical language, over the right base theory, allows us to deal with syntax, deductions and computations, hence, yielding Gödel's Incompleteness theorems and a later avalanche of theorems on algorithmic undecidability of decision problems. But also they understood more: arithmetical formulas can represent approximations of theorems from calculus, geometry, mathematical physics etc., and arithmetical axiomatic systems can prove these formulas. (For example, Alan Turing's most famous paper [20] on computable numbers is an early manifesto of constructivism and arithmetisation of calculus.) One can arithmetise calculus, mathematical physics, geometry, much of algebra and all that is usually considered "separable mathematics". Here is an exercise for the reader: write a first-order arithmetical formula that says $\frac{\sin x}{x} \to_{x \to 0} 1$, and then convert it into a member of the Atlas.

So, members of the Atlas don't only represent the literal 'natural number theory' in the sense of literal meanings of first-order arithmetical formulas, but all possible statements from most of mathematics, approximated or encoded and re-cast in the pure language of prefixed polynomials. Here, in the scope of this note and this project, we are not particularly interested in *theorems* from these various subjects (why substitute meaningful discoveries by unintelligible writing?) or *negations of* these *theorems*, but in members of metamathematically non-trivial EFA-equivalence classes (unprovable statements).

Arithmetical strength of axiomatic systems

We could of course reason about members of the Atlas (or, more generally, transform members of the Atlas) using various axiomatic systems, or rules of transformation, not just EFA. What is needed from a random axiomatic system T, in an arbitrary language, is that the first-order language of arithmetic is interpreted in T. Literally, T may be talking about imaginary wombles or sepulki or 'sets', but there should be a good translation from formulas of the language of arithmetic into formulas in the language of T.

Definition 7. Arithmetical strength
Given a theory T, we define T^a, its arithmetical strength, to be the set of all polynomial expressions with quantifiers, whose translations into the language of T are provable in T. Given two axiomatic systems, T_1 and T_2, we set $T_1 \trianglelefteq T_2$ ("T_1 is arithmetically not stronger than T_2") if $T_1^a \subseteq T_2^a$, and $T_1 \triangleleft T_2$ ("T_2 is strictly arithmetically stronger than T_1") if T_1^a is a proper subset of T_2^a. A theory T_2 is called arithmetically conservative over T_1 if $T_1^a = T_2^a$ (we especially care about the case of $T_2 \supseteq T_1$ in this definition).

For example $\text{ZFC}^a = \text{ZF}^a$ and $\text{ACA}_0^a = \text{PA}^a =$ polynomial translations of theorems of PA.

The hierarchy of classical 'foundational' theories starts as a linear order as we add more induction principles, set-existence axioms (comprehensions, separations, replacements, collections, powerset, etc) and transfinite recursion principles. However, it is absolutely unclear what happens after we exhaust the "obviously true" reasoning principles. Here is a picture of some classic axiomatic theories, ordered by arithmetical strength.

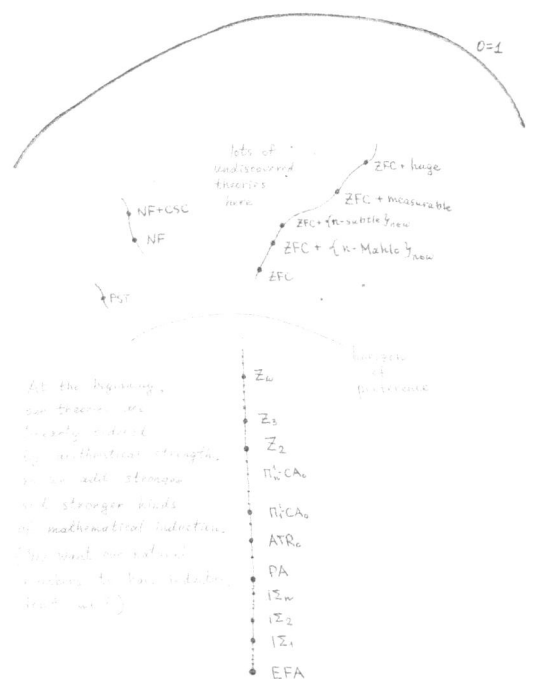

Arithmetical Splitting

The question of discovering genuinely alternative, equally good (without any way to form preference) axiomatic systems T_1 and T_2 that contradict each other on first-order arithmetical formulas is wide open, very controversial and has fascinated the first author for the last quarter-a-century. Although for third-order arithmetical statements (like CH or AD) this is nowadays a relatively non-controversial issue, and splitting is considered, although not by everyone, as an innocent philosophical possibility (in some universes CH holds, in others it doesn't), raising the same possibility for members of the Atlas invites many objections: some horrified by the sheer revolutionary nature of this unimaginable possibility ("isn't there the true set of natural numbers?") and some accompanied by mathematical arguments pointing towards an explanation as to why first-order Arithmetical Splitting should be impossible. Some people appeal to the second-order definition of "arithmetical truth", others appeal to its ordinal iterative approximation, via say, the vague phantom ω_1^{CK}, others confuse the issue with its one-quantifier particular case and point out the compelling reasons for Π_1^0-realism.

We shall not discuss Arithmetical Splitting in this note. An up-to-date discussion will appear in [5] and other articles. Let us just mention that once Arithmetical Splitting is discovered, the members of the Atlas on which the future controversial theories T_1 and T_2 disagree, would be of utmost metamathematical interest. We believe that so far logic is in its infancy and many more axiomatic systems, representing as-yet-unimaginable mathematical phenomena, will be discovered in the future. In particular, Arithmetical Splitting will be discovered and will take central stage in all future metamathematical thinking, beyond mere Gödel Incompleteness.

How to imagine the Atlas as concentric clusters

Every EFA-equivalence class can be thought of as surrounded by other classes that EFA fails to identify, but $I\Sigma_1$ does. Then come the further classes which PA identifies with our class, then stronger theories, etc. The stronger theory we use, the more blind it will be to the important metamathematical differences we care about. So, some classes are closer to each other than others, depending on the strength of a theory, if any, that makes them equivalent.

But also, keep in mind the disconnectedness of the Atlas: if the pair $(\varphi \longleftrightarrow \psi)$ and $\neg(\varphi \longleftrightarrow \psi)$ witnesses Arithmetical Splitting then there will be no eventual "true" theory to prove that φ and ψ belong to the same class. So, we don't expect that the concentric clusters eventually fuse together to form two opposite clusters (the "true" and the "false" statements).

2 First three examples to guide us

Let us start off by briefly glancing at three particular distinct elements of the Atlas. These three elements belong to three different EFA-equivalence classes. It is important to note at this stage: the three expressions below can be re-written into (or proved to be EFA-equivalent to) billions of other polynomial expressions of comparable lengths, so there is no need to concentrate attention on any features of these particular expressions, but think of them as mere generic representatives of their equivalence classes.

Unprovability by primitive recursive means

Theorem 1.
The following statement Φ is equivalent to the Paris-Harrington Principle for pairs, hence to the 1-consistency of $I\Sigma_1$ and, hence, is unprovable in $I\Sigma_1$.

$$\forall\, m\, e\ \ \exists\, N\ \ \forall\, a\, b\ \ \exists\, c\, d\ A\ X\ \forall\, x\, y\ \ \exists\, BCF\ \exists\, hijk\ell npqrst$$

$$x\cdot\Big(y+B-x\Big)\cdot\Big(A+m+B-y\Big)\cdot\Big((A+h-d)^2+((d+1)\cdot i+A-c)^2+(B+n-dx)^2+$$
$$+((dx+1)\cdot j+B-c)^2++(C+r-dy)^2+((dy+1)\cdot k+C-c)^2+(B+s+1-C)^2+$$
$$+(C+t-N)^2+(F+p-b(B+C^2))^2+(a-\ell b(B+C^2)-F-\ell)^2+(X-F+eq)^2\Big)=0$$

Proof.
See the big manuscript [4] for the proof. The proof goes by demonstrating equivalence with PH^2, the Paris-Harrington Principle for pairs, so this statement is unprovable in $I\Sigma_1$. □

This statement, as written, has quantifier complexity Π_6^0. However, the last four blocks of quantifiers can be bounded by some exponential expressions, which can be struggled with and eliminated using the methods from Chapter 6 of [13], so the formula is really equivalent to a Π_2^0 formula. We didn't do any of it because it would blow up the size of the resulting polynomial. However, a lot of this work has been later done by Aran Nayebi in [15] and we highly recommend his article as a continuation of our big manuscript [4]. With our method of counting length, the polynomial above has size **131**. The number of variables is **25**. Here, there are probably many possibilities to simplify the polynomial, perhaps to half the current size, by reusing variables and clever combinatorial equivalences during the proof. Also the choice of coding tricks and the way to arrange the colouring can transform the resulting polynomial. No attempts have yet been made to simplify it. And there are no reasons to think that we are anywhere near the shortest member (the seed) of its equivalence class. Also, it of course should be possible to have a much shorter and simpler expression equivalent to the expression above if we allow additional symbols: exponentiation or both exponentiation and logarithm. It is interesting to see how much better we can do with these extra means.

Unprovability in two-quantifier-induction arithmetic

Theorem 2.
The following statement Φ_2 is equivalent to PH^3 and, hence, to the 1-consistency of $I\Sigma_2$, and thus is not provable in $I\Sigma_2$.

$$\forall\, m\, e\ \exists\, N\ \forall\, a\, b\ \exists\, c\, d\ A\ X\ \forall\, xyz\ \exists\, BCD\ F\ \exists\, hijk\ell npqrstuvw$$

$$x\cdot\Big(y+B-x\Big)\cdot\Big(z+B-y\Big)\cdot\Big(A+m+B-z\Big)\cdot\Big((A+h-d)^2+(c-A-(d+1)\cdot i)^2+$$

$$+(B+n-dx)^2+(c-B-(dx+1)\cdot j)^2+(C+r-dy)^2+(c-C-(dy+1)\cdot k)^2+(D+t-dz)^2+$$
$$+(c-D-(dz+1)\cdot u)^2+(B+s+1-C)^2+(C+v+1-D)^2+(D+w-N)^2+$$
$$+(F+p-b\cdot(B+C^2+D^3))^2+(a-F-\ell\cdot b\cdot(B+C^2+D^3)-\ell)^2+(F-X+qe)^2\Big) = 0.$$

Proof.
For the proof, see our big manuscript [4]. Again, we show that the statement Φ_2 above is equivalent to PH^3, the Paris-Harrington Principle for triples. □

Again, the statement Φ_2 is Π_6^0 but is actually EFA-equivalent to a much longer Π_2^0 formula. The statement has **30** variables. With our method of counting, the polynomial is of size **175**. Also, we know that there are no Σ_1^0 or Π_1^0 expressions in this equivalence class because the 1-consistency of $I\Sigma_2$ or the Paris-Harrington Principle for triples are strictly Π_2^0. The same remark applies to all our examples. For each equivalence class in the Atlas, there is the smallest quantifier-complexity of its members and we often know it from metamathematical considerations.

Unprovability in full Peano Arithmetic

Theorem 3.

Consider the following statement $\Phi(\mathbf{n})$ with one free variable $\mathbf{n}$. For every $\mathbf{n} > 1$, the statement $\Phi(\mathbf{n})$ is equivalent to the Paris-Harrington Principle in dimension $\mathbf{n}$ and hence to the 1-consistency of $I\Sigma_{\mathbf{n}-1}$. In particular, $\forall \mathbf{n}\ \Phi(\mathbf{n})$ is equivalent to PH and hence to the to 1-consistency of Peano Arithmetic. Therefore $\forall \mathbf{n}\ \Phi(\mathbf{n})$ is unprovable in Peano Arithmetic.

$$\forall\, m\, e\, \exists\, N\, \forall\, a\, b\, u\, v\, \exists\, str\, STR\, \alpha\beta\gamma\delta\varepsilon\rho\tau\, \forall\, i\, j$$

$$\exists k\ell BC\sigma\Sigma pPQU\, \Omega Mxyzw\Delta E\, FGK\, HLZ\, XW$$

$$\Big\{[s-t-r-1]^2+[S-T-R-1]^2+\Big[i\cdot(i-p-\mathbf{n}-1)\cdot\Big([s-\sigma-\Sigma(it+1)]^2+[\sigma+w-it]^2+$$
$$+[S-C-z(iT+1)]^2+[C+\Omega-iT]^2+[\sigma+B+1-C]^2\Big)\Big]\cdot\Big([a-\alpha-\beta((t+s^2)b+1)]^2+$$
$$+[\alpha+\gamma-(t+s^2)b]^2+[a-\delta-\varepsilon((T+S^2)b+1)]^2+[\delta+\rho-(T+S^2)b]^2+[\alpha+\tau+1-\delta]^2\Big]\Big\}\cdot$$
$$\cdot\Big\{i\cdot(j+B-i)\cdot(r+m+C-j)\cdot\Big[[s-r-x(t+1)]^2+[r+\ell-t]^2+[s-B-\sigma(ti+1)]^2+$$
$$+[B+w-ti]^2+[s-C-M(tj+1)]^2+[C+y-tj]^2+[B+z+1-C]^2+[C+\Omega-N]^2\Big]^2+$$

$$\left[i \cdot (\mathbf{n}+p+1-i) \cdot k \cdot (r+m+E-k) \cdot (u+\Sigma-v) \cdot \left[[u-F-G(iv+1)]^2 + [F+K-iv]^2 + \right.\right.$$
$$+[s-H-L(tk+1)]^2 + [H+Z-tk]^2 + [u-P-Q((i+1)v+1)]^2 + [P+U-(i+1)v]^2 +$$
$$+[P+X-F] \cdot [(H-P)^2 - W - 1] \Big] \cdot \Big[[a-\alpha-\beta((v+u^2)b+1)]^2 + [\alpha+\gamma-(v+u^2)b]^2 +$$
$$\left.\left. +[\alpha - S - \tau \cdot e]^2 \right]\right]^2 \Big\} \;=\; 0.$$

Proof.
For the proof, see our full manuscript [4]. We prove that for all **n**, the expression above is equivalent to the Paris-Harrington principle in dimension **n**. □

Again, all quantifiers after the first two blocks of quantifiers can be made bounded by some exponential functions, and the famous battle against the bounded quantifier (Chapter 6 of [13]) can reduce the statement to its true Π_2^0 shape, although at the cost of losing the current small size. Here, we have one free variables **n** and **48** bound variables. The polynomial expression $\Phi(n)$ is of size **386**, with n counted as a variable. For each concrete **n**, the size of $\Phi(\mathbf{n})$ is **384+ 2n**.

Corollary 4.
For every **n**, there is a prefixed polynomial equation of length $\leq$ **384 + 2n** that is equivalent to 1-Con($I\Sigma_\mathbf{n}$).

So, the sizes of seeds of 1-Con($I\Sigma_n$) are bounded by a linear function of n, which should in the future yield consequences, including pigeonhole kind of consequences.

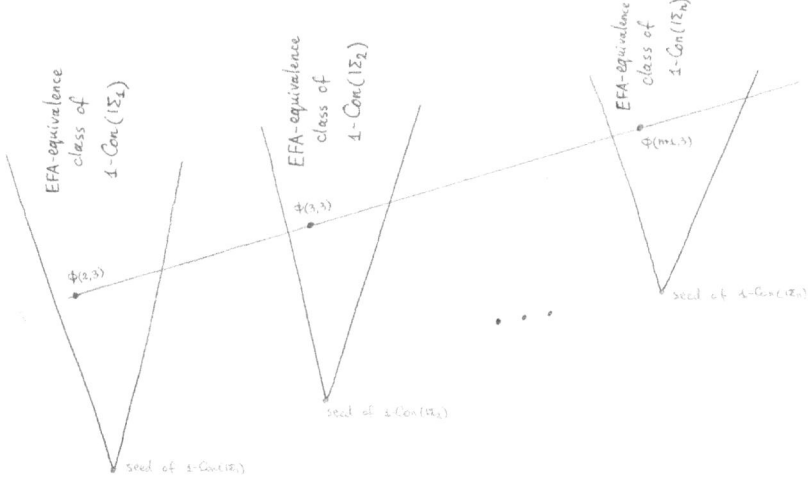

3 Beyond predicative mathematics

The readers could have thought for a moment that the three relatively neat examples above are due to pure luck or to some feature of $I\Sigma_n$ and that in reality it is much harder to reach high impredicative equivalence classes. We were full of doubts on this issue for a while until proving the following theorems in October 2009.

An expression equivalent to the Finite Kruskal Theorem

Theorem 5. The following polynomial equation with quantifiers is equivalent to the Finite Kruskal Theorem and hence is unprovable in predicative mathematics, for example in the theory ATR_0:

$\forall\, K\, \exists\, M\, \forall\, ab\, \exists\, ijcdefhk\, \forall\, lmnpq\, A\, \exists\, grst\, BFGIJLOPQWXYZ\ \alpha\beta\gamma\delta\zeta\eta\theta\kappa\lambda\mu\nu\xi\pi\rho\sigma\tau$

$\forall\, uvxyz\, CDHNT\, \exists ERS\, \forall\, U\, \exists\, V$

$[(i-c-1)^2+(i+d-M)^2+(w+1-t)^2+(t+X-q)^2+(g+1-s)^2+(s+Y+1-r)^2+(r+Z-q)^2+$

$+((p+l^2-bi-1-B)\cdot(l+B-p)\cdot((biA+A-a+p+l^2)^2-B-1)\cdot((K+i-q)^2-B-1)\cdot$

$\cdot(u-pr-1-E)\cdot((prC+C-l+u)^2-E-1)\cdot(v-ps-1-E)\cdot((psD+D-l+v)^2-E-1)\cdot$

$\cdot(x-pt-1-E)\cdot((ptH+H-l+x)^2-E-1)\cdot(u+E-v)\cdot v\cdot(q+E+1-z)\cdot$

$\cdot(((vN-u)^2-E-1)^2+(vR-x)^2+(uS-x)^2)\cdot(y-pz-1-E)\cdot((pzT+T-l+y)^2-E-1)\cdot$

$\cdot((ER-u)^2+(ES-v)^2+((EU-y)^2-V-1)^2))^2]\cdot[mni(m-n)\cdot(K+i+r+1-m)\cdot$

$\cdot(K+j+r+1-m)\cdot(j+r-i)\cdot(M+r+1-j)\cdot((f+e^2+r-bi)^2+(bis+s-a+f+e^2)^2+$

$+(k+h^2+t-bj)^2+(bjW+W-a+k+h^2)^2+(k+X+1-h)^2+(F+Y-fm)^2+$

$+(fmZ+Z-e+F)^2+(F+F^2G^2+g-dm)^2+(dmB+B-c+F+F^2G^2)^2+(kOR+R-h+G)^2+$

$+(G+E-kO)^2+(S+1-OIPQ(e-f))^2+(O+V-K-j)^2+(J+\alpha-fn)^2+(fn\beta+\beta-e+J)^2+$

$+(dn\delta+\delta-c+J+J^2L^2)^2+(J+J^2L^2+\gamma-dn)^2+((L-G)^2-\zeta-1)^2+(P+P^2Q^2+\eta-dI)^2+$

$+(dI\theta+\theta-c+P+P^2Q^2)^2+(I+\kappa-K-i)^2+(P\lambda-F)^2+(P\mu-J)^2+(P-F\nu+J\xi)^2+(Q\pi-G)^2+$

$+(Q\rho-L)^2+(Q-G\tau+L\sigma)^2)]=0.$

This polynomial is of size **648**. It has **66** variables.

A phase transition polynomial between EFA-provability and predicative unprovability

Consider the following quantified polynomial equation $A(\mathbf{m}, \mathbf{n})$, with the two free variables $\mathbf{m}$ and $\mathbf{n}$, which we show in bold font:

$\forall K \exists M \forall ab \exists ijcdefhk \, \phi\chi \, \forall lmnpq \, A \, \Gamma\Delta \, \exists grst \, BFGIJLOPQWXYZ \, \alpha\beta\gamma\delta\zeta\eta\theta\kappa\lambda\mu\nu\xi\pi\rho\sigma\tau\varphi\psi\omega$

$\forall uvxyz \, CDHNT \, \Theta \, \exists ERS \, \Lambda\Upsilon\Phi\Psi\Omega \, k_*l_*m_*n_*o_*p_* \, \forall U \, \exists V$

$[(((\Gamma-i)^2 - \varphi - 1) \cdot ((\Delta - \phi)^2 - \varphi - 1))^2 + (((\Gamma-j)^2 - \psi - 1) \cdot ((\Delta - \chi)^2 - \psi - 1))^2] \cdot [(\Gamma^2 + \Delta^2) \cdot ((1 + \Upsilon - \omega)^2 +$
$+ (\omega\Phi + \Phi - \psi + 1)^2 + (\varphi + \Psi - \omega\Delta - \omega)^2 + (\omega\Delta\Omega + \omega\Omega + \Omega - \psi + \varphi)^2 + (\Delta + k_* + 1 - i)^2 \cdot ((\Lambda + k_* - \omega\Theta)^2 +$
$+ (\omega\Theta l_* + l_* - \psi + \Lambda)^2 + (2\Lambda + m_* - \omega\Theta - \omega)^2 + (\omega\Theta n_* + \omega n_* + n_* - \psi + 2\Lambda)^2) + (\varphi + o_* - \Gamma)^2 +$
$+ (\Gamma + p_* + 1 - 2\varphi)^2))] + [(i - c - 1)^2 + (i + d - M)^2 + (w + 1 - t)^2 + (t + X - q)^2 + (g + 1 - s)^2 + (s + Y + 1 - r)^2 +$
$+ (r + Z - q)^2 + ((p + l^2 - bi - 1 - B) \cdot (l + B - p) \cdot ((biA + A - a + p + l^2)^2 - B - 1) \cdot ((\mathbf{m}K + \mathbf{n}\phi - \mathbf{m}q)^2 - B - 1) \cdot$
$\cdot (u - pr - 1 - E) \cdot ((prC + C - l + u)^2 - E - 1) \cdot (v - ps - 1 - E) \cdot ((psD + D - l + v)^2 - E - 1) \cdot (x - pt - 1 - E) \cdot$
$\cdot ((ptH + H - l + x)^2 - E - 1) \cdot (u + E - v) \cdot (q + E + 1 - z) \cdot v \cdot (((vN - u)^2 - E - 1)^2 + (vR - x)^2 +$
$+ (uS - x)^2) \cdot (y - pz - 1 - E) \cdot ((pzT + T - l + y)^2 - E - 1) \cdot ((ER - u)^2 + ((EU - y)^2 - V - 1)^2 + (ES - v)^2))^2] \cdot$
$\cdot [mni(m - n) \cdot (\mathbf{m}K + \mathbf{n}\phi + r + 1 - \mathbf{m}m) \cdot (\mathbf{m}K + \mathbf{n}\chi + r + 1 - \mathbf{m}m) \cdot (j + r - i) \cdot (M + r + 1 - j) \cdot ((f + e^2 + r - bi)^2 +$
$+ (bis + s - a + f + e^2)^2 + (k + h^2 + t - bj)^2 + (bjW + W - a + k + h^2)^2 + (k + X + 1 - h)^2 + (fmZ + Z - e + F)^2 +$
$+ (F + F^2G^2 + g - dm)^2 + (F + Y - fm)^2 + (dmB + B - c + F + F^2G^2)^2 + (kOR + R - h + G)^2 + (G + E - kO)^2 +$
$+ (S + 1 - OIPQ(e - f))^2 + (mO + V - \mathbf{m}K - \mathbf{n}\chi)^2 + (J + \alpha - fn)^2 + (fn\beta + \beta - e + J)^2 +$
$+ (dn\delta + \delta - c + J + J^2L^2)^2 + (J + J^2L^2 + \gamma - dn)^2 + (dI\theta + \theta - c + P + P^2Q^2)^2 + (P + P^2Q^2 + \eta - dI)^2 +$
$+ ((L - G)^2 - \zeta - 1)^2 + (\mathbf{m}I + \kappa - \mathbf{m}K - \mathbf{n}\phi)^2 + (P\lambda - F)^2 + (P\mu - J)^2 + (P - F\nu + J\xi)^2 + (Q\pi - G)^2 +$
$(Q\rho - L)^2 + (Q - G\tau + L\sigma)^2)] = 0.$

Theorem 6. There exists a real number w such that:

1. if $\frac{\mathbf{n}}{\mathbf{m}} \leq w$ then $I\Delta_0 + \exp$ proves $A(\mathbf{m}, \mathbf{n})$;

2. if $\frac{\mathbf{n}}{\mathbf{m}} > w$ then ATR_0 does not prove $A(\mathbf{m}, \mathbf{n})$.

The number w is the real number introduced by Andreas Weiermann in [21] and is defined as follows: $w = \frac{1}{\log \alpha}$, where α is the Otter's tree constant (the inverse of the radius of convergence of the generating series for unordered trees), $w \approx 0.63957768999472013311\ldots$. The number w is of course primitive recursively computable.

This theorem is a new type of result within Andreas Weiermann's programme of phase transitions between provability and unprovability. More on Weiermann's programme can be found in [22].

The Graph Minor Theorem

The Graph Minor Theorem is a famous example of a former long-standing mathematical conjecture, resolved in the 1980s, that turned out to possess a lot of metamathematical strength (i.e., is unprovable in a rather strong theory). See [8] for the discussion and the proofs.

Now, consider the following polynomial expression with a quantifier-prefix:

$\forall K \exists N \forall ab \exists ij\ def\ ghm\ no\ xy\ ABC\ u\ \forall klpz\ Y \exists cqrvw\ DEF\ GHI\ \forall s \exists t \forall L \exists J\ MX\ OPQ$
$\forall RWZ\ \zeta\theta \exists STUV\ \alpha\beta\gamma\delta\epsilon\eta\kappa\lambda\mu\nu\xi\pi\rho\sigma\tau\phi\chi\varphi\psi\omega\ \Gamma\Delta\Theta\Lambda\Phi\Psi\Xi\Omega\Upsilon\ x_1 x_2 x_3\ \forall y_3 y_4 \exists x_4 x_5 x_6 x_7 x_8 x_9\ y_1 y_2$

$[(i+u+1-N)^2 + ((k+k^2 l^2 + k^3 l^3 p^3 - bi - b - 1 - D) \cdot ((biY + bY + Y - a + k + k^2 l^2 + k^3 l^3 p^3)^2 - D - 1) \cdot$
$\cdot (K+i+D-p) \cdot ((D+2-v)^2 + (((w-c)^2 - E - 1)^2 + ((w-q)^2 - F - 1)^2 + ((w-r)^2 - G - 1)^2) \cdot$
$\cdot ((w+E+1-p)^2 + (s+F-lw-l)^2 + (lwG+lG+G-k+s)^2 + (vH-s)^2) + ((c-q)^2 - I - 1)^2 +$
$+ ((q-r)^2 - J - 1)^2 + ((r-c)^2 - M - 1)^2)^2] \cdot [(i+u+1-j)^2 + (d+d^2 e^2 + d^3 e^3 f^3 + \beta - bi - b)^2 +$
$+ (j+\beta+1-N)^2 + (bi\gamma + b\gamma + \gamma - a + d^2 e^2 + d^3 e^3 f^3)^2 + (g + g^2 h^2 + g^3 h^3 m^3 + \epsilon - bj - b)^2 +$
$+ (bj\delta + b\delta + \delta - a + g + g^2 h^2 + g^3 h^3 m^3)^2 + ((p-1) \cdot p \cdot ((z-k)^2 \cdot (z-l)^2 + ((f+\eta+1-z) \cdot (q-ez-e-1-\eta) \cdot$
$\cdot ((ez\zeta + e\zeta + \zeta - d + q)^2 - \eta - 1) \cdot ((\rho\theta - q)^2 - \eta - 1))^2) + ((\eta + 2 - r)^2 + (v + \kappa - yk - y)^2 + (yk\lambda + y\lambda + \lambda - x + w)^2 +$
$+ (w+\mu-yl-y)^2 + (yl\nu + y\nu + \nu - x + w)^2 + (((s-v)^2 - \pi - 1)^2 + ((s-w)^2 - \rho - 1)^2) \cdot ((s+\pi+1-f)^2 +$
$+ (r\rho - t)^2 + (t+\sigma - os - s)^2 + (os\tau + o\tau + \tau - n + t)^2))^2 + ((C+\xi-k) \cdot ((BL\phi + B\phi + \phi - A + J)^2 +$
$+ (J+\xi - BL - B)^2 + (((L-k)^2 - \varphi - 1) \cdot (D + D^2 E^2 + D^3 E^3 F^3 - J))^2 + (((L-k-1)^2 - \chi - 1) \cdot$
$\cdot (G + G^2 H^2 + G^3 H^3 I^3 - J))^2 + (((L-1)^2 - \psi - 1) \cdot (g + g^2 h^2 + g^3 h^3 m^3 - J))^2 + (((L-G)^2 - \omega - 1) \cdot$
$\cdot (n + n^2 o^2 + n^3 o^3 f^3 - J))^2 + (\Gamma + 1 - M)^2 + (P + \Delta - EM)^2 + (EM\Theta + \Theta - D + P)^2 +$
$+ (U+\Lambda - HR)^2 + (HR\Psi + \Psi - G + U)^2 + (S+\Phi - ER)^2 + (ER\Xi + \Xi - D + S)^2 + (T+\Omega - ER - E)^2 +$
$+ (ER\Upsilon + \Upsilon - D + T)^2 + (R \cdot (I + x_1 + 1 - R) \cdot ((x_1 + 2 - O)^2 + (M + x_2 + 1 - X)^2 + (X + x_3 - F)^2 +$
$+ (Q + x_4 - EX)^2 + (EXx_5 + x_5 - D + Q)^2 + (((I-F+1)^2 + (Ox_6 - P)^2 + (Ox_7 - Q)^2 + (((R-M)^2 +$
$+ (U - PQ)^2) \cdot ((X + x_8 - R)^2 + (U - T)^2) \cdot ((R + x_8 - X)^2 + (U - S)^2 + ((R - M)^2 - x_9 - 1)^2)) \cdot$
$\cdot ((I-F)^2 + (Ox_6 - P)^2 + (Ox_7 - Q)^2 + (O - Px_8 + Qx_9)^2) + ((R-M)^2 + (U-V)^2 + (OV-Q)^2) \cdot$
$\cdot ((R-X)^2 + (U-V)^2 + (OV-Q)^2) \cdot (((R-X)^2 - y_1 - 1)^2 + ((R-M)^2 - y_2 - 1)^2 + (U-S)^2))) \cdot$
$\cdot ((M + x_1 - F)^2 + (I - F + 1)^2 + (\alpha + x_2 - EW)^2 + (EWx_3 + x_3 - D + \alpha)^2 +$
$+ (((Zy_3 - P)^2 - x_4 - 1) \cdot ((Zy_4 - \alpha)^2 - x_4 - 1) \cdot (M - W) \cdot (I - x_4 + 1 - W) \cdot W \cdot (Z - 1))^2 +$
$+ ((R + x_5 + 1 - M)^2 + (U - S)^2) \cdot ((M + x_5 - R)^2 + (U - T)^2)) \cdot ((I - F)^2 + (U - S)^2)^2]^2 = 0.$

Theorem 7.
The statement above is equivalent to the finite Graph Minor Theorem, and, hence, is unprovable in at least Π_1^1-CA_0, and, actually, in a stronger theory.

This polynomial expression is of size **1067**.

Phase transition for the planar graph minor theorem

Very much in the spirit of the previous polynomial expression, we could also write a phase transition expression for the graph minor theorem restricted to planar graphs, based on the article [3]. There is no exact phase transition result for the full Graph Minor Theorem (the number of graphs grows faster than exponentiation) but there is a neat threshold result for the much smaller (exponential) class of planar graphs, with the constant $\frac{1}{\log_2 \gamma}$ separating provable and unprovable instances of the planar Graph Minor Theorem. Here, γ is a classical constant, "the planar graph constant" from the graph enumeration theory, $29.06 < \gamma < 32$. Omitting the minors $K_{3,3}$ and K_5 is easy to express, and the rest of the polynomial is similar to the Graph Minor polynomial expression above.

Other minor-omitting classes of graphs should also be metamathematically interesting.

4 Final remarks

Generation of the Atlas

Passing from first-order arithmetical formulas to pure polynomials adds concreteness and certain clarity to all our logicians' talk about 'arithmetical truth', etc. Since each size-level of the Atlas is finite, we can start implementing the Atlas by hands, perhaps distributing the first few thousand expressions among a team of calculators. Here are the first few members: $0 = 0$, $1 = 0$, $\forall x \, (x = 0)$, $\exists x \, (x = 0)$, $2 = 0$, $\forall x \, (x - 1 = 0)$, $\exists x \, (x - 1 = 0)$, $\forall x \, \exists y \, (x - y = 0)$, $\exists x \, \forall y \, (x - y = 0)$, ..., $\exists xyz \, (x^2 + y^2 - z^2 = 0)$, ..., etc. Clearly, the first few size-levels of the Atlas will only consist of EFA-provable and EFA-refutable expressions. An interesting question would be: at which level (of what size) will the first open problem occur?

Since there is no algorithm to check whether two members of the Atlas are EFA-equivalent, the structure of the Atlas cannot be automatically 'decided' by some algorithm. Instead, the equivalence classes of the Atlas have to be generated separately, starting from already known members and stored in a provisional database. As the generation of members continues, once a proof of equivalence of two expressions is stumbled upon, their provisional classes get permanently identified as one.

Choosing the right language

The language of polynomials may seem pure, primal and most concrete you can ever get, but it definitely isn't the language to store mathematical knowledge and not the language in which our reasoning and practical generation of equivalence classes of the Atlas should proceed.

Dealing with pure polynomials, we are forced to clutter our expressions with encodings and approximations, which takes resources and eventually clouds and encrypts the meanings of the mathematical phenomena we are trying to express. If we are to build a useful chunk of the Atlas practically, we should be able to talk about mathematical objects directly, without spending resources on encryption and losing meaningfullness on the way. Perhaps we should maximise our arithmetical language, by including more concrete functions, operations on finite sets, sequences, trees, relations, cardinalities, other explicit combinatorial notational symbols and even analytical tools.

Distances within a class and between classes

The Atlas is not a list of hard to understand cloud of amorphous and mostly useless "mathematical facts" – far from it. The whole point is that it goes so much beyond metamathematically-unaware projects of automatic theorem proving and proof-verification. What is important for us here is the metamathematical structure of the Atlas, especially the classes that contain unprovable statements (and hence consist of unprovable statements).

Notions of distance can be introduced on members of the same class. Some members of the same class are transformed into each other quickly (and, hence, are near each other), some others would necessarily require a long sequence of transformations to establish their equivalence. Such notions would be of course very sensitive to the choice of the language of the Atlas and to the allowed manipulations that transform one member into another. Short-cuts, notations and extra rules of automatic reasoning would influence such notions of distance within a class. However, unprovable things will stay unprovable, and the Atlas stays rigid in terms of mapping non-identifiable, genuinely different equivalence classes.

Notions of distance between classes should be developed too. Two classes are closer to each other than another two if a weaker extension of EFA can prove equivalence of their members. So, equivalence classes come in ever-larger clusters as we increase the deductive strength that may identify classes.

Feasibility and the need for a normal form

Another important issue is feasibility: we may not have enough space and computational resources to fully generate and store a considerable useful chunk of the Atlas, including the information about equivalence-class structure, distances between classes and distances within a class. We can't store all members and we can't store all proofs. This is a tricky practical task. Perhaps some 'normal form' of expressions can be developed in such a way that it is sufficient to only store the normal forms, while billions of neighbouring elements of that normal form can be reconstructed on demand when needed. (For a trivial example, when a universal formula is proved, there is no need to store any of its instances.)

It was popular some 30-40 years ago to routinely mention non-feasibility of projects like the Atlas. Also, not expecting anything unprovable other than consistency statements, it was common to claim that even single examples of explicitly written unprovable statements are beyond reach. In that era, even the most qualified specialists could fall into the trap of routinely claiming, by inertia, that something must be unfeasible. For example, in [12], the authors are predicting on their first page that our future expression Φ from the Abstract and from Theorem 1 would not be feasible to achieve. They start off with saying that Con_{PA} is very complicated and the Paris-Harrington Principle is perhaps a simpler example but "still very complicated if written arithmetically". Clearly, our current note refutes this claim.

The last 40 years of development of metamathematics, including reverse mathematics and the enormous amount of Harvey Friedman's discoveries, showed that the non-feasibility philosophy and the mindless lip-service to it that we hear from different corners, is, perhaps, out-of-date. On top of that, the computational and storage capacities of modern computers increased so much that practical implementation of a project like the Atlas seems well within reach.

Harvey Friedman's strong statements

In the big manuscript [4], we went further, and produced a crude polynomial expression with quantifiers that is equivalent to Proposition E of Harvey Friedman's Boolean Relation Theory [9]. The expression we got is of size **4620**. The point of this exercise was to show that any strength is within reach, once we have a combinatorial statement of that strength. Since 2009, Harvey Friedman produced much simpler unprovable statements, equivalent to consistency of ZFC + {"there is an n-SRP cardinal"}$_{n \in \omega}$, so the size **4620** can be considerably reduced, using these new statements, perhaps to below **300** symbols in the language of pure polynomials, and much shorter if extra symbols are allowed. Notice that the new statements are equivalent to Π_1^0 formulas, so their equivalence class (if the equivalence can be conducted in EFA which is not always the case for model-theoretic reasons that make equivalence indirect) will contain a statement of non-solvability of a certain Diophantine equation, which we may never see, since practically lowering quantifier-complexity often explodes the polynomial so much.

Other ways to find non-trivial classes

We are hunting for members of the Atlas that belong to non-trivial (unprovable) equivalence classes. Here is what else we can do.

1. Translate more unprovable statements, with the hope of getting elements far from the ones we already obtained: the statement of totality of the Ackermann function (unprovable in $I\Sigma_1$), totality of superexponentiation (unprovable in EFA), Finite Ramsey Theorem (unprovable in EFA), termination of Goodstein sequences, termination of hydra battles, miniaturisations of the Open Ramsey Theorem (unprovable in ATR_0) and many other old known unprovable statements.

   To learn unprovability proofs, and for extensive bibliography, see [2]. For a larger introduction containing an extensive catalogue of examples of increasing strength, we highly recommend the first chapter of Harvey Friedman's monumental monograph [9].

2. A new generation of unprovability results is coming from modern Ramsey theory, especially infinite-dimensional Ramsey theory in the spirit of [18] and [19]. This seems to be a source of many new ideas and many new unprovabilities.

3. The Jones polynomial expression is a famous example of universality. You can find it and the whole discussion of universality and an extensive bibliography of related results of the era in the original papers by James Jones [10], [11] and in, say, [12].

   The papers contain many tricks and methods which we did not study and did not use, but any future implementors of the Atlas should learn and be ready to use, as well as the new tricks that appeared since then. It would take long to list those of them which we know and indicate which ones are more useful for our purposes.

   Also, some knowledge of modern number theory would be necessary. We predict that some new ideas can be extracted from Craig Smoryński's monograph [17].

4. The article [14] by Boris Moroz and Merlin Carl should be of interest.

5. Some extremely general open problems in mathematics (for example on distributions of prime values of polynomials) already have pieces of unprovability

proofs built in their formulations, so are asking for an unprovability proof. One of the most glaring examples is Schinzel's Hypothesis H in number theory. (Its possible unprovability, along the lines of the usual indicator theory proofs, was first suggested by Alan Woods.)

Seed sizes

Whatever language for the Atlas is chosen (like pure polynomials in this note), every equivalence class will have a seed (the shortest member), or, possibly, several seeds of the same size. We can view seeds as the "most compressed" representatives of their equivalence classes, carrying and encoding the phenomenon the class represents in the most efficient way.

We know the two trivial seeds: $0 = 0$ ("the seed of truth"), the seed belonging to the equivalence class of all EFA-provable stuff and $1 = 0$ ("the seed of lies"), the seed of all EFA-refutable stuff. Finding other seeds, of non-trivial classes, would be a very interesting task. Our raw coarse polynomials in this note have the following sizes, giving rough upper bounds on the sizes of their seeds:

	statement equivalent to	unprovable in theory	current size
Theorem 1	PH^2 or 1-Con($I\Sigma_1$)	$I\Sigma_1$	**131**
Theorem 2	PH^3 or 1-Con($I\Sigma_2$)	$I\Sigma_2$	**175**
Theorem 3	PH or 1-Con(PA)	PA	**386**
Theorem 5	Kruskal Theorem	$> ATR_0$	**648**
Theorem 7	Graph Minor Theorem	$> \Pi_1^1\text{-}CA_0$	**1067**
Theorem $*$	1-Con(MAH)	MAH	**4620**

It seems, although we don't have enough evidence to be sure yet, that seeds of equivalence classes of true quantifier-complexities Π_2^0 and Π_3^0 will appear earlier than seeds of classes possessing a Π_1^0-member. The explanation could be that essentially raising quantifier-complexity gives us enormous coding power that allows us to reduce the size of the polynomial. The examples of members of various equivalence classes that we have at the moment were obtained from combinatorial statements. These are the polynomials burdened and cluttered with parts that are responsible for the Gödel coding, and other non-polynomial, in a sense non-arithmetical sections that do the encryption work. In this way we hit members of various equivalence classes that are not likely to be anywhere near the seeds of their classes (at least we don't know any reasons why they should be).

We shouldn't exclude the possibility that some open problems in mathematics may turn out to be equivalent to non-trivial seeds or be close to seeds.

Seed size as a measure of complexity different from strength

Notice that classes like that of Con(ZFC + {n-subtle cardinals}$_{n\in\omega}$) or 1-Con(ZFC + {n-Mahlo cardinals}$_{n\in\omega}$) may appear in the Atlas earlier than, say, Con(ZFC). In this sense, the size of the seed, although somewhat correlated with logical strength, really measures 'complexity of the simplest description' of the class, and the mathematical phenomenon it represents, not its logical strength. There may be weak theories representing phenomena of complex description and strong theories representing a phenomenon that can be easily "wrapped" in a short expression.

References

[1] Jeremy Avigad (2003). Number theory and elementary arithmetic. *Philosophia Mathematica*, (3), 11, pp. 257–284.

[2] Andrey Bovykin (2009). Brief introduction to unprovability. *Logic Colloquium 2006*. Lecture Notes in Logic, pp. 38–64.

[3] Andrey Bovykin (2010). Unprovability threshold for the planar graph minor theorem. *Annals of Pure and Applied Logic*, 162 (3), pp. 175–181.

[4] Andrey Bovykin and Michiel De Smet (2011). A study of the Atlas of all polynomial equations with quantifier-prefixes and the structure of the provable-equivalence classes. Manuscript. 81 pages. Available online.

[5] Andrey Bovykin and Davide Crippa (2017). Five Kinds of Impossibility. Draft book manuscript, 100 pages.

[6] Michiel De Smet (2011). Unprovability and phase transitions in Ramsey theory. PhD thesis. University of Gent, Belgium.

[7] Costas Dimitracopoulos and Haim Gaifman (1980). Fragments of Peano's arithmetic and the MRDP theorem. *Logic and algorithms*, Enseign. Math. 30, pp. 187–206.

[8] Harvey Friedman, Neil Robertson, Paul Seymour (1987). The metamathematics of the graph minor theorem. *Contemporary Mathematics Series of the AMS*, 65, pp. 229–261.

[9] Harvey Friedman (2010). Boolean Relation Theory. Association for Symbolic Logic Series. Cambridge University Press.

[10] James P. Jones (1978). Three universal representations of recursively enumerable sets. *Journal of Symbolic Logic*, 43 (2), pp. 335–351.

[11] James P. Jones (1982). Universal Diophantine equation. *Journal of Symbolic Logic*, 47 (3), pp. 549–571.

[12] James P. Jones, John C. Shepherdson, Verena H. Dyson (1982). Some Diophantine forms of Gödel's Theorem, *Archiv fur Math. Logik und Grundlagen.*, 22, pp. 51–60.

[13] Yuri Matiyasevich (1993). Hilbert's Tenth Problem. The MIT Press.

[14] Boris Moroz and Merlin Carl (2013). A polynomial encoding provability in pure mathematics (outline of an explicit construction). *Bulletin of the Belgian Mathematical So-*

ciety, 20 (1), pp. 181–187.

[15] Aran Nayebi (2013). Exponential prefixed polynomial equations. Manuscript. 19 pages. Abstract published in *Bulletin of Symbolic Logic* 20 (2), p. 252. Full text available online at http://arxiv.org/abs/1303.5777

[16] Bjorn Poonen (2008). Undecidability in number theory. *Notices of the American Mathematical Society*, 55 (3), pp. 344–350.

[17] Craig Smoryński (1991). Logical Number Theory. Springer-Verlag.

[18] Stevo Todorcevic (2005). High-dimensional Ramsey theory and Banach space geometry. In the book with Argyros "Ramsey Methods in Analysis". Birkhäuser.

[19] Stevo Todorcevic (2010). Introduction to Ramsey Spaces. Annals of Mathematical Studies 147. Princeton University Press.

[20] Alan Turing (1937). On computable numbers with an application to the entscheidungsproblem. *Proceedings of the London Mathematical Society* (2), 42, pp. 230–265. Also in: Turing's Collected Works, ed. Gandy, Yates. North-Holland, Elsevier, 2001.

[21] Andreas Weiermann (2003). An application of graphical enumeration to PA. *Journal of Symbolic Logic*, 68 (1), pp. 5–16.

[22] Andreas Weiermann (2005). Analytic combinatorics, proof-theoretic ordinals, and phase transitions for independence results. *Annals of Pure and Applied Logic*, 136, pp. 189–218.

On Families of Anticommuting Matrices

Pavel Hrubeš
Institute of Mathematics of AVCR, Prague
pahrubes@gmail.com

Abstract

Let $e_1, \dots, e_k$ be complex $n \times n$ matrices such that $e_i e_j = -e_j e_i$ whenever $i \neq j$. We conjecture that $\mathrm{rk}(e_1^2) + \mathrm{rk}(e_2^2) + \cdots + \mathrm{rk}(e_k^2) \leq O(n \log n)$. We show that:

(i). $\mathrm{rk}(e_1^n) + \mathrm{rk}(e_2^n) + \cdots + \mathrm{rk}(e_k^n) \leq O(n \log n)$,

(ii). if $e_1^2, \dots, e_k^2 \neq 0$ then $k \leq O(n)$,

(iii). if $e_1, \dots, e_k$ have full rank, or at least $n - O(n/\log n)$, then $k \leq O(\log n)$.

(i) implies that the conjecture holds if $e_1^2, \dots, e_k^2$ are diagonalizable (or if $e_1, \dots, e_k$ are). (ii) and (iii) show it holds when their rank is sufficiently large or sufficiently small.

1 Introduction

Consider a family $e_1, \dots, e_k$ of complex $n \times n$ matrices which pairwise anticommute; i.e., $e_i e_j = -e_j e_i$ whenever $i \neq j$. A standard example is a representation of a Clifford algebra, which gives an anticommuting family of $2\log_2 n + 1$ invertible matrices, if n is a power of two (see Example 1 in Section 3). This is known to be tight: if all the matrices $e_1, \dots, e_k$ are invertible then k is at most $2\log_2 n + 1$. (see [10] and Theorem 1 below). However, the situation is much less understood when the matrices are singular. As an example, consider the following problem:

Question 1. *Assume that every e_i has rank at least $2n/3$. Is k at most $O(\log n)$?*

The research leading to these results has received funding from the European Research Council under the European Union's Seventh Framework Program (FP7/2007-2013) / ERC grant agreement no. 339691.

This paper is republished, with permission, having first appeard as: Pavel Hrubeš, On families of anticommuting matrices, *Linear Algebra and its Applications*, Volume 493, 2016, Pages 494–507.

We expect the answer to be positive. However, we can solve such a problem only under some extra assumptions. In [6], it was shown that an anticommuting family of diagonalisable matrices can be "decomposed" into representations of Clifford algebras. This indeed affirmatively answers Question 1 if the e_i's are diagonalisable. In this paper, we formulate a conjecture which relates the size of an anticommuting family with the rank of matrices in the family. We prove some partial results in this direction. In Theorem 3, we show that the situation is clear when the matrices are diagonalisable, or their squares are diagonalisable, or even $\mathrm{rk}(e_i^2) = \mathrm{rk}(e_i^3)$. However, we can say very little about the case when the matrices are nilpotent. In Theorem 2, we show that, in Question 1, we have $k \leq O(n)$. Theorem 6 implies that $k \leq O(\log n)$ whenever the rank of every e_i is almost full.

One motivation for this study is to understand sum-of-squares composition formulas. A sum-of-squares formula is an identity

$$(x_1^2 + x_2^2 + \cdots + x_k^2) \cdot (y_1^2 + y_2^2 + \cdots + y_k^2) = f_1^2 + f_2^2 + \cdots + f_n^2, \tag{1}$$

where $f_1, \ldots, f_n$ are bilinear complex[1] polynomials. We want to know how large must n be in terms of k so that such an identity exists. This problem has a very interesting history, and we refer the reader to the the monograph [10] for details. A classical result of Hurwitz [3] states that $n = k$ can be achieved only for $k \in \{1, 2, 4, 8\}$. Hence, n is strictly larger than k for most values of k, but it is not known how much larger. In particular, we do not known whether[2] $n = \Omega(k^{1+\epsilon})$ for some $\epsilon > 0$. In [1], it was shown that such a lower bound would resolve an open problem in arithmetic complexity theory (while the authors obtained an $\Omega(n^{7/6})$ lower bound on *integer* composition formulas in [2]). We point out that our conjecture about anticommuting families implies $n = \Omega(k^2/\log k)$, which would be asymptotically tight. This connection is hardly surprising: already Hurwitz's theorem, as well as the more general Hurwitz-Radon theorem [4, 9], can be proved by reduction to an anticommuting system.

2 Preliminaries

A family $e_1, \ldots, e_k$ of $n \times n$ complex matrices will be called *anticommuting* if $e_i e_j = -e_j e_i$ holds for every *distinct* $i, j \in \{1, \ldots, k\}$. We conjecture that the following holds ($\mathrm{rk}(A)$ is the rank of the matrix A):

---

[1]The problem is often phrased over $\mathbb{R}$ when the bilinearity condition is automatic.
[2]Recall that $f(k) = \Omega(g(k))$ if there exists $c > 0$ such that $f(k) \geq cg(k)$ holds for every sufficiently large k.

Conjecture 1. *Let $e_1, \ldots, e_k$ be an anticommuting family of $n \times n$ matrices. Then*
$$\sum_{i=1}^{k} \mathrm{rk}(e_i^2) \leq O(n \log n).$$

The main motivation is the following theorem:

Theorem 1 ([10]). *Let $e_1, \ldots, e_k$ be an anticommuting family of $n \times n$ invertible matrices. Then $k \leq 2 \log_2 n + 1$. The bound is achieved if n is a power of two.*

Under the assumption that e_i^2 are scalar diagonal matrices, this appears in [7] (though it may have been known already to Hurwitz). As stated, it can be found in [10] (Proposition 1.11 and Exercise 12, Chapter 1). There, an exact bound is given (see also Proposition 9 below):

$$k \leq 2q + 1, \text{ if } n = m2^q \text{ with } m \text{ odd}. \tag{2}$$

Theorem 1 shows, first, that the Conjecture holds for invertible matrices and, second, that the purported upper bound cannot be improved: taking $2 \log_2 n + 1$ full rank matrices gives $\sum \mathrm{rk}(e_i^2) = (2 \log_2 n + 1)n$.

A key aspect of Conjecture 1 is that $\sum \mathrm{rk}(e_i^2)$ is bounded in terms of a function of n only. This would fail, had we counted $\sum \mathrm{rk}(e_i)$ instead. For consider 2×2 matrices

$$e_i = \begin{pmatrix} 0 & a_i \\ 0 & 0 \end{pmatrix}, \ a_i \neq 0. \tag{3}$$

They trivially anticommute (as $e_i e_j = e_j e_i = 0$), but $\sum_{i=1}^{k} \mathrm{rk}(e_i) = k$, which can be arbitrarily large. However, we also have $e_i^2 = 0$ and this example is vacuous when counting $\sum \mathrm{rk}(e_i^2)$. The minimum requirement of the Conjecture is that every anticommuting family with non-zero squares is finite. Indeed, we prove that this is the case:

Theorem 2. *Let $e_1, \ldots, e_k$ be an anticommuting family of $n \times n$ matrices with $e_1^2, \ldots, e_k^2 \neq 0$. Then $k \leq O(n)$.*

In Theorem 10, we will show that $k \leq 2n - 3$ if n is sufficiently large, which is tight. Theorem 2 implies

$$\sum_{i=1}^{k} \mathrm{rk}(e_i^2) \leq O(n^2).$$

We will also show:

Theorem 3. *Let $e_1, \ldots, e_k$ be an anticommuting family of $n \times n$ matrices. Then*

$$\sum_{i=1}^{k} \mathrm{rk}(e_i^n) \leq (2 \log_2 n + 1) n\,.$$

This implies that Conjecture 1 holds whenever $\mathrm{rk}(e_i^2) = \mathrm{rk}(e_i^3)$ for every e_i (this is guaranteed if e_i^2 is diagonalisable). Note that if already $e_1, \ldots, e_k$ are diagonalisable, we obtain $\sum_{i=1}^{k} \mathrm{rk}(e_i) \leq (2 \log_2 n + 1) n$.

We will also generalise Theorem 1. In Theorem 6, we show that the assumption that e_i have full rank can be replaced by the assumption that they have almost full rank. This, together with Theorem 2 shows that Conjecture 1 holds if the e_i^2 have either rank close to n or close to $\log n$. Moreover, note that the Conjecture implies positive answer to Question 1: if $\mathrm{rk}(e_i) \geq 2n/3$ then $\mathrm{rk}(e_i^2) \geq n/3$ and so we must have $k \leq O(\log n)$.

Irreducible families In order to prove Theorem 3, we will discuss general structure of anticommuting families. One way to obtain such a family is via a direct sum of simpler families, which we now describe. If $A_1 \in \mathbb{C}^{r_1 \times r_1}$ and $A_2 \in \mathbb{C}^{r_2 \times r_2}$, let $A_1 \oplus A_2$ be the $(r_1 + r_2) \times (r_1 + r_2)$ matrix

$$A_1 \oplus A_2 = \begin{pmatrix} A_1 & 0 \\ 0 & A_2 \end{pmatrix}.$$

A family $e_1, \ldots, e_k \in C^{n \times n}$ will be called *reducible*, if there exists an invertible V such that

$$V e_i V^{-1} = e_i(1) \oplus e_i(2)\,,\ i \in [k], \quad (4)$$

where $e_1(1), \ldots, e_k(1) \in \mathbb{C}^{r_1 \times r_1}$, $e_1(2), \ldots, e_k(2) \in \mathbb{C}^{r_2 \times r_2}$, with $0 < r_1 < n$ and $r_1 + r_2 = n$. If no such decomposition exists, the family will be called *irreducible*. (Note that the similarity transformation $V e_1 V^{-1}, \ldots, V e_k V^{-1}$ preserves anticommutativity and rank, and that $e_1, \ldots, e_k$ anticommutes iff both $e_1(1), \ldots, e_k(1)$ and $e_1(2), \ldots, e_k(2)$ do.)

Hence, irreducible families form basic building blocks of anticommuting families, and it is enough to understand the structure of the irreducibles. We obtain the following result:

Theorem 4. *Let $e_1, \ldots, e_k$ be an irreducible anticommuting family. Then every e_i is either invertible or nilpotent.*

In other words, an irreducible family consists of two parts: the invertible matrices and the nilpotent ones. As in Theorem 1, invertible matrices are quite well-understood. In contrast, we know almost nothing about nilpotent anticommuting matrices. One exception is a nice theorem of Jacobson [5], see also [8]: a family of anticommuting nilpotent matrices is simultaneously upper triangularisable. A more refined description of irreducible families will be give in Proposition 8.

Notation and organization $[k] := \{1, \ldots, k\}$. $\mathbb{C}^{n \times m}$ will denote the set of $n \times m$ complex matrices. For a matrix A, $\text{rk}(A)$ is its the rank. Spectrum of a square matrix A, $\Lambda(A)$, is the set of its eigenvalues. A is nilpotent if $A^r = 0$ for some r (or equivalently, $A^n = 0$, or $\Lambda(A) = \{0\}$).

In Section 3, we give examples of anticommuting families. In Section 4, we prove Theorem 2, and give a generalization of Theorem 1. In Section 5, we prove Theorems 3 and 4. The latter is a corollary of Proposition 8 which describes structure of irreducible families in a greater detail. In Section 6, we reprove (2) and determine the bound from Theorem 2 exactly. In Section 7, we outline the connection between Conjecture 1 and the sums-of-squares problem.

We note that our main results hold in any field of characteristic different from two (Theorem 4 also assumes the field to be algebraically closed).

3 Examples of anticommuting families

One example of an anticommuting family is the one given in (3), consisting of matrices that square to zero. We give two other examples of anticommuting families. Each achieves optimal parameters within its class. Example 1 gives the largest anticommuting family of invertible matrices (Theorem 1), Example 2 the largest family of anticommuting matrices with non-zero squares if $n > 4$ (Theorem 10). It can be checked that both examples are irreducible.

Example 1 – Invertible Matrices Suppose that $e_1, \ldots, e_k \in \mathbb{C}^{n \times n}$ are anticommuting matrices. Then the following is a family of $k+2$ anticommuting matrices of dimension $2n \times 2n$:

$$\begin{pmatrix} I_n & 0 \\ 0 & -I_n \end{pmatrix}, \begin{pmatrix} 0 & I_n \\ -I_n & 0 \end{pmatrix}, \begin{pmatrix} 0 & e_1 \\ e_1 & 0 \end{pmatrix}, \ldots, \begin{pmatrix} 0 & e_k \\ e_k & 0 \end{pmatrix}. \quad (5)$$

Starting with a single non-zero 1×1 matrix, this construction can be applied iteratively to construct a family of $2 \log_2 n + 1$ anticommuting invertible $n \times n$ matrices whenever n is a power of two. Moreover, each matrix is diagonalizable. If n is not a power of two but rather of the form $m2^q$ with m odd, we instead obtain $2q+1$ such matrices.

Example 2 – Nilpotent Matrices If $n \geq 2$, consider $n \times n$ matrices of the form

$$e_i = \begin{pmatrix} 0 & u_i & 0 \\ & & v_i^t \\ & & 0 \end{pmatrix},$$

where $u_i, v_i \in \mathbb{C}^{n-2}$ are row-vectors. (As usual, the unspecified entries are zero). Then

$$e_i e_j = \begin{pmatrix} 0 & 0 & u_i v_j^t \\ & & 0 \\ & & 0 \end{pmatrix},$$

and so $e_i e_j = -e_j e_i$ iff $u_i v_j^t = -u_j v_i^t$ and $e_i^2 \neq 0$ iff $u_i v_i^t \neq 0$. Setting $r := n-2$, it is easy to construct row vectors $u_1, \ldots, u_{2r}, v_1, \ldots, v_{2r} \in \mathbb{C}^r$ such that for every $i, j \in [2r]$

$$u_i v_i^t \neq 0, \ u_i v_j^t = -u_j v_i^t \text{ if } i \neq j.$$

This gives an anticommutung family

$$e_1, \ldots, e_{2n-4} \in \mathbb{C}^{n \times n},$$

where every e_i is nilpotent but satisfies $e_i^2 \neq 0$. One can add one more matrix to the family: the diagonal matrix

$$e_0 := \begin{pmatrix} -1 & & \\ & I_{n-2} & \\ & & -1 \end{pmatrix}.$$

This gives $2n-3$ anticommuting matrices with non-zero squares.

4 Arguments from linear independence

In this section, we prove Theorem 2 and Theorem 6, the latter being a generalization of Theorem 1. We first observe the following:

Remark. *If $e_1, \ldots, e_k$ anticommute and $e_1^2, \ldots, e_k^2 \neq 0$ then they are linearly independent.*

To see this, assume that $e_1 = \sum_{j>1}^{k} a_j e_j$. Since e_1 anticommutes with every e_j, $j > 1$, we have $e_1^2 = e_1(\sum a_j e_j) = -(\sum a_j e_j)e_1 = -e_1^2$ and hence $e_1^2 = 0$.

This means that $k \leq n^2$ if $e_1, \ldots, e_k \in \mathbb{C}^{n \times n}$. We now show that k is actually $O(n)$. In Theorem 10, we will see that the correct bound is $2n - 3$ if $n > 4$, which is tight.

Theorem 2 (restated). *Let $e_1, \ldots, e_k \in \mathbb{C}^{n \times n}$ be an anticommuting family with $e_1^2, \ldots, e_k^2 \neq 0$. Then $k \leq O(n)$.*

Proof. First, there exist row-vectors $u, v \in \mathbb{C}^n$ such that $u e_i^2 v^t \neq 0 \in \mathbb{C}$ for every $i \in [k]$. This is because we can view $u e_i^2 v^t$ as a polynomial in the $2n$-coordinates of u and v. If $e_i^2 \neq 0$, the polynomial is non-trivial, and so a generic u, v satisfies $u e_i^2 v^t \neq 0$ for every $i \in [k]$.

Let us define the $k \times k$ matrix M by

$$M_{ij} := \{u e_i e_j v^t\}_{i,j \in [k]}.$$

Then $\mathrm{rk}(M) \leq n$. This is because M can be factored as $M = L \cdot R$, where L is $k \times n$ matrix with i-th row equal to $u e_i$ and R is $n \times k$ with j-th column equal to $e_j v^t$. On the other hand, we have $\mathrm{rk}(M) \geq k/2$. This is because $M_{ii} \neq 0$ and, since $e_i e_j = -e_j e_i$, $M_{ij} = -M_{ji}$ whenever $j \neq i$. Hence $M + M^t$ is a diagonal matrix with non-zero entries on the diagonal, $\mathrm{rk}(M + M^t) = k$ and so $\mathrm{rk}(M) \geq k/2$. This gives $k/2 \leq \mathrm{rk}(M) \leq n$ and so $k \leq 2n$. $\square$

The Remark can be generalised. For $A = \{i_1, \ldots, i_r\} \subseteq [k]$ with $i_1 < \cdots < i_r$, let e_A be the matrix $e_{i_1} e_{i_2} \cdots e_{i_r}$. $|A|$ denotes the cardinality of A.

Lemma 5. *Let $e_1, \ldots, e_k$ be anticommuting matrices. For $p \leq k$, assume that for every $A \subseteq \{1, \ldots, k\}$ with $|A| \leq p$ we have $\prod_{i \in A} e_i^2 \neq 0$. Then the matrices e_A, with $|A| \leq p$ and $|A|$ even, are linearly independent (similarly with odd $|A|$).*

Proof. Suppose that we have a non-trivial linear combination $\sum_{A \text{ even}} a_A e_A = 0$. Let A_0 be a largest A with $a_A \neq 0$. We will show that $\prod_{i \in A_0} e_i^2 = 0$ holds. This implies the statement of the lemma for even A's; the odd case is analogous. The proof is based on the following observations. First, e_i and e_j^2 always commute. Second, if $i \notin A$ then $e_i e_A = (-1)^{|A|} e_A e_i$, i.e., e_A and e_i commute or anticommute depending on the parity of $|A|$.

Without loss of generality, assume that $A_0 = \{1, \ldots, q\}$. For $r \leq q$ and $z \in \mathbb{N}$ let $S_r(z) := \{A \subseteq \{r+1, \ldots, k\} : |A| = z \bmod 2\}$. We will show that for every $0 \leq r \leq q$,

$$e_1^2 \cdots e_r^2 \left(\sum_{A \in S_r(r)} a_{[r] \cup A} e_A \right) = 0. \tag{6}$$

If $r = 0$, (6) is just the equality $\sum_{A \text{ even}} a_A e_A = 0$. Assume (6) holds for some $r < q$, and we want to show it holds for $r + 1$. Collecting terms that contain e_{r+1} and those that do not, (6) can be rewritten as

$$e_1^2 \cdots e_r^2 e_{r+1} \left(\sum_{A \in S_{r+1}(r+1)} a_{[r+1] \cup A} e_A \right) = -e_1^2 \cdots e_r^2 \left(\sum_{B \in S_{r+1}(r)} a_{[r] \cup B} e_B \right).$$

Let f and g be the left and right hand side of the last equality. Since A range over sets of parity $(r+1) \mod 2$ and B over sets with parity $r \mod 2$, we have $e_{r+1} f = (-1)^{r+1} f e_{r+1}$ and $e_{r+1} g = (-1)^r g e_{r+1}$. As $f = g$, this gives $e_{r+1} f = -f e_{r+1} = 0$ and so $e_{r+1} f = 0$. Hence,

$$e_1^2 \cdots e_r^2 e_{r+1}^2 \sum_{A \in S_{r+1}(r+1)} a_{[r+1] \cup A} e_A = 0,$$

as required in (6). Finally, if we set $r := q$ in (6), we obtain $e_1^2 \cdots e_q^2 \cdot a_{A_0} = 0$ (recall that A_0 is maximal) and so $e_1^2 \cdots e_q^2 = 0$, as required. □

Part (ii) of the following theorem is a generalisation of Theorem 1. Note that part (i) gives $k \leq O(\log n)$ whenever $r \geq n - O(n/\log n)$.

Theorem 6. *Let $e_1, \ldots, e_k$ be anticommuting matrices in $\mathbb{C}^{n \times n}$ and $r := \min_{i \in [k]} \operatorname{rk}(e_i)$.*

(i). *If $r > n(1 - 1/2c)$ with $c \in \mathbb{N}$ then $k \leq cn^{2/c}$.*

(ii). *If $r > n \left(1 - \frac{1}{4(\log_2 n + 1)}\right)$ then $k \leq 2\log_2 n + 1$.*

Proof. (i). By Sylvester's inequality, we have $\operatorname{rk}(\prod_{i \in A} e_i^2) > n - |A|n/c$ for every $A \subseteq [k]$. Hence, $\prod_{i \in A} e_i^2 \neq 0$ whenever $|A| \leq c$. By Lemma 5, the matrices e_A, $A \subseteq [k], |A| = c$, are linearly independent. This implies $\binom{k}{c} \leq n^2$ and the statement follows from the estimate $\binom{k}{c} \geq (k/c)^c$.

In (ii), assume that $k > 2\log_2 n + 1$ and, without loss of generality, $k \leq 2\log_2 n + 2$. As in (i), we conclude $e_1^2 \cdots e_k^2 \neq 0$. The lemma shows that the products e_A, with $|A|$ even, are linearly independent. This gives $2^{k-1} \leq n^2$ and so $k \leq 2\log_2 n + 1$, a contradiction. □

5 Structure of irreducible families

We will now prove Theorems 3 and 4. We start with the following lemma.

Lemma 7. *Let A and B be square matrices of the form*

$$A = \begin{pmatrix} A_1 & 0 \\ 0 & A_2 \end{pmatrix}, \quad B = \begin{pmatrix} B_1 & B_3 \\ B_4 & B_2 \end{pmatrix},$$

where $A_1, B_1 \in \mathbb{C}^{n \times n}$, $A_2, B_2 \in \mathbb{C}^{m \times m}$. If $AB = -BA$, the following hold:

(i). *if there is no λ such that $\lambda \in \Lambda(A_1)$ and $-\lambda \in \Lambda(A_2)$ then $B_3 = 0$ and $B_4 = 0$,*

(ii). *if $\Lambda(A_1) = \{\lambda_1\}$ and $\Lambda(A_2) = \{\lambda_2\}$ for some $\lambda_1, \lambda_2 \neq 0$ then $B_1, B_2 = 0$.*

Proof. We first note the following:

Claim. *Let $X \in \mathbb{C}^{p \times p}$, $Y \in \mathbb{C}^{q \times q}$ and $Z \in \mathbb{C}^{p \times q}$ be such that $XZ = ZY$. If $\Lambda(X) \cap \Lambda(Y) = \emptyset$ then $Z = 0$.*

Proof of the Claim. Without loss of generality, we can assume that Y is upper triangular with its eigenvalues $\lambda_1, \ldots, \lambda_q$ on the diagonal. Let $v_1, \ldots, v_q$ be the columns of Z, and assume that some v_i is non-zero. Taking the first such v_i gives $Xv_i = \lambda_i v_i$ – contradiction with $\lambda_i \notin \Lambda(X)$. □

Anticommutativity of A and B gives $A_1 B_3 = -B_3 A_2$ and $A_2 B_4 = -B_4 A_1$. If A_1, A_2 satisfy the assumption of (i), we have $\Lambda(A_1) \cap \Lambda(-A_2) = \emptyset$ and so $B_3, B_4 = 0$ by the Claim. We also have $A_1 B_1 = -A_1 B_1$. If A_1 is as in (ii), we have $\Lambda(A_1) \cap \Lambda(-A_1) = \emptyset$ and so $B_1 = 0$; similarly for B_2. □

Assume that A is a block-diagonal matrix such that each block has exactly one eigenvalue and the eigenvalues are distinct for different blocks. This can be achieved by converting A to Jordan normal form and regrouping the Jordan blocks according to their eigenvalues. Then Lemma 5 determines block-structure of B, if A and B anticommute. For example, assume that A is block-diagonal

$$A = \begin{pmatrix} A_1 & & & \\ & A_{-1} & & \\ & & A_2 & \\ & & & A_0 \end{pmatrix},$$

where $\Lambda(A_z) = \{z\}$ for $z \in \{1, -1, 2, 0\}$. If A and B anticommute, B must have block-structure

$$B = \begin{pmatrix} 0 & B_1 & & \\ B_2 & 0 & & \\ & & 0 & \\ & & & B_3 \end{pmatrix}.$$

Similar considerations give:

Proposition 8. Let $e_1, \dots, e_k \in \mathbb{C}^{n \times n}$ be an irreducible anticommuting family. Then:

(i). For every e_i, $\Lambda(e_i) \subseteq \{\lambda_i, -\lambda_i\}$ for some $\lambda_i \in \mathbb{C}$. Equality holds whenever at least two matrices in the family are non-zero.

(ii). If at least two of the matrices are invertible then n is even and the multiplicity of λ_i is exactly $n/2$ in an invertible e_i. Moreover, e_i is similar to $\begin{pmatrix} e'_i & 0 \\ 0 & -e'_i \end{pmatrix}$ for some $e'_i \in \mathbb{C}^{n/2 \times n/2}$ with $\Lambda(e'_i) = \{\lambda_i\}$.

(iii). Assume that $e_1 = \begin{pmatrix} e'_1 & 0 \\ 0 & e''_1 \end{pmatrix}$ is invertible with $\Lambda(e'_1) = \{\lambda_1\}$, $\Lambda(e''_1) = \{-\lambda_1\}$. Then the family is of the form

$$e_1 = \begin{pmatrix} e'_1 & 0 \\ 0 & e''_1 \end{pmatrix}, \quad e_j = \begin{pmatrix} 0 & e'_j \\ e''_j & 0 \end{pmatrix}, \, j \in \{2, \dots, k\}, \tag{7}$$

where e'_1, e''_1, e'_j, e''_j are matrices in $\mathbb{C}^{r \times r}$, $\mathbb{C}^{(n-r) \times (n-r)}$, $\mathbb{C}^{r \times (n-r)}$ and $\mathbb{C}^{(n-r) \times r}$, respectively, and r is the multiplicity of λ_1 in e_1. Moreover, the family $e_2 e_3, e_2 e_4, \dots, e_2 e_k$ is anticommuting and reducible if $0 < r < n$.

Proof. (i). Assume that there is some e_i with eigenvalues λ, λ' such that $\lambda \neq -\lambda'$. After a suitable similarity transformation, we can assume that

$$e_i = \begin{pmatrix} e'_i & 0 \\ 0 & e''_i \end{pmatrix},$$

where $e'_i \in \mathbb{C}^{r \times r}$ $e''_i \in \mathbb{C}^{(n-r) \times (n-r)}$ are such that $\Lambda(e'_i) \subseteq \{\lambda, -\lambda'\}$ and $\Lambda(e''_i) \cap \{\lambda, -\lambda'\} = \emptyset$, for some $0 < r < n$. Lemma 7 part (i) gives that every e_j is of the form

$$e_j = \begin{pmatrix} e'_j & 0 \\ 0 & e''_j \end{pmatrix}$$

and hence the family is reducible. If $\Lambda(e_i) = \{\lambda\}$ with $\lambda \neq 0$, the lemma gives $e_j = 0$ for every $j \neq i$.

(7) in (iii) follows from from Lemma 7 part (ii). Moreover, since $e_2, \dots, e_k$ anticommute then so do $e_2 e_3, \dots, e_2 e_k$. The latter family is of the form

$$e_2 e_j = \begin{pmatrix} e'_2 e''_j & 0 \\ 0 & e''_2 e'_j \end{pmatrix}, \, j \in \{3, \dots, k\}, \tag{8}$$

and hence it is reducible.

For (ii), assume that e_1 and e_2 are non-singular. By (i), we have $\Lambda(e_1) \subseteq \{\lambda_1, -\lambda_1\}$. We can assume that e_1 is as in (iii) with r equal to the multiplicity of λ_1, and hence e_2 is as in (7). Since e'_2, e''_2 are of dimension $r \times (n-r)$ resp. $(n-r) \times r$, the rank of e_2 is at most the minimum of $2r$ and $2(n-r)$. We must therefore have $r = n/2$. Moreover, anticommutativity of e_1 and e_2 gives $e'_1 e'_2 = -e'_2 e''_1$. Since e'_2 must be invertible, we have $e''_1 = -e'^{-1}_2 e'_1 e'_2$. Hence, e'_1 and $-e''_1$ are similar and e_1 is similar to $\begin{pmatrix} e'_1 & 0 \\ 0 & -e'_1 \end{pmatrix}$. □

Part (i) of the proposition immediately implies Theorem 4. This in turn gives:

Theorem 3 (restated). *Let $e_1, \ldots, e_k \in \mathbb{C}^{n \times n}$ be an anticommutative family. Then $\sum_{i=1}^{k} \mathrm{rk}(e_i^n) \leq (2 \log_2 n + 1) n$.*

Proof. Argue by induction on n. If $n = 1$, the statement is clear. If $n > 1$, assume first that the family is irreducible. By Theorem 4, every e_i is either invertible or nilpotent. If e_i is nilpotent then $e_i^n = 0$ and it contributes nothing to the rank. On the other hand, Theorem 1 asserts that there can be at most $2 \log_2 n + 1$ anticommuting invertible matrices and so indeed $\sum_{i=1}^{k} \mathrm{rk}(e_i^n) \leq (2 \log_2 n + 1) n$.

If the family is reducible, consider the decomposition in (4). By the inductive assumption, $\sum \mathrm{rk}(e_i(z)^n) \leq \sum \mathrm{rk}(e_i(z)^{r_z}) \leq (2 \log_2 r_z + 1) r_z$ for both $z \in \{1, 2\}$. Since $\mathrm{rk}(e_i^n) = \mathrm{rk}(e_i(1)^n) + \mathrm{rk}(e_i(2)^n)$, we obtain

$$\sum_{i=1}^{k} \mathrm{rk}(e_i^n) \leq \sum_{i=1}^{k} \mathrm{rk}(e_i(1)^{r_1}) + \sum_{i=1}^{k} \mathrm{rk}(e_i(2)^{r_2}) \leq$$
$$\leq (2 \log_2 r_1 + 1) r_1 + (2 \log_2 r_2 + 1) r_2 \leq (2 \log_2 n + 1)(r_1 + r_2) =$$
$$= (2 \log_2 n + 1) n.$$

□

6 Some exact bounds

For completeness, we now sketch a proof of (2) from Section 2. We then prove the exact bound in Theorem 2.

Proposition 9. *Let $e_1, \ldots, e_k$ be an anticommutative family of invertible $n \times n$ matrices, where $n = m 2^q$ and m is odd. Then $k \leq 2q + 1$.*

The bound is achieved by Example 1.

Proof of Proposition 9. Argue by induction on n. If $n > 1$, it is enough to consider the case when the family is irreducible. If $k > 1$, Proposition 8 part (ii) shows that λ_1 has multiplicity $n/2$ in e_1. By part (iii), we can assume that the family is of the form (7), where every $e_i', e_i'' \in \mathbb{C}^{n/2 \times n/2}$, $i \in [k]$, is invertible. If $e_2, \ldots, e_k$ anticommute then so do the $k-2$ matrices $e_2 e_3, e_2 e_4, \ldots, e_2 e_k$. Writing $e_2 e_j$ as in (8), we obtain that $e_2' e_3'', \ldots, e_2' e_k''$ is a family of $k-2$ invertible anticommuting matrices in $\mathbb{C}^{n/2 \times n/2}$. The inductive assumption gives $k - 2 \leq 2(q-1) + 1$ and so $k \leq 2q + 1$ as required. $\square$

For a natural number n, let $\alpha(n)$ denote the largest k so that there exists an anticommuting family $e_1, \ldots, e_k \in \mathbb{C}^{n \times n}$ with $e_1^2, \ldots, e_k^2 \neq 0$.

Theorem 10.
$$\alpha(n) = \begin{cases} 2n - 1, & \text{if } n \in \{1, 2\} \\ 2n - 2, & \text{if } n \in \{3, 4\} \\ 2n - 3, & \text{if } n > 4 \end{cases}$$

The rest of this section is devoted to proving the theorem.

Lemma 11. *If $n > 1$, $\alpha(n)$ equals the maximum of the following quantities: a) $2n - 3$, b) $\max_{0 < r < n}(\alpha(r) + \alpha(n-r))$, c) $2 + \alpha(n/2)$ (where we set $\alpha(n/2) := -1$ if n is odd).*

Proof. That $\alpha(n)$ is at least the maximum is seen as follows. $\alpha(n) \geq$ a) is Example 2. $\alpha(n) \geq 2 + \alpha(n/2)$ is seen from (5) in Example 1. For b), suppose we have two anticommuting families $e_1(z), \ldots, e_{k_z}(z) \in \mathbb{C}^{r_z \times r_z}$, $z \in \{1, 2\}$. Then the following is an anticommuting family of $(r_1 + r_2) \times (r_1 + r_2)$ matrices: $e_1(1) \oplus 0, \ldots, e_{k_1} \oplus 0, 0 \oplus e_1(2), \ldots, 0 \oplus e_{k_2}(2)$ (with $0 \in \mathbb{C}^{r_1 \times r_1}, \mathbb{C}^{r_2 \times r_2}$ respectively).

We now prove the opposite inequality. Let $e_1, \ldots, e_k \in \mathbb{C}^{n \times n}$ be an anticommuting family with $e_1^2, \ldots, e_k^2 \neq 0$. We first give two claims.

Claim 1. *If all the e_i's are nilpotent then $k \leq 2(n-2)$.*

Proof of Claim 1. By the theorem of Jacobson [5], a family of anticommuting nilpotent matrices is simultaneously upper triangularisable. So let us assume that $e_1, \ldots, e_k$ are upper-triangular with zero diagonal, and proceed as in the proof of Theorem 2. For M as defined in the proof, it is enough to show that $\mathrm{rk}(M) \leq n - 2$, which gives $k \leq 2(n-2)$. If the e_i's are upper triangular with zero diagonal, we can see that the first column of L and the last row of R are zero. This shows that $\mathrm{rk}(M) = \mathrm{rk}(LR) \leq n - 2$. $\square$

Claim 2. *If e_1, e_2 are invertible then $k \leq 2 + \alpha(n/2)$.*

Proof of Claim 2. As in the proof of Theorem 9, we can assume that the family is of the form (7) where e_2', e_2'' are invertible and $r = n/2$. Considering the family $e_2 e_3, \ldots, e_2 e_k$, written as in (8), we obtain that $e_2' e_3'', \ldots, e_2' e_k''$ is an anticommuting family of $k - 2$ matrices in $\mathbb{C}^{n/2 \times n/2}$. If we show that $(e_2' e_j'')^2 \neq 0$ for every $j \in \{3, \ldots, k\}$, we obtain $k - 2 \leq \alpha(n/2)$ as required.

Let $j \in \{3, \ldots, k\}$. Anticommutativity of e_2 and e_j gives $e_2' e_j'' = -e_j' e_2''$ and $e_2'' e_j' = -e_j'' e_2'$. Hence

$$(e_2' e_j'')^2 = e_2' e_j'' e_2' e_j'' = e_2'(e_j'' e_2') e_j'' = -e_2' e_2'' e_j' e_j'',$$
$$= e_2' e_j''(e_2' e_j'') = -e_2' e_j'' e_j' e_2''.$$

If $(e_2' e_j'')^2 = 0$, the first equality gives $e_j' e_j'' = 0$ and the second $e_j'' e_j' = 0$ (recall that e_2', e_2'' are invertible). But since $e_j^2 = e_j' e_j'' \oplus e_j'' e_j'$, this gives $e_j^2 = 0$ – contrary to the assumption $e_j^2 \neq 0$. □

To prove the Lemma, assume first that $e_1, \ldots, e_k$ is irreducible. Then the e_i's are either invertible or nilpotent. If there is at most one invertible e_i, Claim 1 gives $k - 1 \leq 2(n - 2)$, as in a). If at least two e_i's are invertible, Claim 2 gives $k \leq 2 + \alpha(n/2)$, as in b). If the family is reducible, write it as in (4). For $z \in \{1, 2\}$, let $A_z := \{i \in [k] : e_i(z)^2 \neq 0\}$. Then $A_1 \cup A_2 = [k]$ and so $k \leq \alpha(r_1) + \alpha(r_2)$, as in c). □

Proof of Theorem 10. Using the Lemma, it is easy to verify that the theorem holds for $n \leq 4$. If $n > 4$, the lemma gives $\alpha(n) \geq 2n - 3$ and it suffices to prove the opposite inequality. Assume that n is the smallest $n > 4$ such that $\alpha(n) > 2n - 3$. This means that for every $n' < n$, $\alpha(n') = 2n' - \epsilon(n')$ where $\epsilon(n') = 1$ if $n' \in \{1, 2\}$ and $\epsilon(n') > 1$ otherwise. Then either $\alpha(r) + \alpha(n - r) > 2n - 3$ for some $0 < r < n$, or $2 + \alpha(n/2) > 2n - 3$. The first case is impossible: we have $\alpha(r) + \alpha(n - r) = 2n - \epsilon(r) - \epsilon(n - r)$. But $\epsilon(r) + \epsilon(n - r) < 3$ implies $r, (n - r) \in \{1, 2\}$ and so $n \leq 4$. If $2 + \alpha(n/2) > 2n - 3$, we have $2 + 2(n/2) - 2\epsilon(n/2) > 2n - 3$ and so $n < 5 - 2\epsilon(n/2) \leq 3$. □

7 Sum-of-squares formulas

We now briefly discuss the sum-of-squares problem. Let $\sigma(k)$ be the smallest n so that there exists a sum-of-squares formula as in (1) from the Introduction. The following can be found in Chapter 0 of [10]:

Lemma 12. *$\sigma(k)$ is the smallest n such that there exists $k \times n$ matrices $A_1, \ldots A_k$ which satisfy*

$$A_i A_i^t = I_k, \quad A_i A_j^t = -A_j A_i^t, \quad \text{if } i \neq j,$$

for every $i,j \in [k]$.

The matrices from the lemma can be converted to anticommuting matrices, which provides a connection between the sum-of-squares problem and Conjecture 1, as follows.

Proposition 13. *If $\sigma(k) = n$, there exists an anticommuting family $e_1, \ldots, e_k \in \mathbb{C}^{(n+2k) \times (n+2k)}$ such that $\mathrm{rk}(e_1^2) = \cdots = \mathrm{rk}(e_k^2) = k$.*

Proof. Take the $(2k+n) \times (2k+n)$ matrices (with $0 \in \mathbb{C}^{k \times k}$)

$$e_i := \begin{pmatrix} 0 & A_i & 0 \\ & & A_i^t \\ & & 0 \end{pmatrix}, \, i \in [k].$$

The matrices have the required properties as seen from

$$e_i e_j = \begin{pmatrix} 0 & 0 & A_i A_j^t \\ & & 0 \\ & & 0 \end{pmatrix}.$$

□

Proposition 13 shows that Conjecture 1 implies

$$\sigma(k) = \Omega(k^2 / \log k),$$

which would be tight. We can see that the constructed matrices are nilpotent which is exactly the case of Conjecture 1 we do not know how to handle. Furthermore, they satisfy some additional properties: $e_1^2 = e_2^2 = \cdots = e_k^2$ and $e_1^3, \ldots, e_k^3 = 0$. Hence, in order to understand the sum-of-squares composition formulas, we may as well focus on Conjecture 1 with these additional assumptions. Finally, let us note that Proposition 13 is too generous if $\sigma(k) = k$. In this case, we can actually obtain $k-1$ invertible anticommuting matrices in $\mathbb{C}^{k \times k}$. Again following [10], let

$$e_1 := A_1 A_k^t, \; e_2 := A_2 A_k^t, \ldots, \; e_{k-1} := A_{k-1} A_k^t.$$

The matrices anticommute, as seen from $A_i A_k^t A_j A_k^t = -A_i A_k^t A_k A_j^t = -A_i A_j^t$ (note that $A_k A_k^t = I$ implies $A_k^t A_k = I$ for square matrices). This is one way how to obtain Hurwitz's $\{1,2,4,8\}$-theorem: if $\sigma(k) = k$, we have $k-1$ invertible anticommuting matrices in $\mathbb{C}^{k \times k}$. By Theorem 1, this gives $k - 1 \leq 2 \log_2 k + 1$ and hence $k \leq 8$. Furthermore, the precise bound in (2) rules out the k's which are not a power of two.

References

[1] P. Hrubeš, A. Wigderson, and A. Yehudayoff. Non-commutative circuits and the sum of squares problem. *J. Amer. Math. Soc.*, 24:871–898, 2011.

[2] P. Hrubeš, A. Wigderson, and A. Yehudayoff. An asymptotic bound on the composition number of integer sums of squares formulas. *Canadian Mathematical Bulletin*, 56:70–79, 2013.

[3] A. Hurwitz. Über die Komposition der quadratischen Formen von beliebigvielen Variabeln. *Nach. Ges. der Wiss. Göttingen*, pages 309–316, 1898.

[4] A. Hurwitz. Über die Komposition der quadratischen Formen. *Math. Ann.*, 88:1–25, 1923.

[5] N. Jacobson. *Lie Algebras*. Interscience, New York, 1962.

[6] Y. Kumbasar and A. H. Bilge. Canonical forms for families of anti-commuting diagonalizable operators. *ArXiv*, 2011.

[7] M. H. A. Newman. Note on an algebraic theorem of Eddington. *J. London Math. Soc*, 7:93–99, 1932.

[8] H. Radjavi. The Engel-Jacobson theorem revisited. *J.Algebra*, 111:427–430, 1987.

[9] J. Radon. Lineare scharen orthogonalen Matrizen. *Abh. Math. Sem. Univ. Hamburg*, 1(2-14), 1922.

[10] D. B. Shapiro. *Compositions of quadratic forms*. De Gruyter expositions in mathematics 33, 2000.

Worms and Spiders:
Reflection Calculi and Ordinal Notation Systems

David Fernández-Duque
Institute de Recherche en Informatique de Toulouse, Toulouse University, France.
Department of Mathematics, Ghent University, Belgium.
`david.fernandez@irit.fr`

To the memory of Professor Grigori Mints.

Abstract

We give a general overview of ordinal notation systems arising from reflection calculi, and extend them to represent impredicative ordinals defined using Buchholz-style collapsing functions.

1 Introduction

I had the honor of receiving the Gödel Centenary Research Prize in 2008 based on work directed by my doctoral advisor, Grigori 'Grisha' Mints. The topic of my dissertation was *dynamic topological logic,* and while this remains a research interest of mine, in recent years I have focused on studying polymodal provability logics. These logics have proof-theoretic applications and give rise to ordinal notation systems, although previously only for ordinals below the Feferman-Shütte ordinal, Γ_0. I last saw Professor Mints in the *First International Wormshop* in 2012, where he asked if we could represent the Bachmann-Howard ordinal, $\psi(\varepsilon_{\Omega+1})$, using provability logics. It seems fitting for this volume to once again write about a problem posed to me by Professor Mints.

Notation systems for $\psi(\varepsilon_{\Omega+1})$ and other 'impredicative' ordinals are a natural step in advancing Beklemishev's Π_1^0 ordinal analysis[1] to relatively strong theories

---

This work was partially funded by ANR-11-LABX-0040-CIMI within the program ANR-11-IDEX-0002-02.

[1]The Π_1^0 ordinal of a theory T is a way to measure its 'consistency strength'. A different measure, more widely studied, is its Π_1^1 ordinal; we will not define either in this work, but the interested reader may find details in [4] and [33], respectively.

of second-order arithmetic, as well as systems based on Kripke-Platek set theory. Indeed, Professor Mints was not the only participant of the Wormshop interested in representing impredicative ordinals within provability algebras. Fedor Pakhomov brought up the same question, and we had many discussions on the topic. At the time, we each came up with a different strategy for addressing it. These discussions inspired me to continue reflecting about the problem the next couple of years, eventually leading to the ideas presented in the latter part of this manuscript.

1.1 Background

The Gödel-Löb logic GL is a modal logic in which $\Box\varphi$ is interpreted as 'φ *is derivable in T*', where T is some fixed formal theory such as Peano arithmetic. This may be extended to a polymodal logic GLP_ω with one modality $[n]$ for each natural number n, as proposed by Japaridze [27]. The modalities $[n]$ may be given a natural proof-theoretic interpretation by extending T with new axioms or infinitary rules. However, GLP_ω is not an easy modal logic to work with, and to this end Dashkov [14] and Beklemishev [7, 6] have identified a particularly well-behaved fragment called the *reflection calculus* (RC), which contains the dual modalities $\langle n \rangle$, but does not allow one to define $[n]$.

Because of this, when working within RC, we may simply write n instead of $\langle n \rangle$. With this notational convention in mind, of particular interest are *worms*, which are expressions of the form
$$m_1 \ldots m_n \top,$$
which can be read as

> *It is m_1-consistent with T that it is m_2 consistent with T that ... that T is m_n-consistent.*

In [26], Ignatiev proved that the set of worms of GLP_ω is well-ordered by consistency strength and computed their order-type. Beklemishev has since shown that trasfinite induction along this well-order may be used to give an otherwise finitary proof of the consistency of Peano arithmetic [4].

Indeed, the order-type of the set of worms in RC_ω is ε_0, an ordinal which already appeared in Gentzen's earlier proof of the consistency of PA [21]. Moreover, as Beklemishev has observed [5], worms remain well-ordered if we instead work in RC_Λ (or GLP_Λ), where Λ is an arbitrary ordinal. The worms of RC_Λ give a notation system up to the Feferman-Schütte ordinal Γ_0, considered the upper bound of predicative mathematics.

This suggests that techniques based on reflection calculi may be used to give a proof-theoretic analysis of theories of strength Γ_0, the focus of an ongoing research project. However, if worms only provide notations for ordinals below Γ_0, then these techniques cannot be applied to 'impredicative' theories, such as Kripke-Platek set theory with infinity, whose proof-theoretic ordinal is much larger and is obtained by 'collapsing' an uncountable ordinal.

1.2 Goals of the article

The goal of this article is to give a step-by-step and mostly self-contained account of the ordinal notation systems that arise from reflection calculi. Sections 2-5 are devoted to giving an overview of known, 'predicative' notation systems, first for ε_0 and then for Γ_0. However, our presentation is quite a bit different from those available in the current literature. In particular, it is meant to be 'minimalist', in the sense that we only prove results that are central to our goal of comparing the reflection-based ordinal notations to standard proof-theoretic ordinals. Among other things, we sometimes do not show that the notation systems considered are computable.

The second half presents new material, providing impredicative notation systems based on provability logics. We first introduce *impredicative worms*, which give a representation system for $\psi(e^{\Omega+1}1)$, an ordinal a bit larger than the Bachmann-Howard ordinal. Then we introduce *spiders*, which are used to represent ordinals up to $\psi_0\Omega^\omega 1$ in Buchholz-style notation [11]. Here, $\Omega^\omega 1$ is the first fixed point of the aleph function; unlike the predicative systems discussed above, these notation systems also include notations for several uncountable ordinals. The latter are then 'collapsed' in order to represent countable ordinals much larger than Γ_0.

Although our focus is on notations arising from the reflection calculi and not on proof-theoretic interpretations of the provability operators, we precede each notation system with an informal discussion on such interpretations. These discussions are only given as motivation; further details may be found in the references provided. We also go into detail discussing the 'traditional' notation systems for each of the proof-theoretical ordinals involved before discussing the reflection-based version, and thus this text may also serve as an introduction of sorts to ordinal notation systems.

1.3 Layout of the article

§2: Review of the basic definitions and properties of the reflection calculus RC and the transfinite provability logic GLP.

§3: Introduction to *worms* and their order-theoretic properties.

§4: Computation of the order-type of worms with finite entries, and a brief overview of their interpretation in the language of Peano arithmetic.

§5: Computation of the order-type of worms with ordinal entries, and an overview of their interpretation in the language of second-order arithmetic.

§6: Introduction and analysis of impredicative worms, obtained by introducing an uncountable modality and its collapsing function.

§7: Introduction to *spiders,* variants of worms interpreted using the aleph function and its collapses.

§8: Concluding remarks.

2 The reflection calculus

Provability logics are modal logics for reasoning about Gödel's provability operator and its variants [10]. One uses $\Box \varphi$ to express 'φ *is provable in* T'; here, T may be Peano arithmetic, or more generally, any sound extension of elementary arithmetic (see Section 4.1 below). The dual of $\Box$ is $\Diamond = \neg \Box \neg$, and we may read $\Diamond \varphi$ as 'φ *is consistent with* T'. This unimodal logic is called *Gödel-Löb logic,* which Japaridze extended to a polymodal variant with one modality $[n]$ for each natural number in [28], further extended by Beklemishev to allow one modality for each ordinal in [5].

The resulting polymodal logics have some nice properties; for exmample, they are decidable, provided the modalities range over some computable linear order. However, there are also some technical difficulties when working with these logics; most notoriously, they are incomplete for their relational semantics, and their topological semantics are quite complex [9, 18, 15, 25].

Fortunately, Dashkov [14] and Beklemishev [6, 7] have shown that for proof-theoretic applications, it is sufficient to restrict to a more manageable fragment of Japaridze's logic called the *Reflection Calculus* (RC). Due to its simplicity relative to Japaridze's logic, we will perform all of our modal reasoning directly within RC.

2.1 Ordinal numbers and well-orders

(Ordinal) reflection calculi are polymodal systems whose modalities range over a set or class of ordinal numbers, which are canonical representatives of well-orders. Recall that if A is a set (or class), a *preorder* on A is a trasitive, reflexive relation $\preccurlyeq \, \subseteq A \times A$. The preorder $\preccurlyeq$ is *total* if, given $a, b \in A$, we always have that $a \preccurlyeq b$ or $b \preccurlyeq a$, and *antisymmetric* if whenever $a \preccurlyeq b$ and $b \preccurlyeq a$, it follows that $a = b$. A total, antisymmetric preorder is a *linear order*. We say that $\langle A, \preccurlyeq \rangle$ is a *pre-well-order* if $\preccurlyeq$ is a total preorder and every non-empty $B \subseteq A$ has a minimal element (i.e., there is $m \in B$ such that $m \preccurlyeq b$ for all $b \in B$). A *well-order* is a pre-well-order that is also linear. Note that pre-well-orders are not the same as well-quasiorders (the latter need not be total). Pre-well-orders will be convenient to us because, as we will see, worms are pre-well-ordered but not linearly ordered.

Define $a \prec b$ by $a \preccurlyeq b$ but $b \not\preccurlyeq a$, and $a \approx b$ by $a \preccurlyeq b$ and $b \preccurlyeq a$. The next proposition may readily be checked by the reader:

Proposition 2.1. *Let $\langle A, \preccurlyeq \rangle$ be a total preorder. Then, the following are equivalent:*

1. *$\preccurlyeq$ is a pre-well-order;*

2. *if $a_0, a_1, \ldots \subseteq A$ is any infinite sequence, then there are $i < j$ such that $a_i \preccurlyeq a_j$;*

3. *there is no infinite descending sequence*

$$a_0 \succ a_1 \succ a_2 \succ \ldots \subseteq A;$$

4. *if $B \subseteq A$ is such that for every $a \in A$,*

$$(\forall b \prec a \, (b \in B)) \to a \in B,$$

then $B = A$.

We use the standard interval notation for preorders: $(a, b) = \{x : a \prec x \prec b\}$, $(a, \infty) = \{x : a \prec x\}$, etc. With this, we are ready to introduce ordinal numbers as a special case of a well-ordered set. Their formal definition is as follows:

Definition 2.2. *Say that a set A is* transitive *if whenever $B \in A$, it follows that $B \subseteq A$. Then, a set ξ is an* ordinal *if ξ is transitive and $\langle \xi, \in \rangle$ is a strict well-order.*

When ξ, ζ are ordinals, we write $\xi < \zeta$ instead of $\xi \in \zeta$ and $\xi \leq \zeta$ if $\xi < \zeta$ or $\xi = \zeta$. The class of ordinal numbers will be denoted Ord. We will rarely appeal to Definition 2.2 directly; instead, we will use some basic structural properties of the class of ordinal numbers as a whole. First, observe that Ord is itself a (class-sized) well-order:

Lemma 2.3. *The class* Ord *is well-ordered by $\leq$, and if $\Theta \subseteq$ Ord is a set, then Θ is an ordinal if and only if Θ is transitive.*

Thus if ξ is any ordinal, then $\xi = \{\zeta \in \mathsf{Ord} : \zeta < \xi\}$, and $0 = \varnothing$ is the least ordinal. For $\xi \in \mathsf{Ord}$, define $\xi + 1 = \xi \cup \{\xi\}$; this is the least ordinal greater than ξ. It follows from these observations that any natural number is an ordinal, but there are infinite ordinals as well; the set of natural numbers is itself an ordinal and denoted ω. More generally, new ordinals can be formed by taking successors and unions:

Lemma 2.4.

1. *If ξ is any ordinal, then $\xi + 1$ is also an ordinal. Moreover, if $\zeta < \xi + 1$, it follows that $\zeta \leq \xi$.*

2. *If Θ is a set of ordinals, then $\lambda = \bigcup \Theta$ is an ordinal. Moreover, if $\xi < \lambda$, it follows that $\xi < \theta$ for some $\theta \in \Theta$.*

These basic properties will suffice to introduce the reflection calculus, but later in the text we will study ordinals in greater depth. A more detailed introduction to the ordinal numbers may be found in a text such as [29].

2.2 The reflection calculus

The modalities of reflection calculi are indexed by elements of some set of ordinals Λ. Alternately, one can take Λ to be the class of all ordinals, obtaining a class-sized logic. Formulas of RC_Λ are built from the grammar

$$\top \mid \phi \wedge \psi \mid \langle \lambda \rangle \phi,$$

where $\lambda < \Lambda$ and ϕ, ψ are formulas of RC_Λ; we may write $\lambda \phi$ instead of $\langle \lambda \rangle \phi$, particularly since RC_Λ does not contain expressions of the form $[\lambda]\phi$. The set of formulas of RC_Λ will be denoted $\mathcal{L}_\Lambda$, and we will simply write $\mathcal{L}_{\mathsf{RC}}$ and RC instead of $\mathcal{L}_{\mathsf{Ord}}$, $\mathsf{RC}_{\mathsf{Ord}}$. Propositional variables may also be included, but we will omit them since they are not needed for our purposes. Note that this strays from convention, since the variable-free fragment is typically denoted RC^0. Reflection calculi derive *sequents* of the form $\phi \Rightarrow \psi$, using the following rules and axioms:

$$\phi \Rightarrow \phi \qquad \phi \Rightarrow \top \qquad \frac{\phi \Rightarrow \psi \quad \psi \Rightarrow \theta}{\phi \Rightarrow \theta}$$

$$\phi \wedge \psi \Rightarrow \phi \qquad \phi \wedge \psi \Rightarrow \psi \qquad \frac{\phi \Rightarrow \psi \quad \phi \Rightarrow \theta}{\phi \Rightarrow \psi \wedge \theta}$$

$$\lambda\lambda\phi \Rightarrow \lambda\phi \qquad \frac{\phi \Rightarrow \psi}{\lambda\phi \Rightarrow \lambda\psi}$$

$$\lambda\phi \Rightarrow \mu\phi \qquad \text{for } \mu \leq \lambda;$$

$$\lambda\phi \wedge \mu\psi \Rightarrow \lambda(\phi \wedge \mu\psi) \qquad \text{for } \mu < \lambda.$$

Let us write $\phi \equiv \psi$ if $\mathsf{RC}_\Lambda \vdash \phi \Rightarrow \psi$ and $\mathsf{RC}_\Lambda \vdash \psi \Rightarrow \phi$. Then, the following equivalence will be useful to us:

Lemma 2.5. *Given formulas ϕ and ψ and ordinals $\mu < \lambda$,*

$$(\lambda\phi \wedge \mu\psi) \equiv \lambda(\phi \wedge \mu\psi).$$

Proof. The left-to-right direction is an axiom of RC. For the other direction we observe that $\lambda\mu\psi \Rightarrow \mu\psi$ is derivable using the axioms $\lambda\mu\psi \Rightarrow \mu\mu\psi$ and $\mu\mu\psi \Rightarrow \mu\psi$, from which the desired derivation can easily be obtained. □

Reflection calculi enjoy relatively simple relational semantics, where formulas have truth values on some set of points X, and each expression $\lambda\varphi$ is evaluated using an accessibility relation $\succ_\lambda$ on X.

Definition 2.6. *An RC_Λ-frame is a structure $\mathfrak{F} = \langle X, \langle \succ_\lambda \rangle_{\lambda < \Lambda} \rangle$ such that for all $x, y, z \in X$ and all $\mu < \lambda < \Lambda$,*

(i) *if $x \succ_\mu y \succ_\mu z$ then $x \succ_\mu z$,*

(ii) *if $z \succ_\mu x$ and $z \succ_\lambda y$ then $y \succ_\mu x$, and*

(iii) *if $x \succ_\lambda y$ then $x \succ_\mu y$.*

The valuation *on $\mathfrak{F}$ is the unique function* $\llbracket \cdot \rrbracket_{\mathfrak{F}} : \mathcal{L}_\Lambda \to 2^X$ *such that*

$$\llbracket \bot \rrbracket_{\mathfrak{F}} = \varnothing$$

$$\llbracket \neg \phi \rrbracket_{\mathfrak{F}} = X \setminus \llbracket \phi \rrbracket_{\mathfrak{F}}$$

$$\llbracket \phi \wedge \psi \rrbracket_{\mathfrak{F}} = \llbracket \phi \rrbracket_{\mathfrak{F}} \cap \llbracket \psi \rrbracket_{\mathfrak{F}}$$

$$\llbracket \lambda \phi \rrbracket_{\mathfrak{F}} = \{ x \in X : \exists y \prec_\lambda x \, (y \in \llbracket \phi \rrbracket_{\mathfrak{F}}) \}.$$

We may write $(\mathfrak{F}, x) \models \psi$ *instead of* $x \in \llbracket \psi \rrbracket_{\mathfrak{F}}$. *As usual,* ϕ *is* satisfied *on $\mathfrak{F}$ if* $\llbracket \phi \rrbracket_{\mathfrak{F}} \neq \varnothing$, *and* true *on $\mathfrak{F}$ if* $\llbracket \phi \rrbracket_{\mathfrak{F}} = X$.

Theorem 2.7. *For any class or set of ordinals Λ, RC_Λ is sound for the class of RC_Λ-frames.*

Proof. The proof proceeds by a standard induction on the length of a derivation and we omit it. □

In fact, Dashkov proved that RC_ω is also complete for the class of RC_ω-frames [14];[2] it is very likely that his result can be generalized to full RC over the ordinals, either by adapting his proof or by applying reduction techniques as in [8]. However, we remark that only soundness will be needed for our purposes.

2.3 Transfinite provability logic

The reflection calculus was introduced as a restriction of Japaridze's logic GLP_ω [27], which itself was extended by Beklemishev to full GLP [5], containing one modality for each ordinal number. Although we will work mostly within the reflection calculus, for historical reasons it is convenient to review the logic GLP.

The (variable-free) language of GLP is defined by the following grammar:

$$\top \mid \bot \mid \phi \wedge \psi \mid \phi \to \psi \mid \langle \lambda \rangle \phi.$$

Note that in this language we can define negation (as well as other Boolean connectives), along with $[\lambda]\phi = \neg \langle \lambda \rangle \neg \phi$.

The logic GLP_Λ is then given by the following rules and axioms:

(i) all propositional tautologies,

(ii) $[\lambda](\phi \to \psi) \to ([\lambda]\phi \to [\lambda]\psi)$ for all $\lambda < \Lambda$,

---

[2] Beware that RC_ω in our notation is not the same as $\mathsf{RC}\omega$ in [7].

(iii) $[\lambda]([\lambda]\phi \to \phi) \to [\lambda]\phi$ for all $\lambda < \Lambda$,

(iv) $[\mu]\phi \to [\lambda]\phi$ for $\mu < \lambda < \Lambda$,

(v) $\langle\mu\rangle\phi \to [\lambda]\langle\mu\rangle\phi$ for $\mu < \lambda < \Lambda$,

(vi) modus ponens and

(vii) necessitation for each $[\xi]$.

The reader may recognize axiom (iii) as Löb's axiom [32], ostensibly absent from RC; it is simply not expressible there. However, it was proven by Dashkov that GLP is conservative over RC, in the following sense:

Theorem 2.8. *If $\phi, \psi \in \mathcal{L}_{\mathsf{RC}}$, then $\mathsf{RC} \vdash \phi \Rightarrow \psi$, if and only if $\mathsf{GLP} \vdash \phi \to \psi$.*

Proof. That $\mathsf{RC} \vdash \phi \Rightarrow \psi$ implies $\mathsf{GLP} \vdash \phi \to \psi$ is readily proven by induction on the length of a derivation; one need only verify that, for $\mu < \lambda$,

$$\mathsf{GLP} \vdash \langle\lambda\rangle\phi \wedge \langle\mu\rangle\psi \to \langle\lambda\rangle(\phi \wedge \langle\mu\rangle\psi),$$

using the GLP axiom (v).

The other direction was proven for RC_ω by Dashkov in [14]. To extend to modalities over the ordinals, assume that $\mathsf{GLP} \vdash \phi \to \psi$. Then, there are finitely many modalities appearing in the derivation of $\vdash \phi \to \psi$, hence $\mathsf{GLP}_\Theta \vdash \phi \to \psi$ for some finite set Θ. But GLP_Θ readily embeds into GLP_ω (see [8]), and thus we can use the conservativity of GLP_ω over RC_ω to conclude that $\mathsf{RC} \vdash \phi \Rightarrow \psi$. □

As we have mentioned, full GLP (with propositional variables), or even GLP_2, is incomplete for its relational semantics. Without propositional variables, Ignatiev has built a relational model in which every consistent formula of variable-free GLP_ω is satisfied [26], and Joosten and I extended this to variable-free GLP over the ordinals. However, these models are infinite, and even $1\top$ cannot be satisfied on any finite relational model validating variable-free GLP. On the other hand, every worm has a relatively small RC-model, as we will see below.

3 Worms and consistency orderings

Worms are expressions of RC (or GLP) representing iterated consistency assertions. Ignatiev first observed that the worms in GLP_ω are well-founded [26]. The order-types of worms in GLP_2 were then studied by Boolos [10], and in full GLP by Beklemishev [5] and further by Joosten and I in [20], this time working in RC. Moreover, this particular well-order has surprising proof-theoretical applications: Beklemishev has used transfinite induction along the RC_ω worms to prove the consistency of Peano arithmetic and compute its Π^0_1 ordinal [4].

In this section we will review the ordering between worms and show that it is well-founded. Let us begin with some preliminaries.

3.1 Basic definitions

Definition 3.1. *A* worm *is any* RC *formula of the form*

$$\mathfrak{w} = \lambda_1 \ldots \lambda_n \top,$$

with each λ_i an ordinal and $n < \omega$ (including the 'empty worm', $\top$). The class of worms is denoted $\mathbb{W}$.

If Λ is a set or class of ordinals and each $\lambda_i \in \Lambda$, we write $\mathfrak{w} \sqsubset \Lambda$. The set of worms $\mathfrak{v}$ such that $\mathfrak{v} \sqsubset \Lambda$ is denoted $\mathbb{W}_\Lambda$.

'Measuring' worms is the central theme of this work. Let us begin by giving notation for some simple measurements, such as the length and the maximum element of a worm.

Definition 3.2. *If $\mathfrak{w} = \lambda_1 \ldots \lambda_n \top$, then we set $\#\mathfrak{w} = n$ (i.e., $\#\mathfrak{w}$ is the* length *of $\mathfrak{w}$). Define $\min \mathfrak{w} = \min_{i \in [1,n]} \lambda_i$, and similarly $\max \mathfrak{w} = \max_{i \in [1,n]} \lambda_i$. The class of worms $\mathfrak{w}$ such that $\mathfrak{w} = \top$ or $\mu \leq \min \mathfrak{w}$ will be denoted $\mathbb{W}_{\geq \mu}$. We define $\mathbb{W}_{>\mu}$ analogously.*

These give us some idea of 'how big' a worm is, but what we are truly interested in is in ordering worms by their *consistency strength:*

Definition 3.3. *Given an ordinal λ, we define a relation $\lhd_\lambda$ on $\mathbb{W}$ by $\mathfrak{v} \lhd_\lambda \mathfrak{w}$ if and only if $\mathsf{RC} \vdash \mathfrak{w} \Rightarrow \lambda \mathfrak{v}$. We also define $\mathfrak{v} \trianglelefteq_\mu \mathfrak{w}$ if $\mathfrak{v} \lhd_\mu \mathfrak{w}$ or $\mathfrak{v} \equiv \mathfrak{w}$.*

Instead of $\lhd_0, \trianglelefteq_0$ we may simply write $\lhd, \trianglelefteq$. As we will see, these orderings have some rather interesting properties. Let us begin by proving some basic facts about them:

Lemma 3.4. *Let $\mu \leq \lambda$ be ordinals and $\mathfrak{u}, \mathfrak{v}, \mathfrak{w}$ be worms. Then:*

1. *if $\mathfrak{w} \neq \top$ and $\mu < \min \mathfrak{w}$, then $\top \lhd_\mu \mathfrak{w}$,*

2. *if $\mathfrak{v} \lhd_\lambda \mathfrak{w}$, then $\mathfrak{v} \lhd_\mu \mathfrak{w}$, and*

3. *if $\mathfrak{u} \lhd_\mu \mathfrak{v}$ and $\mathfrak{v} \lhd_\mu \mathfrak{w}$, then $\mathfrak{u} \lhd_\mu \mathfrak{w}$.*

Proof. For the first item, write $\mathfrak{w} = \lambda \mathfrak{v}$, so that $\lambda \geq \mu$. Then, $\mathfrak{v} \Rightarrow \top$ is an axiom of RC, from which we can derive $\lambda \mathfrak{v} \Rightarrow \lambda \top$ and from there use the axiom $\lambda \top \Rightarrow \mu \top$.

For the second item, if $\mathfrak{v} \lhd_\lambda \mathfrak{w}$, then by definition, $\mathfrak{w} \Rightarrow \lambda \mathfrak{v}$ is derivable. Using the axiom $\lambda \mathfrak{v} \Rightarrow \mu \mathfrak{v}$, we see that $\mathfrak{w} \Rightarrow \mu \mathfrak{v}$ is derivable as well, that is, $\mathfrak{v} \lhd_\mu \mathfrak{w}$.

Transitivity simply follows from the fact that $\mathsf{RC} \vdash \mu \mu \mathfrak{u} \Rightarrow \mu \mathfrak{u}$, so that if $\mathfrak{u} \lhd_\mu \mathfrak{v}$ and $\mathfrak{v} \lhd_\mu \mathfrak{w}$, we have that $\mathsf{RC} \vdash \mathfrak{w} \Rightarrow \mu \mathfrak{v} \Rightarrow \mu \mu \mathfrak{u} \Rightarrow \mu \mathfrak{u}$, so $\mathfrak{u} \lhd_\mu \mathfrak{w}$. $\square$

3.2 Computing the consistency orders

The definition of $\mathfrak{v} \lhd_\lambda \mathfrak{w}$ does not suggest an obvious algorithm for deciding whether it holds or not. Fortunately, it can be reduced to computing the ordering between smaller worms; in this section, we will show how this is done. Let us begin by proving that $\lhd_\mu$ is always irreflexive. To do this, we will use the following frames.

Definition 3.5. *Let* $\mathfrak{w} = \lambda_n \ldots \lambda_0 \top$ *be any worm (note that we are using a different enumeration from that in Definition 3.1). Define a frame* $\mathfrak{F}(\mathfrak{w}) = \langle X, \langle \succ_\lambda \rangle_{\lambda < \Lambda} \rangle$ *as follows.*

First, set $X = [0, n+1] \subseteq \mathbb{N}$. *To simplify notation below, let* $\lambda_{n+1} = 0$. *Then, define* $x \succ_\eta y$ *if and only if:*

1. $x > y$ *and for all* $i \in [y, x)$, $\lambda_i \geq \eta$, *or*

2. $x \leq y$ *and for all* $i \in [x, y]$, $\lambda_i > \eta$.

Although this might not be obvious from the definition, these frames are indeed RC-frames.

Lemma 3.6. *Given any worm* $\mathfrak{w}$, $\mathfrak{F}(\mathfrak{w})$ *is an* RC*-frame.*

Proof. We must check that $\mathfrak{F}(\mathfrak{w})$ satisfies each item of Definition 2.6.

(i) Suppose that $x \succ_\eta y \succ_\eta z$. If $x > y$, consider three sub-cases.

   a. If $y > z$, from $x \succ_\eta y \succ_\eta z$ we see that for all $i \in [z, y) \cup [y, x) = [z, x)$, $\lambda_i \geq \eta$, so that $x \succ_\eta z$.

   b. If $z \in [y, x)$, from $[z, x) \subseteq [y, x)$ and $x \succ_\eta y$ we obtain $\lambda_i \geq \eta$ for all $i \in [z, x)$, so $x \succ_\eta z$.

   c. If $z \geq x$, from $[x, z] \subseteq [y, z]$ and $y \succ_\eta z$ we obtain $\lambda_i > \eta$ for all $i \in [x, z]$, hence $x \succ_\eta z$.

The cases where $x \leq y$ are analogous.

(ii). As in the previous item, we must consider several cases. Suppose that $\mu < \eta$, $z \succ_\mu x$ and $z \succ_\eta y$. If $z > x$, we consider three subcases.

  a. If $y \leq x$, then from $z \succ_\eta y$ and $[y, x] \subseteq [y, z]$ we obtain $\lambda_i \geq \eta > \mu$ for all $i \in [y, x]$, hence $y \succ_\mu x$.

  b. If $y \in (x, z]$, then from $[x, y) \subseteq [x, z)$ and $z \succ_\mu x$ we obtain $\lambda_i \geq \mu$ for all $i \in [x, y)$, hence $y \succ_\mu x$.

  c. If $y > z$, then from $z \succ_\mu x$ we we have that $\lambda_i \geq \mu$ for all $i \in [x, z)$, while from $z \succ_\eta y$ it follows that for all $i \in [z, y)$, $\lambda_i > \eta > \mu$, giving us $y \succ_\mu x$.

Cases where $z \leq x$ are similar.

(iii). That $\succ_\mu$ is monotone on μ is obvious from its definition. $\square$

Thus to prove that $\lhd_\mu$ is irreflexive, it suffices to show that there is $x \in [0, n+1]$ such that $(\mathfrak{F}(\mathfrak{w}), x) \models \lambda_{n-1} \ldots \lambda_0 \top$ but $(\mathfrak{F}(\mathfrak{w}), x) \not\models \lambda_n \ldots \lambda_0 \top$, as then by setting $\mu = \lambda_n$ and $\mathfrak{v} = \lambda_{n-1} \ldots \lambda_0 \top$ we see that $\mathfrak{v} \not\lhd_\mu \mathfrak{v}$. The following lemma will help us find such an x.

Lemma 3.7. *Let $\mathfrak{w} = \lambda_n \ldots \lambda_0 \top$ be a worm, and for any $i \in [0, n+1]$, define $\mathfrak{w}[i]$ recursively by $\mathfrak{w}[0] = \top$ and $\mathfrak{w}[i+1] = \lambda_i \mathfrak{w}[i]$. Then:*

1. $(\mathfrak{F}(\mathfrak{w}), i) \models \mathfrak{w}[i]$, *and*

2. *if $x \in [0, i)$, then $(\mathfrak{F}(\mathfrak{w}), x) \not\models \mathfrak{w}[i]$.*

Proof. The first claim is easy to check from the definition of $\mathfrak{F}(\mathfrak{w})$, so we focus on proving the second by induction on i. The base case is vacuously true as $[0, 0) = \varnothing$. Otherwise, assume the claim for i, and consider $x \in [0, i+1)$; we must show that $(\mathfrak{F}(\mathfrak{w}), x) \not\models \mathfrak{w}[i+1] = \lambda_i \mathfrak{w}[i]$, which means that for all $y \prec_{\lambda_i} x$, $(\mathfrak{F}(\mathfrak{w}), y) \not\models \mathfrak{w}[i]$. Note that we cannot have that $y \in [i, n+1]$, as in this case $y \geq i \geq x$; but obviously $\lambda_i \not> \lambda_i$, so that $y \not\prec_{\lambda_i} x$. It follows that $y \in [0, i)$, and we can apply the induction hypothesis to $\mathfrak{w}[i]$. $\square$

Lemma 3.8. *Given any ordinal μ and any worm $\mathfrak{v}$, we have that $\mathfrak{v} \not\lhd_\mu \mathfrak{v}$.*

Proof. Let μ be any ordinal, $\mathfrak{v}$ be any worm, and consider the RC-frame $\mathfrak{F}(\mu \mathfrak{v})$. If $n = \#\mathfrak{v}$, observe that $(\mu \mathfrak{v})[n] = \mathfrak{v}$, hence by Lemma 3.7, $(\mathfrak{F}(\mu \mathfrak{v}), n) \models \mathfrak{v}$ but $(\mathfrak{F}(\mu \mathfrak{v}), n) \not\models \mu \mathfrak{v}$; it follows from Theorem 2.7 that $\mathfrak{v} \not\lhd_\mu \mathfrak{v}$. $\square$

Thus the worm orderings are irreflexive. Next we turn our attention to a useful operation between worms. Specifically, worms can be regarded as strings of symbols, and as such we can think of concatenating them.

Definition 3.9. *Let* $\mathfrak{v} = \xi_1 \ldots \xi_n \top$ *and* $\mathfrak{w} = \zeta_1 \ldots \zeta_m \top$ *be worms. Then, define*
$$\mathfrak{v}\mathfrak{w} = \xi_1 \ldots \xi_n \zeta_1 \ldots \zeta_m \top$$

Often we will want to put an extra ordinal between the worms, and we write $\mathfrak{v} \, \lambda \, \mathfrak{w}$ for $\mathfrak{v}(\lambda \mathfrak{w})$.

Lemma 3.10. *If* $\mathfrak{w}, \mathfrak{v}$ *are worms and* $\mu < \min \mathfrak{w}$, *then* $\mathfrak{w} \, \mu \, \mathfrak{v} \equiv \mathfrak{w} \wedge \mu \mathfrak{v}$.

Proof. By induction on $\#\mathfrak{w}$. If $\mathfrak{w} = \top$, the claim becomes $\mu\mathfrak{v} \equiv \top \wedge \mu\mathfrak{v}$, which is obviously true. Otherwise, we write $\mathfrak{w} = \lambda \mathfrak{u}$ with $\lambda > \mu$, and observe that by Lemma 2.5,
$$\lambda \mathfrak{u} \wedge \mu \mathfrak{v} \equiv \lambda(\mathfrak{u} \wedge \mu \mathfrak{v}) \stackrel{\text{IH}}{\equiv} \lambda(\mathfrak{u} \, \mu \, \mathfrak{v}) = \mathfrak{w} \, \mu \, \mathfrak{v}. \qquad \square$$

Thus we may "pull out" the initial segment of a worm, provided the following element is a lower bound for this initial segment. In general, for any ordinal λ, we can pull out the maximal initial segment of $\mathfrak{w}$ which is bounded below by λ; this segment is the λ-*head* of $\mathfrak{w}$, and what is left over (if anything) is its λ-*body*.

Definition 3.11. *Let* λ *be an ordinal and* $\mathfrak{w} \in \mathbb{W}_{\geq \lambda}$. *We define* $h_\lambda(\mathfrak{w})$ *to be the maximal initial segment of* $\mathfrak{w}$ *such that* $\lambda < \min h_\lambda(\mathfrak{w})$, *and define* $b_\lambda(\mathfrak{w})$ *as follows: if* λ *appears in* $\mathfrak{w}$, *then we set* $b_\lambda(\mathfrak{w})$ *to be the unique worm such that* $\mathfrak{w} = h_\lambda(\mathfrak{w}) \lambda b_\lambda(\mathfrak{v})$. *Otherwise, set* $b_\lambda(\mathfrak{w}) = \top$.

We may write h, b instead of h_0, b_0. We remark that our notation is a variant from that used in [20], where our h_λ would be denoted $h_{\lambda+1}$.

Lemma 3.12. *Given a worm* $\mathfrak{w} \neq \top$ *and an ordinal* $\mu \leq \min \mathfrak{w}$,

1. $h_\mu(\mathfrak{w}) \in \mathbb{W}_{>\mu}$,

2. $\#h_\mu(\mathfrak{w}) \leq \#\mathfrak{w}$, *with equality holding only if* $\mu < \min \mathfrak{w}$, *in which case* $h_\mu(\mathfrak{w}) = \mathfrak{w}$;

3. $\#b_\mu(\mathfrak{w}) < \#\mathfrak{w}$, *and*

4. $\mathfrak{w} \equiv h_\mu(\mathfrak{w}) \wedge \mu b_\mu(\mathfrak{w})$.

Proof. The first two claims are immediate from the definition of h_μ. For the third, this is again obvious in the case that μ occurs in $\mathfrak{w}$, otherwise we have that $b_\mu(\mathfrak{w}) = \top$ and by the assumption that $\mathfrak{w} \neq \top$ we obtain $\#b_\mu(\mathfrak{w}) < \#\mathfrak{w}$.

The fourth claim is an instance of Lemma 3.10 if μ appears in $\mathfrak{w}$, otherwise $\mathfrak{w} = h_\mu(\mathfrak{w})$ and we use Lemma 3.4 to see that $\mathfrak{w} \Rightarrow \mu\top = \mu b_\mu(\top)$ is derivable. □

With this we can reduce relations between worms to those between their heads and bodies.

Lemma 3.13. *If $\mathfrak{w}, \mathfrak{v} \neq \top$ are worms and $\mu \leq \min \mathfrak{w}\mathfrak{v}$, then*

1. *$\mathfrak{w} \triangleleft_\mu \mathfrak{v}$ whenever*

    (a) *$\mathfrak{w} \trianglelefteq_\mu b_\mu(\mathfrak{v})$, or*

    (b) *$b_\mu(\mathfrak{w}) \triangleleft_\mu \mathfrak{v}$ and $h_\mu(\mathfrak{w}) \triangleleft_{\mu+1} h_\mu(\mathfrak{v})$, and*

2. *$\mathsf{RC} \vdash \mathfrak{v} \Rightarrow \mathfrak{w}$ whenever $b_\mu(\mathfrak{w}) \triangleleft_\mu \mathfrak{v}$ and $\mathsf{RC} \vdash h_\mu(\mathfrak{v}) \Rightarrow h_\mu(\mathfrak{w})$.*

Proof. For the first claim, if $\mathfrak{w} \trianglelefteq_\mu b_\mu(\mathfrak{v})$, then by Lemma 3.12.4 we have that $\mathfrak{v} \Rightarrow \mu b_\mu(\mathfrak{v})$, that is, $b_\mu(\mathfrak{v}) \triangleleft_\mu \mathfrak{v}$. By transitivity we obtain $\mathfrak{w} \triangleleft_\mu \mathfrak{v}$. If $b_\mu(\mathfrak{w}) \triangleleft_\mu \mathfrak{v}$ and $h_\mu(\mathfrak{w}) \triangleleft_{\mu+1} h_\mu(\mathfrak{v})$, reasoning in RC we have that

$$\mathfrak{v} \Rightarrow h_\mu(\mathfrak{v}) \wedge \mathfrak{v} \Rightarrow \langle\mu+1\rangle h_\mu(\mathfrak{w}) \wedge \mu b_\mu(\mathfrak{w}) \equiv \langle\mu+1\rangle h_\mu(\mathfrak{w})\mu b_\mu(\mathfrak{w}) \Rightarrow \mu\mathfrak{w},$$

and $\mathfrak{w} \triangleleft_\mu \mathfrak{v}$, as needed.

For the second, if $b_\mu(\mathfrak{w}) \triangleleft_\mu \mathfrak{v}$ and $\mathsf{RC} \vdash h_\mu(\mathfrak{v}) \Rightarrow h_\mu(\mathfrak{w})$, we have that

$$\mathfrak{v} \Rightarrow h_\mu(\mathfrak{v}) \wedge \mathfrak{v} \Rightarrow h_\mu(\mathfrak{w}) \wedge \mu b_\mu(\mathfrak{w}) \equiv \mathfrak{w}.$$ □

As we will see, Lemma 3.13 gives us a recursive way to compute $\triangleleft_\mu$. This recursion will allow us to establish many of the fundamental properties of $\triangleleft_\mu$, beginning with the fact that it defines a total preorder.

Lemma 3.14. *Given worms $\mathfrak{v}, \mathfrak{w}$ and $\mu \leq \min(\mathfrak{w}\mathfrak{v})$, exactly one of $\mathfrak{w} \trianglelefteq_\mu \mathfrak{v}$ or $\mathfrak{v} \triangleleft_\mu \mathfrak{w}$ occurs.*

Proof. That they cannot simultaneously occur follows immediately from Lemma 3.8, since $\triangleleft_\mu$ is irreflexive.

To show that at least one occurs, proceed by induction on $\#\mathfrak{w} + \#\mathfrak{v}$. To be precise, assume inductively that whenever $\#\mathfrak{w}' + \#\mathfrak{v}' < \#\mathfrak{w} + \#\mathfrak{v}$ and $\mu \leq \min(\mathfrak{w}'\mathfrak{v}')$ is arbitrary, then either $\mathfrak{w}' \trianglelefteq_\mu \mathfrak{v}'$ or $\mathfrak{v}' \trianglelefteq_\mu \mathfrak{w}'$. If either $\mathfrak{v} = \top$ or $\mathfrak{w} = \top$, then the claim is immediate from Lemma 3.4.

Otherwise, let $\lambda = \min(\mathfrak{w}\mathfrak{v})$, so that $\lambda \geq \mu$. If $\mathfrak{w} \trianglelefteq_\lambda b_\lambda(\mathfrak{v})$, then by Lemma 3.13, $\mathfrak{w} \triangleleft_\lambda \mathfrak{v}$, and similarly if $\mathfrak{v} \trianglelefteq_\lambda b_\lambda(\mathfrak{w})$, then $\mathfrak{v} \triangleleft_\lambda \mathfrak{w}$. On the other hand, if neither occurs then by the induction hypothesis we have that $b_\lambda(\mathfrak{v}) \triangleleft_\lambda \mathfrak{w}$ and $b_\lambda(\mathfrak{w}) \triangleleft_\lambda \mathfrak{v}$.

Since λ appears in either $\mathfrak{w}$ or $\mathfrak{v}$, by Lemma 3.12.2 we have that

$$\#h_\lambda(\mathfrak{w}) + \#h_\lambda(\mathfrak{v}) < \#\mathfrak{w} + \#\mathfrak{v},$$

so that by the induction hypothesis, either $h_\lambda(\mathfrak{w}) \triangleleft_\mu h_\lambda(\mathfrak{v})$, $h_\lambda(\mathfrak{w}) \equiv h_\lambda(\mathfrak{v})$, or $h_\lambda(\mathfrak{v}) \triangleleft_\lambda h_\lambda(\mathfrak{w})$. If $h_\lambda(\mathfrak{w}) \triangleleft_\lambda h_\lambda(\mathfrak{v})$, we may use Lemma 3.13.1 to see that $\mathfrak{w} \triangleleft_\lambda \mathfrak{v}$, so that by Lemma 3.4, $\mathfrak{w} \triangleleft_\mu \mathfrak{v}$. Similarly, if $h_\lambda(\mathfrak{v}) \triangleleft_\lambda h_\lambda(\mathfrak{w})$, we obtain $\mathfrak{v} \triangleleft_\mu \mathfrak{w}$. If $h_\lambda(\mathfrak{w}) \equiv h_\lambda(\mathfrak{v})$, then Lemma 3.13.2 yields both $\mathfrak{w} \Rightarrow \mathfrak{v}$ and $\mathfrak{w} \Rightarrow \mathfrak{v}$, i.e., $\mathfrak{w} \equiv \mathfrak{v}$. □

Corollary 3.15. *If* $\mathsf{RC} \vdash \mathfrak{w} \Rightarrow \mathfrak{v}$, *then* $\mathfrak{v} \trianglelefteq \mathfrak{w}$.

Proof. Towards a contradiction, suppose that $\mathsf{RC} \vdash \mathfrak{w} \Rightarrow \mathfrak{v}$ but $\mathfrak{v} \not\trianglelefteq \mathfrak{w}$. By Lemma 3.14, $\mathfrak{w} \triangleleft \mathfrak{v}$. Hence $\mathfrak{v} \Rightarrow \mathfrak{w} \Rightarrow 0\mathfrak{v}$, and $\mathfrak{v} \triangleleft \mathfrak{v}$, contradicting the irreflexivity of $\triangleleft$. □

Moreover, the orderings $\triangleleft_\lambda$, $\triangleleft_\mu$ coincide on $\mathbb{W}_{\geq \max\{\lambda,\mu\}}$:

Lemma 3.16. *Let* $\mathfrak{w}, \mathfrak{v}$ *be worms and* $\mu, \lambda \leq \min(\mathfrak{w}\mathfrak{v})$. *Then,* $\mathfrak{w} \triangleleft_\mu \mathfrak{v}$ *if and only if* $\mathfrak{w} \triangleleft_\lambda \mathfrak{v}$.

Proof. Assume without loss of generality that $\mu \leq \lambda$. One direction is already in Lemma 3.4. For the other, assume towards a contradiction that $\mathfrak{w} \triangleleft_\mu \mathfrak{v}$ but $\mathfrak{w} \not\triangleleft_\lambda \mathfrak{v}$. Then, by Lemma 3.14, $\mathfrak{v} \trianglelefteq_\lambda \mathfrak{w}$ and thus $\mathfrak{v} \trianglelefteq_\mu \mathfrak{w}$, so that $\mathfrak{v} \trianglelefteq_\mu \mathfrak{w} \triangleleft_\mu \mathfrak{v}$, contradicting the irreflexivity of $\triangleleft_\mu$ (Lemma 3.8). □

With this we can give an improved version of Lemma 3.13, that will be more useful to us later.

Theorem 3.17. *The relation* $\trianglelefteq_\lambda$ *is a total preorder on* $\mathbb{W}_{\geq \lambda}$, *and for all* $\mu \leq \lambda$ *and* $\mathfrak{w}, \mathfrak{v} \in \mathbb{W}_{\geq \lambda}$ *with* $\mathfrak{w}, \mathfrak{v} \neq \top$,

1. $\mathfrak{w} \triangleleft_\mu \mathfrak{v}$ *if and only if*

   (a) $\mathfrak{w} \trianglelefteq_\mu b_\lambda(\mathfrak{v})$, *or*

   (b) $b_\lambda(\mathfrak{w}) \triangleleft_\mu \mathfrak{v}$ *and* $h_\lambda(\mathfrak{w}) \triangleleft_\mu h_\lambda(\mathfrak{v})$, *and*

2. $\mathfrak{w} \trianglelefteq_\mu \mathfrak{v}$ *if and only if*

   (a) $\mathfrak{w} \trianglelefteq_\mu b_\lambda(\mathfrak{v})$, *or*

   (b) $b_\lambda(\mathfrak{w}) \triangleleft_\mu \mathfrak{v}$ *and* $h_\lambda(\mathfrak{w}) \trianglelefteq_\mu h_\lambda(\mathfrak{v})$.

Proof. Totality is Lemma 3.14. Let us prove item 2; the proof of item 1 is similar. If (2a) holds, then by Lemma 3.16, $\mathfrak{w} \trianglelefteq_\lambda b_\lambda(\mathfrak{v})$, so that by Lemma 3.13.1, $\mathfrak{w} \triangleleft_\lambda \mathfrak{v}$, and once again by Lemma 3.16, $\mathfrak{w} \triangleleft_\mu \mathfrak{v}$. If (2b) holds, then by Lemma 3.16 we obtain $b_\lambda(\mathfrak{w}) \triangleleft_\lambda \mathfrak{v}$ and $h_\lambda(\mathfrak{w}) \trianglelefteq_{\lambda+1} h_\lambda(\mathfrak{v})$. If $h_\lambda(\mathfrak{w}) \triangleleft_{\lambda+1} h_\lambda(\mathfrak{v})$, we may use Lemma 3.13.1 to obtain $\mathfrak{w} \triangleleft_\lambda \mathfrak{v}$. Otherwise, by Lemma 3.13.2, we see that $\mathsf{RC} \vdash \mathfrak{v} \Rightarrow \mathfrak{w}$, which by Corollary 3.15 gives us $\mathfrak{w} \triangleleft_\lambda \mathfrak{v}$. In either case, $\mathfrak{w} \triangleleft_\mu \mathfrak{v}$.

For the other direction, assume that (2a) and (2b) both fail. Then by Lemma 3.14 together with Lemma 3.16, we have that $b_\lambda(\mathfrak{v}) \triangleleft_\lambda \mathfrak{w}$ and either $\mathfrak{v} \trianglelefteq_\lambda b_\lambda(\mathfrak{w})$ or $h_\lambda(\mathfrak{v}) \triangleleft_{\lambda+1} h_\lambda(\mathfrak{w})$. In either case $\mathfrak{v} \triangleleft_\lambda \mathfrak{w}$, and thus $\mathfrak{w} \not\triangleleft_\mu \mathfrak{v}$. □

Before continuing, it will be useful to derive a few straightforward consequences of Theorem 3.17.

Corollary 3.18. *Every $\phi \in \mathcal{L}_{\mathsf{RC}}$ is equivalent to some $\mathfrak{w} \in \mathbb{W}$. Moreover, we can take $\mathfrak{w}$ so that every ordinal appearing in $\mathfrak{w}$ already appears in ϕ.*

Proof. By induction on the complexity of ϕ. We have that $\top$ is a worm and for $\phi = \lambda \psi$, by induction hypothesis we have that $\psi \equiv \mathfrak{v}$ for some worm $\mathfrak{v}$ with all modalities appearing in ψ and hence $\phi \equiv \lambda \mathfrak{v}$.

It remains to consider an expression of the form $\psi \wedge \phi$. Using the induction hypothesis, there are worms $\mathfrak{w}, \mathfrak{v}$ equivalent to ϕ, ψ, respectively, so that $\psi \wedge \phi \equiv \mathfrak{w} \wedge \mathfrak{v}$. We proceed by a secondary induction on $\#\mathfrak{w} + \#\mathfrak{v}$. Note that the claim is trivial if either $\mathfrak{w} = \top$ or $\mathfrak{v} = \top$, so we assume otherwise.

Let μ be the least ordinal appearing either in $\mathfrak{w}$ or in $\mathfrak{v}$, so that

$$\psi \wedge \phi \equiv (h_\mu(\mathfrak{w}) \wedge h_\mu(\mathfrak{v})) \wedge (\mu b_\mu(\mathfrak{w}) \wedge \mu b_\mu(\mathfrak{v})).$$

By induction hypothesis, $h_\mu(\mathfrak{w}) \wedge h_\mu(\mathfrak{v}) \equiv \mathfrak{u}_1$ for some $\mathfrak{u}_1 \in \mathbb{W}_{\mu+1}$ with all modalities occurring in $\phi \wedge \psi$. Meanwhile, either $b_\mu(\mathfrak{w}) \triangleleft_\mu b_\mu(\mathfrak{v})$, $b_\mu(\mathfrak{w}) \equiv b_\mu(\mathfrak{v})$ or $b_\mu(\mathfrak{v}) \triangleleft_\mu b_\mu(\mathfrak{w})$. In the first case,

$$\mu b_\mu(\mathfrak{v}) \Rightarrow \mu \mu b_\mu(\mathfrak{w}) \Rightarrow \mu b_\mu(\mathfrak{w}),$$

and in the second $\mu b_\mu(\mathfrak{v}) \Rightarrow \mu b_\mu(\mathfrak{w})$; in either case, $\mu b_\mu(\mathfrak{w}) \wedge \mu b_\mu(\mathfrak{v}) \equiv \mu b_\mu(\mathfrak{v})$. Similarly, if $b_\mu(\mathfrak{v}) \triangleleft_\mu b_\mu(\mathfrak{w})$, then $\mu b_\mu(\mathfrak{w}) \wedge \mu b_\mu(\mathfrak{v}) \equiv \mu b_\mu(\mathfrak{w})$. In either case,

$$\mu b_\mu(\mathfrak{w}) \wedge \mu b_\mu(\mathfrak{v}) \equiv \mu b_\mu(\mathfrak{u}_0)$$

for some worm $\mathfrak{u}_0 \in \{\mathfrak{w}, \mathfrak{v}\}$, and thus

$$\phi \wedge \psi \equiv (h_\mu(\mathfrak{w}) \wedge h_\mu(\mathfrak{v})) \wedge (\mu b_\mu(\mathfrak{w}) \wedge \mu b_\mu(\mathfrak{v})) \equiv \mathfrak{u}_1 \wedge \mu \mathfrak{u}_0 \equiv \mathfrak{u}_1 \mu \mathfrak{u}_0. \quad □$$

Below, we remark that $\mathfrak{w} \sqsubset \mu$ is equivalent to $\max \mathfrak{w} < \mu$.

Corollary 3.19. *Let μ be an ordinal and $\top \neq \mathfrak{w} \in \mathbb{W}$. Then,*

1. *if $\mathfrak{w} \neq \top$ and $\mu < \max \mathfrak{w}$ then $\mu\top \lhd \mathfrak{w}$,*

2. *if $\mathfrak{w} \neq \top$ and $\mu \leq \max \mathfrak{w}$ then $\mu\top \unlhd \mathfrak{w}$, and*

3. *if $\mathfrak{w} \sqsubset \mu$ then $\mathfrak{w} \lhd \mu\top$.*

Proof. For the first claim, proceed by induction on $\#\mathfrak{w}$. Write $\mathfrak{w} = \lambda\mathfrak{v}$ and consider two cases. If $\lambda \leq \mu$, by induction on length, $\mu\top \lhd \mathfrak{v}$, so $\mu\top \lhd \mathfrak{v} \lhd \mathfrak{w}$. Otherwise, $\lambda > \mu$, so from $\mathfrak{v} \Rightarrow \top$, $\lambda\top \Rightarrow \mu\top$, and Lemma 3.10 we obtain

$$\mathfrak{w} \Rightarrow \lambda\top \wedge \mu\top \Rightarrow \lambda\mu\top \Rightarrow 0\mu\top.$$

The second claim is similar. Again, write $\mathfrak{w} = \lambda\mathfrak{v}$. If $\mu > \lambda$, we have inductively that $\mu\top \unlhd \mathfrak{v} \lhd \mathfrak{w}$. Otherwise, $\mu \leq \lambda$, in which case

$$\mathfrak{w} \Rightarrow \lambda\top \Rightarrow \mu\top,$$

and we may use Corollary 3.15.

For the third, we proceed once again by induction on $\#\mathfrak{w}$. The case for $\mathfrak{w} = \top$ is obvious. Otherwise, let $\eta = \min \mathfrak{w}$. Then, by the induction hypothesis, $h_\eta(\mathfrak{w}) \lhd \mu\top = h_\eta(\mu\top)$, while also by the induction hypothesis $b_\eta(\mathfrak{w}) \lhd \mu\top$, hence $\mathfrak{w} \lhd \mu\top$ by Theorem 3.17. □

3.3 Well-orderedness of worms

We have seen that $\unlhd_\mu$ is a total preorder, but in fact we have more; it is a pre-well-order. We will prove this using a Kruskal-style argument [31]. It is very similar to Beklemishev's proof in [5], although he uses normal forms for worms. Here we will use our 'head-body' decomposition instead.

Theorem 3.20. *For any ordinal λ and any $\eta \leq \lambda$, $\lhd_\eta$ is a pre-well-order on $\mathbb{W}_{\geq \lambda}$.*

Proof. We have already seen that $\mathbb{W}_\lambda$ is total in Theorem 3.17, so it remains to show that there are no infinite $\lhd_\eta$-descending chains. We will prove this by contradiction, assuming that there is such a chain.

Let $\mathfrak{w}_0$ be any worm such that $\mathfrak{w}_0$ is the first element of some infinite descending chain $\mathfrak{w}_0 \rhd_\eta \mathfrak{v}_1 \rhd_\eta \mathfrak{v}_2 \rhd_\eta \ldots$ and $\#\mathfrak{w}_0$ is minimal among all worms that can be the first element of such a chain. Then, for $i > 0$, choose $\mathfrak{w}_i$ recursively by letting it be a worm such that there is an infinite descending chain

$$\mathfrak{w}_0 \rhd_\eta \mathfrak{w}_1 \rhd_\eta \ldots \rhd_\eta \mathfrak{w}_i \rhd_\eta \mathfrak{v}_{i+1} \rhd_\eta \ldots,$$

and such that $\#\mathfrak{w}_i$ is minimal among all worms with this property (where $\mathfrak{w}_j$ is already fixed for $j < i$). Let $\vec{\mathfrak{w}}$ be the resulting chain.

Now, let $\mu \geq \eta$ be the least ordinal appearing in $\vec{\mathfrak{w}}$, and define $h(\vec{\mathfrak{w}})$ to be the sequence

$$h_\mu(\mathfrak{w}_0), h_\mu(\mathfrak{w}_1), \ldots, h_\mu(\mathfrak{w}_i), \ldots$$

Let j be the first natural number such that μ appears in $\mathfrak{w}_j$. By Lemma 3.12.2, $h_\mu(\mathfrak{w}_i) = \mathfrak{w}_i$ for all $i < j$, while $\#h_\mu(\mathfrak{w}_j) < \#\mathfrak{w}_j$, so by the minimality of $\#\mathfrak{w}_j$, $h(\vec{\mathfrak{w}})$ is not an infinite decreasing chain. Hence for some k, $h_\mu(\mathfrak{w}_k) \unrhd_\eta h_\mu(\mathfrak{w}_{k+1})$.

Next, define $b(\vec{\mathfrak{w}})$ to be the sequence

$$\mathfrak{w}_0, \ldots, \mathfrak{w}_{k-1}, b_\mu(\mathfrak{w}_k), \mathfrak{w}_{k+2}, \mathfrak{w}_{k+3}, \ldots$$

In other words, we replace $\mathfrak{w}_k$ by $b_\mu(\mathfrak{w}_k)$ and skip $\mathfrak{w}_{k+1}$. By the minimality of $\#\mathfrak{w}_k$, this cannot be a decreasing sequence, and hence $b_\mu(\mathfrak{w}_k) \unlhd_\eta \mathfrak{w}_{k+2} \lhd_\eta \mathfrak{w}_{k+1}$.

It follows from Theorem 3.17 that $\mathfrak{w}_k \unlhd_\eta \mathfrak{w}_{k+1}$, a contradiction. We conclude that there can be no decreasing sequence, and $\lhd_\eta$ is well-founded, as claimed. $\square$

One consequence of worms being pre-well-ordered is that we can assign them an ordinal number measuring their order-type. In the next section we will make this precise.

3.4 Order-types on a pre-well-order

As we have mentioned, any well-order may be canonically represented using an ordinal number. To do this, if $\mathfrak{A} = \langle A, \preccurlyeq \rangle$ is any pre-well-order, for $a \in A$ define

$$o(a) = \bigcup_{b \prec a} (o(b) + 1).$$

Observe that o is strictly increasing, in the following sense:

Definition 3.21. *Let $\langle A, \preccurlyeq_A \rangle, \langle B, \preccurlyeq_B \rangle$ be preorders, and $f \colon A \to B$. We say that f is* stricty increasing *if*

1. *for all $x, y \in A$, $x \preccurlyeq_A y$ implies $f(x) \preccurlyeq_B f(y)$, and*

2. *for all $x, y \in A$, $x \prec_A y$ implies $f(x) \prec_B f(y)$.*

We note that if $\prec_A$ is total, then there are other equivalent ways of defining strictly increasing maps:

Lemma 3.22. *If $\langle A, \preccurlyeq_A \rangle, \langle B, \preccurlyeq_B \rangle$ are total preorders and $f: A \to B$, then the following are equivalent:*

1. *f is strictly increasing;*

2. *for all $x, y \in A$, $x \preccurlyeq_A y$ if and only if $f(x) \preccurlyeq_B f(y)$;*

3. *for all $x, y \in A$, $x \prec_A y$ if and only if $f(x) \prec_B f(y)$.*

Proof. Straightforward, using the fact that $a \prec_A b$ if and only if $b \not\preccurlyeq_A a$, and similarly for $\prec_B$. □

Then, the map o can be characterized as the only strictly increasing, initial map $f: A \to \mathsf{Ord}$, where $f: A \to B$ is *initial* if whenever $b \prec_B f(a)$, it follows that $b = f(a')$ for some $a' \prec_A a$:

Lemma 3.23. *Let $\langle A, \preccurlyeq \rangle$ be a pre-well-order. Then,*

1. *for all $x, y \in A$, $x \prec y$ if and only if $o(x) < o(y)$, and*

2. *$o: A \to \mathsf{Ord}$ is an initial map.*

The proof proceeds by transfinite induction along $\prec$ and we omit it, as is the case of the proof of the following:

Lemma 3.24. *Let $\langle A, \preccurlyeq \rangle$ be a pre-well-order. Suppose that $f: A \to \mathsf{Ord}$ satisfies*

1. *$x \prec y$ implies that $f(x) < f(y)$,*

2. *$x \preccurlyeq y$ implies that $f(x) \leq f(y)$, and*

3. *if $\xi \in f[A]$ then $\xi \subseteq f[A]$.*

Then, $f = o$.

Observe that $o(a) = o(b)$ implies that $a \preccurlyeq b$ and $b \preccurlyeq a$, i.e. $a \approx b$. Let us state this explicitly for the case of worms.

Lemma 3.25. *If $\mathfrak{w}, \mathfrak{v}$ are worms such that $o(\mathfrak{w}) = o(\mathfrak{v})$, then $\mathfrak{w} \equiv \mathfrak{v}$.*

Proof. Reasoning by contrapositive, assume that $\mathfrak{w} \not\equiv \mathfrak{v}$. Then by Lemma 3.14, either $\mathfrak{w} \lhd \mathfrak{v}$, which implies that $o(\mathfrak{w}) < o(\mathfrak{v})$, or $\mathfrak{v} \lhd \mathfrak{w}$, and hence $o(\mathfrak{v}) < o(\mathfrak{w})$. In either case, $o(\mathfrak{w}) \neq o(\mathfrak{v})$. □

Computing $o(\mathfrak{w})$ will take some work, but it is not too difficult to establish some basic relationships between $o(\mathfrak{w})$ and the ordinals appearing in $\mathfrak{w}$.

Lemma 3.26. *Let* $\mathfrak{w} \neq \top$ *be a worm and* μ *an ordinal. Then,*

1. *if* $\mu \leq \max \mathfrak{w}$, *then* $\mu \leq o(\mu\top) \leq o(\mathfrak{w})$, *and*

2. *if* $\max \mathfrak{w} < \mu$, *then* $o(\mathfrak{w}) < o(\mu\top)$.

Proof. First we proceed by induction on μ to show that $\mu \leq o(\mu\top)$. Suppose that $\eta < \mu$. Then by Corollary 3.19, $\eta\top \lhd \mu\top$, while by the induction hypothesis $\eta \leq o(\eta\top)$, and hence $\eta \leq o(\eta\top) < o(\mu\top)$. Since $\eta < \mu$ was arbitrary, $\mu \leq o(\mu\top)$. That $o(\mu\top) \leq o(\mathfrak{w})$ if $\mu \leq \max \mathfrak{w}$ follows from Corollary 3.19, since $\mu\top \trianglelefteq \mathfrak{w}$.

The second claim is immediate from Corollary 3.19.3. $\square$

Let us conclude this section by stating a useful consequence of the fact that $o \colon \mathbb{W} \to \mathsf{Ord}$ is initial.

Corollary 3.27. *For every ordinal* ξ *there is a worm* $\mathfrak{w} \trianglelefteq \xi\top$ *such that* $\xi = o(\mathfrak{w})$.

Proof. By Lemma 3.26, $\xi \leq o(\xi\top)$, so this is a special case of Lemma 3.23.2. $\square$

4 Finite worms

In the previous section we explored some basic properties of o, but they are not sufficient to compute $o(\mathfrak{w})$ for a worm $\mathfrak{w}$. In this section we will provide an explicit calculus for $o \restriction \mathbb{W}_\omega$ (where $\restriction$ denotes domain restriction). $\mathbb{W}_\omega$ is a particularly interesting case-study in that it has been used by Beklemishev for a Π^0_1 ordinal analysis of Peano arithmetic. Before we continue, it will be illustrative to sketch the relationship between $\mathbb{W}_\omega$ and PA.

4.1 First-order arithmetic

Expressions of RC_ω have a natural proof-theoretical interpretation in first-order arithmetic. We will use the language Π_ω of first-order arithmetic containing the signature

$$\{0, 1, +, \cdot, 2^\cdot, =\}$$

so that we have symbols for addition, multiplication, and exponentiation, as well as Boolean connectives and quantifiers ranging over the natural numbers. Elements of Π_ω are *formulas*. The set of all formulas where all quantifiers are *bounded*, that is, of the form $\forall x{<}t\ \phi$ or $\exists x{<}t\ \phi$ (where t is any term), is denoted Δ_0. A formula of the form $\exists x_n \forall x_{n-1} \ldots \delta(x_1, \ldots, x_n)$, with $\delta \in \Delta_0$, is Σ_n, and a formula of the form $\forall x_n \exists x_{n-1} \ldots \delta(x_1, \ldots, x_n)$ is Π_n. These classes are extended modulo provable equivalence, so that every formula falls into one of them. Note that the negation of a Σ_n formula is Π_n and vice-versa.

To simplify notation we may assume that some additional function symbols are available, although these are always definable from the basic arithmetical operations. In particular, we assume that we have for each n a function $\langle x_1, \ldots, x_n \rangle$ coding a sequence as a single natural number.

In order to formalize provability within arithmetic, we fix some Gödel numbering mapping a formula $\psi \in \Pi_\omega$ to its corresponding Gödel number $\ulcorner \psi \urcorner$, and similarly for terms and sequences of formulas, which can be used to represent derivations. We also define the *numeral* of $n \in \mathbb{N}$ to be the term

$$\bar{n} = 0 + \underbrace{1 + \ldots + 1}_{n \text{ times}}.$$

In order to simplify notation, we will often identify ψ with $\ulcorner \psi \urcorner$.

We will assume that every theory T contains classical predicate logic, is closed under modus ponens, and that there is a Δ_0 formula $\texttt{Proof}_T(x, y)$ which holds if and only if x codes a derivation in T of a formula coded by y. Using Craig's trick, any theory with a computably enumerable set of axioms is deductively equivalent to one in this form, so we do not lose generality by these assumptions.

If ϕ is a natural number (supposedly coding a formula), we use $\Box_T \phi$ as shorthand for $\exists y \, \texttt{Proof}_T(y, \bar{\phi})$. We also write $\Box_T \phi(\dot{x}_0, \ldots, \dot{x}_n)$ as short for $\exists \psi \, (\psi = \phi(\bar{x}_0, \ldots, \bar{x}_n) \wedge \Box_T \psi)$. To get started on proving theorems about arithmetic, we need a minimal 'background theory'. This will use Robinson's arithmetic Q enriched with axioms for the exponential; call the resulting theory Q^+. To be precise, Q^+ is axiomatized by classical first-order logic with equality, together with the following:

- $\forall x \, (x + 0 = x)$
- $\forall x \, (x \neq 0 \leftrightarrow \exists y \, x = y + 1)$
- $\forall x \forall y \, (x + 1 = y + 1 \rightarrow x = y)$
- $\forall x \forall y \, (x + (y + 1) = (x + y) + 1)$
- $\forall x \, (x \times 0 = 0)$
- $\forall x \forall y \, (x \times (y + 1) = (x \times y) + y)$
- $2^0 = 1$
- $\forall x \, (2^{x+1} = 2^x + 2^x)$

Aside from these basic axioms, the following schemes will be useful in axiomatizing many theories of interest to us. Let Γ to denote a set of formulas. Then, the induction schema for Γ is defined by

$$\text{I}\Gamma\text{:} \quad \phi(0) \wedge \forall x (\phi(x) \rightarrow \phi(x + 1)) \rightarrow \forall x \phi(x), \quad \text{where } \phi \in \Gamma.$$

Elementary arithmetic is the first-order theory

$$\text{EA} = \text{Q}^+ + \text{I}\Delta_0,$$

and *Peano arithmetic* is the first-order theory

$$\text{PA} = \text{Q}^+ + \text{I}\Pi_\omega.$$

As usual, $\Diamond_T \phi$ is defined as $\neg \Box_T \neg \varphi$, and this will be used to interpret the RC-modality 0. Other modalities can be interpreted as stronger notions of consistency. For this purpose it is very useful to consider the provability predicates $[n]_T$, where $[n]_T$ is a natural first-order formalization of "provable from the axioms of T together with some true Π_n sentence". More precisely, let True_{Π_n} be the standard partial truth-predicate for Π_n formulas, which is itself of complexity Π_n (see [24] for information about partial truth definitions within EA). Then, we define

$$[n]_T \varphi \leftrightarrow \exists \pi \, (\text{True}_{\Pi_n}(\pi) \wedge \Box_T(\pi \to \varphi)).$$

Definition 4.1. *Given a theory T, we then define $\cdot_T \colon \mathcal{L}_{\mathsf{RC}} \to \Pi_\omega$ given recursively by*

(i) $\top_T = \top$,

(ii) $(\phi \wedge \psi)_T = \phi_T \wedge \psi_T$, *and*

(iii) $(n\phi)_T = \langle n \rangle_T \phi_T$.

The next theorem follows from the arithmetical completeness of GLP_ω proven by Ignatiev [26] together with the conservativity of GLP_ω over RC_ω (Theorem 2.8).

Theorem 4.2. *Let T be any sound, representable extension of PA. Given a formula ϕ of RC_ω, $\mathsf{RC}_\omega \vdash \phi$ if and only if $T \vdash \phi_T$.*

We remark that Japaridze first proved a variant of this result, where $[n]_T$ is defined using iterated ω-rules [27]. A similar interpretation will be discussed in Section 5.2 in the context of second-order arithmetic. However, the interpretation we have sketched using proof predicates has been used by Beklemishev to provide a consistency proof of Peano arithmetic as well as a Π^0_1 ordinal analysis. Here we will briefly sketch the consistency proof; for details, see [4].

The first step is to represent Peano arithmetic in terms of n-consistency:

Theorem 4.3. *It is provable in EA that*

$$\text{PA} \equiv \text{EA} + \{\langle n \rangle_{\text{EA}} \top : n < \omega\}.$$

This is a reformulation of a result of Kreisel and Lévy [30], although they used *primitive recursive arithmetic* in place of EA. The variant with EA is due to Beklemishev.

The consistency proof will be realized mostly within a 'finitary base theory', EA^+, which is only a bit stronger than EA. To describe it, first define the *superexponential*, denoted 2_m^n, to be the function given recursively by (i) $2_0^n = 2^n$ and (ii) $2_{m+1}^n = 2^{2_m^n}$. Thus, 2_m^1 denotes an exponential tower of m 2's. Then, we let EA^+ be the extension of EA with an axiom stating that the superexponential function is total. With this, we may enunciate Beklemishev's *reduction rule*:

Theorem 4.4. *If $\mathfrak{w} \sqsubset \omega$ is any worm, then EA^+ proves that*

$$\big(\forall \mathfrak{v} \lhd \mathfrak{w} \,(\Diamond_{EA} \mathfrak{v}_{EA})\big) \to \Diamond_{EA} \mathfrak{w}_{EA}.$$

This extends a previous result by Schmerl [35]. Meanwhile, the reader may recognize this as the premise of the *transfinite induction scheme* for worms. To be precise, if $\phi(x), x \prec y$ are arithmetical formulas, then the transfinite induction scheme for ϕ along $\prec$ is given by:

$$\mathtt{TI}_\prec(\phi) = \Big(\forall x \,((\forall y \prec x \,\phi(y)) \to \phi(x))\Big) \to \forall x \,\phi(x).$$

If Γ is a set of formulas, then $\mathtt{TI}_\prec(\Gamma)$ is the scheme $\{\mathtt{TI}_\prec(\phi) : \phi \in \Gamma\}$.

Observe that $\Diamond_{EA}\phi \in \Pi_1$ independently of ϕ; with this in mind, we obtain the following as an immediate consequence of Theorem 4.4:

Theorem 4.5. $EA^+ + \mathtt{TI}_{\lhd \upharpoonright \mathbb{W}_\omega}(\Pi_1) \vdash \Diamond_{PA} \top.$

In words, we can prove the consistency of Peano arithmetic using EA^+ and transfinite induction along $\langle \mathbb{W}_\omega, \lhd \rangle$. In fact, we use only one instance of transfinite induction for a predicate $\phi(x)$ expressing "$x \sqsubset \omega$ and $\Diamond_{EA} x_{EA}$".

Compare this to Gentzen's work [21], where he proves the consistency of Peano arithmetic with transfinite induction up to the ordinal ε_0. In the remainder of this section, we will see how finite worms and ε_0 are closely related.

4.2 The ordinal ε_0

The ordinal ε_0 is naturally defined by extending the arithmetical operations of addition, multiplication and exponentiation to the transfinite. In view of Lemma 2.4, we may have to consider not only successor ordinals, but also unions of ordinals. Fortunately, these operations are exhaustive.

Lemma 4.6. *Let ξ be an ordinal. Then, exactly one of the following occurs:*

(i) $\xi = 0$;

(ii) there exists ζ such that $\xi = \zeta + 1$, in which case we say that ξ is a successor; or

(iii) $\xi = \bigcup_{\zeta < \xi} \zeta$, in which case we say that ξ is a limit.

Thus we may recursively define operations on the ordinals if we consider these three cases. For example, ordinal addition is defined as follows:

Definition 4.7. *Given ordinals ξ, ζ, we define $\xi + \zeta$ by recursion on ζ as follows:*

1. $\xi + 0 = \xi$

2. $\xi + (\zeta + 1) = (\xi + \zeta) + 1$

3. $\xi + \zeta = \bigcup_{\vartheta < \zeta} (\xi + \vartheta)$, *for ζ a limit ordinal.*

Ordinal addition retains some, but not all, of the properties of addition on the natural numbers; it is associative, but not commutative. For example, $1 + \omega = \omega < \omega + 1$, and more generally $1 + \xi = \xi < \xi + 1$ whenever ξ is infinite. We also have a form of subtraction, but only on the left:

Lemma 4.8. *If $\zeta < \xi$ are ordinals, there exists a unique η such that $\zeta + \eta = \xi$.*

The proof follows by a standard transfinite induction on ξ. We will denote this unique η by $-\zeta + \xi$. It will be convenient to spell out some of the basic properties of left-subtraction:

Lemma 4.9. *Let α, β, γ be ordinals. Then:*

(i) $-0 + \alpha = \alpha$ and $-\alpha + \alpha = 0$;

(ii) if $\alpha \leq \beta$ and $-\alpha + \beta \leq \gamma$ then $-\alpha + (\beta + \gamma) = (-\alpha + \beta) + \gamma$;

(iii) if $\alpha + \beta \leq \gamma$ then $-\beta + (-\alpha + \gamma) = -(\alpha + \beta) + \gamma$;

(iv) if $\alpha \leq \beta \leq \alpha + \gamma$ then $-\beta + (\alpha + \gamma) = -(-\alpha + \beta) + \gamma$.

Proof. These properties are proven using the associativity of addition and the fact that $-\mu + \lambda$ is unique. We prove only (iii) as an example. Observe that

$$(\alpha + \beta) + (-\beta + (-\alpha + \gamma)) = \alpha + (\beta + (-\beta + (-\alpha + \gamma)))$$
$$= \alpha + (-\alpha + \gamma) = \gamma;$$

but $-(\alpha + \beta) + \gamma$ is the unique η such that $(\alpha + \beta) + \eta = \gamma$, so we conclude that (iii) holds. The other properties are proven similarly. □

The definition of addition we have given can be used as a template to generalize other arithmetical operations. Henceforth, if $\langle \mu_\xi \rangle_{\xi < \lambda}$ is an increasing sequence of ordinals, we will write $\lim_{\xi < \lambda} \mu_\xi$ instead of $\bigcup_{\xi < \lambda} \mu_\xi$.

Definition 4.10. *Given ordinals ξ, ζ, we define $\xi \cdot \zeta$ by recursion on ζ as follows:*

1. $\xi \cdot 0 = 0$,
2. $\xi \cdot (\zeta + 1) = \xi \cdot \zeta + \xi$, *and*
3. $\xi \cdot \zeta = \lim_{\vartheta < \zeta} \xi \cdot \vartheta$, *for ζ a limit ordinal.*

Similarly, we define ξ^ζ by:

1. $\xi^0 = 1$,
2. $\xi^{\zeta+1} = \xi^\zeta \cdot \xi$, *and*
3. $\xi^\zeta = \lim_{\vartheta < \zeta} \xi^\vartheta$, *for ζ a limit ordinal.*

Addition, multiplication and exponentiation give us our first examples of *normal functions*. These are functions that are increasing and continuous, in the following sense:

Definition 4.11. *A function $f \colon \mathsf{Ord} \to \mathsf{Ord}$ is normal if:*

1. *whenever $\xi < \zeta$, it follows that $f(\xi) < f(\zeta)$, and*
2. *whenever λ is a limit ordinal, $f(\lambda) = \lim_{\xi < \lambda} f(\xi)$.*

Normal functions are particularly nice to work with. Among other things, they have the following property, proven by an easy transfinite induction:

Lemma 4.12. *If $f \colon \mathsf{Ord} \to \mathsf{Ord}$ is normal, then for every ordinal ξ, $\xi \leq f(\xi)$.*

Of course this does not rule out the possibility that $\xi = f(\xi)$, and in fact the identity function is an example of a normal function. As we have mentioned, the elementary arithmetical functions give us further examples:

Lemma 4.13. *Let α be any ordinal. Then, the functions $f, g, h \colon \mathsf{Ord} \to \mathsf{Ord}$ given by*

1. $f(\xi) = \alpha + \xi$,
2. $g(\xi) = (1 + \alpha) \cdot \xi$,

3. $h(\xi) = (2+\alpha)^\xi$

are all normal.

Note, however, that the function $\xi \mapsto \xi + \alpha$ is not normal in general, and neither are $\xi \mapsto 0 \cdot \xi$, $\xi \mapsto 1^\xi$. But $\xi \mapsto \omega^\xi$ is normal, and this function is of particular interest, since it is the basis of the Cantor normal form representation of ordinals (similar to a base-n representation of natural numbers), where we write

$$\xi = \omega^{\alpha_n} + \ldots + \omega^{\alpha_0}$$

with the α_i's non-decreasing. Moreover, the ordinals of the form ω^β are exactly the *additively indecomposable* ordinals; that is, non-zero ordinals that cannot be written as the sum of two smaller ordinals. Let us summarize some important properties of this function:

Lemma 4.14. *Let $\xi \neq 0$ be any ordinal. Then:*

1. *There are ordinals α, β such that $\xi = \alpha + \omega^\beta$. The value of β is unique.*

2. *We can take $\alpha = 0$ if and only if, for all $\gamma, \delta < \xi$, we have that $\gamma + \delta < \xi$.*

We call this the *Cantor decomposition* of ξ. Cantor decompositions can often be used to determine whether $\xi < \zeta$:

Lemma 4.15. *Given ordinals $\xi = \alpha + \omega^\beta$ and $\zeta = \gamma + \omega^\delta$,*

1. *$\xi < \zeta$ if and only if*

    (a) $\xi \leq \gamma$, or

    (b) $\alpha < \zeta$ and $\beta < \delta$, and

2. *$\xi \leq \zeta$ if and only if*

    (a) $\xi \leq \gamma$, or

    (b) $\alpha < \zeta$ and $\beta \leq \delta$.

Note, however, that this decomposition is only useful when $\beta < \xi$ or $\gamma < \zeta$, which as we will see is not always the case. In particular, the ordinal ε_0 is the first ordinal such that $\varepsilon_0 = \omega^{\varepsilon_0}$. Roughly, it is defined by beginning with 0 and closing under the operation $\langle \alpha, \beta \rangle \mapsto \alpha + \omega^\beta$. Since many proof-theoretical ordinals are defined by taking the closure under a family of functions, it will be convenient to formalize such a closure with some generality.

The general scheme is to consider a family of ordinal functions $f_1, \ldots, f_n$, then considering the least ordinal ξ such that $f_i(\alpha_1, \ldots, \alpha_m) < \xi$ whenever each $\alpha_i < \xi$. To simplify our presentation, let us make a few preliminary observations:

1. The functions f_i may be partial or total. Since a total function is a special case of a partial function, we may in general consider $f_i \colon \mathsf{Ord}^m \dashrightarrow \mathsf{Ord}$ (where $f \colon A \dashrightarrow B$ indicates that f is a partial function).

2. We may have functions with fixed or variable arity. Given a class A, let $A^{<\omega}$ denote the class of finite sequences $\langle a_1, \ldots, a_m \rangle$ with $m < \omega$ and each $a_i \in A$. An ordinal function with fixed arity m may be regarded as a partial function on $\mathsf{Ord}^{<\omega}$, whose domain is $\mathsf{Ord}^m \subseteq \mathsf{Ord}^{<\omega}$. Thus without loss of generality, we may assume that all partial functions have variable arity.

3. We may represent the family $f_1, \ldots, f_n$ as a single function by setting
$$f(i, \alpha_1, \ldots, \alpha_m) = f_i(\alpha_1, \ldots, \alpha_m).$$
Note that this idea can also be used to represent infinite families of functions as a single function.

Thus we may restrict our discussion to ordinals closed under a single partial function of variable arity, and will do so in the next definition.

Definition 4.16. *Let $f \colon \mathsf{Ord}^{<\omega} \dashrightarrow \mathsf{Ord}$ be a partial function. Given a set of ordinals Θ, define $f[\Theta]$ to be the set of all ordinals λ such that there exist $\mu_1, \ldots, \mu_n \in \Theta$ (possibly with $n=0$) such that $\lambda = f(\mu_1, \ldots, \mu_n)$.*

For $n < \omega$, define inductively $\Theta_0^f = \Theta$ and $\Theta_{n+1}^f = \Theta_n^f \cup f[(\Theta_n^f)]$. Then, define
$$\overline{\Theta}^f = \bigcup_{n<\omega} \Theta_n^f.$$

The set $\overline{\Theta}^f$ is the *closure of Θ under f*, and indeed behaves like a standard closure operation:

Lemma 4.17. *Let $f \colon \mathsf{Ord}^{<\omega} \dashrightarrow \mathsf{Ord}$ and let Θ be any set of ordinals. Then,*

1. *$\Theta \cup f[(\overline{\Theta}^f)] \subseteq \overline{\Theta}^f$,*

2. *if $\Theta \cup \overline{\Xi}^f \subseteq \Xi$ then $\overline{\Theta}^f \subseteq \Xi$, and*

3. *for any ordinal λ, $\lambda \in (\overline{\Theta}^f) \setminus \Theta$ if and only if there are $\mu_1, \ldots, \mu_n \in \overline{\Theta}^f \setminus \{\lambda\}$ with $\lambda = f(\mu_1, \ldots, \mu_n)$.*

Proof. For the first item, note that if $\lambda_1, \ldots, \lambda_n \in \overline{\Theta}^f$ then $\lambda_1, \ldots, \lambda_n \in \Theta_m^f$ for m large enough and hence $f(\lambda_1, \ldots, \lambda_n) \in \Theta_{m+1}^f \subseteq \overline{\Theta}^f$. The second follows by showing indutively that $\Theta_n^f \subseteq \Xi$ for all n, hence $\overline{\Theta}^f \subseteq \Xi$. For the third, assume otherwise, and consider $\Xi = \overline{\Theta}^f \setminus \{\lambda\}$. One can readily verify that $\Theta \cup f[\Xi] \subseteq \Xi$, contradicting the previous item. □

With this, we are ready to define the ordinal ε_0. Below, recall that we are following the standard set-theoretic convention that $1 = \{0\}$.

Definition 4.18. *Define* $\mathrm{Cantor} \colon \mathrm{Ord}^2 \to \mathrm{Ord}$ *by* $\mathrm{Cantor}(\alpha, \beta) = \alpha + \omega^\beta$. *Then, we define*
$$\varepsilon_0 = \overline{1}^{\mathrm{Cantor}}.$$

As promised, ε_0 is the first fixed-point of the function $\xi \mapsto \omega^\xi$:

Theorem 4.19. *The set ε_0 is an ordinal and satisfies the identity $\varepsilon_0 = \omega^{\varepsilon_0}$. Moreover, if $0 < \xi < \varepsilon_0$, there are $\alpha, \beta < \xi$ such that $\xi = \alpha + \omega^\beta$.*

Proof. First we will show that if $0 < \xi \in \varepsilon_0$, then there are $\alpha, \beta < \xi$ such that $\xi = \alpha + \omega^\beta$. By Lemma 4.17.3, there are $\alpha, \beta \in \varepsilon_0$ with $\alpha, \beta \neq \xi$ and such that $\xi = \alpha + \omega^\beta$. Since $\omega^\beta > 0$ it follows that $\alpha < \xi$, and since $\beta \leq \omega^\beta \leq \xi$ it follows that $\beta \leq \xi$; but $\beta \neq \xi$, so $\beta < \xi$.

Now, since every element of ε_0 is an ordinal, in view of Lemma 2.3, in order to show that ε_0 is also an ordinal it suffices to show that if $\xi < \zeta \in \varepsilon_0$, then $\xi \in \varepsilon_0$. We proceed by induction on ζ with a secondary induction on ξ. Write $\zeta = \alpha + \omega^\beta$ and $\xi = \gamma + \omega^\delta$ with $\alpha, \beta \in \varepsilon_0 \cap \zeta$. Since $\xi < \zeta$, by Lemma 4.15, we have that either $\xi \leq \alpha$ or $\gamma < \zeta$ and $\delta < \beta$. In the first case, our induction hypothesis applied to $\alpha < \zeta$ gives us $\xi \in \varepsilon_0$, in the second the secondary induction hypothesis on $\gamma < \xi$ gives us $\gamma \in \varepsilon_0$ and the induction hypothesis on $\beta < \zeta$ gives us $\delta \in \varepsilon_0$, hence $\xi = \alpha + \omega^\beta \in \varepsilon_0$. □

4.3 Order-types of finite worms

Our work on elementary ordinal operations and the ordinal ε_0 will suffice to compute the order-types of 'finite' worms, i.e., worms where every entry is finite. In order to give a calculus for these order-types, we will need to consider, in addition to concatenation, 'promotion' ($\uparrow$) and 'demotion' ($\downarrow$) operations on worms. Below, let us write $\mathcal{L}_{\geq \lambda}$ for the sublanguage of $\mathcal{L}_{\mathsf{RC}}$ which only contains modalities $\xi \geq \lambda$.

Definition 4.20. *Let $\phi \in \mathcal{L}_{\mathsf{RC}}$ and λ be an ordinal. We define $\lambda \uparrow \phi$ to be the result of replacing every ordinal ξ appearing in ϕ by $\lambda + \xi$. Formally, $\lambda \uparrow \top = \top$, $\lambda \uparrow (\phi \wedge \psi) = (\lambda \uparrow \psi) \wedge (\lambda \uparrow \psi)$, and $\lambda \uparrow \langle\mu\rangle\phi = \langle\lambda + \mu\rangle(\lambda \uparrow \phi)$.*

If $\phi \in \mathcal{L}_{\geq \lambda}$, we similarly define $\lambda \downarrow \phi$ by replacing every occurrence of ξ by $-\lambda + \xi$.

The relationship between ↑ and ↓ is analogous to that between ordinal addition and subtraction. The following are all straightforward consequences of Lemma 4.9 and we omit the proofs.

Lemma 4.21. *Let α, β be ordinals and $\phi \in \mathcal{L}_{\mathsf{RC}}$. Then,*

(i) $0 \uparrow \phi = \phi$;

(ii) $\alpha \uparrow (\beta \uparrow \phi) = (\alpha + \beta) \uparrow \phi$;

(iii) if $\phi \in \mathcal{L}_{\geq \beta + \alpha}$ then $\alpha \downarrow (\beta \downarrow \phi) = (\beta + \alpha) \downarrow \phi$;

(iv) if $\alpha \leq \beta$ then $\alpha \downarrow (\beta \uparrow \phi) = (-\alpha + \beta) \uparrow \phi$, *and*

(v) if $\alpha \leq \beta$ and $\phi \in \mathcal{L}_{\geq -\alpha + \beta}$ then $\alpha \uparrow \phi \in \mathcal{L}_{\geq \beta}$ and

$$\beta \downarrow (\alpha \uparrow \phi) = (-\alpha + \beta) \downarrow \phi.$$

The operation $\phi \mapsto \lambda \uparrow \phi$ is particularly interesting in that it provides a sort of self-embedding of RC:

Lemma 4.22. *Let α, β be ordinals and $\phi, \psi \in \mathcal{L}_{\mathsf{RC}}$. If $\phi \Rightarrow \psi$ is derivable in RC, then so is $(\lambda \uparrow \phi) \Rightarrow (\lambda \uparrow \psi)$.*

Proof. By induction on the length of a derivation of $\phi \Rightarrow \psi$; intuitively, one replaces every formula θ appearing in the derivation by $\lambda \uparrow \theta$. The details are straightforward and left to the reader. □

The promotion operator gives us an order-preserving transformation on the class of worms:

Lemma 4.23. *Given a worm $\mathfrak{w} \in \mathbb{W}_{\geq \mu}$ and an ordinal λ, the following are equivalent:*

(i) $\mathfrak{w} \lhd_\mu \mathfrak{v}$;

(ii) $\lambda \uparrow \mathfrak{w} \lhd_\mu \lambda \uparrow \mathfrak{v}$, *and*

(iii) $\lambda \uparrow \mathfrak{w} \lhd_\lambda \lambda \uparrow \mathfrak{v}$.

Proof. The equivalence between (ii) and (iii) is immediate from Lemma 3.16, so we focus on the equivalence between (i) and (iii).

If $\mathfrak{w} \lhd_\mu \mathfrak{v}$, then $\mathfrak{w} \lhd \mathfrak{v}$, so RC derives $\mathfrak{v} \Rightarrow 0\mathfrak{w}$. By Lemma 4.22, RC also derives $(\lambda \uparrow \mathfrak{v}) \Rightarrow \lambda(\lambda \uparrow \mathfrak{w})$, that is, $(\lambda \uparrow \mathfrak{w}) \lhd_\lambda (\lambda \uparrow \mathfrak{v})$.

Conversely, if $\lambda \uparrow \mathfrak{w} \lhd_\lambda \lambda \uparrow \mathfrak{v}$, assume towards a contradiction that $\mathfrak{w} \not\lhd_\mu \mathfrak{v}$, so that by Lemma 3.14, $\mathfrak{v} \unlhd_\mu \mathfrak{w}$. Again by Lemma 4.22, $(\lambda \uparrow \mathfrak{v}) \unlhd_\lambda (\lambda \uparrow \mathfrak{w})$, so $(\lambda \uparrow \mathfrak{v}) \unlhd_\lambda (\lambda \uparrow \mathfrak{w}) \lhd_\lambda (\lambda \uparrow \mathfrak{v})$, contradicting irreflexivity. $\square$

Lemma 4.23 is useful for comparing worms; if we wish to settle whether $\lambda \uparrow \mathfrak{w} \lhd \lambda \uparrow \mathfrak{v}$, then it suffices to check whether $\mathfrak{w} \lhd \mathfrak{v}$. More generally, we obtain the following variant of Theorem 3.17. Below, recall that we write h, b instead of h_0, b_0.

Lemma 4.24. *Given worms* $\mathfrak{w}, \mathfrak{v} \neq \top$,

1. $\mathfrak{w} \lhd \mathfrak{v}$ *if and only if*

    (a) $\mathfrak{w} \unlhd b(\mathfrak{v})$, *or*

    (b) $b(\mathfrak{w}) \lhd \mathfrak{v}$ *and* $1 \downarrow h(\mathfrak{w}) \lhd 1 \downarrow h(\mathfrak{v})$;

2. $\mathfrak{w} \unlhd \mathfrak{v}$ *if and only if*

    (a) $\mathfrak{w} \unlhd b(\mathfrak{v})$, *or*

    (b) $b(\mathfrak{w}) \lhd \mathfrak{v}$ *and* $1 \downarrow h(\mathfrak{w}) \unlhd 1 \downarrow h(\mathfrak{v})$.

If all entries of $\mathfrak{v} \neq \top$ are natural numbers, $1 \downarrow h(\mathfrak{w})$ will be 'smaller' than $\mathfrak{w}$. To be precise, it will have a smaller 1-*norm*, defined as follows:

Definition 4.25. *We define* $\|\cdot\|_1 : \mathbb{W}_\omega \to \omega$ *recursively by*

1. $\|\top\|_1 = 0$;

2. *if* $\mathfrak{w} \neq \top$ *and* $\min \mathfrak{w} = 0$,
$$\|\mathfrak{w}\|_1 = \|h(\mathfrak{w})\|_1 + \|b(\mathfrak{w})\|_1 + 1;$$

3. *if* $\mathfrak{w} \neq \top$ *and* $\min \mathfrak{w} > 0$,
$$\|\mathfrak{w}\|_1 = \|1 \downarrow \mathfrak{w}\|_1 + 1.$$

Recall that we use h and b as shorthands for h_0, b_0.

Lemma 4.26. *For every worm* $\mathfrak{w} \sqsubset \omega$ *with* $\mathfrak{w} \neq \top$,

1. $\|b(\mathfrak{w})\|_1 < \|\mathfrak{w}\|_1$, *and*

2. $\|1 \downarrow h(\mathfrak{w})\|_1 < \|\mathfrak{w}\|_1$.

Proof. For the first claim, note that if 0 appears in $\mathfrak{w}$ then $\|b(\mathfrak{w})\|_1 + 1 \leq \|\mathfrak{w}\|_1$. If 0 does not appear, $\|b(\mathfrak{w})\|_1 = 0 < \|\mathfrak{w}\|_1$.

For the second, if $h(\mathfrak{w}) = \top$ then once again $\|1 \downarrow h(\mathfrak{w})\|_1 = 0 < \|\mathfrak{w}\|_1$, and if $h(\mathfrak{w}) \neq \top$ then
$$\|1 \downarrow h(\mathfrak{w})\|_1 + 1 = \|h(\mathfrak{w})\|_1 \leq \|\mathfrak{w}\|_1,$$
so $\|1 \downarrow h(\mathfrak{w})\|_1 < \|h(\mathfrak{w})\|_1 \leq \|\mathfrak{w}\|_1$. □

We remark that there are other possible ways to define $\|\cdot\|_1$ that would also satisfy Lemma 4.26; for example, we can define $\|\mathfrak{w}\|_1' = \#\mathfrak{w} + \max \mathfrak{w}$, or

$$\|m_1 \ldots m_n \top\|_1'' = \sum_{i=1}^{n}(m_i + 1).$$

However, these definitions do not generalize well to worms with transfinite entries, which will be the focus of Section 5. On the other hand, our norm $\|\cdot\|_1$ can be applied to transfinite worms with only a minor modification.

Our goal now is to give an explicit calculus for computing $o(\mathfrak{w})$ if $\mathfrak{w} \sqsubset \omega$. In view of Lemma 3.24, it is sufficient to propose a candidate function for o and show that it has the required properties. Now, if we compare Lemma 4.24 with Lemma 4.15, we observe that the clauses for checking whether $\mathfrak{w} \triangleleft \mathfrak{v}$ in terms of

$$b(\mathfrak{w}), 1 \downarrow h(\mathfrak{w}), b(\mathfrak{v}), 1 \downarrow h(\mathfrak{v})$$

are analogous to the clauses for checking whether $\alpha + \omega^\beta < \gamma + \omega^\delta$ in terms of $\alpha, \beta, \gamma, \delta$, respectively. This suggests that

$$o(\mathfrak{w}) = ob(\mathfrak{w}) + \omega^{o(1 \downarrow h(\mathfrak{w}))}, \tag{1}$$

and we will use this idea to define our 'candidate function'.

Definition 4.27. *Let $\mathfrak{v}, \mathfrak{w}$ be worms and α an ordinal.*
Then, define a map $\acute{o} \colon \mathbb{W}_\omega \to \mathsf{Ord}$ by

1. *$\acute{o}(\top) = 0$, and*

2. *if $\mathfrak{w} \neq \top$ then $\acute{o}(\mathfrak{w}) = \acute{o}(b(\mathfrak{w})) + \omega^{\acute{o}(1 \downarrow h(\mathfrak{w}))}$.*

First, let us check that $\acute{o}$ is indeed a function:

Lemma 4.28. *The map $\acute{o}$ is well-defined.*

Proof. This follows from an easy induction on $\|\mathfrak{w}\|_1$ using Lemma 4.26. □

It remains to check that ó is strictly increasing and initial. Let us begin with the former:

Lemma 4.29. *The map* $ó\colon \mathbb{W}_\omega \to \mathsf{Ord}$ *is strictly increasing.*

Proof. We will prove by induction on $\|\mathfrak{w}\|_1 + \|\mathfrak{v}\|_1$ that $\mathfrak{w} \lhd \mathfrak{v}$ if and only if $ó(\mathfrak{w}) < ó(\mathfrak{v})$. Note that $\mathfrak{w} \lhd \top$ is never true, nor is $\xi < ó(\top) = 0$, so we may assume that $\mathfrak{v} \neq \top$. Then, if $\mathfrak{w} = \top$ it follows that $ó(\top) = 0$, so both sides are true. Hence we may also assume that $\mathfrak{w} \neq \top$.

By Lemma 4.24, $\mathfrak{w} \lhd \mathfrak{v}$ if and only if either $\mathfrak{w} \trianglelefteq b(\mathfrak{v})$ or $b(\mathfrak{w}) \lhd \mathfrak{v}$ and $1 \downarrow h(\mathfrak{w}) \lhd 1 \downarrow h(\mathfrak{v})$. Observe that, by the induction hypothesis,

1. $\mathfrak{w} \trianglelefteq b(\mathfrak{v})$ if and only if $ó(\mathfrak{w}) \leq ób(\mathfrak{v})$, since

$$\|\mathfrak{w}\|_1 + \|b(\mathfrak{v})\|_1 < \|\mathfrak{w}\|_1 + \|\mathfrak{v}\|_1\,;$$

2. $b(\mathfrak{w}) \lhd \mathfrak{v}$ if and only if $ób(\mathfrak{w}) < ó(\mathfrak{v})$, since

$$\|b(\mathfrak{w})\|_1 + \|\mathfrak{v}\|_1 < \|\mathfrak{w}\|_1 + \|\mathfrak{v}\|_1\,,$$

and

3. $1 \downarrow h(\mathfrak{w}) \lhd 1 \downarrow h(\mathfrak{v})$ if and only if $ó(1 \downarrow h(\mathfrak{w})) < ó(1 \downarrow h(\mathfrak{v}))$, since

$$\|1 \downarrow h(\mathfrak{w})\|_1 + \|1 \downarrow h(\mathfrak{v})\|_1 < \|\mathfrak{w}\|_1 + \|\mathfrak{v}\|_1\,.$$

This implies that $\mathfrak{w} \lhd \mathfrak{v}$ if and only if either $ó(\mathfrak{w}) \leq ób(\mathfrak{v})$, or $ób(\mathfrak{w}) < ó(\mathfrak{v})$ and $ó(1 \downarrow h(\mathfrak{w})) < ó(1 \downarrow h(\mathfrak{v}))$. But by Lemma 4.15.1, the latter is equivalent to

$$ób(\mathfrak{w}) + \omega^{ó(1\downarrow h(\mathfrak{w}))} < ób(\mathfrak{v}) + \omega^{ó(1\downarrow h(\mathfrak{v}))},$$

i.e., $ó(\mathfrak{w}) < ó(\mathfrak{v})$. □

It remains to check that the range of ó is ε_0. We will use the following lemma:

Lemma 4.30. *For all* $m < \omega$, $ó(m\top) < \varepsilon_0$.

Proof. By induction on n; if $n = 0$ then $ó(0\top) = 0 + \omega^0 = 1 < \varepsilon_0$. Otherwise, by induction hypothesis $ó(n\top) < \varepsilon_0$, so

$$ó(\langle n+1\rangle\top) = \omega^{ó(n\top)} < \varepsilon_0,$$

as claimed. □

Lemma 4.31. *An ordinal ξ lies in the range of $\acute{o}$ if and only if $\xi < \varepsilon_0$.*

Proof. First, assume that $\xi < \varepsilon_0$; we must find $\mathfrak{w} \sqsubset \omega$ such that $\xi = \acute{o}(\mathfrak{w})$. Proceed by induction on ξ. If $\xi = 0$, then $\xi = \acute{o}(\top)$. Otherwise, by Theorem 4.19, $\xi = \alpha + \omega^\beta$ for some $\alpha, \beta < \xi$. By the induction hypothesis, there are worms $\mathfrak{u}, \mathfrak{v}$ such that $\alpha = \acute{o}(\mathfrak{u})$ and $\beta = \acute{o}(\mathfrak{v})$, thus

$$\acute{o}((1 \uparrow \mathfrak{v}) \, 0 \, \mathfrak{u}) = \acute{o}(\mathfrak{u}) + \omega^{\acute{o}(\mathfrak{v})} = \alpha + \omega^\beta = \xi.$$

Next we check that if $\mathfrak{w} \sqsubset \omega$, then $\acute{o}(\mathfrak{w}) < \varepsilon_0$. Fix $M > \max \mathfrak{w}$; then, by Corollary 3.19.3, $\mathfrak{w} \lhd M\top$, so that $\acute{o}(\mathfrak{w}) \lhd \acute{o}(M\top)$. But by Lemma 4.30, $\acute{o}(M\top) < \varepsilon_0$, as claimed. □

We now have all the necessary ingredients to show that $\acute{o} = o$.

Lemma 4.32. *For all $\mathfrak{w} \sqsubset \omega$, $o(\mathfrak{w}) = \acute{o}(\mathfrak{w})$.*

Proof. By Lemma 4.28, $\acute{o}$ is well-defined on $\mathbb{W}_\omega$, and by Lemmas 4.29 and 4.31, it is strictly increasing and initial. By Lemma 3.24, $o = \acute{o}$ on $\mathbb{W}_\omega$. □

Let us conclude this section by summarizing our main results:

Theorem 4.33. *The map $o \colon \mathbb{W}_\omega \to \varepsilon_0$ is surjective and satisfies*

1. $o(\top) = 0$, *and*

2. $o((1 \uparrow \mathfrak{v}) \, 0 \, \mathfrak{w}) = o(\mathfrak{w}) + \omega^{o(\mathfrak{v})}$.

Proof. Immediate from Lemma 4.32 and the definition of $\acute{o}$. □

5 Transfinite worms

We have now seen that finite worms give a notation for ε_0, the proof-theoretic ordinal of Peano arithmetic. However, stronger theories, including many important theories of reverse mathematics, have much larger proof-theoretic strength, suggesting that RC_ω is not suitable for their Π_1^0 ordinal analysis. Fortunately, Theorem 3.20 is valid even when worms have arbitrary ordinal entries. In this section, we will extend Theorem 4.33 to all of $\mathbb{W}$.

5.1 Subsystems of second-order arithmetic

Let us begin by discussing proof-theoretic interpretations of RC_Λ with $\Lambda > \omega$. It will be convenient to pass to the language Π^1_ω of second-order arithmetic. This language extends that of first-order arithmetic with new variables $X, Y, Z, \ldots$ denoting sets of natural numbers, along with new atomic formulas $t \in X$ and second-order quantifiers $\forall X, \exists X$. As is standard, we may define $X \subseteq Y$ by $\forall x(x \in X \to x \in Y)$, and $X = Y$ by $X \subseteq Y \wedge Y \subseteq X$.

When working in a second-order context, we write Π^0_n instead of Π_n (note that these formulas could contain second-order parameters, but no quantifiers over sets). The classes Σ^1_n, Π^1_n are defined analogously to their first-order counterparts, but using alternating second-order quantifiers and setting $\Sigma^1_0 = \Pi^1_0 = \Delta^1_0 = \Pi^0_\omega$. It is well-known that every second-order formula is equivalent to another in one of the above forms.

When axiomatizing second-order arithmetic, the focus passes from induction to *comprehension;* that is, axioms stating the existence of sets whole elements satisfy a prescribed property. Some important axioms and schemes are:

Γ-**CA:** $\exists X \forall x \, (x \in X \leftrightarrow \phi(x))$, where $\phi \in \Gamma$ and X is not free in ϕ;

Δ^0_1-**CA:** $\forall x(\pi(w) \leftrightarrow \sigma(x)) \to \exists X \forall x \, (x \in X \leftrightarrow \sigma(x))$, where $\sigma \in \Sigma^0_1$, $\pi \in \Pi^0_1$, and X is not free in σ or π;

Ind: $0 \in X \wedge \forall x \, (x \in X \to x + 1 \in X) \to \forall x \, (x \in X)$.

We mention one further axiom that requires a more elaborate setup. We may represent well-orders in second-order arithmetic as pairs of sets $\Lambda = \langle |\Lambda|, \leq_\Lambda \rangle$, and define

$$\mathtt{WO}(\Lambda) = \mathtt{linear}(\Lambda) \wedge \forall X \subseteq |\Lambda| \, (\exists x \in X \to \exists y \in X \forall z \in X \, y \leq_\Lambda z),$$

where $\mathtt{linear}(\Lambda)$ is a formula expressing that Λ is a linear order.

Given a set X whose elements we will regard as ordered pairs $\langle \lambda, n \rangle$, let $X_{<_\Lambda \lambda}$ be the set of all $\langle \mu, n \rangle$ with $\mu <_\Lambda \lambda$. With this, we define the *transfinite recursion* scheme by

$$\mathtt{TR}_\phi(X, \Lambda) = \forall \lambda \in |\Lambda| \, \forall n \, (n \in X \leftrightarrow \phi(n, X_{<_\Lambda \lambda})).$$

Intuitively, $\mathtt{TR}_\phi(X, \Lambda)$ states that X is made up of "layers" indexed by elements of Λ, and the elements of the λth layer are those natural numbers n satisfying $\phi(n, X_{<_\Lambda \lambda})$, where $X_{<_\Lambda \lambda}$ is the union of all previous layers. If Γ is a set of formulas, we denote the Γ-*transfinite recursion* scheme by

$$\Gamma\text{-}\mathtt{TR} = \big\{ \forall \Lambda (\mathtt{WO}(\Lambda) \to \exists X \, \mathtt{TR}_\phi(X, \Lambda)) : \phi \in \Gamma \big\}.$$

Now we are ready to define some important theories:

$$\begin{aligned} \text{ECA}_0 &: \quad Q^+ + \text{Ind} + \Delta_0^0\text{-CA}; \\ \text{RCA}_0^* &: \quad Q^+ + \text{Ind} + \Delta_1^0\text{-CA}; \\ \text{RCA}_0 &: \quad Q^+ + I\Sigma_1^0 + \Delta_1^0\text{-CA}; \\ \text{ACA}_0 &: \quad Q^+ + \text{Ind} + \Sigma_1^0\text{-CA}; \\ \text{ATR}_0 &: \quad Q^+ + \text{Ind} + \Pi_\omega^0\text{-TR}; \\ \Pi_1^1\text{-CA}_0 &: \quad Q^+ + \text{Ind} + \Pi_1^1\text{-CA}. \end{aligned}$$

These are listed from weakest to strongest. The theories RCA_0, ACA_0 ATR_0 and $\Pi_1^1\text{-CA}_0$, together with the theory of *weak König's lemma*, WKL_0, are the 'Big Five' theories of reverse mathematics, where RCA_0 functions as a 'constructive base theory', and the stronger four theories are all equivalent to many well-known theorems in mathematical analysis. For a detailed treatment of these and other subsystems of second-order arithmetic, see [36].

ECA_0 (the theory of *elementary comprehension*) is the second-order analogue of elementary arithmetic, and is a bit weaker than the more standard RCA_0^*. Meanwhile, *arithmetical comprehension* (ACA_0) is essentially the second-order version of PA, and has the same proof-theoretic ordinal, ε_0. Thus the next milestone in the Π_1^0 ordinal analysis program is naturally ATR_0, the theory of *arithmetical transfinite recursion*. Appropriately, the constructions we will use to interpret the modalities $\langle \lambda \rangle$ for countable $\lambda > \omega$ may be carried out within ATR_0.

5.2 Iterated ω-rules

If we wish to interpret $[\lambda]_T \phi$ for transfinite λ, we need to consider a notion of provability that naturally extends beyond ω. One such notion, which is well-studied in proof theory (see, e.g., [33]), considers infinitary derivations with the ω-*rule*. Intuitively, this rule has the form

$$\frac{\phi(\bar{0}) \quad \phi(\bar{1}) \quad \phi(\bar{2}) \quad \phi(\bar{3}) \quad \phi(\bar{4}) \quad \ldots}{\forall x \, \phi(x)}$$

The parameter λ in $[\lambda]_T \phi$ denotes the nesting depth of ω-rules that may be used for proving ϕ. The notion of λ-provability is defined as follows:

Definition 5.1. *Let T be a theory of second-order arithmetic and $\phi \in \Pi_\omega^1$. For an ordinal λ, we define $[\lambda]_T \phi$ recursively if either*

(i) $\Box_T \phi$, *or*

(ii) *there are an ordinal $\mu < \lambda$ and a formula $\psi(x)$ such that*

(a) for all $n < \omega$, $[\mu]_T \psi(\bar{n})$, and

(b) $\Box_T(\forall x \psi(x) \to \phi)$.

This notion can be formalized by representing ω-proofs as infinite trees, as presented by Arai [2] and Girard [22]. Here we will instead use the formalization of Joosten and I [19]. We use a set P as an *iterated provability class*, whose elements are codes of pairs $\langle \lambda, \varphi \rangle$, with λ a code for an ordinal and φ a code for a formula. The idea is that we want P to be a set of pairs $\langle \lambda, \varphi \rangle$ satisfying Definition 5.1 if we set $[\lambda]_T \varphi \leftrightarrow \langle \lambda, \varphi \rangle \in P$. Thus we may write $[\lambda]_P \varphi$ instead of $\langle \lambda, \varphi \rangle \in P$.

Definition 5.2. *Fix a well-order Λ on $\mathbb{N}$. Say that a set P of natural numbers is an* iterated provability class *for Λ if it satisfies the expression*

$$[\lambda]_P \varphi \leftrightarrow \Big(\Box_T \varphi \vee \exists \psi \exists \xi <_\Lambda \lambda \, \big(\forall n \, [\xi]_P \psi(\dot{n}) \wedge \Box_T(\forall x \psi(x) \to \varphi)\big)\Big).$$

Let $\mathrm{IPC}_T^\Lambda(P)$ be a Π^0_ω formula stating that P is an iterated provabiltiy class for Λ. Then, define

$$[\lambda]_T^\Lambda \phi := \forall P \, (\mathrm{IPC}_T^\Lambda(P) \to [\lambda]_P \phi).$$

Note that $[\lambda]_T^\Lambda$ is a Π^1_1 formula. Alternately, one could define $[\lambda]_T^\Lambda$ as a Σ^1_1 formula, but the two definitions are equivalent due to the following.

Lemma 5.3.

1. *It is provable in ACA_0 that if Λ is a countable well-order and P, Q are both iterated provability classes for Λ, then $P = Q$.*

2. *It is provable in ATR_0 that if Λ is a countable well-order, then there exists an iterated provability class for Λ.*

The first claim is proven by considering two IPC's P, Q and showing by transfinite induction on λ that $[\lambda]_P \phi \leftrightarrow [\lambda]_Q \phi$; this induction is readily available in ACA_0 since the expression $[\lambda]_P \phi$ is arithmetical. For the second, we simply observe that the construction of an IPC is a special case of arithmetical transfinite recursion. See [19] for more details.

If we fix a computable well-order Λ and a theory T in the language of second-order arithmetic, we can readily define $\cdot_T^\Lambda \colon \mathcal{L}_\Lambda \to \Pi^1_\omega$ as in Definition 4.1, but setting $(\lambda \phi)_T^\Lambda = \langle \bar{\lambda} \rangle_T^\Lambda \phi_T$ We then obtain the following:

Theorem 5.4. *Let Λ be a computable well-order and T be a theory extending ACA_0 such that it is provable in T that Λ is well-ordered, and that there is a set P satisfying $\mathrm{IPC}_T^\Lambda(P)$.*

Then, for any sequent $\phi \Rightarrow \psi$ of $\mathcal{L}_\Lambda$, $\mathrm{RC} \vdash \phi \Rightarrow \psi$ if and only if $T \vdash \phi_T^\Lambda \to \psi_T^\Lambda$.

Proof. This is proven in [19] with GLP_Λ in place of RC_Λ, and this version is obtained by observing that GLP_Λ is conservative over RC_Λ by Theorem 2.8. □

The computability condition in Λ is included due to the fact that in the proof of Theorem 5.4, we need to be able to prove properties about Λ within T; for example, we need for
$$\forall x \, \forall y \, (x \leq_\Lambda y \to \Box_T(\dot x \leq_\Lambda \dot y))$$
to hold. However, we can drop this condition if we allow an *oracle* for Λ; or, more generally, for any set of natural numbers. To do this, we add a set-constant O to the language of second-order arithmetic in order to 'feed' information about any set of numbers into T.

To be precise, given a theory T and $A \subseteq \mathbb{N}$, define $T|A$ to be the theory whose rules and axioms are those of T together with all instances of $\bar n \in O$ for $n \in X$, and all instances of $\bar n \not\in O$ for $n \not\in X$. Then, for any formula ϕ, we define
$$[\lambda|X]^\Lambda_T \phi = [\lambda]^\Lambda_{T|X} \phi.$$

Its dual, $\langle \lambda|X \rangle^\Lambda_T \phi$, is defined in the usual way. With this, we obtain an analogue of Theorem 4.3 for ATR_0, proven by Cordón-Franco, Joosten, Lara-Martín and myself in [13]:

Theorem 5.5. $\mathrm{ATR}_0 \equiv \mathrm{ECA}_0 + \forall \Lambda \, \forall X \, \langle \lambda|X \rangle^\Lambda_T \top.$

This result may well be the first step in a consistency proof of ATR_0 in the style of Theorem 4.5. Moreover, the proof-theoretic strength of ATR_0 is measured by the Feferman-Schütte ordinal, Γ_0. In the rest of this section, we will see how the worm ordering relates to this ordinal.

5.3 Ordering transfinite worms

Let us extend our calculus for computing o to worms that may contain transfinite entries. In Section 4, we used the operations b, h and $1 \downarrow$ to simplify worms and compute their order-types. However, this will not suffice for transfintie worms. For example, if $\mathfrak{w} = \omega 0 \omega \top$, we have that $h(\mathfrak{w}) = \omega \top$ while $b(\mathfrak{w}) = \omega \top$, both of which are shorter than ω. However,
$$1 \downarrow (\omega \top) = \langle -1 + \omega \rangle \top = \omega \top;$$
thus, demoting by 1 will not get us anywhere. Instead, we could demote by ω, and obtain $\omega \downarrow (\omega \top) = 0 \top$, which is indeed 'simpler'. As we will see, this is the appropriate way to decompose infinite worms:

Lemma 5.6. *Given a worm* $\mathfrak{w} \not= \top$, *there exist unique* $\mu < \Lambda$ *and worms* $\mathfrak{w}_1, \mathfrak{w}_0$ *such that either* $\mathfrak{w}_1 = \top$ *or* $0 < \min \mathfrak{w}_1$ *and*

$$\mathfrak{w} = \mu \uparrow (\mathfrak{w}_1 \, 0 \, \mathfrak{w}_0).$$

Proof. Take $\mu = \min \mathfrak{w}$, $\mathfrak{w}_1 = h(\mu \downarrow \mathfrak{w})$ and $\mathfrak{w}_0 = b(\mu \downarrow \mathfrak{w})$; evidently these are the only possible values that satisfy the desired equation. □

With this we may define the norm of a worm $\mathfrak{w}$, which roughly corresponds to the number of operations of 0-concatenation and μ-promotion needed to construct $\mathfrak{w}$.

Definition 5.7. *For* $\mathfrak{w} \sqsubset \mathsf{Ord}$ *we define* $\|\mathfrak{w}\|$ *inductively by*

1. $\|\top\| = 0$;

2. *if* $\mathfrak{w} \not= \top$ *and* $\min \mathfrak{w} = 0$, *set*

$$\|\mathfrak{w}\| = \|h(\mathfrak{w})\| + \|b(\mathfrak{w})\| + 1;$$

3. *otherwise, let* $\mu = \min \mathfrak{w} > 0$, *and set*

$$\|\mathfrak{w}\| = \|\mu \downarrow \mathfrak{w}\| + 1.$$

The following is obvious from Definition 5.7 and Lemma 5.6:

Lemma 5.8. *For every worm* $\mathfrak{w}$, $\|\mathfrak{w}\| \in \mathbb{N}$ *is well-defined. Moreover, if* $\mathfrak{w} = \alpha \uparrow (\mathfrak{w}_1 \, 0 \, \mathfrak{w}_0)$ *with* $0 < \min \mathfrak{w}_1$, *then* $\|\mathfrak{w}_1\|, \|\mathfrak{w}_0\| < \|\mathfrak{w}\|$.

Thus we may try to compute $o(\mathfrak{w})$ by recursion on $\|\mathfrak{w}\|$. Assuming that the identity $o(\mathfrak{w}) = ob(\mathfrak{w}) + \omega^{o(1 \downarrow h(\mathfrak{w}))}$ remains valid for transfinite worms, we only have to find a way to compute $o(\mu \uparrow \mathfrak{w})$ in terms of $o(\mathfrak{w})$. Fortunately, the map $o(\mathfrak{w}) \mapsto o(\mu \uparrow \mathfrak{w})$ is well-defined; let us denote it by σ^μ.

Lemma 5.9. *There exists a unique family of functions* $\vec{\sigma} = \langle \sigma^\xi \rangle_{\xi \in \mathsf{Ord}}$ *such that* $\sigma^\xi \colon \mathsf{Ord} \to \mathsf{Ord}$ *and, for every ordinal* ξ *and every worm* $\mathfrak{w}$, $\sigma^\xi o(\mathfrak{w}) = o(\xi \uparrow \mathfrak{w})$.

Proof. Given ordinals ξ, ζ, we need to see that there exists a unique ordinal ϑ such that $\vartheta = o(\xi \uparrow \mathfrak{w})$ whenever $\zeta = o(\mathfrak{w})$.

First observe that, by Corollary 3.27, there is some worm $\mathfrak{w}_*$ such that $\zeta = o(\mathfrak{w}_*)$. Since by Theorem 3.20, the class of worms is well-ordered, $o(\xi \uparrow \mathfrak{w}_*)$ is well-defined. It remains to check that if $\mathfrak{w}$ is an arbitrary worm such that $o(\mathfrak{w}) = \xi$, then also $o(\xi \uparrow \mathfrak{w}) = o(\xi \uparrow \mathfrak{w}_*)$. But if $o(\mathfrak{w}) = o(\mathfrak{w}_*)$, by Lemma 3.25 we have that $\mathfrak{w} \equiv \mathfrak{w}_*$, and thus by Lemma 4.22, $\xi \uparrow \mathfrak{w} \equiv \xi \uparrow \mathfrak{w}_*$. The latter implies that $o(\xi \uparrow \mathfrak{w}) = o(\xi \uparrow \mathfrak{w}_*)$, as needed. □

Lemma 5.10. *The family of functions $\vec{\sigma}$ has the following properties:*

1. *σ^α is strictly increasing for all α;*

2. *$\sigma^0 \xi = \xi$, and*

3. *$\sigma^{\alpha+\beta} = \sigma^\alpha \sigma^\beta$.*

Proof. For item 1, suppose that $\xi < \zeta$. If $\xi = o(\mathfrak{w})$ and $\zeta = o(\mathfrak{v})$, then by Lemma 3.23, $\mathfrak{w} \triangleleft \mathfrak{v}$, so that by Lemma 4.23, $\alpha \uparrow \mathfrak{w} \triangleleft \alpha \uparrow \mathfrak{v}$ and thus $o(\alpha \uparrow \mathfrak{w}) < o(\alpha \uparrow \mathfrak{v})$; a similar argument shows that if $o(\mathfrak{w}) \trianglelefteq o(\mathfrak{v})$, then $o(\alpha \uparrow \mathfrak{w}) \trianglelefteq o(\alpha \uparrow \mathfrak{v})$. Item 2 follows from the fact that $0 \uparrow \mathfrak{w} = \mathfrak{w}$ for all $\mathfrak{w}$, so if $\zeta = o(\mathfrak{w})$ we have that $\sigma^0 \zeta = o(0 \uparrow \mathfrak{w}) = \zeta$.

Item 3 is immediate from Lemma 4.21.(ii), since if $o(\mathfrak{w}) = \zeta$ then $o(\beta \uparrow \mathfrak{w}) = \sigma^\beta(\zeta)$, which means that

$$o(\alpha \uparrow (\beta \uparrow \mathfrak{w})) = \sigma^\alpha \sigma^\beta(\zeta).$$

But, on the other hand, $\alpha \uparrow (\beta \uparrow \mathfrak{w}) = (\alpha + \beta) \uparrow \mathfrak{w}$, and

$$o((\alpha + \beta) \uparrow \mathfrak{w}) = \sigma^{\alpha+\beta} \zeta,$$

and we conclude that $\sigma^{\alpha+\beta} \zeta = \sigma^\alpha \sigma^\beta \zeta$. □

Observe also that if $\zeta < \varepsilon_0$, then by Theorem 4.33, there is $\mathfrak{w} \sqsubset \omega$ such that $\zeta = o(\mathfrak{w})$, and hence by Theorem 4.33, $\sigma^1 \zeta = o(1 \uparrow \mathfrak{w}) = -1 + \omega^{o(\mathfrak{w})}$ (where we subtract 1 to account for the case $\mathfrak{w} = \top$). Thus for $\zeta < \varepsilon_0$, $\sigma^1 \zeta = -1 + \omega^\zeta$. It is thus natural to conjecture that $\sigma^1 \zeta = -1 + \omega^\zeta$ for all ζ. In the next section we will discuss how a family of ordinal functions satisfying these properties can be constructed, and show that they are closely related to the Feferman-Schütte ordinal Γ_0.

5.4 Hyperations and the Feferman-Schütte ordinal

Beklemishev has shown how provability algebras give rise to a notation system for Γ_0. Such ordinals are usually presented using Veblen progressions [37], but alternatively they may be defined through *hyperations*, which are more convenient in our present context.

Definition 5.11. *Let f be a normal function. Then, we define the* hyperation *of f to be the unique family of normal functions $\langle f^\zeta \rangle_{\zeta \in \mathsf{On}}$ such that*

(i) $f^1 = f$

(ii) $f^{\alpha+\beta} = f^\alpha f^\beta$ for all ordinals α, β

(iii) $\langle f^\zeta \rangle_{\zeta \in \mathsf{On}}$ is pointwise minimal amongst all families of normal functions satisfying the above clauses[3].

It is not obvious that such a family of functions exists, but a detailed construction is given by Joosten and myself in [17]. It is also shown there that they may be computed by the following recursion:

Lemma 5.12. *Let f be a normal function such that $f(0) = 0$. Then, given ordinals λ, μ,*

(i) $f^0 \mu = \mu$;

(ii) $f^{\lambda+1} \mu = f^\lambda f \mu$;

(iii) *if μ is a limit,* $f^\lambda \mu = \lim_{\xi < \mu} f^\lambda \xi$;

(iv) *if λ is a limit,* $f^\lambda (\mu + 1) = \lim_{\xi < \lambda} f^\xi (f^\lambda(\mu) + 1)$.

Although each function f^ξ is normal, the function $\xi \mapsto f^\xi \mu$ typically is not, even when $\mu = 0$, since if $f(0) = 0$ then it follows that $f^\xi 0 = 0$ for all ξ. However, when $f(0) > 0$ then $\xi \mapsto f^\xi 0$ is normal, and more generally, we have the following:

Lemma 5.13. *Assume that $f \colon \mathsf{Ord} \to \mathsf{Ord}$ is normal and suppose that μ is the least ordinal such that $f(\mu) > \mu$ (if it exists).*

Then, the function $\xi \mapsto f^\xi \mu$ is normal, and for all ξ, $f^\xi \upharpoonright \mu$ is the identity (where $\upharpoonright$ denotes domain restriction).

We omit the proof which proceeds by transfinite induction using Lemma 5.12. We are particularly interested in hyperating $e(\xi) = -1 + \omega^\xi$; the family of functions $\langle e^\xi \rangle_{\xi \in \mathsf{Ord}}$ are the *hyperexponentials*. Observe that, in view of Lemma 5.13, $e^\xi 0 = 0$ for all ξ and the function $\xi \mapsto e^\xi 1$ is normal. Aside from the clauses mentioned above, we remark that to entirely determine the value of $e^\lambda \mu$ we need the additional clause

$$e^1(\mu + 1) = \lim_{n < \omega} ((1 + e^1 \mu) \cdot n);$$

this follows directly from the definitions of ordinal exponentiation and the function e.

Aguilera and I proved the following in [1]:

---

[3]That is, if $\langle g^\zeta \rangle_{\zeta \in \mathsf{On}}$ is a family of functions satisfying conditions (i) and (ii), then for all ordinals ξ, ζ, $f^\zeta \xi \leq g^\zeta \xi$.

Proposition 5.14. *For every ordinal $\xi > 0$, there exist unique ordinals α, β such that β is 1 or additively decomposable and $\xi = e^\alpha \beta$.*

We call α above the *degree of indecomposability* of ξ; in particular, if ξ is already additively decomposable, then $\alpha = 0$. More generally, $e^\alpha \beta$ is always additively indecomposable if $\alpha, \beta > 0$, since

$$e^\alpha \beta = e e^{-1+\alpha} \beta = -1 + \omega^{e^{-1+\alpha}\beta} = \omega^{e^{-1+\alpha}\beta}.$$

Note that by writing β as a sum of indecomposables we may iterate this lemma and thus write any ordinal in terms of $e, +, 0$ and 1. This form is unique if we do not allow sums of the form $\xi + \eta$ where $\xi + \eta = \eta$.

We will not review Veblen progressions here; however, as these are more standard than hyperexponentials, we remark that notations using hyperexponentials or Veblen functions can be easily translated from one to the other using the following proposition. Below, φ_α denotes the Veblen functions as defined in [33].

Proposition 5.15. *Given ordinals α, β,*

1. $e^\alpha(0) = 0$,

2. $e^1(1 + \beta) = \varphi_0(1 + \beta)$,

3. $e^{\omega^{1+\alpha}}(1 + \beta) = \varphi_{1+\alpha}(\beta)$.

The proof can be found in [17]. We have seen that every ordinal $\xi < \varepsilon_0$ can be written as a sum of the form $\alpha + \omega^\beta$ with $\alpha, \beta < \xi$. In general, it is desirable in any ordinal notation system that, if we have a notation for an additively indecomposable ξ, then we also have notations for ordinals $\alpha, \beta < \xi$ such that $\alpha + \beta = \xi$. If instead ξ is additively indecomposable, it is also convenient to have notations for α, β such that $\xi = e^\alpha \beta$ (although we cannot always guarantee that $\alpha < \xi$). The following definition captures these properties.

Definition 5.16. *Let Θ be a set of ordinals.*

1. *We say that Θ is additively reductive if whenever ξ is additively decomposable, we have that $\xi \in \Theta$ if and only if there are $\alpha, \beta \in \xi \cap \Theta$ such that $\xi = \alpha + \beta$.*

2. *We say that Θ is hyperexponentailly reductive if whenever $\xi > 1$ is additively indecomposable, we have that $\xi \in \Theta$ if and only if there are $\alpha, \beta \in \Theta$ such that $\beta < \xi$ and $\xi = e^\alpha \beta$.*

3. *We say that Θ is reductive if it is additively and hyperexponentially reductive.*

Additively reductive sets of ordinals always contain Cantor decompositions of their elements and are closed under left subtraction by *arbitrary* ordinals:

Lemma 5.17. *Let Θ be an additively reductive set of ordinals such that $0 \in \Theta$. Then:*

1. *If $0 \neq \xi \in \Theta$ is arbitrary, there are ordinals α, β such that $\alpha, \omega^\beta \in \Theta$ and $\xi = \alpha + \omega^\beta$.*

2. *If $\beta \in \Theta$ and $\alpha < \beta$ (not necessarily a member of Θ), then $-\alpha + \beta \in \Theta$.*

Proof. For the first claim, if ξ is additively indecomposable there is nothing to do, since we already have that $\xi = \omega^\beta$ for some β. Otherwise, using the assumption that Θ is additively reductive, write $\xi = \gamma + \delta$ with $\gamma, \delta \in \xi \cap \Theta$.

By the induction hypothesis applied to δ, there are η, β such that $\eta, \omega^\beta \in \Theta$ and $\delta = \eta + \omega^\beta$. Again using the assumption that Θ is additively reductive, we may set $\alpha = \gamma + \eta \in \Theta$, and see that $\xi = \alpha + \omega^\beta$.

Now we prove the second item by induction on ξ. We may assume that β is additively indecomposable, since otherwise $-\alpha + \beta \in \{0, \beta\} \subseteq \Theta$. Thus we may write $\beta = \gamma + \delta$ with $\gamma, \delta \in \beta \cap \Theta$. If $\alpha \leq \gamma$, by the induction hypothesis $-\alpha + \gamma \in \Theta$, and thus $-\alpha + \beta = (-\alpha + \gamma) + \delta \in \Theta$. Otherwise, also by the induction hypothesis applied to $\delta < \xi$,
$$-\alpha + \beta = -(-\gamma + \alpha) + \delta \in \Theta. \qquad \square$$

Meanwhile, hyperexponentially reductive sets of ordinals always contain hyperexponential normal forms for their elements:

Lemma 5.18. *If Θ contains 0 and is hyperexponentially reductive, then for every $\xi \in \Theta$, there are $\alpha, \beta \in \Theta$ such that $\beta = 1$ or is additively decomposable, and $\xi = e^\alpha \beta$.*

Proof. By induction on ξ; if ξ is additively decomposable or 1 then $\xi = e^0 \xi$, otherwise there are $\alpha', \beta' \in \Theta$ with $\beta < \xi$ such that $\xi = e^{\alpha'} \beta'$. By induction hypothesis there are $\gamma, \beta \in \Theta$ such that $\beta = 1$ or is additively decomposable and $\beta' = e^\gamma \beta$. Setting $\alpha = \alpha' + \gamma$, we see that $\xi = e^\alpha \beta$, as desired. $\qquad \square$

The ordinal Γ_0 can be constructed by closing $\{0, 1\}$ under addition and hyperexponentiation, or more succinctly by the function $\alpha, \beta, \gamma \mapsto e^\alpha(\beta + \gamma)$. In fact, Γ_0 is the least *hyperexponentially perfect* set, in the sense of the following definition:

Definition 5.19. *Define a function* $\mathrm{HE}\colon \mathrm{Ord}^3 \to \mathrm{Ord}$ *by*
$$\mathrm{HE}(\alpha, \beta, \gamma) = e^\alpha(\beta + \gamma).$$
Given a set of ordinals Θ, say that Θ is hyperexponentially closed *if $2 \cup \mathrm{HE}[\Theta] \subseteq \Theta$. We say that Θ is* hyperexponentially perfect *if it is reductive and hyperexponentially closed.*

It is easy to see that Θ is hyperexponentially perfect if and only if it is reductive and $0, 1 \in \Theta$. Note also that hyperexponentially closed sets are closed under both addition and hyperexponentiation:

Lemma 5.20. *If $0 \in \Theta$ and $\alpha, \beta \in \Theta$, then $\alpha + \beta, e^\alpha \beta \in \mathrm{HE}[\Theta]$.*

Proof. If Θ is hyperexponentially closed then by definition we have that $0 \in \Theta$, hence if $\alpha, \beta \in \Theta$, $\alpha + \beta = e^0(\alpha + \beta) \in \mathrm{HE}[\Theta]$ and $e^\alpha \beta = e^\alpha(\beta + 0) \in \mathrm{HE}[\Theta]$. □

With this, we are ready to define the ordinal Γ_0:

Theorem 5.21. *Let $\Gamma_0 = \overline{2}^{\mathrm{HE}}$. Then, Γ_0 is an ordinal and for every $\xi < \Gamma_0$ with $\xi > 1$, there are ordinals $\alpha, \beta, \gamma < \xi$ such that $\xi = e^\alpha(\beta + \gamma)$.*

Proof. The proof closely mimics that of Theorem 4.19. First we will show that if $1 < \xi \in \Gamma_0$, then there are $\alpha, \beta, \gamma < \xi$ such that $\xi = e^\alpha(\beta + \gamma)$. By Lemma 4.17.3, there are $\alpha, \beta, \gamma \in \Gamma_0$ with $\alpha, \beta, \gamma \neq \xi$ and such that $\xi = e^\alpha(\beta + \gamma)$. Since $\xi \neq 0$ it follows that $\beta + \gamma \geq 1$, and since the function e^α is normal, $\beta + \gamma \leq \xi$, from which we obtain $\beta, \gamma < \xi$. Similarly, $\alpha \leq \xi$ since $e^\alpha(\beta + \gamma) \geq e^\alpha 1$ and the function $\alpha \mapsto e^\alpha 1$ is normal. Thus we also have $\alpha < \xi$.

Next we show that Γ_0 is transitive. We proceed by induction on ζ with a secondary induction on ξ to show that $\xi < \zeta \in \Gamma_0$ implies that $\xi \in \Gamma_0$. We may without loss of generality assume that $\xi, \zeta > 1$. Write $\zeta = e^\alpha(\beta + \gamma)$ with $\alpha, \beta, \gamma \in \Gamma_0 \cap \zeta$. Then, using Proposition 5.14, write $\xi = e^\lambda \mu$ with $\mu = 1$ or additively decomposable.

Now consider two cases. If $\lambda = 0$, we have that $\mu = \xi > 1$, hence ξ is additively decomposable and we can write $\xi = \nu + \eta$, with $\nu, \eta < \xi$. By the secondary induction hypothesis, $\nu, \eta \in \Gamma_0$, hence $\xi = e^0(\nu + \eta) \in \Gamma_0$.

Otherwise, $\lambda > 0$, and we consider two subcases. If $\alpha \geq \lambda$, by the induction hypothesis applied to $\alpha < \zeta$, $\lambda \in \Gamma_0$. But $\xi > 1$ and is additively indecomposable, while $\mu \leq \xi$ is 1 or additively decomposable, so $\mu < \xi$. By the secondary induction hypothesis, $\mu \in \Gamma_0$, hence $\xi = e^\lambda \mu \in \Gamma_0$. If instead $\alpha < \lambda$, we observe that $e^\lambda \mu = e^\alpha e^{-\alpha + \lambda} \mu$, and by normality of e^α, $e^{-\alpha + \lambda} \mu < \beta + \gamma$. Since $e^{-\alpha + \lambda} \mu$ is additively indecomposable, it follows that $e^{-\alpha + \lambda} \mu \leq \max\{\beta, \gamma\}$, so that by the induction hypothesis applied to $\max\{\beta, \gamma\} < \zeta$, we have that $e^{-\alpha + \lambda} \mu \in \Gamma_0$. Since $\alpha \in \Gamma_0$, $\xi = e^\alpha e^{-\alpha + \lambda} \mu \in \Gamma_0$. □

Thus Γ_0 can be characterized as the least hyperexponentially closed ordinal, or alternatively the least hyperexponentially perfect ordinal. Later we will see that it can also be obtained using worms, by closing under o.

5.5 Order-types of transfinite worms

As in Section 4.3, our strategy for giving a calculus for computing o will be to guess a candidate function and prove that it has the required properties. Let us assume that Theorem 4.33 remains true for transfinite worms. Moreover, note that the functions e^ξ satisfy all desired properties of our functions o^ξ. Thus we will conjecture that $e^\xi = o^\xi$ for every ordnal ξ, and propose the following candidate:

Definition 5.22. *Let Λ be an ordinal, $\mathfrak{v}, \mathfrak{w} \in \mathbb{W}^\Lambda$ be worms and $\alpha < \Lambda$ an ordinal. Then, define*

1. $\hat{o}(\top) = 0$,

2. $\hat{o}(\mathfrak{w}) = \hat{o}b(\mathfrak{w}) + \omega^{\hat{o}(1 \downarrow h(\mathfrak{w}))}$ *if $\mathfrak{w} \neq \top$ and $\min \mathfrak{w} = 0$,*

3. $\hat{o}(\mathfrak{w}) = e^\mu \hat{o}(\mu \downarrow \mathfrak{w})$ *if $\mathfrak{w} \neq \top$ and $\mu = \min \mathfrak{w} > 0$.*

The next few lemmas establish that $\hat{o}$ behaves as it should.

Lemma 5.23. *If $\mathfrak{w} \neq \top$ is any worm, then $\hat{o}(\mathfrak{w}) \neq 0$.*

Proof. If $\min \mathfrak{w} = 0$, this is obvious since $\omega^\xi > 0$ independently of ξ. Otherwise, $\hat{o}(\mathfrak{w}) = e^\mu \hat{o}(\mu \downarrow \mathfrak{w})$ with $\mu = \min \mathfrak{w} > 0$. But $\min \mu \downarrow \mathfrak{w} = 0$, so by the previous case $\hat{o}(\mu \downarrow \mathfrak{w}) \neq 0$ and hence $\hat{o}(\mathfrak{w}) = e^\mu \hat{o}(\mu \downarrow \mathfrak{w}) \neq 0$. □

Lemma 5.24. *For any worm $\mathfrak{w} \neq \top$, $\hat{o}(\mathfrak{w}) = \hat{o}b(\mathfrak{w}) + \omega^{\hat{o}(1 \downarrow h(\mathfrak{w}))}$.*

Proof. If $\min \mathfrak{w} = 0$, there is nothing to prove. Otherwise, $\min \mathfrak{w} > 0$, so we can write $\min \mathfrak{w} = 1 + \eta$ for some η. Moreover, $h(\mathfrak{w}) = \mathfrak{w}$ and $b(\mathfrak{w}) = \top$, so $\hat{o}h(\mathfrak{w}) = \hat{o}(\mathfrak{w}) \neq 0$ and $\hat{o}b(\mathfrak{w}) = 0$. Meanwhile, $(1 + \eta) \downarrow \mathfrak{w} \neq \top$, so $\hat{o}((1 + \eta) \downarrow \mathfrak{w}) \neq 0$ and thus $e^\eta \hat{o}((1 + \eta) \downarrow \mathfrak{w}) > 0$, from which it follows that

$$-1 + \omega^{e^\eta(\hat{o}((1+\eta)\downarrow \mathfrak{w}))} = \omega^{e^\eta(\hat{o}((1+\eta)\downarrow \mathfrak{w}))}. \tag{2}$$

Finally, observe that

$$\hat{o}(1 \downarrow h(\mathfrak{w})) = e^\eta \hat{o}(\eta \downarrow (1 \downarrow h(\mathfrak{w}))) = e^\eta \hat{o}((1 + \eta) \downarrow h(\mathfrak{w})). \tag{3}$$

Putting all of this together,

$$\begin{aligned}
\hat{o}(\mathfrak{w}) &= e^{1+\eta}\hat{o}((1+\eta)\downarrow\mathfrak{w}) & \text{by definition}\\
&= 0 + ee^{\eta}(\hat{o}((1+\eta)\downarrow\mathfrak{w})) & \text{since } e^{1+\eta} = ee^{\eta}\\
&= \hat{o}(b(\mathfrak{w})) + ee^{\eta}(\hat{o}((1+\eta)\downarrow\mathfrak{w})) & \text{since } \hat{o}(b(\mathfrak{w})) = 0\\
&= \hat{o}(b(\mathfrak{w})) + (-1 + \omega^{e^{\eta}(\hat{o}((1+\eta)\downarrow\mathfrak{w}))}) & \text{by definition of } e\\
&= \hat{o}(b(\mathfrak{w})) + \omega^{e^{\eta}(\hat{o}((1+\eta)\downarrow\mathfrak{w}))} & \text{by (2)}\\
&= \hat{o}(b(\mathfrak{w})) + \omega^{\hat{o}(1\downarrow h(\mathfrak{w}))} & \text{by (3),}
\end{aligned}$$

as claimed. □

Lemma 5.25. *For any worm $\mathfrak{w}$ and ordinal λ, $\hat{o}(\lambda\uparrow\mathfrak{w}) = e^{\lambda}\hat{o}(\mathfrak{w})$.*

Proof. If $\mathfrak{w} = \top$, then

$$\hat{o}(\lambda\uparrow\top) = \hat{o}(\top) = 0 = e^{\lambda}0 = e^{\lambda}\hat{o}(\top).$$

Otherwise, $\mathfrak{w} \neq \top$. If $\lambda = 0$ the lemma follows from the fact that $0\uparrow\mathfrak{w} = \mathfrak{w}$ and e^0 is the identity, and if $\min\mathfrak{w} = 0$ then $\min(\lambda\uparrow\mathfrak{w}) = \lambda$ and

$$\hat{o}(\lambda\uparrow\mu) = e^{\lambda}\hat{o}(\lambda\downarrow(\lambda\uparrow\mathfrak{w})) = e^{\lambda}\hat{o}(\mathfrak{w}).$$

If not, let $\mu = \min\mathfrak{w} > 0$, so that $\hat{o}(\mathfrak{w}) = e^{\mu}(\mu\downarrow\mathfrak{w})$. Observe that $\min(\lambda\uparrow\mathfrak{w}) = \lambda + \mu$. Hence,

$$\begin{aligned}
\hat{o}(\lambda\uparrow\mathfrak{w}) &= e^{\lambda+\mu}((\lambda+\mu)\downarrow(\lambda\uparrow\mathfrak{w}))\\
&= e^{\lambda+\mu}(\mu\downarrow(\lambda\downarrow(\lambda\uparrow\mathfrak{w})))\\
&= e^{\lambda}e^{\mu}(\mu\downarrow\mathfrak{w})\\
&= e^{\lambda}\hat{o}(\mathfrak{w}),
\end{aligned}$$

as claimed. □

With this we can prove that $\hat{o}$ is strictly increasing and initial.

Lemma 5.26. *The map $\hat{o}\colon \mathbb{W} \to \mathsf{Ord}$ is strictly increasing.*

Proof. We proceed by induction on $\|\mathfrak{w}\| + \|\mathfrak{v}\|$ to show that $\mathfrak{w} \triangleleft \mathfrak{v}$ if and only if $\hat{o}(\mathfrak{w}) < \hat{o}(\mathfrak{v})$. If $\mathfrak{w} = \top$ the claim is immediate from Lemma 5.23, so we assume otherwise. Note that in this case $\mathfrak{w} \triangleright \top$ and $\hat{o}(\mathfrak{w}) > \hat{o}(\top)$, so we may also assume that $\mathfrak{v} \neq \top$.

Thus we consider $\mathfrak{w}, \mathfrak{v} \neq \top$, and define $\mu = \min(\mathfrak{w}\mathfrak{v})$. If $\mu = 0$, we observe that either $\|h(\mathfrak{w})\| < \|\mathfrak{w}\|$ or $\|h(\mathfrak{v})\| < \|\mathfrak{v}\|$, and we can proceed exactly as in the proof of Lemma 4.29. Thus we consider only the case for $\mu > 0$.

Note that in this case we have that

$$\|\mu \downarrow \mathfrak{w}\| + \|\mu \downarrow \mathfrak{v}\| < \|\mathfrak{w}\| + \|\mathfrak{v}\|,$$

so we may apply the induction hypothesis to $\mu \downarrow \mathfrak{w}$ and $\mu \downarrow \mathfrak{v}$. Hence we obtain:

$$\begin{aligned}
\mathfrak{w} \triangleleft \mathfrak{v} &\Leftrightarrow (\mu \downarrow \mathfrak{w}) \triangleleft (\mu \downarrow \mathfrak{v}) &&\text{by Lemma 4.23}\\
&\Leftrightarrow \hat{o}(\mu \downarrow \mathfrak{w}) < \hat{o}(\mu \downarrow \mathfrak{v}) &&\text{by induction hypothesis}\\
&\Leftrightarrow e^\mu \hat{o}(\mu \downarrow \mathfrak{w}) < e^\mu \hat{o}(\mu \downarrow \mathfrak{v}) &&\text{by normality of } e^\mu\\
&\Leftrightarrow \hat{o}\mathfrak{w} < \hat{o}\mathfrak{v} &&\text{by Lemma 5.25,}
\end{aligned}$$

as needed. □

Lemma 5.27. *The map $\hat{o} \colon \mathbb{W} \to \mathsf{Ord}$ is surjective.*

Proof. Proceed by induction on $\xi \in \mathsf{Ord}$ to show that there is $\mathfrak{w}$ with $\hat{o}(\mathfrak{w}) = \xi$. For the base case, $\xi = 0 = \hat{o}(\top)$. Otherwise, by Proposition 5.14, ξ can be written in the form $e^\alpha \beta$ with β additively decomposable or 1. Write $\beta = \gamma + \omega^\delta$, so that $\gamma, \delta < \beta \leq \xi$. By the induction hypothesis, there are worms $\mathfrak{u}, \mathfrak{v}$ such that $\hat{o}(\mathfrak{u}) = \gamma$ and $\hat{o}(\mathfrak{v}) = \delta$. Then, $\xi = \hat{o}(\alpha \uparrow ((1 \uparrow \mathfrak{v}) \, 0 \, \mathfrak{u}))$, as needed. □

Lemma 5.28. *For every worm $\mathfrak{w}$, $\hat{o}(\mathfrak{w}) = o(\mathfrak{w})$.*

Proof. Immediate from Lemmas 5.26 and 5.27 using Lemma 3.24. □

Before giving the definitive version of our calculus, let us show that the clasue for $\mathfrak{w} \, 0 \, \mathfrak{v}$ can be simplified somewhat.

Lemma 5.29. *Given arbitrary worms $\mathfrak{w}$, $\mathfrak{v}$, $o(\mathfrak{w} \, 0 \, \mathfrak{v}) = o(\mathfrak{v}) + 1 + o(\mathfrak{w})$.*

Proof. Observe that by Lemma 5.25 together with Lemma 5.28, we have that for any worm $\mathfrak{u}$, $o(1 \uparrow \mathfrak{u}) = eo(\mathfrak{u}) = -1 + \omega^{o(\mathfrak{u})}$, so that

$$\omega^{o(\mathfrak{u})} = 1 + o(1 \uparrow \mathfrak{u}). \tag{4}$$

With this in mind, proceed by induction on $\#\mathfrak{v} + \#\mathfrak{w}$ to prove the lemma. First consider the case where $0 < \min \mathfrak{v}$. In this case, $h(\mathfrak{v} \, 0 \, \mathfrak{w}) = \mathfrak{v}$, so that

$$o(\mathfrak{v} \, 0 \, \mathfrak{w}) = o(\mathfrak{w}) + \omega^{o(1 \downarrow \mathfrak{v})} = o(\mathfrak{w}) + 1 + o(\mathfrak{v}),$$

where the first equality is by Defintion 5.22 and the second follows from (4).

If $\mathfrak{v}$ does contain a zero, we have that $\mathfrak{v} = h(\mathfrak{v})\, 0\, b(\mathfrak{v})$, so that

$$\mathfrak{v}\, 0\, \mathfrak{w} = h(\mathfrak{v})\, 0\, b(\mathfrak{v})\, 0\, \mathfrak{w}.$$

This means that $h(\mathfrak{v}\, 0\, \mathfrak{w}) = h(\mathfrak{v})$ and $b(\mathfrak{v}\, 0\, \mathfrak{w}) = b(\mathfrak{v})\, 0\, \mathfrak{w}$. Applying the induction hypothesis to $b(\mathfrak{v})\, 0\, \mathfrak{w}$, we obtain

$$ob(\mathfrak{v}\, 0\, \mathfrak{w}) = o(\mathfrak{w}) + 1 + ob(\mathfrak{v}),$$

and thus

$$o(\mathfrak{v}\, 0\, \mathfrak{w}) = ob(\mathfrak{w}\, 0\, \mathfrak{v}) + \omega^{o(1\downarrow h(\mathfrak{v}))}$$
$$\stackrel{\text{IH}}{=} o(\mathfrak{w}) + 1 + ob(\mathfrak{v}) + \omega^{o(1\downarrow h(\mathfrak{v}))} = o(\mathfrak{w}) + 1 + o(\mathfrak{v}),$$

as needed. □

Let us put our results together to give our definitive calculus for o.

Theorem 5.30. *Let $\mathfrak{v}, \mathfrak{w}$ be worms and α be an ordinal. Then,*

1. $o(\top) = 0$,

2. $o(\mathfrak{v}\, 0\, \mathfrak{w}) = o(\mathfrak{w}) + 1 + o(\mathfrak{v})$, *and*

3. $o(\alpha \uparrow \mathfrak{w}) = e^{\alpha} o(\mathfrak{w})$.

Proof. The first item is immediate from Definition 5.22, the second from Lemma 5.29, and the third from Lemma 5.25, respectively, using the fact that $o = \hat{o}$ by Lemma 5.28. □

Note that Theorem 5.30 can be applied to any worm $\mathfrak{w}$, and hence it gives a complete calculus for computing o. Next, let us see how this gives rise to a notation system for Γ_0.

5.6 Beklemishev's predicative worms

Now we review results from [5] showing that Γ_0 is the least set definable by iteratively taking order-types of worms. Let us begin by discussing the properties of sets of worms obtained from additively reductive sets of ordinals. Recall that $\mathfrak{w} \sqsubset \Theta$ means that every ordinal appearing in $\mathfrak{w}$ belongs to Θ.

Lemma 5.31. *Let Θ be an additively reductive set of ordinals such that $0 \in \Theta$, and let $\mathfrak{w} \sqsubset \Theta$. Then,*

1. If $\mu \in \Theta$, $\mu \uparrow \mathfrak{w} \sqsubset \Theta$, and

2. if $\mu \leq \mathfrak{w}$ is arbitrary, then $\mu \downarrow \mathfrak{w} \sqsubset \Theta$.

Proof. Suppose that $\mathfrak{w} = \lambda_1 \ldots \lambda_n \top \sqsubset \Theta$. If $\mu \in \Theta$, using the fact that Θ is closed under addition, for each $i \in [1, n]$ we have that $\mu + \lambda_i \in \Theta$. Thus $\mu \uparrow \mathfrak{w} \sqsubset \Theta$.

Similarly, by Lemma 5.17.2, if μ is arbitrary then $-\mu + \lambda_i \in \Theta$ for each $i \in [1, n]$, so $\mu \downarrow \mathfrak{w} \sqsubset \Theta$. □

Now, let us make the notion of "closing under o" precise.

Definition 5.32. *Observe that o may be regarded as a function $o\colon \mathsf{Ord}^{<\omega} \to \mathsf{Ord}$ by setting*
$$o(\mu_1, \ldots, \mu_n) = o(\mu_1 \ldots \mu_n \top).$$
Then, given a set of ordinals Θ, if $o[\Theta] \subseteq \Theta$ we say that Θ is worm-closed, *and if $\Theta = o[\Theta]$ we say that Θ is* worm-perfect.

Even when Θ is not worm-perfect, sets of the form $o[\Theta]$ are rather well-behaved:

Lemma 5.33. *If Θ is any set of ordinals, then $0 \in o[\Theta]$. If moreover $0 \in \Theta$, then also $1 \in o[\Theta]$, and $o[\Theta]$ is additively reductive.*

Proof. Observe that $0 = o(\top)$, and $\top \sqsubset \Theta$ since $\top$ contains no ordinals, so $0 \in o[\Theta]$. Similarly, $1 = o(0\top)$, and $0\top \sqsubset \Theta$ if $0 \in \Theta$.

Let us see that $o[\Theta]$ is additively reductive when $0 \in \Theta$. First assume that $\alpha, \beta \in o[\Theta]$. Then, there are worms $\mathfrak{u}, \mathfrak{v} \sqsubset \Theta$ such that $\alpha = o(\mathfrak{u})$ and $\beta = o(\mathfrak{v})$. If $\beta \geq \omega$, then
$$o(\mathfrak{v}\, 0\, \mathfrak{u}) = o(\mathfrak{u}) + 1 + o(\mathfrak{v}) = \alpha + 1 + \beta = \alpha + \beta,$$
otherwise
$$o(\langle 0 \rangle^\beta \mathfrak{u}) = o(\mathfrak{u}) + \beta = \alpha + \beta,$$
where we define $\langle \lambda \rangle^n = \underbrace{\langle \lambda \rangle \ldots \langle \lambda \rangle}_{n \text{ times}}$. Both $\mathfrak{v}\, 0\, \mathfrak{u}, \langle 0 \rangle^\beta \mathfrak{u} \sqsubset \Theta$, so $\alpha + \beta \in o[\Theta]$.

Conversely, if $\xi \in o[\Theta]$ is additively decomposable, write $\xi = o(\mathfrak{w})$. Then, $\xi = ob(\mathfrak{w}) + 1 + oh(\mathfrak{w})$, and since $1 + oh(\mathfrak{w})$ is additively indecomposable, we have that $\xi \neq 1 + oh(\mathfrak{w})$ and hence $ob(\mathfrak{w}), 1 + oh(\mathfrak{w}) < \xi$. Clearly $ob(\mathfrak{w}) \in o[\Theta]$, while $1 + oh(\mathfrak{w})$ is either 1 or $oh(\mathfrak{w})$, both of which belong to $o[\Theta]$. □

Lemma 5.34. *Let Θ be any set of ordinals. Then, Θ is worm-perfect if and only if it is hyperexponentially perfect.*

Proof. Assume first that Θ is worm-perfect. By Lemma 5.33, $0 \in \Theta$, thus also $1 \in \Theta$ and Θ is additively reductive. It remains to prove that $\operatorname{HE}[\Theta] \subseteq \Theta$ and that Θ is hyperexponentially reductive.

To show that $\operatorname{HE}[\Theta] \subseteq \Theta$, it suffices to check that $e^\alpha \beta \in \Theta$ whenever $\alpha, \beta \in \Theta$, given that we already know that Θ is closed under addition. If $\alpha, \beta \in \Theta$, since Θ is worm-perfect, there is $\mathfrak{w} \sqsubset \Theta$ such that $o(\mathfrak{w}) = \beta$. By Lemma 5.31, $\alpha \uparrow \mathfrak{w} \sqsubset \Theta$, and by Theorem 5.30, $e^\alpha \beta = o(\alpha \uparrow \mathfrak{w}) \in \Theta$.

Next we show that if $1 < \xi \in \Theta$, there are $\alpha, \beta \in \Theta$ such that $\xi = e^\alpha \beta$ and $\beta < \xi$. Since Θ is worm-perfect, $\xi = o(\mathfrak{w})$ for some $\mathfrak{w} \sqsubset \Theta$. We proceed by induction on $\|\mathfrak{w}\|$ to find suitable $\alpha, \beta \in \Theta$. We may assume that $\mathfrak{w} \neq \top$ since $\xi > 0$, and we set $\mu = \min \mathfrak{w}$. If $\mu = 0$, then $h(\mathfrak{w}), b(\mathfrak{w}) \sqsubset \Theta$, and since Θ is worm-perfect, $ob(\mathfrak{w}), oh(\mathfrak{w}) \in \Theta$. Now, if $oh(\mathfrak{w}) = \xi$, by induction on $\|h(\mathfrak{w})\|$ we see that there exist suitable $\alpha, \beta \in \Theta$. If instead $oh(\mathfrak{w}) < \xi$, this means that $\xi = ob(\mathfrak{w}) + 1 + oh(\mathfrak{w})$ is additively decomposable, contrary to our assumption.

Now consider $\mu > 0$. By Lemma 5.31, $\mu \downarrow \mathfrak{w} \sqsubset \Theta$. Hence by induction on $\|\mu \downarrow \mathfrak{w}\| < \|\mathfrak{w}\|$, we have that $o(\mu \downarrow \mathfrak{w}) = e^\eta \beta$ for some $\eta, \beta \in \Theta$ with $\beta < o(\mu \downarrow \mathfrak{w})$. It follows that

$$o(\mathfrak{w}) = e^\mu o(\mu \downarrow \mathfrak{w}) = e^\mu e^\eta \beta = e^{\mu+\eta} \beta,$$

and since Θ is closed under addition, we may set $\alpha = \mu + \eta \in \Theta$.

For the other direction, assume that Θ is hyperexponentially perfect. To show that $o[\Theta] \subseteq \Theta$, we will prove by induction on $\|\mathfrak{w}\|$ that if $\mathfrak{w} \sqsubset \Theta$, then $o(\mathfrak{w}) \in \Theta$. For the base case, if $\mathfrak{w} = \top$, then $o(\mathfrak{w}) = 0 \in \Theta$. Otherwise, let $\mu = \min \mathfrak{w}$.

If $\mu = 0$, then by induction hypothesis $oh(\mathfrak{w}), ob(\mathfrak{w}) \in \Theta$. Since also $1 \in \Theta$, then $o(\mathfrak{w}) = ob(\mathfrak{w}) + 1 + oh(\mathfrak{w}) \in \Theta$. Otherwise, $\|\mu \downarrow \mathfrak{w}\| < \|\mathfrak{w}\|$, and as before, $\mu \downarrow \mathfrak{w} \sqsubset \Theta$. It follows by the induction hypothesis that $o(\mu \downarrow \mathfrak{w}) \in \Theta$. Moreover, since μ appears in $\mathfrak{w}$ we must have that $\mu \in \Theta$, thus $o(\mathfrak{w}) = e^\mu o(\mu \downarrow \mathfrak{w}) \in \Theta$, using the fact that Θ is hyperexponentially closed.

Finally, we show that $\Theta \subseteq o[\Theta]$. We prove by induction on ξ that if $\xi \in \Theta$, then $\xi = o(\mathfrak{w})$ for some $\mathfrak{w} \sqsubset \Theta$. If $\xi = 0$ we may take $\mathfrak{w} = \top$. If not, using the fact that Θ is hyperexponentially perfect, write $\xi = e^\alpha \beta$ with $\alpha, \beta \in \Theta$ and $\beta = 1$ or additively decomposable. If $\beta = 1$, then $\xi = e^\alpha 1 = o(\alpha \top)$. Otherwise, since Θ is additively reductive, we may write $\beta = \gamma + \delta'$ with $\gamma, \delta' \in \beta \cap \Theta$. Using Lemma 5.17 we see that $\delta = -1 + \delta' \in \Theta$. By the induction hypothesis, there are worms $\mathfrak{u}, \mathfrak{v} \sqsubset \Theta$ such that $\gamma = o(\mathfrak{u})$, $\delta = o(\mathfrak{v})$, and thus

$$\beta = \gamma + \delta' = \gamma + 1 + \delta = o(\mathfrak{u}) + 1 + o(\mathfrak{v}) = o(\mathfrak{v} \, 0 \, \mathfrak{u}).$$

But $\mathfrak{v} \, 0 \, \mathfrak{u} \sqsubset \Theta$, and thus by Lemma 5.31, $\alpha \uparrow (\mathfrak{v} \, 0 \, \mathfrak{u}) \sqsubset \Theta$, and $o(\alpha \uparrow (\mathfrak{v} \, 0 \, \mathfrak{u})) = e^\alpha \beta$, as needed. □

With this, we obtain our worm-based characterization of Γ_0:

Theorem 5.35. *Γ_0 is the least worm-perfect set of ordinals.*

Proof. Γ_0 is the least hyperexponentially perfect set, and since it is transitive and closed under addition, it is additively reductive. Hence Γ_0 is also worm-perfect, and since any worm-perfect set is hyperexponentially perfect, there can be no smaller worm-perfect set. □

5.7 Autonomous worms and predicative ordinal notations

The map $o\colon \mathbb{W} \to \mathsf{Ord}$ suggests that worms could themselves be used as modalities. This gives rise to Beklemishev's *autonomous worms* [5]:

Definition 5.36. *We define the set of autonomous worms W to be the least set such that $\top \in \mathsf{W}$ and, if $\mathsf{w}, \mathsf{v} \in \mathsf{W}$, then $(\mathsf{w})\mathsf{v} \in \mathsf{W}$.*

The idea is to interpret autonomous worms as regular worms using o:

Definition 5.37. *We define a map $\cdot^o \colon \mathsf{W} \to \mathbb{W}$ given recursively by*

1. $\top^o = \top$
2. $\big((\mathsf{w})\mathsf{v}\big)^o = \langle o(\mathsf{w}^o) \rangle \mathsf{v}^o$.

We then define $\mathsf{o}\colon \mathsf{W} \to \mathsf{Ord}$ by setting $\mathsf{o}(\mathsf{w}) = o(\mathsf{w}^o)$.

As Beklemishev has noted, autonomous worms give notations for any ordinal below Γ_0.

Theorem 5.38. *If γ is any ordinal, then $\gamma < \Gamma_0$ if and only if there is $\mathsf{w} \in \mathsf{W}$ such that $\gamma = \mathsf{o}(\mathsf{w})$.*

Proof. To see that $\Gamma_0 \subseteq \mathsf{o}[\mathsf{W}]$, it suffices in view of Theorem 5.35 to observe that $\mathsf{o}[\mathsf{W}]$ is worm-perfect by construction.

To see that $\mathsf{o}[\mathsf{W}] \subseteq \Gamma_0$, one proves by induction on the number of parentheses in w that if Θ contains 0 and is worm-closed, then $\mathsf{o}(\mathsf{w}) \in \Theta$. In particular, $\mathsf{o}(\mathsf{w}) \in \Gamma_0$. □

6 Impredicative worms

Now we turn to a possible solution to Mints' and Pakhomov's problem of representing the Bachmann-Howard ordinal using worms. This ordinal is related to *inductive definitions*, that is, least fixed points of monotone operators $F\colon 2^{\mathbb{N}} \to 2^{\mathbb{N}}$. Let us begin by reviewing these operators and their fixed points.

1	()	ω	(())	ε_0	((()))
ε_1	((()))((()))	$\varepsilon_\omega + \varepsilon_0$	((()))()(()(()))	$e^{e^{e^{e^{1}1}}}1$	((((()))))

Figure 1: Some ordinals represented as autonomous worms. We use the identity $\varepsilon_\xi = e^\omega(1+\xi)$, which is a special case of Proposition 5.15.

6.1 Inductive definitions

Let $F: 2^{\mathbb{N}} \to 2^{\mathbb{N}}$. We say that F is *monotone* if $F(X) \subseteq F(Y)$ whenever $X \subseteq Y$. For example, if $f: \mathbb{N}^{<\omega} \to \mathbb{N}$, we obtain a monotone operator by setting $F(X) = f[X]$; as we have seen in Lemma 4.17, we can reach a fixed point for such an F by iterating it ω-many times and taking the union of these iterations. More generally, any monotone operator has a least fixed point:

Definition 6.1. *Let $F: 2^{\mathbb{N}} \to 2^{\mathbb{N}}$ be monotone. We define μF to be the unique set such that:*

1. *$\mu F = F(\mu F)$, and*
2. *If $X \subseteq \mathbb{N}$ is such that $F(X) \subseteq X$, then $\mu F \subseteq X$.*

The Knaster-Tarski theorem states that the set μF is always well-defined [23]; it can always be reached "from below" by iterating F, beginning from the empty set. However, in general, we may need to iterate F far beyond ω.

Definition 6.2. *Let $F: 2^{\mathbb{N}} \to 2^{\mathbb{N}}$. For an ordinal ξ, we define an operator $F^\xi: 2^{\mathbb{N}} \to 2^{\mathbb{N}}$ inductively by*

1. *$F^0(X) = X$,*
2. *$F^{\xi+1}(X) = F(F^\xi(X))$,*
3. *$F^\lambda(X) = \bigcup_{\xi<\lambda} F^\xi(X)$ for λ a limit ordinal.*

These iterations eventually become constant, but the ordinal at which they stabilize can be rather large; in principle, our only guarantee is that it is countable, since at each stage before reaching a fixed point we must add at least one natural number. Below, recall that ω_1 denotes the first uncountable cardinal.

Lemma 6.3. *If $F: 2^{\mathbb{N}} \to 2^{\mathbb{N}}$ is monotone, then there is $\lambda < \omega_1$ such that $F^\lambda(\varnothing) = \mu F$.*

We omit the proof, which follows from cardinality considerations. Alternately, it is possible to construct least fixed points 'from above', by taking the intersection of all F-closed sets.

Lemma 6.4. *If $F\colon 2^{\mathbb{N}} \to 2^{\mathbb{N}}$ is monotone, then*
$$\mu F = \bigcap \{Y \subseteq \mathbb{N} : F(Y) \subseteq Y\}.$$

Monotone operators and their fixed points can be formalized in second-order arithmetic, provided they are definable. Any formula $\phi(n, X) \in \Pi^1_\omega$ (with no other free variables) can be regarded as an operator on $2^{\mathbb{N}}$ given by $X \mapsto \{n \in \mathbb{N} : \phi(n, X)\}$. Say that a formula ϕ is in *negation normal form* if it contains no instances of $\to$, and $\neg$ occurs only on atomic formulas. It is well-known that every formula is equivalent to one in negation normal form, obtained by applying De Morgan's rules iteratively.

Definition 6.5. *Let ϕ be a formula in negation normal form and X a set-variable. We say ϕ is positive on X if ϕ contains no occurrences of $t \notin X$.*

Positive formulas give rise to monotone operators, due to the following:

Lemma 6.6. *Given a formula $\phi(n, X)$ that is positive on X, it is provable in ECA_0 that*
$$\forall X \ \forall Y \ \Big(X \subseteq Y \to \forall n \ (\phi(n, X) \to \phi(n, Y))\Big).$$

Thus if we define $F_\phi \colon 2^{\mathbb{N}} \to 2^{\mathbb{N}}$ by $F_\phi(X) = \{n \in \mathbb{N} : \phi(n, X)\}$, F_ϕ will be monotone on X whenever ϕ is positive on X. Moreover, if ϕ is arithmetical, Lemma 6.4 may readily be formalized in $\Pi^1_1\text{-}\mathrm{CA}_0$, by defining
$$M = \Big\{n \in \mathbb{N} : \forall X \big(\forall m (\phi(m, X) \to m \in X) \to n \in X\big)\Big\}.$$

Thus we arrive at the following:

Lemma 6.7. *Let $\phi(n, X)$ be arithmetical and positive on X. Then, it is provable in $\Pi^1_1\text{-}\mathrm{CA}_0$ that there is a least set M such that, for all n,*
$$n \in M \leftrightarrow \phi(n, M).$$

We will denote this set M by $\mu X.\phi$.

With these tools in mind, we are now ready to formalize ω-logic in second-order arithmetic.

6.2 Formalizing full ω-logic

We have discussed before how the ω-rule can be iterated along a well-order. However, we may also consider full ω-logic based on a theory T; that is, the set of formulas that can be derived using the ω-rule and reasoning in T, regardless of the nesting depth of these ω-rules. Let us write $[\infty]_T \phi$ if ϕ is derivable in this fashion. To be precise, we want $[\infty]_T \phi$ to hold whenever:

(i) $\Box_T \phi$,

(ii) $\phi = \forall x \psi(x)$ and for all n, $[\infty]_T \psi(\bar{n})$, or

(iii) there is ψ such that $[\infty]_T \psi$ and $[\infty]_T (\psi \to \phi)$.

In words, $[\infty]_T$ is closed under T and the ω-rule. This notion may be formalized using ω-trees to represent infinite derivations, as in [2, 22]. We follow a different approach, using a fixed-point construction as in [16].

Definition 6.8. *Fix a theory T, possibly with oracles. Let $\mathsf{SPC}_T(Q)$ be a Π_1^1 formula naturally expressing that Q is the least set such that $\phi \in Q$ whenever (i) $\Box_T \phi$ holds, (ii) $\phi = \forall v \psi(v)$ and for all n, $\psi(\bar{n}) \in Q$, or (iii) there exists $\psi \in Q$ such that $\psi \to \phi \in Q$.*

Then, define
$$[\infty]_T \phi \equiv \forall Q (\mathsf{SPC}_T(Q) \to \phi \in Q).$$

In view of Lemma 6.3, this fixed point is reached after some countable ordinal, which gives us the following:

Proposition 6.9. *Given a theory T and $\phi \in \Pi_\omega^1$, $[\infty]_T \phi$ holds if and only if $[\xi]_T \phi$ holds for some $\xi < \omega_1$.*

As before, we may also consider saturated provabiltiy operators with oracles, and we write $[\infty | A]_T \phi$ instead of $[\infty]_{T|A} \phi$. Since these provability operators are defined via a least fixed point, in view of Lemma 6.7, their existence can be readily proven in Π_1^1-CA_0.

Lemma 6.10. *Let T be any theory, possibly with oracles. Then, it is provable in Π_1^1-CA_0 that there exists a set Q such that $\mathsf{SPC}_T(Q)$ holds.*

This notion of provability allows us to represent Π_1^1-CA_0 in terms of a strong consistency assertion, in the spirit of Theorems 4.3 and 5.5. The following is proven in [16]:

Theorem 6.11. Π_1^1-$\mathsf{CA}_0 \equiv \mathsf{ECA}_0 + \forall X \, \langle \infty | X \rangle_T \top$.

This suggests that studying worms which contain the modality $\langle\infty\rangle$ may be instrumental in studying theories capable of reasoning about least fixed points. In view of Proposition 6.9, we may identify $\langle\infty\rangle$ with $\langle\Omega\rangle$ for some ordinal Ω large enough so that $[\infty]_T\phi$ is equivalent to $[\Omega]_T\phi$; we can take $\Omega = \omega_1$, for example, but a large enough countable ordinal will do. In the next section, we will see how adding uncountable ordinals to our notation system allows us to provide notations for much larger countable ordinals as well.

6.3 Beyond the Bachmann-Howard ordinal

It is not hard to see that ε_0 and Γ_0 are countable; for example, it is an easy consequence of Theorem 5.38. With a bit of extra work, one can see that they are computable as well, for example representing elements of Γ_0 as in Theorem 5.38. However, this does not mean that uncountable ordinals cannot appear as a "detour" in defining proof-theoretic ordinals. Indeed, the Bachmann-Howard ordinal precisely arises by adding a symbol for an uncountable ordinal. Before continuing, let us recall a few basic properties of cardinals and cardinalities.

Definition 6.12. *Given a set A, we define $|A|$ to be the least ordinal κ such that there is a bijection $f\colon A \to \kappa$. If $\kappa = |\kappa|$, we say that κ is a* cardinal.

The following properties are well-known and discussed in detail, for example, in [29].

Lemma 6.13. *Let A, B be sets. Then,*

1. *$|A \cup B| \leq \max\{\omega, |A|, |B|\}$;*

2. *if at least one of A, B is infinite, then $|A \cup B| = \max\{|A|, |B|\}$;*

3. *$|A \times B| \leq \max\{\omega, |A|, |B|\}$,*

4. *if one of A, B is infinite and both are non-empty, $|A \times B| = \max\{|A|, |B|\}$, and*

5. *if $\{A_i : i \in I\}$ is a family of sets, then*

$$\left|\bigcup_{i \in I} A_i\right| \leq \max\left\{\omega, \sup_{i \in I} |A_i|, |I|\right\}.$$

These results readily allow us to compute the cardinalities of ordinals obtained using addition and multiplication.

Lemma 6.14. *Let α, β be ordinals. Then,*

1. $|\alpha + \beta| \leq \max\{\omega, |\alpha|, |\beta|\}$;
2. $|\alpha + \beta| = \max\{|\alpha|, |\beta|\}$ *if one of the two is infinite;*
3. $|\alpha\beta| \leq \max\{\omega, |\alpha|, |\beta|\}$, *and*
4. $|\alpha\beta| = \max\{|\alpha|, |\beta|\}$ *if one of the two is infinite and both are non-zero.*

Proof. These claims are immediate from Lemma 6.13 if we observe that $\alpha + \beta$ is the disjoint union of α with $[\alpha, \alpha + \beta)$, and $|[\alpha, \alpha + \beta)| = |\beta|$, while $\alpha\beta$ is in bijection with $\alpha \times \beta$ (via the map $\alpha\xi + \zeta \mapsto (\zeta, \xi) \in \alpha \times \beta$). □

Similar claims hold for the hyperexponential function:

Lemma 6.15. *Let α, β be arbitrary ordinals. Then, $|e^\alpha \beta| \leq \max\{\omega, |\alpha|, |\beta|\}$. If moreover $\beta > 0$ and $\max\{\alpha, \beta\} \geq \omega$, then $|e^\alpha \beta| = \max\{|\alpha|, |\beta|\}$.*

Proof. To bound $|e^\alpha \beta|$, we proceed by induction on α with a secondary induction on β to show that $|e^\alpha \beta| \leq \max\{\omega, |\alpha|, |\beta|\}$. We consider several cases, using Lemma 5.12. If $\alpha = 0$, then $e^0 \beta = \beta$, so the claim is obviously true. If $\beta = 0$, we see that $e^\alpha 0 = 0$, so the claim holds as well. For $\alpha = 1$ and $\beta = \gamma + 1$,

$$e(\gamma + 1) = \lim_{n < \omega} (1 + e\gamma) \cdot n \stackrel{\text{IH}}{\leq} \max\{\omega, |\alpha|, |\beta|\}.$$

If α is a limit and $\beta = 1$,

$$e^\alpha 1 = \lim_{\gamma < \alpha} e^\gamma 1 \stackrel{\text{IH}}{\leq} \max\{\omega, |\alpha|, |\beta|\}.$$

For $\alpha = \gamma + 1$ with $\gamma > 0$ we obtain

$$e^{\gamma+1}\beta = e^\gamma e\beta \stackrel{\text{IH}}{\leq} \max\{|\alpha|, |e\beta|\}$$
$$\stackrel{\text{IH}}{\leq} \max\{\omega, |\alpha|, \max\{\omega, |\alpha|, |\beta|\}\} = \max\{\omega, |\alpha|, |\beta|\}.$$

If β is a limit, then we obtain

$$e^\alpha \beta = \lim_{\gamma < \beta} e^\alpha \gamma \stackrel{\text{IH}}{\leq} \max\{\omega, |\alpha|, |\beta|\}.$$

Finally, for limit α and $\beta = \delta + 1$ we obtain

$$e^\alpha(\delta + 1) = \lim_{\gamma < \alpha} e^\gamma(e^\alpha(\delta) + 1) \stackrel{\text{IH}}{\leq} \max\{\omega, |\alpha|, |\beta|\}.$$

Since this covers all cases, the result follows.

For the second claim, if $\beta > 0$, then $e^\alpha \beta \geq \max\{\alpha, \beta\}$, so $|e^\alpha \beta| \geq \max\{|\alpha|, |\beta|\}$ and we obtain the desired equality if one of the two is infinite. □

Corollary 6.16. *If κ is an uncountable cardinal, then κ is additively indecomposable and $e^\kappa 1 = \kappa$.*

Proof. We know that $e^\kappa 1 \geq \kappa$. However, from Lemma 6.15, $|e^\xi 1| < \kappa$ whenever $\xi < \kappa$, so that $e^\xi 1 < \kappa$. But $e^\kappa 1 = \lim_{\xi < \kappa} e^\xi 1$, so $e^\kappa 1 = \kappa$, from which it also follows that $\kappa = \omega^\kappa$ and thus is additively indecomposable. □

We have a simiar situation with worms; it is very easy to infer the cardinality of $o(\mathfrak{w})$ by looking at the entries in $\mathfrak{w}$.

Lemma 6.17. *If $\mathfrak{w} \in \mathbb{W}$ then $|o(\mathfrak{w})| \leq |\max \omega \mathfrak{w}|$. If moreover $\mathfrak{w} \neq \top$ and $\max \mathfrak{w} \geq \omega$, then $|o(\mathfrak{w})| = |\max \mathfrak{w}|$.*

Proof. We prove by induction on $\|\mathfrak{w}\|$ that $|o(\mathfrak{w})| \leq |\max \omega \mathfrak{w}|$. For $\mathfrak{w} = \top$ this is obvious. Otherwise, let $\mu = \min \mathfrak{w}$. If $\mu = 0$, then $o(\mathfrak{w}) = ob(\mathfrak{w}) + 1 + oh(\mathfrak{w})$, so that by Lemma 6.15,

$$|o(\mathfrak{w})| = |ob(\mathfrak{w}) + 1 + oh(\mathfrak{w})| \leq \max\{\omega, |ob(\mathfrak{w})|, 1, |oh(\mathfrak{w})|\}.$$

By the induction hypothesis $|oh(\mathfrak{w})| \leq |\max \omega h(\mathfrak{w})| \leq |\max \omega \mathfrak{w}|$ and similarly for $|ob(\mathfrak{w})|$, so we obtain $|o(\mathfrak{w})| \leq |\max \omega \mathfrak{w}|$.

If $\mu > 0$, then $o(\mathfrak{w}) = e^\mu (\mu \downarrow \mathfrak{w})$. Since $\mu, \max(\mu \downarrow \mathfrak{w}) \leq \max \mathfrak{w}$ and $\|\mu \downarrow \mathfrak{w}\| < \|\mathfrak{w}\|$, we use the induction hypothesis and Lemma 6.15 once again to see that

$$|o(\mathfrak{w})| = |e^\mu o(\mu \downarrow \mathfrak{w})| \leq \max\{\omega, |\mu|, |\max(\mu \downarrow \mathfrak{w})|\} \leq |\max \omega \mathfrak{w}|.$$

The claim follows.

For the second claim, if $\mathfrak{w} \neq \top$ and $\max \mathfrak{w} \geq \omega$, then by Lemma 3.26.1, $o(\mathfrak{w}) \geq \max \mathfrak{w}$, so

$$|o(\mathfrak{w})| \geq |\max \mathfrak{w}| = |\max \omega \mathfrak{w}|,$$

and thus we obtain equality. □

Similarly, closure under a function f does not produce many more ordinals than we had to begin with:

Lemma 6.18. *If $f: \mathrm{Ord}^{<\omega} \dashrightarrow \mathrm{Ord}$ and Θ is a set of ordinals, then*

$$|\Theta| \leq \left|\overline{\Theta}^f\right| \leq \max\{\omega, |\Theta|\}.$$

Proof. We inductively check that

$$|\Theta| \leq \left|\Theta_n^f\right| \leq \max\{\omega, |\Theta|\}, \tag{5}$$

from which the lemma follows using the fact that $\overline{\Theta}^f = \bigcup_{n<\omega} \Theta_n^f$.

We have that $\Theta_0^f = \Theta$, so (5) holds. Now, assume inductively that (5) holds for n. Then, $\Theta_{n+1}^f = \Theta_n^f \cup f[\Theta_n^f]$; by the induction hypothesis,

$$|\Theta| \leq \left|\Theta_n^f\right| \leq \left|\Theta_{n+1}^f\right|.$$

Now, elements of $f[\Theta]$ are of the form $f(\xi_1, \ldots, \xi_m)$ with $\xi_1, \ldots, \xi_m \in \Theta_n^f$; but there are at most $\max\{\omega, \left|\Theta_n^f\right|\}$ of these, so

$$\left|f[\Theta_n^f]\right| \leq \max\left\{\omega, \left|\Theta_n^f\right|\right\} \stackrel{\text{IH}}{\leq} \max\{\omega, |\Theta|\},$$

from which it follows that

$$\left|\Theta_{n+1}^f\right| = \left|\Theta_n^f \cup f[\Theta_n^f]\right| \leq \max\left\{\omega, \left|\Theta_n^f\right|, \left|f[\Theta_n^f]\right|\right\} \stackrel{\text{IH}}{\leq} \max\{\omega, |\Theta|\}. \qquad \square$$

This tells us that none of the ordinal operations we have discussed so far will give rise to any uncountable ordinals. So, we may add one directly; we can then use it to produce more countable ordinals using *collapsing functions*. We shall present them using hyperexponentials rather than Veblen functions, although this change is merely cosmetic as the two define the same ordinals. It is standard to use Ω to denote a 'big' ordinal, which for convenience may be assumed to be ω_1. However, we mention that, with some additional technical work, one can take $\Omega = \omega_1^{CK}$, the first non-computable ordinal [34].

Definition 6.19. *Let Ω, ξ be ordinals. We simultaneously define the sets $C(\xi)$ and the ordinals $\psi(\xi)$ by induction on ξ as follows:*

1. *$C(\xi)$ is the least set such that*

    (a) *$\Omega \in C(\xi)$,*

    (b) *$C(\xi)$ is hyperexponentially closed, and*

    (c) *if $\alpha \in C(\xi)$ and $\alpha < \xi$ then $\psi(\alpha) \in C(\xi)$.*

2. *$\psi(\xi)$ is the least λ such that $\lambda \notin C(\xi)$.*

In the notation of Definition 4.16, let BH_ξ be the pair of functions $\{\mathrm{HE}, \psi \restriction \xi\}$. Then,
$$C(\xi) = \overline{\{0, 1, \Omega\}}^{\mathrm{BH}_\xi}.$$

Thus our previous work on closures under ordinal functions readily applies to the sets $C(\xi)$. The function ψ appears in the ordinal analysis of systems such as ID_1 and Kripke-Platek set-theory with infinity [33].

Lemma 6.20. *If ξ is any ordinal, then $\psi(\xi)$ is additively indecomposable and $\psi(\xi) = e^{\psi(\xi)} 1$.*

Proof. To see that $\psi(\xi)$ is additively indecomposable, we will assume otherwise and reach a contradiction. Hence, suppose that $\psi(\xi) = \alpha + \beta$ with $\alpha, \beta < \psi(\xi)$. By definition of $\psi(\xi)$ we have that $\alpha, \beta \in C(\xi)$, hence $\psi(\xi) = \alpha + \beta \in C(\xi)$, contradicting its definition.

Next we show that $\psi(\xi) = e^{\psi(\xi)} 1$. By Proposition 5.14, there are α, β with β either 1 or additively decomposable such that $\psi(\xi) = e^\alpha \beta$. Since $\psi(\xi)$ is additively indecomposable we have that $\beta \ne \psi(\xi)$, and since e^α is normal, we have that $\beta < \psi(\xi)$. Now, towards a contradiction, assume that $\alpha < \psi(\xi)$; then $\alpha, \beta \in C(\xi)$ so $\psi(\xi) \in C(\xi)$, contrary to its definition. We conclude that $\alpha = \psi(\xi)$, and again since $e^{\psi(\xi)}$ is normal and $e^{\psi(\xi)} 1 \ge \psi(\xi)$, that $\beta = 1$. $\square$

We remark that the above lemma already tells us that the *countable* ordinals we can construct using ψ are much bigger than Γ_0; indeed, we already have that $\Gamma_0 = \psi(0)$, and this is only scratching the surface of our notation system: ordinals such as $\psi(\Omega)$ or $\psi(e^\omega(\Omega + 1))$ are much larger. The latter is the Howard-Bachmann ordinal $\psi(\varepsilon_{\Omega+1})$, as one can readily check that $e^\omega \xi = \varepsilon_\xi$ for all ξ using Proposition 5.15.

Lemma 6.21. *Assume that Ω is such that $\Omega = e^\Omega 1$. If ξ is any ordinal, then $C(\xi)$ is hyperexponentially perfect.*

Proof. We already know that $C(\xi)$ is hyperexponentially closed, so it remains to show that it is reductive. Let $\zeta \in C(\xi)$. By Lemma 4.17.3, either $\zeta \in \{0, 1, \Omega\}$, there are $\alpha, \beta, \gamma \ne \zeta$ with $\zeta = e^\alpha(\beta + \gamma)$, or $\zeta = \psi(\alpha)$ for some $\alpha \in C(\xi) \cap \xi$. If $\zeta < 2$, there is nothing to prove, so we assume otherwise.

First assume that $\zeta = e^\alpha(\beta + \gamma)$. If ζ is additively decomposable, by Lemma 6.20, we cannot have that $\alpha > 0$, so we conclude that $\zeta = e^0(\beta + \gamma) = \beta + \gamma$, as needed. If it is additively indecomposable, since $\beta + \gamma \in C(\xi)$, then we already have that $\zeta = e^\alpha(\beta + \gamma)$ with $\alpha, \beta + \gamma \in C(\xi)$. In all other cases, ζ must be additively indecomposable. If $\zeta = \Omega$, then $\zeta = e^\Omega 1$ and $\Omega, 1 \in C(\xi)$, and if $\zeta = \psi(\alpha)$, by Lemma 6.20, $\zeta = e^\zeta 1$, with $\zeta, 1 \in C(\xi)$. □

The intention of the function ψ is to produce new countable ordinals from possibly uncountable ones. Let us see that this is the case:

Lemma 6.22. *Let ξ be any ordinal and $\Omega = \omega_1$. Then, $C(\xi)$ is countable and $\psi(\xi) < \Omega$.*

Proof. The first claim is an instance of Lemma 6.18, while the second is immediate from the first. □

Observe that $\sup C(\xi) = \Gamma_{\Omega+1}$, the first hyperexponentially closed ordinal which is greater than Ω, and thus the smallest ordinal not contained in any $C(\xi)$ is $\psi(\Gamma_{\Omega+1})$. However, our worm notation will give slightly smaller ordinals. Thus it will be convenient to consider a "cut-off" version of the sets $C(\xi)$. Let us see that these cut-off versions maintain a restricted version of the minimality property of $C(\xi)$.

Lemma 6.23. *If $\mu \leq \lambda$ are ordinals such that $\Omega < \lambda$, then $C(\mu) \cap \lambda$ is the least set D such that:*

(i) $0, 1, \Omega \in D$;

(ii) if $\alpha, \beta, \gamma \in D$ and $e^\alpha(\beta+\gamma) < \lambda$ then $e^\alpha(\beta+\gamma) \in D$, and

(iii) if $\alpha \in D \cap \mu$ then $\psi(\alpha) \in D$.

Proof. First we observe that $C(\mu) \cap \lambda$ indeed satisfies (i)-(iii), where for the first item we use the assumption that $\Omega < \lambda$ and for the third we use Lemma 6.22 to see that $\psi(\alpha) < \Omega < \lambda$. Now, let D be the least set satisfying (i)-(iii), and consider

$$D' = D \cup (C(\mu) \setminus \lambda).$$

One readily verifies that $0, 1, \Omega \in D'$, and that if $\alpha, \beta, \gamma \in D'$ then $e^\alpha(\beta + \gamma) \in D'$ (using the fact that $D \subseteq C(\mu) \cap \lambda \subseteq C(\mu)$ by minimality of D). Finally, if $\alpha < \mu$ and $\alpha \in D'$, then since $\mu \leq \lambda$ we have that $\alpha \in D$, and since D satisfies (iii) we have that $\psi(\alpha) \in D \subseteq D'$. But by definition $C(\mu)$ is the least set with these properties, so we obtain $C(\mu) \subseteq D'$, and hence

$$C(\mu) \cap \lambda \subseteq D' \cap \lambda = D,$$

as was to be shown. □

We remark that the ordinal $\psi(\Gamma_{\Omega+1})$ is computable, meaning that it is isomorphic to an ordering $\langle A, \preceq \rangle$, where $A \subseteq \mathbb{N}$ and both A and $\preceq$ are Δ^0_1-definable; however, we will not go into details here, and instead refer the reader to a text such as [33].

6.4 Collapsing uncountable worms

Now let us turn our attention to uncountable worms. The general idea is as follows. We have seen in Theorem 5.38 that worms give us a notation system for Γ_0 if we interpret $\langle\mathfrak{w}\rangle$ as $\langle o(\mathfrak{w})\rangle$. Meanwhile, now we have a new modality $\langle\infty\rangle$, which we can regard as $\langle\omega_1\rangle$. Note that, by Corollary 6.16,

$$o(\langle\omega_1\rangle\top) = e^{\omega_1}o(\langle 0\rangle\top) = \omega_1.$$

Thus if we add the new symbol Ω representing $\langle\omega_1\rangle$ to Beklemishev's autonomous worms, we see inductively that

$$\langle\omega_1\rangle\top = \Omega^o = (\Omega)^o = ((\Omega))^o\ldots$$

Moreover, if such operations are to be interpreted proof-theoretically using iterated ω-rules, then in view of Proposition 6.9 we have that $\langle\omega_1\rangle\top \equiv \langle\omega_1 + \xi\rangle\top$ for any ordinal ξ. Thus we also would have, for example,

$$\langle\omega_1\rangle\top = (\Omega)^o = (()\Omega)^o = (\Omega\Omega)^o\ldots$$

This would lead to quite a wasteful notation system! Thus we will adopt the following rule: when writing an autonomous worm $(\mathbf{w})\top$, if $o(\mathbf{w})$ is countable, then we will take it at face-value and interpret $(\mathbf{w})\top$ as $\langle o(\mathbf{w})\rangle\top$. However, if $o(\mathbf{w})$ is uncountable, we will first "project" it to a countable ordinal, in order to represent large countable worms.

Of course, projections will be very similar to collapsing functions; however, given that countable ordinals are taken at face value, these projections will have the property that $\pi = \pi \circ \pi$ (thus their name). Other than that, their construction is very similar to that of ψ:

Definition 6.24. *Given a worm $\mathfrak{w} \in \mathbb{W}$ and an ordinal Ω, we define $U(\mathfrak{w}) \subseteq \mathsf{Ord}$ and a map $\pi\colon \mathbb{W} \to \mathsf{Ord}$ by induction on $\mathfrak{w}$ along $\lhd$ as follows.*

1. *Let $U(\mathfrak{w})$ be the least set of ordinals such that*

    (a) $\Omega \in U(\mathfrak{w})$,

    (b) *if $\mathfrak{u} \sqsubset U(\mathfrak{w})$ and $\mathfrak{u} \lhd \mathfrak{w}$ then $\pi(\mathfrak{u}) \in U(\mathfrak{w})$.*

2. *Then, set*

    (a) $\pi(\mathfrak{w}) = o(\mathfrak{w})$ *if $\mathfrak{w} \sqsubset \Omega$,*

    (b) *otherwise, set $\pi(\mathfrak{w})$ to be the least ordinal μ such that $\mu \notin U(\mathfrak{w})$.*

We will write $\pi(\mathfrak{w})$ or $\pi\mathfrak{w}$ indistinctly. Once again, we can write Definition 6.24 in the terminology of Definition 4.16 by setting

$$U(\mathfrak{w}) = \overline{\{\Omega\}}^{\pi\restriction\{\mathfrak{v}:\mathfrak{v}\triangleleft\mathfrak{w}\}}.$$

Thus Lemma 6.18 gives us the following:

Lemma 6.25. *For every worm $\mathfrak{w}$, $U(\mathfrak{w})$ and $\pi\mathfrak{w}$ are countable.*

Throughout this section we will assume that $\Omega = \omega_1$, so that from Lemma 6.25 we obtain $\pi\mathfrak{w} < \Omega$ for all worms $\mathfrak{w}$. As was the case for defining ψ, with some extra technical work we can take $\Omega = \omega_1^{CK}$ instead.

Note that $U(\mathfrak{w})$ itself is not worm-closed, as it does not contain, for example, the ordinal $\Omega + 1 = o(0\Omega\top)$. However, its countable part is indeed worm-perfect. The next lemmas will establish this fact. First, we show that it is worm-closed.

Lemma 6.26. *For any worm $\mathfrak{v}$ with $o(\mathfrak{v}) \geq \Omega$, $U(\mathfrak{v}) \cap \Omega$ is worm-closed.*

Proof. By Corollary 3.19.3, if $\mathfrak{w} \sqsubset U(\mathfrak{v}) \cap \Omega$, then $\mathfrak{w} \triangleleft \Omega\top \triangleleft \mathfrak{v}$, so that $o(\mathfrak{w}) = \pi(\mathfrak{w}) \in U(\mathfrak{v})$. But by Lemma 6.17, $o(\mathfrak{w}) < \Omega$, so $o(\mathfrak{w}) \in U(\mathfrak{v}) \cap \Omega$ as needed. □

Recall that Lemma 6.20 states that $\psi(\xi) = e^{\psi(\xi)}1$. Next, we show that π enjoys a similar property.

Lemma 6.27. *If $o(\mathfrak{w}) \geq \Omega$, then $o(\langle\pi\mathfrak{w}\rangle\top) = \pi\mathfrak{w}$.*

Proof. Suppose not. Then, by Lemma 3.26.1, $\pi\mathfrak{w} < o(\langle\pi\mathfrak{w}\rangle\top)$, so that by Corollary 3.27, there is a worm $\mathfrak{v}$ such that $o(\mathfrak{v}) = \pi\mathfrak{w}$. Since $o(\mathfrak{v}) < o(\langle\pi\mathfrak{w}\rangle\top)$, by Lemma 3.26.1 once again, we must have that $\mathfrak{v} \sqsubset \pi\mathfrak{w} \subseteq U(\mathfrak{w})$. But by Lemma 6.26, $\pi\mathfrak{w} = o(\mathfrak{v}) \in U(\mathfrak{w})$, contradicting the definition of $\pi\mathfrak{w}$. □

Lemma 6.28. *For any worm $\mathfrak{w}$, $U(\mathfrak{w}) \cap \Omega$ is worm-perfect and*

$$U(\mathfrak{w}) \cap \Omega = U(\mathfrak{w}) \setminus \{\Omega\}.$$

Proof. For the first claim, in view of Lemma 6.26, it remains to show that if $\xi \in U(\mathfrak{w}) \cap \Omega$, then $\xi = o(\mathfrak{v})$ for some $\mathfrak{v} \sqsubset U(\mathfrak{w}) \cap \Omega$. By definition of $U(\mathfrak{w})$, if $\xi \in U(\mathfrak{w}) \cap \Omega$, then $\xi = \pi\mathfrak{u}$ for some $\mathfrak{u} \sqsubset U(\mathfrak{w})$. If $\mathfrak{u} \sqsubset \Omega$, then $\xi = \pi\mathfrak{u} = o\mathfrak{u}$. Otherwise, by Lemma 6.27, $\xi = \pi\mathfrak{u} = o(\langle\pi\mathfrak{u}\rangle\top)$.

The second claim is immediate from Lemma 6.25 and the assumption that $\Omega = \omega_1$, since $\pi\mathfrak{w} < \Omega$ for every worm $\mathfrak{w}$. □

However, as we have mentioned, $U(\mathfrak{w})$ itself is not worm-closed, and neither is $o[U(\mathfrak{w})]$. Nevertheless, the latter does satisfy a bounded form of hyperexponential closure:

Lemma 6.29. *Given any worm $\mathfrak{w}$ and ordinals α, β, if $\alpha, \beta \in o[U(\mathfrak{w})]$ and $e^\alpha \beta < e^{\Omega+1}1$ then $e^\alpha \beta \in o[U(\mathfrak{w})]$.*

Proof. If $\alpha, \beta \in o[U(\mathfrak{w})]$ and $e^\alpha \beta < e^{\Omega+1}1$, we may assume without loss of generality that $\beta > 0$ (since otherwise $e^\alpha \beta = 0$), so by the assumption that $o(\mathfrak{w}) < e^{\Omega+1}1 = e^\Omega \omega$, we see by monotonicity that either $\alpha < \Omega$, or $\alpha = \Omega$ and $\beta < \omega$.

First assume that $\alpha < \Omega$, and let

$$\mathfrak{v} = \lambda_1 \ldots \lambda_n \top \sqsubset U(\mathfrak{w})$$

be such that $\beta = o(\mathfrak{v})$. In view of Lemma 6.28, for each $\lambda \in [1, n]$, either $\lambda_i = \Omega$, in which case $\alpha + \lambda_i = \lambda_i$, or $\lambda_i \in U(\mathfrak{w}) \cap \Omega$, which since $U(\mathfrak{w}) \cap \Omega$ is worm-perfect (Lemma 6.28) gives us $\alpha + \lambda_i \in U(\mathfrak{w}) \cap \Omega \subseteq U(\mathfrak{w})$ (Lemma 5.33). Thus $\alpha + \lambda_i \in U(\mathfrak{w})$ for each i, hence $\alpha \uparrow \mathfrak{v} \sqsubset U(\mathfrak{w})$, and

$$o(\alpha \uparrow \mathfrak{v}) = e^\alpha o(\mathfrak{v}) = e^\alpha \beta.$$

Otherwise, $\alpha = \Omega$, so $\beta < \omega$ and we see that $o(\langle \Omega \rangle^\beta \top) = e^\alpha \beta$. In either case, it follows that $e^\alpha \beta \in o[(U(\mathfrak{w}))]$. □

Lemma 6.30. *Suppose that $\Omega = \omega_1$. Then, given any worm $\mathfrak{w}$,*

$$e^{\Omega+1}1 = \sup \left\{ o(\mathfrak{v}) : \exists \mathfrak{w} \ (\mathfrak{v} \sqsubset U(\mathfrak{w})) \right\}.$$

Proof. Let

$$\Lambda = \sup \left\{ o(\mathfrak{v}) : \exists \mathfrak{w} \ (\mathfrak{v} \sqsubset U(\mathfrak{w})) \right\}.$$

We have that

$$e^{\Omega+1}1 = e^\Omega \omega = \lim_{n < \omega} e^\Omega n.$$

But, $e^\Omega n = o(\langle \Omega \rangle^n \top)$, so $e^{\Omega+1}1 \leq \Lambda$.

To see that $\Lambda \leq e^{\Omega+1}1$, proceed by induction on $\|\mathfrak{v}\|$ to show that if $\mathfrak{v} \sqsubset U(\mathfrak{w})$ for some $\mathfrak{w}$, then $o(\mathfrak{v}) < e^{\Omega+1}1$.

If $\mathfrak{v} = \top$ there is nothing to prove, and if $\min \mathfrak{v} = 0$ then by the induction hypothesis, $oh(\mathfrak{v}), ob(\mathfrak{v}) < e^{\Omega+1}1$. Since the latter is additively indecomposable,

$$o(\mathfrak{v}) = ob(\mathfrak{v}) + 1 + oh(\mathfrak{v}) < e^{\Omega+1}1.$$

Finally, if $\mu = \min \mathfrak{v} > 0$, then $o(\mathfrak{v}) = e^\mu o(\mu \downarrow \mathfrak{v})$. Consider two cases. If $\mu < \Omega$, then since by the induction hypothesis $o(\mu \downarrow \mathfrak{v}) < e^{\Omega+1}1$, we obtain

$$e^\mu o(\mu \downarrow \mathfrak{v}) < e^\mu e^{\Omega+1}1 = e^{\mu+\Omega+1}1 = e^{\Omega+1}1.$$

Otherwise, $\mu = \Omega$, but this means that $\mu \downarrow \mathfrak{v} = 0^n\top$ for some n, hence $o(\mathfrak{v}) = e^\Omega n < e^{\Omega+1}1$. □

The above results tell us that π behaves a lot like a version of ψ that is restricted to $e^{\Omega+1}1$. Let us see that this is, in fact, the case.

Lemma 6.31. *For every worm $\mathfrak{w}$ with $o(\mathfrak{w}) \in [\Omega, e^{\Omega+1}1]$,*

1. $C(-\Omega + o(\mathfrak{w})) \cap e^{\Omega+1}1 = o[U(\mathfrak{w})]$, *and*

2. $\pi(\mathfrak{w}) = \psi(-\Omega + o(\mathfrak{w}))$.

Proof. We prove both claims by induction on $o(\mathfrak{w})$. Set $C = C(-\Omega + o(\mathfrak{w}))$. First let us show that
$$C \cap e^{\Omega+1}1 \subseteq o[U(\mathfrak{w})].$$
Note that by Lemma 6.23, $C \cap e^{\Omega+1}1$ is the least set containing $0, 1, \Omega$, closed under $\alpha, \beta, \gamma \mapsto e^\alpha(\beta+\gamma)$ below $e^{\Omega+1}1$, and closed under $\psi \restriction (-\Omega + o(\mathfrak{w}))$. But by Lemma 5.33, $o[U(\mathfrak{w})]$ is closed under addition and by Lemma 6.29, by hyperexponentiation below $e^{\Omega+1}1$, so we only need to check that it is closed under $\psi \restriction (-\Omega + o(\mathfrak{w}))$.

If $\alpha \in o[U(\mathfrak{w})]$ and $\alpha < o(-\Omega + o(\mathfrak{w}))$, then by Lemma 5.33 we have that $\Omega + \alpha = o(\mathfrak{u})$ for some $\mathfrak{u} \sqsubset U(\mathfrak{w})$. Then, by the induction hypothesis,
$$\psi(\alpha) = \psi(-\Omega + o(\mathfrak{u})) = \pi(\mathfrak{u}) \in U(\mathfrak{w}),$$
so that $\pi(\mathfrak{u})\top \sqsubset U(\mathfrak{w})$ and by Lemma 6.27, $\pi(\mathfrak{u}) = o(\pi(\mathfrak{u})\top)$, as needed. Thus by the minimality of $C \cap e^{\Omega+1}1$, we conclude that $C \cap e^{\Omega+1}1 \subset o[U(\mathfrak{w})]$.

Next we check that
$$o[U(\mathfrak{w})] \subseteq C \cap e^{\Omega+1}1.$$
By Lemma 6.30, $o[U(\mathfrak{w})] \subseteq e^{\Omega+1}1$, so we only need to prove that $o[U(\mathfrak{w})] \subseteq C$. But, in view of Lemmas 6.21 and Lemma 5.34, C is worm-perfect. Thus to show that $o[U(\mathfrak{w})] \subseteq C$, it suffices to prove that $U(\mathfrak{w}) \subseteq C$. As before, we show that C satisfies the inductive definition of $U(\mathfrak{w})$.

Let $\mathfrak{v} \sqsubset C$ be such that $\mathfrak{v} \lhd \mathfrak{w}$. Once again by Lemma 6.21, we have that $o(\mathfrak{v}) \in C$. Now, if $o(\mathfrak{v}) < \Omega$, then this gives us $\pi\mathfrak{v} = o(\mathfrak{v}) \in C$. Otherwise, $-\Omega + o(\mathfrak{v}) < -\Omega + o(\mathfrak{w})$, and thus $\psi(-\Omega + o(\mathfrak{v})) \in C$. But, by the induction

3341

hypothesis, $\psi(-\Omega+o(\mathfrak{v})) = \pi\mathfrak{v}$, so that $\pi\mathfrak{v} \in C$, as needed. By minimality of $U(\mathfrak{w})$, we conclude that $U(\mathfrak{w}) \subseteq C$ and thus $o[U(\mathfrak{w})] \subseteq C$.

Since we have shown both inclusions, we conclude that

$$o[U(\mathfrak{w})] = C \cap e^{\Omega+1}1.$$

Moreover, $\psi(-\Omega+o(\mathfrak{w}))$ is defined as the least ordinal not in $C = C(-\Omega+o(\mathfrak{w}))$, and since C is countable it is also the least ordinal not in $C \cap \Omega$. Similarly, $\pi\mathfrak{w}$ is the least ordinal not in $U(\mathfrak{w}) \cap \Omega = o[U(\mathfrak{w})] \cap \Omega$. Since these two sets are equal, it follows also that $\psi(-\Omega + o(\mathfrak{w})) = \pi\mathfrak{w}$. □

Corollary 6.32. $\pi(\langle\Omega+1\rangle\top) = \psi(e^{\Omega+1}1)$.

Proof. Immediate from Lemma 6.31 using the fact that

$$e^{\Omega+1}1 = e^{\Omega+1}o(\langle 0\rangle\top) = o(\langle e^{\Omega+1}\rangle\top).$$ □

6.5 Impredicative worm notations

Now let us extend Beklemishev's autonomous worms with the new modality Ω and projections of uncountable worms. Aside from the addition of Ω, the presentation is very similar to that of Section 5.7.

Definition 6.33. *Define the set of impredicative autonomous worms to be the least set W_Ω such that*

(i) $\top \in \mathsf{W}_\Omega$, *and*

(ii) if $\mathtt{w}, \mathtt{v} \in \mathsf{W}_\Omega$, *then*

    (a) $(\mathtt{w})\mathtt{v} \in \mathsf{W}_\Omega$, *and*   *(b)* $\Omega\mathtt{v} \in \mathsf{W}_\Omega$.

As before, the intention is for impredicative autonomous worms to be interpreted as standard worms. We do this via the following translation:

Definition 6.34. *We define a map* $\cdot^\pi \colon \mathsf{W}_\Omega \to \mathbb{W}$ *given by*

1. $\top^\pi = \top$,

2. $((\mathtt{w})\mathtt{v})^\pi = \langle\pi(\mathtt{w}^\pi)\rangle\mathtt{v}^\pi$, *and*

3. $(\Omega\mathtt{v})^\pi = \langle\Omega\rangle\mathtt{v}^\pi$.

Every ordinal in $U(\langle\Omega+1\rangle\top)\cap\Omega$ can be represented as an autonomous worm. Below, define $\mathsf{pw} = \pi(\mathsf{w}^\pi)$.

Lemma 6.35. *If $\Omega = \omega_1$, then for every ordinal $\xi \in U(\langle\Omega+1\rangle\top)\cap\Omega$ there is $\mathsf{w}\in\mathsf{W}_\Omega$ such that $\xi = \mathsf{pw}$.*

Proof. Using the notation of Definition 4.16, we prove by induction on n that if $\xi \in \{\Omega\}_n^\pi \cap \Omega$, then there is $\mathsf{w}\in\mathsf{W}_\Omega$ such that $\xi = \mathsf{pw}$. If $n=0$ there is nothing to prove, so we may assume that $n=k+1$. Write $\xi = \pi(\mathfrak{v})$ with $\mathfrak{v} \sqsubset \{\Omega\}_k^\pi$. If $\mathfrak{v} = \top$, then $\xi = 0 = \mathsf{p}\top$. Otherwise, we can write $\mathfrak{v} = \lambda\mathfrak{u}$ for some worm $\mathfrak{u}$. By a secondary induction on the length of $\mathfrak{v}$, we have that $\mathfrak{u} = \mathsf{u}^\pi$ for some $\mathsf{u} \in \mathsf{W}_\Omega$; meanwhile, either $\lambda = \Omega$, and $\mathsf{v} = \Omega\mathsf{u} \in \mathsf{W}_\Omega$ satisfies

$$\mathsf{pv} = \pi(\mathsf{v}^\pi) = \pi(\langle\Omega\rangle\mathsf{u}^\pi) = \pi(\langle\Omega\rangle\mathfrak{u}) = \pi(\mathfrak{v}) = \xi,$$

or $\lambda < \Omega$, which means that $\lambda \in \{\Omega\}_k^\pi$, so by the induction hypothesis, $\lambda = \mathsf{pw}$ for some $\mathsf{w} \in \mathsf{W}_\Omega$. It follows that $\xi = \mathsf{p}(\mathsf{w})\mathsf{v}$, as desired. □

Just as autonomous worms gave us a notation system for Γ_0, impredicative autonomous worms give us a notation system for $\psi(e^{\Omega+1}1)$.

Theorem 6.36. *If $\Omega = \omega_1$, then for every $\xi < \psi(e^{\Omega+1}1)$ there is $\mathsf{w} \in \mathsf{W}_\Omega$ such that $\xi = \mathsf{pw}$.*

Proof. By Corollary 6.32,

$$\psi(e^{\Omega+1}1) = \pi(\langle\Omega+1\rangle\top),$$

and the latter is, by definition, the least ordinal not belonging to $U(\langle\Omega+1\rangle\top)$. Moreover, $\psi(e^{\Omega+1}1)$ is countable by Lemma 6.22, so we have that $\xi < \Omega$. It follows that

$$\psi(e^{\Omega+1}1) \subseteq U(\langle\Omega+1\rangle\top) \cap \Omega;$$

thus we obtain the claim by Lemma 6.35. □

Impredicative autonomous worms may be suitable for a consistency proof in the spirit of Theorem 4.5 for theories with proof-theoretic strength the Bachmann-Howard ordinal (or even slightly more powerful theories). Examples of such theories are the theory ID_1 of non-iterated inductive definitions, Kripke-Platek with infinity, and *parameter-free* Π_1^1-CA_0, where the Π_1^1 comprehension axiom is restricted to formulas without free set variables. However, the proof-theoretical ordinal of unrestricted Π_1^1-CA_0 is quite a bit larger, and obtained by collapsing all of the ordinals $\{\aleph_n : n < \omega\}$.

We remark that our notation system does not take the oracle in $[\infty|X]_T$ into account, and it is possible that autonomous worms with oracles would indeed give us a notation system for the proof-theoretical ordinal of $\Pi^1_1\text{-CA}_0$. However, we will not follow this route; instead, we will pass from worms to *spiders,* which will allow us to obtain notations for this, and much larger, ordinals.

7 Spiders

The problem with using iterated ω-rules to interpret $[\lambda]_T\phi$ is that GLP no longer applies when $\lambda \geq \omega_1$; since we have that $[\omega_1+1]_T\phi$ is equivalent to $[\omega_1]_T\phi$, we cannot expect the GLP axiom $\langle\omega_1\rangle\phi \to [\omega_1+1]\langle\omega_1\rangle\phi$ to hold. So the question naturally arises: what kind of (sound) provability operator could derive all true instances of $\langle\omega_1\rangle\phi$?

Well, we know that $\langle\omega_1\rangle\phi$ is equivalent to $\forall\xi<\omega_1 \langle\xi\rangle\phi$, which gives us a strategy for proving that $\langle\infty\rangle_T\phi$ holds: prove that

$$\langle 0\rangle_T\phi, \langle 1\rangle_T\phi, \langle 2\rangle_T\phi, \ldots, \langle\omega\rangle_T\phi, \ldots, \langle\Gamma_0\rangle_T\phi, \ldots \langle\psi(\varepsilon_{\Omega+1})\rangle_T\phi, \ldots$$

all hold, and more generally, that $\langle\xi\rangle_T\phi$ holds for all $\xi < \omega_1$. Let us sketch some ideas for formalizing this in the language of set-theory. We remark that this material is exploratory, and will be studied in detail in upcoming work.

7.1 $\aleph_\xi$-rules

We use $\mathcal{L}_\in$ to denote the language of first-order set theory whose only relation symbols are $\in$ and $=$. As we did in second-order arithmetic, we use $x \subseteq y$ as a shorthand for $\forall z(z \in x \to z \in y)$. We also use $\exists!x\phi(x)$ as the standard shorthand for "there is a unique". Then, recall that Zermenlo-Fraenkel set theory with choice, denoted ZFC, is the extension of first-order logic axiomatized by the universal closures of the following:

Extensionality: $(x \subseteq y \wedge y \subseteq x) \to y = x$;

Foundation: $\exists x\, \phi(x) \to \exists x\, (\phi(x) \wedge \forall y \in x\, \neg\phi(y))$, where $\phi(x)$ is an arbitrary formula in which y does not occur free;

Pair: $\exists z\, (x \in z \wedge y \in z)$;

Union: $\exists y\, \forall z \in x\, (z \subseteq y)$;

Powerset: $\exists y\, \forall z\, (z \subseteq x \to z \in y)$;

Separation: $\exists y \, \forall z \, (z \in y \leftrightarrow z \in x \wedge \phi(z))$, where y does not occur free in $\phi(z)$,

Collection: $\forall x \in w \, \exists y \, \phi(x, y) \to \exists z \, \forall x \in w \, \exists y \in z \, \phi(x, y)$, where z does not occur free in $\phi(x)$,

Infinity: $\exists w \left(\exists x \, (x \in w \wedge \forall y \, (y \notin x)) \wedge \forall x \in w \, \exists y \in w \, \forall z (z \in y \leftrightarrow z \in x \vee z = x) \right)$, and

Choice: $\forall x \in w \left(\exists y \, (y \in x) \wedge \forall y \in w \, (\exists z (z \in x \wedge z \in y) \to x = y) \right)$
$$\to \exists z \, \forall x \in w \, \exists! y \, (y \in x \wedge y \in z).$$

As we have stated the union and powerset axioms we may obtain sets that are too big, but we can then obtain the desired sets using separation. Observe also that the Foundation scheme states that $\in$ is well-founded; this allows us to simply define an ordinal as a transitive set all of whose elements are transitive as well, obtaining well-foundedness for free.

This set-theoretic context will allow us to define an analogue of the ω-rule which quantifies over all elements of ω_1; more generally, for any cardinal κ we can define the *κ-rule* by
$$\frac{\langle \phi(\xi) \rangle_{\xi < \kappa}}{\forall x < \kappa \, \phi(x)}.$$

Of course, in order to do this we need to have names for all elements of κ, as well as κ itself. To this effect, let $\mathcal{L}_\in^\kappa$ be a (possibly uncountable) extension of $\mathcal{L}_\in$ which contains one constant c_ξ for each $\xi < \kappa$; to simplify notation, we may assume that $c_\xi = \xi$ and simply write the latter. Then, the κ-rule is readily applicable in any language extending $\mathcal{L}_\in^{\kappa+1}$. Similarly, for a theory T over $\mathcal{L}_\in$, let T^κ be the extension of T over $\mathcal{L}_\in^\kappa$ with the axioms $\xi \in \zeta$ whenever $\xi < \zeta \leq \kappa$, and $\xi \notin \zeta$ whenever $\zeta \leq \xi \leq \kappa$.

If T is an extension of ZFC^κ, we may enrich T by operators of the form $\left[\begin{smallmatrix} \lambda \\ \kappa \end{smallmatrix} \right]_T \phi$, meaning that ϕ is provable using κ-rules of depth at most α. Recall that if ξ is an ordinal, then $\aleph_\xi$ denotes the ξth infinite ordinal. Then, any infinite cardinal κ may be represented in the form $\aleph_\beta$ for some β, and we write $\left[\begin{smallmatrix} \xi \\ \beta \end{smallmatrix} \right]_T \phi$ to state that ϕ may be proven by iterating $\aleph_\beta$-rules along ξ.

If we want the $\aleph$ function to be well-defined, we must work within a cardinal that is closed under $\xi \mapsto \aleph_\xi$. Fortunately, $\xi \mapsto \aleph_\xi$ is a normal function, so we may hyperate it, and readily observe that $\aleph^\omega(0)$ is the first ordinal ξ such that $\aleph_\xi = \xi$. Thus we may assume that T is an extension of $\text{ZFC}^{\aleph^\omega(0)}$.

Definition 7.1. Let T be a theory over $\mathcal{L}_\in^{\aleph_\omega(0)}$, α, β be ordinals, and $\phi \in \mathcal{L}_\in^{\aleph_\omega(0)}$. Then, by recursion on β with a secondary recursion on α, we define $[{}^\alpha_\beta]_T \phi$ to hold if either

1. $\Box_T \phi$, or

2. there are a formula $\psi(x)$ and ordinals γ, η such that $\eta \leq \beta$ and either $\eta < \beta$ or $\gamma < \alpha$, and such that

    (a) for each $\delta < \aleph_\eta$, $[{}^\gamma_\eta]_T \psi(\delta)$, and

    (b) $\Box_T ((\forall x < \aleph_\eta \psi(x)) \to \phi)$.

As was the case with ω-rules, we have that for any β, the $\aleph_\beta$-rule saturates by $\aleph_{\beta+1}$:

Theorem 7.2. *If $[{}^\lambda_\eta]_T \phi$ for arbitrary λ, then there is $\lambda' < \aleph_{\eta+1}$ such that $[{}^{\lambda'}_\eta]_T \phi$.*

Proof. By induction on η with a secondary induction on λ. If $\Box_T \phi$ holds then clearly $[{}^0_\eta]_T \phi$. Otherwise, there are a formula $\psi(x)$ and ordinals γ and $\delta \leq \eta$ such that either $\delta < \eta$ or $\gamma < \lambda$, and for each $\xi < \aleph_\delta$, $[{}^\gamma_\delta]_T \psi(\xi)$ and $\Box_T((\forall x < \aleph_\delta \psi(x)) \to \phi)$. By the induction hypothesis, for each $\xi < \aleph_\delta$ there is

$$\lambda_\xi < \aleph_{\delta+1} \leq \aleph_{\eta+1}$$

such that $[{}^{\lambda_\xi}_\delta]_T \psi(\xi)$. By Lemma 6.13, we have that

$$\lambda = \sup_{\xi < \aleph_\delta} \lambda_\xi < \aleph_{\eta+1},$$

and therefore also $\lambda + 1 < \aleph_{\eta+1}$. But then observe that $[{}^{\lambda+1}_\eta]_T \phi$, as desired. $\Box$

Thus we have a similar situation as we had when considering $\langle \omega_1 + \xi \rangle_T \phi$; any expressions of the form $\langle {}^{\aleph_{\beta+1}+\alpha}_\beta \rangle_T \phi$ is equivalent to $\langle {}^{\aleph_{\beta+1}}_\beta \rangle_T \phi$. Moreover, observe that $\langle {}^{\aleph_{\beta+1}}_\beta \rangle_T \phi$ is in turn equivalent to $\langle {}^0_{\beta+1} \rangle_T \phi$; thus we should only be interested in expressions of the form $\langle {}^\alpha_\beta \rangle_T \phi$ in cases when $\alpha < \aleph_{\beta+1}$. Otherwise, as we did for impredicative worms, we may collapse α to an ordinal $\psi_\beta(\alpha) < \aleph_{\beta+1}$.

In Section 7.3 we will review a version of Buchholz's ordinal notation system which achieves exactly that, and in Section 7.4 we will see how these ideas may be applied to *spiders*, which are similar to worms but based on modalities $\langle {}^\alpha_\beta \rangle$. However, before we continue, we remark that working with uncountable languages has some obvious drawbacks. Fortunately, this can be avoided by working with admissible ordinals rather than cardinals.

7.2 Iterated admissibles

If we work with an uncountable language then the usual proof of the validity of

$$\langle {}^0_0\rangle_T \phi \to [{}^1_0]_T \langle {}^0_0\rangle_T \phi$$

will not go through, given that we cannot code all possible derivations as natural numbers. There is more than one way to get around this problem; one can allow only ordinals appearing in ϕ to be used in a derivation of ϕ, for example. Alternately, we can work with admissible ordinals, (many of) which are countable, instead of cardinals.

In the set-theoretical context, a Δ_0 *formula* is any formula ϕ of $\mathcal{L}_\in$ such that all quantifiers appearing in ϕ are either of the form $\forall x \in y$ or $\exists x \in y$. Then, *Kripke-Platek set theory* is the subtheory KP of ZFC in which the axioms of choice, powerset and infinity are removed, and separation and collection are restricted to $\phi \in \Delta_0$.

With this in mind, we say that an ordinal α is *admissible* if $\mathbb{L}_\alpha$ (in Gödel's constructible hierarchy) is a model of KP. Admissible sets are studied in great detail in [3]. Moreover, an analogue of Theorem 7.2 also holds if we define:

(i) $\omega_0^{CK} = \omega$,

(ii) $\omega_{\xi+1}^{CK}$ to be the least admissible α such that $\omega_\xi^{CK} < \alpha$, and

(iii) $\omega_\lambda^{CK} = \lim_{\xi<\lambda} \omega_\xi^{CK}$ for λ a limit ordinal.

This allows us to interpret $[{}^\alpha_\beta]_T$ using a countable language by replacing the $\aleph_\beta$-rule by the ω_β^{CK}-*rule*,

$$\frac{\langle \phi(\xi) \rangle_{\xi < \omega_\beta^{CK}}}{\forall x < \omega_\beta^{CK}\, \phi(x)}.$$

Working with admissibles rather than cardinals makes the properties of collapsing functions more difficult to prove, but this has been done by Rathjen in [34]. For simplicity, in this text we will continue to work with the $\aleph$-function.

7.3 Collapsing the Aleph function

In this section we will review a variant of Buchholz's notation system of ordinal notations based on collapsing the aleph function [11]. The ordinals obtained appear, for example, in the proof-theoretical analysis of the theories ID_ν of iterated inductive definitions [12]. Below, define $\Omega(\xi) = -\omega + \aleph_\xi$; we will continue with this convention throughout the rest of the text.

Definition 7.3. *Given ordinals η, ξ, we simultaneously define the sets $C_\eta(\xi)$ and the ordinals $\psi_\eta(\xi)$ by induction on ξ as follows:*

1. *$C_\eta(\xi)$ is the least set such that*

    (a) *$2 + \Omega(\eta) \subseteq C_\eta(\xi)$;*

    (b) *if $\alpha, \beta, \gamma \in C_\eta(\xi)$ then $e^\alpha(\beta + \gamma) \in C(\xi)$, and*

    (c) *if $\alpha, \beta \in C_\eta(\xi)$ and $\beta < \xi$, then $\psi_\alpha(\beta) \in C_\eta(\xi)$;*

2. *$\psi_\eta(\xi) = \min\{\xi : \xi \notin C_\eta(\xi)\}$.*

Observe that (1a) could be simplified somewhat if we had defined $\Omega(0) = 2$, but our presentation will in turn simplify some expressions later. As before, it is possible to define $C_\eta(\xi)$ using the notation of Definition 4.16 and thus we can apply our previous work to these sets. Aside from the first item, which is easy to check, the following lemma summarizes the analogues of Lemmas 6.20, 6.21, and 6.22. The proofs are essentially the same and we omit them.

Lemma 7.4. *Given ordinals η, μ,*

1. *$\psi_{1+\eta}(0) = \Omega(1 + \eta)$;*

2. *$\psi_\eta(\mu)$ is additively indecomposable and satisfies $e^{\psi_\eta(\mu)}1 = \psi_\eta(\mu)$;*

3. *$C_\eta(\mu)$ is hyperexponentially perfect,*

4. *$|C_\eta(\mu)| = \Omega(\eta)$, and*

5. *$\psi_\eta(\mu) \in [\Omega(\eta), \Omega(\eta + 1))$.*

The first ordinal that we cannot write using indexed collapsing functions is $\psi_0(\Omega^\omega 1)$:

Lemma 7.5. *Given ordinals $\eta < \Omega^\omega 1$ and an arbitrary ordinal μ,*

$$\sup C_\eta(\mu) = \Omega^\omega 1.$$

Proof. To see that $\sup C_\eta(\mu) \leq \Omega^\omega 1$, we observe that $\Omega^\omega 1$ is closed under all of the operations defining $C_\eta(\mu)$:

Since $\eta < \Omega^\omega 1$, we have that $\Omega(\eta) \subseteq \Omega^\omega 1$. By Lemmas 6.14 and 6.15, we see that if $\alpha, \beta, \gamma < \Omega^\omega 1$, then

$$\kappa := |e^\alpha(\beta + \gamma)| \leq \max\{\omega, |\alpha|, |\beta|, |\gamma|\}.$$

We then have that $\kappa < \Omega^\omega 1$, so writing $\kappa = \Omega(\xi)$ for some $\xi < \Omega^\omega 1$, we observe that
$$e^\alpha(\beta + \gamma) < \Omega(\xi + 1) < \Omega^\omega 1.$$
Finally we note that if $\nu, \xi < \Omega^\omega 1$, then by Lemma 7.4.5, $\psi_\nu(\xi) < \Omega(\nu + 1) < \Omega^\omega 1$.

Now, to see that
$$\sup C_\eta(\mu) \geq \Omega^\omega 1,$$
simply consider the sequence $(\pi_n)_{n<\omega}$ given by $\pi_0 = 0$ and $\pi_{n+1} = \psi_{\pi_n}(0) \in C_\eta(\mu)$. By Lemma 7.4.1 we have that $\pi_{n+1} = \Omega(\pi_n)$ which by Lemma 5.12 converges to $\Omega^\omega 1$. $\square$

The ordinal $\psi_0(\Omega^\omega 1)$ is also computable, but we will not prove this here; see e.g. [11] for details. In the next section, we will present a variant of the functions ψ_ν using worm-like notations obtained from iterated $\aleph_\xi$-rules.

7.4 Iterated Alephs and spiders

We have seen in Theorem 5.38 that Beklemishev's autonomous worms give a notation system for all ordinals below the Feferman-Schütte ordinal Γ_0, and in Theorem 6.36 that impredicative worms extend this to all ordinals below $\psi(e^{\Omega+1}1)$ (which becomes $\psi_0(e^{\psi_0(0)+1}1)$ in our version of Buccholz's notation). Now let us introduce spiders, which may be used to give notations for much larger ordinals than we could with worms.

Definition 7.6. *Let Λ be either an ordinal or the class of all ordinals, and $f: \Lambda \to \Lambda$ be a normal function. We define $\langle {\Lambda \atop f} \rangle$ to be the class of all pairs of ordinals $\langle {\lambda \atop \mu} \rangle$ such that $f(\mu) + \lambda < f(\mu + 1)$, and write $\mathbb{S}_f^\Lambda$ for the set of all expressions of the form*
$$\boldsymbol{\lambda}_1 \ldots \boldsymbol{\lambda}_n \top,$$
with each $\boldsymbol{\lambda}_i \in \langle {\Lambda \atop f} \rangle$. We simply write $\mathbb{S}$ instead of $\mathbb{S}_\Omega^{\mathrm{Ord}}$. Elements of $\mathbb{S}$ are called spiders.

We will restrict our attention to the case where $f(\xi) = \Omega(\xi) = -\omega + \aleph_\xi$, although we state Definition 7.6 with some generality to stress that there are other possible choices for f. In a way, spiders are simply a different way to represent worms; to pass from one representation to the other, we introduce two auxiliary functions.

Definition 7.7. *Let α be any ordinal. Then, define*

(i) *$\lfloor \alpha \rfloor$ to be the greatest ordinal such that $\Omega(\lfloor \alpha \rfloor) \leq \alpha$, and*

(ii) $\dot{\alpha} = -\Omega(\lfloor\alpha\rfloor) + \alpha$.

This definition is sound because for any normal function f with $f(0) = 0$ and any ordinal μ, there is always a greatest ordinal ξ such that $f(\xi) \leq \mu$. The 'translation' between worms and spiders is the following:

Definition 7.8. *Define:*

1. $\flat\colon \left\langle{\Omega\atop\Omega}^{\mathsf{Ord}}\right\rangle \to \mathsf{Ord}$ by $\flat\left\langle\lambda\atop\mu\right\rangle = \Omega(\mu) + \lambda$, and set $\left\langle\lambda\atop\mu\right\rangle \leq \left\langle\eta\atop\nu\right\rangle$ if and only if $\flat\left\langle\lambda\atop\mu\right\rangle \leq \flat\left\langle\eta\atop\nu\right\rangle$. If $\mathfrak{X} = \boldsymbol{\lambda}_1 \ldots \boldsymbol{\lambda}_n\top \in \mathbb{S}$, set $\flat\mathfrak{X} = \flat\boldsymbol{\lambda}_1 \ldots \flat\boldsymbol{\lambda}_n\top$.

2. $\sharp\colon \mathsf{Ord} \to \left\langle\Lambda\atop\Omega\right\rangle$ by $\sharp\lambda = \left\langle\dot{\lambda}\atop\lfloor\lambda\rfloor\right\rangle$. If $\mathfrak{w} = \mu_1 \ldots \mu_n\top \in \mathbb{W}$, set $\sharp\mathfrak{w} = \sharp\mu_1 \ldots \sharp\mu_n\top$.

The following is then immediately verified:

Lemma 7.9. *The class functions $\flat$ and $\sharp$ are bijective and inverses of each other.*

With this, we can extend our worm notation to spiders.

Definition 7.10. *If $\mathfrak{X} \in \mathbb{S}$, define*

1. $O(\mathfrak{X}) = o(\flat\mathfrak{X})$,

2. $H(\mathfrak{X}) = h(\flat\mathfrak{X})$ and $B(\mathfrak{X}) = b(\flat\mathfrak{X})$,

3. $\mathfrak{X} \lhd \mathfrak{Y}$ if and only if $\flat\mathfrak{X} \lhd \flat\mathfrak{Y}$, and

4. if μ is any ordinal, $\mu \uparrow \mathfrak{X} = \sharp(\mu \uparrow \flat\mathfrak{X})$.

Alternately, we can define the head and body of a spider without first turning them into worms:

Lemma 7.11. *Given a spider $\mathfrak{X}$, $H(\mathfrak{X})$ is the maximum initial segment*

$$H(\mathfrak{X}) = \left\langle\lambda_1\atop\eta_1\right\rangle \ldots \left\langle\lambda_m\atop\eta_m\right\rangle \top \in \mathbb{S}$$

of $\mathfrak{X}$ such that for all $i \in [1, m]$, either $\lambda_i \neq 0$ or $\eta_i \neq 0$.

If $H(\mathfrak{X}) = \mathfrak{X}$ then $B(\mathfrak{X}) = \top$, otherwise $B(\mathfrak{X})$ is the unique spider such that

$$\mathfrak{X} = H(\mathfrak{X})\left\langle 0\atop 0\right\rangle B(\mathfrak{X}).$$

As was the case with worms, the cardinality of $O(\mathfrak{X})$ is easy to extract from $\mathfrak{X}$:

Lemma 7.12. *If*
$$\mathfrak{X} = \left\langle \begin{matrix} \lambda_1 \\ \eta_1 \end{matrix} \right\rangle \ldots \left\langle \begin{matrix} \lambda_n \\ \eta_n \end{matrix} \right\rangle \top \in \mathbb{S},$$
then

1. *for every $i \in [1, n]$, $\lambda_i, \eta_i \leq O(\mathfrak{X})$, and*
2. *if $|O(\mathfrak{X})| > \omega$, then $|O(\mathfrak{X})| = \Omega(\max_{i \in [1,n]} \eta_i)$.*

Proof. Immediate by applying Lemma 6.17 to $\flat\mathfrak{X}$ and observing that if $\mu > 0$, $|\Omega(\mu) + \lambda| = \Omega(\mu)$ given that $\lambda < \Omega(\mu + 1)$. $\square$

We can also give an analogue of $\sqsubset$ for spiders:

Definition 7.13. *If*
$$\mathfrak{W} = \left\langle \begin{matrix} \lambda_1 \\ \eta_1 \end{matrix} \right\rangle \ldots \left\langle \begin{matrix} \lambda_n \\ \eta_n \end{matrix} \right\rangle \top \in \mathbb{S}$$
and Θ is a set of ordinals, we define $\mathfrak{W} \sqsubset_\Omega \Theta$ if each $\lambda_i, \eta_i \in \Theta$.

With this, we are ready to 'project' spiders.

Definition 7.14. *Given $\mathfrak{X}, \mathfrak{Y} \in \mathbb{S}$, we define $U_\mathfrak{Y}(\mathfrak{X}) \subseteq \mathsf{Ord}$ and an ordinal $\pi_\mathfrak{Y} \mathfrak{X}$ by induction on $\mathfrak{X}$ along $\triangleleft$ as follows.*

1. *Let $U_\mathfrak{Y}(\mathfrak{X})$ be the least set of ordinals such that if*
$$\mathfrak{U}, \mathfrak{V} \sqsubset_\Omega \Omega(O(\mathfrak{Y})) \cup U_\mathfrak{Y}(\mathfrak{X})$$
and $\mathfrak{V} \triangleleft \mathfrak{X}$, then $\pi_\mathfrak{U} \mathfrak{V} \in U_\mathfrak{Y}(\mathfrak{X})$.

2. *For any $\mathfrak{Y} \in \mathbb{S}$,*
   (a) *If $\mathfrak{X} \triangleleft \left\langle \begin{matrix} 0 \\ O(\mathfrak{Y})+1 \end{matrix} \right\rangle \top$, set $\pi_\mathfrak{Y}(\mathfrak{X}) = O(\mathfrak{X})$;*
   (b) *otherwise,*
$$\pi_\mathfrak{Y}(\mathfrak{X}) = \min\{\xi : \xi \notin U_\mathfrak{Y}(\mathfrak{X})\}.$$

In the remainder of this section, we will see that the functions $\pi_\mathfrak{X}$ behave very similarly to the functions ψ_ν. We begin with a simple lemma.

Lemma 7.15. *If $\mathfrak{X}, \mathfrak{Y}$ are spiders with $O(\mathfrak{X}) > 1$, then $0, 1 \in U_\mathfrak{Y}(\mathfrak{X})$.*

Proof. Immediate from observing that $0 = O(\top) = \pi_\top \top$ and $1 = O(\langle {}^0_0 \rangle \top) = \pi_\top \left(\langle {}^0_0 \rangle \top \right)$. $\square$

With the next few lemmas, we show that the elements of $U_{\mathfrak{Y}}(\mathfrak{X}) \cap O(\mathfrak{X})$ can be characterized as the order-types of suitable spiders. In the process, we obtain some useful properties of $\pi_{\mathfrak{Y}}\mathfrak{X}$.

Lemma 7.16. *If* $\mathfrak{X} \sqsubset_\Omega U_{\mathfrak{V}}(\mathfrak{W})$ *and* $\mathfrak{X} \lhd \mathfrak{W}$, *then* $O(\mathfrak{X}) \in U_{\mathfrak{V}}(\mathfrak{W})$.

Proof. Let $\mathfrak{X} \sqsubset_\Omega U_{\mathfrak{V}}(\mathfrak{W})$ be such that $\mathfrak{X} \lhd \mathfrak{W}$. Since Ω is normal, for every ξ we have that $\xi \leq \Omega(\xi)$. In particular,

$$O(\mathfrak{X}) < O(\mathfrak{X}) + 1 \leq \Omega(O(\mathfrak{X}) + 1).$$

It follows that $O(\mathfrak{X}) = \pi_{\mathfrak{X}}\mathfrak{X} \in U_{\mathfrak{V}}(\mathfrak{W})$. □

With this, we can show that $\pi_{\mathfrak{Y}}\mathfrak{X}$ has cardinality $\Omega(O(\mathfrak{Y}))$, provided $O(\mathfrak{X})$ is large enough.

Lemma 7.17. *If* $\mathfrak{X}, \mathfrak{Y}$ *are spiders with* $O(\mathfrak{X}) \geq \Omega(O(\mathfrak{Y})) + 1$, *then*

$$\pi_{\mathfrak{Y}}\mathfrak{X} \in [\Omega(O(\mathfrak{Y})), \Omega(O(\mathfrak{Y}) + 1)).$$

Proof. If $\xi < \Omega(O(\mathfrak{Y}))$, then by Corollary 3.27 we obtain $\mathfrak{w} \sqsubset \Omega(O(\mathfrak{Y}))$ such that $o(\mathfrak{w}) = \xi$ and observe that $\sharp\mathfrak{w} \lhd \mathfrak{X}$, so that by Lemma 7.16, $\xi = O(\sharp\mathfrak{w}) \in U_{\mathfrak{Y}}(\mathfrak{X})$. It follows that $\pi_{\mathfrak{Y}}\mathfrak{X} \geq \Omega(O(\mathfrak{Y}))$. Meanwhile, by Lemma 6.18, $|U_{\mathfrak{Y}}(\mathfrak{X})| \leq \omega + \Omega(O(\mathfrak{Y}))$, so $\pi_{\mathfrak{Y}}\mathfrak{X} < \Omega(O(\mathfrak{Y}) + 1)$. □

Moreover, $\pi_{\mathfrak{Y}}\mathfrak{X}$ satisfies an analogue of Lemma 6.27:

Lemma 7.18. *If* $O(\mathfrak{X}) \geq \Omega(O(\mathfrak{Y})) + 1$, *then* $O\left(\langle \genfrac{}{}{0pt}{}{\pi_{\mathfrak{Y}}\mathfrak{X}}{O(\mathfrak{Y})} \rangle \top \right) = \pi_{\mathfrak{Y}}\mathfrak{X}$.

Proof. Analogous to the proof of Lemma 6.27, except that to reach a contradiction we use Lemma 7.12.1 to obtain a spider $\mathfrak{V}$ such that $O(\mathfrak{V}) = \pi_{\mathfrak{Y}}\mathfrak{X}$ and all of whose entries are strictly bounded by $\pi_{\mathfrak{Y}}\mathfrak{X}$. □

With this we can show that the elements of $U_{\mathfrak{Y}}(\mathfrak{X}) \cap O(\mathfrak{X})$ are the order-types of suitable spiders, as claimed.

Lemma 7.19. *Let* $\mathfrak{X}, \mathfrak{Y}$ *be spiders and* ξ *an ordinal. Then,* $\xi \in U_{\mathfrak{Y}}(\mathfrak{X}) \cap O(\mathfrak{X})$ *if and only if there is* $\mathfrak{W} \sqsubset_\Omega U_{\mathfrak{Y}}(\mathfrak{X}) \cap O(\mathfrak{X})$ *such that* $\xi = O(\mathfrak{W})$.

Proof. One direction is Lemma 7.16. For the other, if $\xi \in U_{\mathfrak{Y}}(\mathfrak{X}) \cap O(\mathfrak{X})$, then there are $\mathfrak{U}, \mathfrak{V} \sqsubset_\Omega U_{\mathfrak{Y}}(\mathfrak{X})$ such that $\mathfrak{U} \lhd \mathfrak{X}$ and $\xi = \pi_{\mathfrak{V}}\mathfrak{U}$. If $\mathfrak{U} \lhd \langle \genfrac{}{}{0pt}{}{0}{O(\mathfrak{V})+1} \rangle \top$, then we already have $\xi = O(\mathfrak{U})$. If not, by Lemma 7.17,

$$O(\mathfrak{V}) \leq \Omega(O(\mathfrak{V})) \leq \xi < O(\mathfrak{X}),$$

so that by Lemma 7.16, $O(\mathfrak{V}) \in U_{\mathfrak{Y}}(\mathfrak{X})$, and hence $\langle {}_{O(\mathfrak{V})}^{\xi} \rangle \top \sqsubset_{\Omega} U_{\mathfrak{Y}}(\mathfrak{X}) \cap O(\mathfrak{X})$.

Meanwhile, by Lemma 7.18,

$$O\left(\left\langle \begin{matrix} \pi_{\mathfrak{Y}}\mathfrak{U} \\ O(\mathfrak{V}) \end{matrix} \right\rangle \top\right) = \pi_{\mathfrak{Y}}\mathfrak{U} = \xi,$$

as needed. $\square$

Lemma 7.19 is useful in showing that $U_{\mathfrak{Y}}(\mathfrak{X})$ is well-behaved. For example, it satisfies a bounded version of additive reducibility.

Lemma 7.20. *Given spiders $\mathfrak{X}, \mathfrak{Y}$ and an additively decomposable ordinal $\xi < O(\mathfrak{X})$, we have that $\xi \in U_{\mathfrak{Y}}(\mathfrak{X})$ if and only if there are $\alpha, \beta \in U_{\mathfrak{Y}}(\mathfrak{X}) \cap \xi$ such that $\xi = \alpha + \beta$.*

Proof. Analogous to the proof of Lemma 5.33. To illustrate, let us check that if $\xi \in U_{\mathfrak{Y}}(\mathfrak{X}) \cap O(\mathfrak{X})$ is additively decomposable, then there are $\alpha, \beta \in U_{\mathfrak{Y}}(\mathfrak{X}) \cap O(\mathfrak{X})$ such that $\xi = \alpha + \beta$. Using Lemma 7.19, write $\xi = O(\mathfrak{W})$ with $\mathfrak{W} \sqsubset_{\Omega} U_{\mathfrak{Y}}(\mathfrak{X})$. Then, by Theorem 5.30,

$$\xi = O(\mathfrak{W}) = o(\flat\mathfrak{W}) = ob(\flat\mathfrak{W}) + 1 + oh(\flat\mathfrak{W}).$$

Set $\alpha = ob(\flat\mathfrak{W})$ and $\beta = 1 + oh(\flat\mathfrak{W})$. Observe that $\alpha < \xi$, while β is additively indecomposable so $\beta \neq \xi$. Hence, $\alpha, \beta < \xi$.

Finally, observe that $H(\mathfrak{W}), B(\mathfrak{W}) \sqsubset_{\Omega} U_{\mathfrak{Y}}(\mathfrak{X})$,

$$\beta = 1 + oh(\flat\mathfrak{W}) = o((\flat H(\mathfrak{W}))0) = O\left(H(\mathfrak{W})\langle {}_0^0 \rangle\right),$$

and $H(\mathfrak{W}) \sqsubset_{\Omega} U_{\mathfrak{Y}}(\mathfrak{X})$; similarly, $\alpha = OB(\mathfrak{W})$, so $\alpha, \beta \in U_{\mathfrak{Y}}(\mathfrak{X})$. $\square$

Note that $U_{\mathfrak{Y}}(\mathfrak{X})$ is not necessarily additively reductive; however, this truncated form of additive reducibility is sufficient to obtain the conclusion of Lemma 5.17:

Lemma 7.21. *Let Θ be a set of ordinals such that $0 \in \Theta$, and λ be an ordinal such that, whenever $\xi < \lambda$ is additively reducible, then $\xi \in \Theta$ if and only if there are $\alpha, \beta < \xi$ such that $\alpha + \beta = \xi$. Then, for any ordinal ξ:*

1. *if $0 \neq \xi \in \Theta \cap \lambda$, there are ordinals α, β such that $\alpha, \omega^{\beta} \in \Theta$ and $\xi = \alpha + \omega^{\beta}$;*

2. *if $\beta \in \Theta \cap \lambda$ and $\alpha < \beta$ (not necessarily a member of Θ), then $-\alpha + \beta \in \Theta$.*

The proof is identical to that of Lemma 5.17 and we omit it. Next we see that the sets $U_{\mathfrak{Y}}(\mathfrak{X})$ are also closed under some operations related to cardinality.

Lemma 7.22. *If $\mathfrak{X}, \mathfrak{Y}$ are worms and $\xi \in U_{\mathfrak{Y}}(\mathfrak{X}) \cap \Omega(\mathfrak{X})$, then:*

1. $\lfloor \xi \rfloor \in U_{\mathfrak{Y}}(\mathfrak{X})$;

2. *if moreover $\Omega(\xi) < O(\mathfrak{X})$, then $\Omega(\xi) \in U_{\mathfrak{Y}}(\mathfrak{X})$.*

Proof. For the first claim, if ξ is at most countable, $\lfloor \xi \rfloor = 0 \in U_{\mathfrak{Y}}(\mathfrak{X})$. If not, by Lemma 7.19, $\xi = O(\mathfrak{W})$ for some $\mathfrak{W} \sqsubset_\Omega U_{\mathfrak{Y}}(\mathfrak{X})$, and by 7.12.2, $\eta = \lfloor \xi \rfloor$ occurs in $\mathfrak{W}$, hence $\eta \in U_{\mathfrak{Y}}(\mathfrak{X})$.

For the second, we observe that $\langle {}^{0}_{O(\mathfrak{W})} \rangle \top \lhd \langle {}^{0}_{O(\mathfrak{W})+1} \rangle \top$, so that

$$\Omega(\xi) = O\left(\left\langle {}^{0}_{O(\mathfrak{W})} \right\rangle \top\right) = \pi_{\mathfrak{W}}\left(\left\langle {}^{0}_{O(\mathfrak{W})} \right\rangle \top\right).$$

If we moreover have $\Omega(\xi) < O(\mathfrak{X})$, this gives us $\Omega(\xi) \in U_{\mathfrak{Y}}(\mathfrak{X})$. □

The following lemmas show that our work on worms can be used to study the sets $U_{\mathfrak{Y}}(\mathfrak{X})$.

Lemma 7.23. *Given spiders $\mathfrak{W}, \mathfrak{X}, \mathfrak{Y}$,*

$$\mathfrak{W} \sqsubset_\Omega U_{\mathfrak{Y}}(\mathfrak{X}) \cap O(\mathfrak{X})$$

if and only if

$$\flat\mathfrak{W} \sqsubset U_{\mathfrak{Y}}(\mathfrak{X}) \cap O(\mathfrak{X}).$$

Proof. Let

$$\mathfrak{W} = \left\langle {}^{\lambda_1}_{\eta_1} \right\rangle \ldots \left\langle {}^{\lambda_n}_{\eta_n} \right\rangle \top \in \mathbb{S}.$$

If $\mathfrak{W} \sqsubset_\Omega U_{\mathfrak{Y}}(\mathfrak{X}) \cap O(\mathfrak{X})$, then each $\lambda_i, \eta_i \in U_{\mathfrak{Y}}(\mathfrak{X})$. By Lemma 7.22, $\Omega(\eta_i) \in U_{\mathfrak{Y}}(\mathfrak{X})$, and by Lemma 7.20, $\Omega(\eta_i) + \lambda_i \in U_{\mathfrak{Y}}(\mathfrak{X})$. Since

$$\Omega(\eta_i), \Omega(\eta_i) + \lambda_i \leq O(\mathfrak{W}) < O(\mathfrak{X}),$$

it follows that $\flat\mathfrak{W} \sqsubset U_{\mathfrak{Y}}(\mathfrak{X}) \cap O(\mathfrak{X})$.

Conversely, if $\flat\mathfrak{W} \sqsubset U_{\mathfrak{Y}}(\mathfrak{X}) \cap O(\mathfrak{X})$, write $\flat\mathfrak{W} = \mu_1 \ldots \mu_n \top$. By Lemma 7.22, $\lfloor \mu_i \rfloor \in U_{\mathfrak{Y}}(\mathfrak{X})$, and by Lemma 7.20 together with Lemma 7.21.2, $\dot{\mu}_i \in U_{\mathfrak{Y}}(\mathfrak{X})$. It follows from Lemma 7.9 that

$$\mathfrak{W} = \sharp\flat\mathfrak{W} \sqsubset_\Omega U_{\mathfrak{Y}}(\mathfrak{X}) \cap O(\mathfrak{X})$$

by observing that $\lfloor \mu_i \rfloor, \dot{\mu}_i \leq \mu_i < O(\mathfrak{X})$. □

With this we see that the sets $U_{\mathfrak{Y}}(\mathfrak{X}) \cap O(\mathfrak{X})$ are almost worm-perfect.

Theorem 7.24. *Given spiders $\mathfrak{X}, \mathfrak{Y}$ and an ordinal $\xi < O(\mathfrak{X})$, $\xi \in U_{\mathfrak{Y}}(\mathfrak{X}) \cap O(\mathfrak{X})$ if and only if there is $\mathfrak{z} \sqsubset U_{\mathfrak{Y}}(\mathfrak{X}) \cap O(\mathfrak{X})$ with $\xi = o(\mathfrak{z})$.*

Proof. Given an ordinal ξ, by Lemma 7.19, $\xi \in U_{\mathfrak{Y}}(\mathfrak{X}) \cap O(\mathfrak{X})$ if and only if there is $\mathfrak{z} \sqsubset_\Omega U_{\mathfrak{Y}}(\mathfrak{X}) \cap O(\mathfrak{X})$ with $\xi = O(\mathfrak{z})$. But by Lemma 7.23, by setting $\mathfrak{z} = \flat \mathfrak{z}$ we see that this is equivalent to there existing $\mathfrak{z} \sqsubset U_{\mathfrak{Y}}(\mathfrak{X}) \cap O(\mathfrak{X})$ with $\xi = o(\mathfrak{z})$. □

As a consequence, we obtain that $U_{\mathfrak{Y}}(\mathfrak{z})$ is closed under bounded hyperexponentiation.

Lemma 7.25. *If $\mathfrak{X}, \mathfrak{Y}$ are worms and $\alpha, \beta \in U_{\mathfrak{Y}}(\mathfrak{X})$ are such that $e^\alpha \beta < O(\mathfrak{X})$, then $e^\alpha \beta \in U_{\mathfrak{Y}}(\mathfrak{X})$.*

Proof. We may assume that $0 < \alpha, \beta < e^\alpha \beta$, so that if $e^\alpha \beta < O(\mathfrak{X})$, then $\alpha, \beta < O(\mathfrak{X})$. By Theorem 7.24, $\beta = o(\mathfrak{v})$ for some $\mathfrak{v} = \lambda_1 \ldots \lambda_n \top \sqsubset U_{\mathfrak{Y}}(\mathfrak{X}) \cap O(\mathfrak{X})$. Since $e^\alpha \beta$ is additively indecomposable, for each $i \in [1, n]$, $\alpha + \lambda_i \leq \alpha + \beta < e^\alpha \beta$, hence by Lemma 7.20, $\alpha + \lambda_i \in U_{\mathfrak{Y}}(\mathfrak{X})$. Thus $\mathfrak{w} = \alpha \uparrow \mathfrak{v} \sqsubset U_{\mathfrak{Y}}(\mathfrak{X})$, and $o(\mathfrak{w}) = e^\alpha \beta$, which by Theorem 7.24 implies that $e^\alpha \beta \in U_{\mathfrak{Y}}(\mathfrak{X})$. □

This tells us that, below $O(\mathfrak{X})$, the sets $U_{\mathfrak{Y}}(\mathfrak{X})$ behave very similar to the sets $C_\eta(\lambda)$. Conversely, we can prove that the sets $C_\eta(\lambda)$ are 'spider-perfect'.

Lemma 7.26. *If η, λ are ordinals and $\mathfrak{W} \sqsubset_\Omega C_\eta(\lambda)$, then $O(\mathfrak{W}) \in C_\eta(\lambda)$.*

Proof. Suppose that $\mathfrak{W} \sqsubset_\Omega C_\eta(\lambda)$ and $\mathfrak{W} \triangleleft \mathfrak{X}$. The set $C_\eta(\lambda)$ is closed under $\Omega(\cdot)$ and addition, so from $\mathfrak{W} \sqsubset_\Omega C$ we obtain $\flat\mathfrak{W} \sqsubset C$. But $C_\eta(\lambda)$ is hyperexponentially perfect, thus by Lemma 5.34 it is worm-perfect. We conclude that $O(\mathfrak{W}) = o(\flat\mathfrak{W}) \in C$. □

Thus the functions $\pi_{\mathfrak{Y}}$ should closely mimic the functions ψ_η. However, a full translation between the two systems would go beyond the scope of the current work. Instead, we conclude with a conjecture.

Conjecture 7.27. $\psi_0 \Omega^\omega 1 = \pi_\top \left(\langle {}^{0}_{\Omega^\omega 1} \rangle \top \right)$.

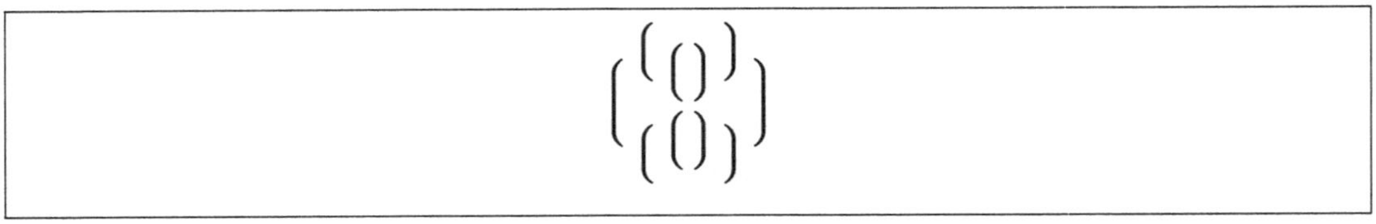

$$\left(\genfrac{}{}{0pt}{}{(\,)}{(\,)}\right)$$

Figure 2: An autonomous spider.

7.5 Autonomous spiders and ordinal notations

We can use autonomous spiders to produce an ordinal notation system, similar to Beklemishev's autonomous worms. We define them as follows:

Definition 7.28. *We define the set of* autonomous spiders, $\mathbb{S}$, *to be the least set such that:*

1. $\top \in \mathbb{S}$;

2. *if* $X, Y, Z \in \mathbb{S}$, *then* $\binom{X}{Y} Z \in \mathbb{S}$.

As with autonomous worms, each autonomous spider can be interpreted as a 'real' spider.

Definition 7.29. *We define a function* $\cdot_O^\pi \colon \mathbb{S} \to \mathbb{S}$ *by*

1. $\top_O^\pi = \top$,

2. $\left(\binom{X}{Y} Z\right)_O^\pi = \langle \genfrac{}{}{0pt}{}{\pi_{Y_O^\pi} X_O^\pi}{O(Y_O^\pi)} \rangle Z_O^\pi$.

For $X, Y \in \mathbb{S}$ *we set* $O(X) = O(X_O^\pi)$ *and* $\mathsf{p}_Y X = \pi_{Y_O^\pi} X_O^\pi$.

We will often omit writing $\top$, so that for example $(\,)$ denotes $\binom{\top}{\top}\top$. The proofs of the following two results are analogous to those of Lemma 6.35 and Theorem 6.36, respectively, and we omit them.

Theorem 7.30. *For any* $\xi \in U_\top \left(\langle \genfrac{}{}{0pt}{}{0}{\Omega^\omega(0)} \rangle \top \right)$, *there exists* $X \in \mathbb{S}$ *such that* $\xi = O(X)$.

Thus assuming Conjecture 7.27, the autonomous spiders indeed provide a notation system for all ordinals below $\psi_0 \Omega^\omega 1$, along with some uncountable ordinals.

8 Concluding remarks

We have developed notation systems for impredicative ordinals based on reflection calculi, thus providing a positive answer to Mints' and Pakhomov's question. These notation systems are obtained by considering strong provability operators extending a theory T. In the process, we have also given a general overview of existing notation systems based on worms.

This work is still exploratory and further developments are required to fully flesh out our proposal. First, no decision procedure is given to determine whether $O(w) < O(v)$ when w, v are impredicative autonomous worms or spiders. While such a decision procedure might be extractable from Theorem 3.17 together with procedures for more standard systems based on ψ, it would be preferable to provide deductive calculi in the style of RC. Second, the set-theoretic interpretations sketched in Section 7 are only tentative and require a rigorous treatment. I'll leave both of these points for future work.

The ultimate goal of the efforts presented here are for the computation of Π^0_1 ordinals of strong theories of second-order arithmetic. There are many more hurdles to overcome before attaining such a goal, but hopefully the ideas presented here will help to lead the way forward.

Acknowledgements

I would like to take this opportunity to express my gratitude to Professor Mints not only for suggesting the topic of this paper, but also for his inspiration and support as a doctoral advisor. His passing was a great personal loss and a great loss to logic. I would also like to thank Fedor Pakhomov for bringing up the same issue and for many enlightening discussions; Lev Beklemishev and Joost Joosten for introducing me to the world of worms, and for many useful comments regarding this manuscript; and Andrés Cordón-Franco, Félix Lara-Martín, as well as my student Juan Pablo Aguilera, for their contributions to the results reviewed here, and Ana Borges for her sharp eye spotting errors in an earlier draft.

Finally, I would like to thank the John Templeton Foundation and the Kurt Gödel Society for the support they have given myself and other logicians through their fellowship program. Their effort is a great boost to logic worldwide; let us hope that it continues to encourage many more generations of logicians.

$$\omega\ \big(()\big)\quad \varepsilon_0\ \Big(\big(()\big)\Big)\quad \Gamma_0\ \Big(^{\big(()\big)}\Big)$$

Figure 3: Some familiar ordinals represented as autonomous spiders.

References

[1] J. P. Aguilera and D. Fernández-Duque. Strong Completeness of Provability Logic for Ordinal Spaces. *ArXiv*, 2015.

[2] T. Arai. Some results on cut-elimination, provable well-orderings, induction and reflection. *Annals of Pure and Applied Logic*, 95(1):93 – 184, 1998.

[3] J. Barwise. *Admissible sets and structures: An approach to definability theory.* Perspectives in mathematical logic. Springer-Verlag, Berlin, New York, 1975.

[4] L. D. Beklemishev. Provability algebras and proof-theoretic ordinals, I. *Annals of Pure and Applied Logic*, 128:103–124, 2004.

[5] L. D. Beklemishev. Veblen hierarchy in the context of provability algebras. In P. Hájek, L. Valdés-Villanueva, and D. Westerståhl, editors, *Logic, Methodology and Philosophy of Science, Proceedings of the Twelfth International Congress*, pages 65–78. Kings College Publications, 2005.

[6] L. D. Beklemishev. Calibrating provability logic. In T. Bolander, T. Braüner, T. S. Ghilardi, and L. Moss, editors, *Advances in Modal Logic*, volume 9, pages 89–94, London, 2012. College Publications.

[7] L. D. Beklemishev. Positive provability logic for uniform reflection principles. *Annals of Pure and Applied Logic*, 165(1):82–105, 2014.

[8] L. D. Beklemishev, D. Fernández-Duque, and J. J. Joosten. On provability logics with linearly ordered modalities. *Studia Logica*, 102(3):541.

[9] L. D. Beklemishev and D. Gabelaia. Topological completeness of the provability logic GLP. *Annals of Pure and Applied Logic*, 164(12):1201–1223, 2013.

[10] G. S. Boolos. *The Logic of Provability.* Cambridge University Press, Cambridge, 1993.

[11] W. Buchholz. A new system of proof-theoretic ordinal functions. *Annals of Pure and Applied Logic*, 32:195 – 207, 1986.

[12] W. Buchholz, S. Feferman, W. Pohlers, and W. Sieg. *Iterated Inductive Definitions and Subsystems of Analysis: Recent Proof-Theoretical Studies*, volume 897 of *Lecture Notes in Mathematics*. Springer-Verlag Berlin Heidelberg, 1981.

[13] A. Cordón Franco, D. Fernández-Duque, J. J. Joosten, and F. Lara Martín. Predicativity through transfinite reflection. *Journal of Symbolic Logic*, 2017.

[14] E. V. Dashkov. On the positive fragment of the polymodal provability logic GLP. *Mathematical Notes*, 91(3-4):318–333, 2012.

[15] D. Fernández-Duque. The polytopologies of transfinite provability logic. *Archive for Mathematical Logic*, 53(3-4):385–431, 2014.

[16] D. Fernández-Duque. Impredicative consistency and reflection. *ArXiv e-prints*, 2015.

[17] D. Fernández-Duque and J. J. Joosten. Hyperations, Veblen progressions and transfinite iteration of ordinal functions. *Annals of Pure and Applied Logic*, 164(7-8):785–801, 2013.

[18] D. Fernández-Duque and J. J. Joosten. Models of transfinite provability logics. *Journal of Symbolic Logic*, 78(2):543–561, 2013.

[19] D. Fernández-Duque and J. J. Joosten. The omega-rule interpretation of transfinite provability logic. *ArXiv*, 1205.2036 [math.LO], 2013.

[20] D. Fernández-Duque and J. J. Joosten. Well-orders in the transfinite Japaridze algebra. *Logic Journal of the IGPL*, 22(6):933–963, 2014.

[21] G. Gentzen. Die Widerspruchsfreiheit der reinen Zahlentheorie. *Mathematische Annalen*, 112:493–565, 1936.

[22] J.-Y. Girard. *Proof theory and logical complexity. Vol. 1.* Studies in proof theory. Bibliopolis, Napoli, 1987.

[23] A. Granas and J. Dugundji. *Fixed Point Theory*. Springer Monographs in Mathematics. Springer-Verlag, New York, 2003.

[24] P. Hájek and P. Pudlák. *Metamathematics of First Order Arithmetic*. Springer-Verlag, Berlin, Heidelberg, New York, 1993.

[25] T. F. Icard III. A topological study of the closed fragment of GLP. *Journal of Logic and Computation*, 21:683–696, 2011.

[26] K. N. Ignatiev. On strong provability predicates and the associated modal logics. *The Journal of Symbolic Logic*, 58:249–290, 1993.

[27] G. Japaridze. The polymodal provability logic. In *Intensional logics and logical structure of theories: material from the Fourth Soviet-Finnish Symposium on Logic*. Metsniereba, Telavi, 1988. In Russian.

[28] G. K. Japaridze. *The modal logical means of investigation of provability*. PhD thesis, Moscow State University, 1986. In Russian.

[29] Thomas Jech. *Set theory, The Third Millenium Edition, Revised and Expanded*. Monographs in Mathematics. Springer, 2002.

[30] G. Kreisel and A. Lévy. Reflection principles and their use for establishing the complexity of axiomatic systems. *Zeitschrift für mathematische Logik und Grundlagen der Mathematik*, 14:97–142, 1968.

[31] J. B. Kruskal. Well-quasi-ordering, the tree theorem, and vazsonyi's conjecture. *Transactions of the American Mathematical Society*, 95(2):210–225, 1960.

[32] M. H. Löb. Solution of a problem of Leon Henkin. *Journal of Symbolic Logic*, 20:115–118, 1955.

[33] W. Pohlers. *Proof Theory, The First Step into Impredicativity*. Springer-Verlag, Berlin Heidelberg, 2009.

[34] Michael Rathjen. How to develop proof-theoretic ordinal functions on the basis of admissible ordinals. *Mathematical Logic Quarterly*, 39(1):47–54, 1993.

[35] U. R. Schmerl. A fine structure generated by reflection formulas over primitive recursive arithmetic. In *Logic Colloquium '78 (Mons, 1978)*, volume 97 of *Stud. Logic Foundations Math.*, pages 335–350. North-Holland, Amsterdam, 1979.

[36] S. G. Simpson. *Subsystems of Second Order Arithmetic*. Cambridge University Press, New York, 2009.

[37] O. Veblen. Continuous increasing functions of finite and transfinite ordinals. *Transactions of the American Mathematical Society*, 9:280–292, 1908.

Recent Progress in Proof Mining in Nonlinear Analysis

Ulrich Kohlenbach
Department of Mathematics
Technische Universität Darmstadt
Schlossgartenstraße 7, 64289 Darmstadt, Germany
kohlenbach@mathematik.tu-darmstadt.de

Dedicated to the memory of Professor Georg Kreisel (1923-2015)

Abstract

During the last two decades the program of 'proof mining' emerged which uses tools from mathematical logic (so-called proof interpretations) to systematically extract explicit quantitative information (e.g. rates of convergence) from prima facie nonconstructive proofs (e.g. convergence proofs). This has been applied particularly successful in the context of nonlinear analysis: fixed point theory, ergodic theory, topological dynamics, convex optimization and abstract Cauchy problems. In this paper we give a survey on some of the results, both on the logical foundation of proof mining as well as its applications in nonlinear analysis, obtained since the monograph [65] appeared.

1 Introduction

Back in the 50's, Georg Kreisel's program of 'unwinding of proofs' asked for a reorientation of proof theory by shifting the historical emphasis on foundational issues of Hilbert's program (consistency proofs) towards applications of proof-theoretic methods for well-defined mathematical goals. Whereas some of these goals, which were still close to foundational concerns, were much developed in the past 50 years (e.g. the classification of provably recursive function(al)s of various formal systems), applications to concrete problems from core mathematics remained rather scattered,

---

The author is grateful to the Kurt Gödel Society (Vienna) and the John Templeton Foundation for being awarded with a 100.000 EUR Gödel Research Prize Fellowship in 2011 which much contributed to the awareness of the proof mining paradigm in core mathematics.

two notably exceptions being C. Delzell's work on Hilbert's 17th problem (see e.g. [34]) and H. Luckhardt's extraction of a polynomial bound on the number of solutions in Roth's theorem from a proof of that theorem due to Esnault and Vieweg (see [103]).

Starting in [58], we engaged in giving Kreisel's ideas for a new form of an applied proof theory a fresh start. This time the focus was on applications in analysis since in this area proof theory can already contribute to the highly nontrivial issue of determining the correct representation of continuous objects in which then quantitative data such as effective bounds could be computed. Back in the 90's, the main emphasis was on the analysis of proofs that used Heine-Borel compactness in the form of the noneffective binary ('weak') König's lemma WKL and the proof-theoretic approach towards the elimination of WKL from classes of proofs. One such class that turned out to be particularly fruitful was that of uniqueness proofs which led to new quantitative results concerning the issue of 'strong uniqueness' in best approximation theory (both w.r.t. the uniform norm, i.e. Chebycheff approximation, as well as w.r.t. the L^1-norm (see [65] for a book treatment of these developments and references to the literature).

In 2000-2003, we started to investigate strong convergence results for iterative procedures of nonexpansive and other classes of mappings in general normed spaces and (with L. Leuştean) hyperbolic spaces and succeeded in the extraction of highly uniform explicit bounds. Here 'highly uniform' refers to the fact that these bounds are essentially independent from the data of the abstract space such as the starting point of the iteration and the mapping used in the iteration, despite of the absence of any compactness assumption, but only depend on some general norm bounds and data from the concrete Polish spaces involved such as $C[0,1]$.

This led in 2003-05 to the discovery of the first so-called 'logical metatheorems' which explain these findings as instances of general proof-theoretic phenomena ([63, 44]). These logical results are called metatheorems since they take as assumption the existence of a proof of a theorem in some formal framework $\mathcal{A}^\omega[X,\ldots]$ and then assert the extractability of an effective uniform bound from such a given proof together with the verification that this bound is correct in any structure X that satisfies the axioms specified in $\mathcal{A}^\omega[X,\ldots]$. These logical metatheorems are based on certain proof-theoretic transformations ('proof interpretations') which are far reaching extensions and modifications of Gödel's famous functional ('Dialectica') interpretation ([65]). The main tool used is the so-called monotone functional interpretation due to [60]. All these proof interpretations prima facie only work for proofs that are based on a constructive ('intuitionistic' in the sense of Brouwer) logic. In order to apply them to proofs using ordinary ('classical') logic one uses - as a pre-processing step - an appropriate so-called negative translation (Gödel) which provides an embedding of the classical reasoning into an intuitionistic system. The crucial feature of the resulting combined interpretation is that important classes of theorems, e.g. theorems of the logical form $\forall x \exists y\, A_\exists(x,y)$, where $A_\exists$ is purely existential, survive this passage (since functional interpretations eliminate the Markov principle even in higher types) and get equipped with an explicit bound $\Phi(x^*)$ on $\exists y$ which only depends on some bounding ('majorizing') data x^* on x. Instead of giving the general type-inductive definition of the majorization relation $x^* \gtrsim x$ we just list the cases we need here: if $x \in \mathbb{N}$, then also $x^* \in \mathbb{N}$ and $x^* \geq x$, if $x \in \mathbb{N}^\mathbb{N}$, then also $x^* \in \mathbb{N}^\mathbb{N}$ and x^* is a nondecreasing upper bound on x. If $x \in X$, where X is an abstract metric space, then $x^* \in \mathbb{N}$ and we define $x^* \gtrsim x := x^* \geq d(a,x)$, where $a \in X$ is some fixed (for the definition of the majorizability relation) reference point. In the case where X is a normed space we always use $a := 0_X$. Finally, if $x \in X^\mathbb{N}$ or $x \in X^X$ resp., then $x^* \in \mathbb{N}^\mathbb{N}$ is nondecreasing with $x^*(n) \geq d(a, x(m))$, whenever $n \geq m$, or $x^*(n) \geq d(a, x(y))$, whenever $n \geq d(a,y)$, resp. While each x in $\mathbb{N}, \mathbb{N}^\mathbb{N}, X, X^\mathbb{N}$ is majorizable, this not always is the case for selfmaps $x \in X^X$. However, important classes of such mappings, e.g. the class of all nonexpansive mappings, have actually very simple majorants.

We now state a special case of our logical metatheorem for the context of abstract normed spaces:

Theorem 1.1 ([63, 44]). *Let P, K be Polish resp. compact metric spaces (explicitly representable in $\mathcal{A}^\omega$), $A_\exists$ be a purely existential formula, $\underline{z} := z_1, \ldots, z_k$ be variables ranging over X, $\mathbb{N} \to X$ or $X \to X$.*

> *From a proof $\mathcal{A}^\omega[X, \|\cdot\|] \vdash \forall x \in P\, \forall y \in K\, \forall \underline{z}\, \exists v^\mathbb{N}\, A_\exists(x, y, \underline{z}, v)$*
>
> *one can extract a computable $\Phi : \mathbb{N}^\mathbb{N} \times \underline{\mathbb{N}}^{(\mathbb{N})} \to \mathbb{N}$ s.t.*
>
> *$\forall y \in K\, \exists v \leq \Phi(r_x, \underline{z}^*)\, A_\exists(x, y, \underline{z}, v)$ holds in every normed space X,*

where $r_x \in \mathbb{N}^\mathbb{N}$ is a representative of $x \in P$, and all $\underline{z}$ and $\underline{z}^* = z_1^*, \ldots, z_k^* \in \mathbb{N}^{(\mathbb{N})}$ s.t. $z_i^* \gtrsim z_i$ for $1 \leq i \leq k$.

The important point here is that Φ does not operate on the X-data $\underline{z}$ (in which case we not even could make sense of the term 'computable' unless X comes equipped with some notion of effectivity) but only on majorants $\underline{z}^*$ of $\underline{z}$. Moreover, Φ is not just 'computable' but of some restricted complexity which depends on the strength of the mathematical principles that are included in the deductive framework $\mathcal{A}^\omega[X,\ldots]$. In the application to a concrete proof, Φ reflects the computational content of that proof.

The development of the proof mining paradigm has been a back-and-forth movement from experimental case studies to the formulation of general theorems as the one above which explain the structure of the findings in these case studies as instances of a logical pattern which can then be used to systematically find new promising areas for case studies which in turn prompt new proof-theoretic results and so on. As discussed in the next section, theorems of the form above have been tailored towards numerous classes of metric and normed structures X and of mappings $T: X \to X$, where the input data often are enriched by appropriate moduli functions $\omega : \mathbb{N} \to \mathbb{N}$ such as (suitable forms) of moduli of uniform convexity or uniform smoothness for X or of uniform continuity of T etc.

In this paper we give a survey on these developments, both w.r.t. the logical foundations as well as to applications in the context of nonlinear analysis, since around 2010.

Notation: Throughout this paper, for $f \in \mathbb{N} \to \mathbb{N}$ we use $f^{(n)}(m)$ to denote the n-th iteration of f starting from m, i.e.

$$f^{(0)}(m) := m, \quad f^{(n+1)}(m) := f(f^{(n)}(m)).$$

For selfmappings $T : C \to C$ of some subset C of a metric or normed space, we simply use $T^n x$ (for $x \in C$) to denote the n-th iteration as here there is no danger to confuse this with the n-th power.

2 New Developments in the Logical Foundation of Proof Mining

In this section we survey the current stage on the logical methodology used in the proof-mining program as it is applied in nonlinear analysis.

2.1 Classes of abstract spaces admissible in proof mining

Whereas the main applications of proof mining to analysis before 2000 where in the context of specific Polish spaces, such as $C[0,1]$ only, since 2001 almost all applications concern situations where the theorem in question refers to some abstract class of spaces X in addition to such concrete ones. E.g. the mean ergodic theorem is a result formulated in the context of arbitrary (not necessarily separable) Hilbert spaces X. As discussed in detail in [65], the extractability of highly uniform bounds depending only on general metric bounds as input data rather than requiring compactness assumptions rests crucially on the fact that the proof being analyzed does not use the separability of X, as the uniform quantitative form of separability is total boundedness (and so in the presence of completeness implies boundedly compactness). In order to deal with such abstract (not assumed to be separable) classes of structures, the approach started in [63] has been to add X as a kind of atom to the formal systems at hand by including X as a new base type. To obtain a logical metatheorem on the extractability of uniform bounds for a class of metric structures X one needs the following requirements:

(a) the axioms used to axiomatize X have (possibly after being enriched with suitable quantitative moduli ω in $\mathbb{N}$ or $\mathbb{N}^\mathbb{N}$) a monotone functional interpretation by simple majorizing functionals,

(b) all the constants used to axiomatize X have effective majorants of low complexity.

These conditions usually follow from the fact that the axioms used to characterize X (when interpreted in the full set-theoretic model $\mathcal{S}^{\omega,X}$) can be written in purely universal form (once the various moduli are given; see [63, 44, 65]) or - more generally - in the form of axioms

$$\Delta :\equiv \forall \underline{a}^{\underline{\delta}} \exists \underline{b} \preceq_{\underline{\sigma}} \underline{r}\underline{a} \forall \underline{c}^{\underline{\gamma}} A_0(\underline{a},\underline{b},\underline{c}),$$

where A_0 is quantifier-free and does not contain any further free variables, $\underline{r}$ is a closed term (of suitable types) of $\mathcal{A}[X,\ldots]$, the types $\underline{\delta},\underline{\sigma},\underline{\gamma}$ satisfy some modest conditions (see [49]; for the finite types over $\mathbb{N}$ only such axioms were already considered in [59]). Here $\preceq$ is pointwise defined where in the normed case one takes $x^X \preceq_X y^X :\equiv \|x\| \leq \|y\|$.[1]

As shown in [49], such axioms cover all normed structures axiomatizable in positive bounded logic, both w.r.t. the normal strict interpretation as well as w.r.t. the weaker approximate satisfaction relation used in that context. Structures axiomatized in this weaker approximate sense are not only closed under ultraproducts but also ultraroots (see [51]). By adding a nonstandard axiom F_-^X of the form Δ, which is not true in the full set-theoretic model but which holds in the model of all strongly majorizable functionals $\mathcal{M}^{\omega,X}$ (in the sense of Bezem [17] extended to

---

[1] This can be adapted to the metric case by taking $x^X \preceq_X y^\mathbb{N} =\equiv d(x,a) \leq y$ for some reference point $a \in X$.

the types over $\mathbb{N}$ and X), one can prove (using quantifier-free choice QF-AC which does not $\mathcal{M}^{\omega,X}$) a uniform boundedness principle Σ_1^0-UB$_-^X$ which implies that each sentence in positive bounded logic is equivalent to its approximations (written as a single sentence) so that the usual validity in a model and approximate validity coincide. This corresponds to the fact that the ultraproduct of X satisfies an axiom in positive bounded logic in the strong sense already when it satisfies it in the approximate sense. Since QF-AC gets eliminated by the functional interpretation and F_-^X can be eliminated from the verification of extractable bounds for $\forall\exists$-sentences by interpreting things in the model $\mathcal{M}^{\omega,X}$, one may freely use the strong reading of Δ when proving a theorem while the extracted bound will also be valid in the - in general - larger class of spaces X which are only required to satisfy the approximate version of Δ (see [49] for all this). In total, the following classes of spaces have been shown so far to satisfy appropriate logical metatheorems on the extractability of effective uniform bounds:

1. metric spaces (see [63] for the bounded case and [44] for the unbounded case),

2. W-hyperbolic and CAT(0)-spaces (see [63] for the bounded case and [44] for the unbounded case),

3. CAT(κ)-spaces for $\kappa > 0$ ([83]),

4. uniformly convex W-hyperbolic spaces with monotone modulus of uniform convexity, so called UCW-spaces (see [97, 100]), where then the bounds depend on a given modulus of uniform convexity,

5. δ-hyperbolic spaces and $\mathbb{R}$-trees (see [97]),

6. normed spaces, uniformly convex normed spaces and inner product spaces (see [63]),

7. uniformly smooth normed spaces (with single-valued and norm-to-norm uniformly continuous normalized duality mapping) (see [80]), where then the bounds depend on an appropriate concept of modulus of uniform smoothness,

8. totally bounded metric spaces (see [65] and [81]), where then the bounds depend on an appropriate notion of modulus of total boundedness,

9. metric completions of the spaces listed so far (see [65]),

10. Banach lattices (see [49]),

11. abstract L^p- and $C(K)$-spaces (see [49]),

12. bands in the $L^p(L^q)$-Banach lattice (see [49]).

If these conditions are satisfied, the contribution of the X-axioms to the extractable bounds mainly consists in the moduli ω referred to above. A further convenient, but not mandatory, requirement is that the axioms should imply the uniform continuity of the constants occurring in the axioms so that their extensionality (w.r.t. $x =_X y :\equiv d_X(x,y) =_\mathbb{R} 0$) can be inferred which must not be included as an =-equality axiom as the uniform quantitative interpretation of extensionality by monotone functional interpretation upgrades extensionality to uniform continuity (on bounded domains). However, whereas in the model-theoretic approaches to metric structures as in Chang and Keisler's continuous ([29]) or in Henson and Iovino's positive bounded logic ([51]), uniform continuity is a necessary part of the framework (used e.g. to define the ultrapower of X, see - however - the recent preprint [31] which aims at weakening this requirement), this is not case in the proof-theoretic treatment where one can avoid to use uniform continuity

(i) if the proof in question can be formalized with a weaker quantifier-free extensionality rule QF-ER instead of the extensionality axiom or

(ii) if a condition weaker than uniform continuity turns out to be sufficient to provide a uniform quantitative form of the special instances of the extensionality axiom used in the proof.

E.g. consider the definition of W-hyperbolic spaces X from [63] which is axiomatized (in addition to being a metric space) by four axioms (W1)-(W4) on a formal convexity operator $W : X \times X \times [0,1] \to X$, where (W4), in particular, expresses the uniform continuity (and hence the extensionality) of W as a function in x, y (for fixed $\lambda \in [0,1]$), whereas (W2) implies the uniform continuity (and hence extensionality) in λ. It turns out that many proofs do not use (W4) and the extraction of effective uniform bounds goes through without problems so that the extracted bounds are valid in the larger class of spaces axiomatized by (W1)-(W3) which coincides with the spaces of hyperbolic type from [46] (see e.g. Remark 3.25 in [76]).

If, however, one is in a situation where all the functions involved satisfy appropriate uniform continuity assumptions and the model theory for metric and normed spaces it applicable, then proof-theoretic bound extraction theorems may be viewed as constructive quantitative versions of qualitative uniformity results obtainable using ultraproducts (see [3]). Whereas for the latter, only the truth of the theorem in question in the respective class of structures is needed and only the existence of a uniform bound follows (though by results in effective model theory recently announced by J. Rute one may even conclude the existence of computable bounds obtained by blind unbounded search through some infinitary term language), the proof-theoretic approach uses the formalizability of a proof of the theorem in some suitable system and then extracts an explicit subrecursive bound reflecting the computational content of the given proof.

2.2 Classes of majorizable functions that are admissible in proof mining

The above comments on extensionality not only apply to the constants used in axiomatizing X but also to the classes $\mathcal{F}$ of functions $T : X \to X$ in the theorem

$$\forall T \in \mathcal{F} \, \exists n \in \mathbb{N} \, A_\exists(T, n)$$

for which we want to extract effective uniform bounds. E.g. if uniform continuity of T follows from $T \in \mathcal{F}$, then full extensionality is for free. Otherwise, the items (i),(ii) may apply: as an instance of the first item, one may refer to the nonlinear ergodic theorem due to [129] for selfmappings $T : K \to K$ of a subset K of a Hilbert space satisfying the Wittmann-condition

$$(W) \, \forall x, y \in K \, (\|Tx + Ty\| \leq \|x + y\|),$$

whose proof only uses QF-ER. The condition (W) trivially implies that T is majorized by $T^* := Id_\mathbb{N}$, while T in general will be discontinuous (see example 3.1 in [129]). In section 3.2 we will discuss the quantitative analysis of Wittmann's theorem given in [116].

Recently, the 2nd item above has been used substantially in the context of metric fixed point theory. There extensionality often is used only in the form

$$p \in Fix(T) \wedge x =_X p \to x \in Fix(T),$$

where $Fix(T)$ denotes the fixed point set of some selfmapping $T : X \to X$ of a metric space (X, d). This is a genuine use of the extensionality axiom that cannot be replaced by QF-ER. However, its uniform quantitative interpretation (as obtained via monotone functional interpretation) is satisfied by moduli $\delta_T, \omega_T : \mathbb{N} \to \mathbb{N}$ such that

$$(*) \begin{cases} \forall x, p \in X \, \forall k \in \mathbb{N} \, \left(d(p, Tp) < \frac{1}{\delta_T(k)+1} \wedge d(x, p) < \frac{1}{\omega_T(k)+1} \right. \\ \qquad \left. \to d(x, Tx) \leq \frac{1}{k+1} \right). \end{cases}$$

This condition (which is introduced in [81] under the name of '(moduli of) uniform closedness') can always be satisfied when T is uniformly continuous with a modulus of uniform continuity Ω_T by defining

$$\omega_T(k) := \max\{4k+3, \Omega_T(4k+3)\} \text{ and } \delta_T(k) := 2k+1$$

(see Lemma 7.1 in [81]), but is also satisfied for important classes of in general discontinuous functions: e.g. T satisfies the so-called condition (E) (introduced in [39] as a generalization of a condition (C) in [124]) if there exists a $\mu \geq 1$ such that

$$\forall x, y \in C \ (d(x, Ty) \leq \mu d(x, Tx) + d(x, y)).$$

It is easy to see that in this case we may take (w.l.o.g. $\mu \in \mathbb{N}$)

$$\omega_T(k) := 4k+3 \text{ and } \delta_T(k) := 2\mu(k+1) - 1$$

(see [81])[2] while T may fail to be continuous. E.g., as shown in [39], $T : [-2, 1] \to [-2, 1]$ defined by

$$T(x) := |x|/2, \text{ for } x \in [-2, 1), \text{ and } := -1/2, \text{ for } x = 1,$$

satisfies the condition (E) with $\mu := 3$.

The condition (E) also implies that T is majorizable (w.r.t. a reference point $a \in X$) by

$$T^*(n) := C_{a,\mu} + n,$$

where $\mathbb{N} \ni C_{a,\mu} \geq \mu \cdot d(a, Ta)$.

In addition to the issue that maybe only partial extensionality is available, the membership relation $T \in \mathcal{F}$ also has to satisfy the requirements $(a), (b)$ on the X-axioms in order to be able to design a bound-extraction theorem for the class $\mathcal{F}$. In the examples just discussed this is trivial by the existence of simple T-majorants and since both the condition (W) as well as (E) (for given μ) are purely universal and hence admissible in logical bound-extraction theorems for these classes of mappings while in the case of (E) the extracted bound will additionally depend on the upper bound $C_{a,\mu}$ for some $a \in X$.

So far the following classes of mappings have been shown to have appropriate bound extraction theorems and have been used in actual unwindings of proofs:

---

[2]Correction to [81]: in the definition of the moduli χ, δ for mappings satisfying the condition (E) one has to use $\lceil \cdot \rceil$ to make the bounds χ, δ_T natural numbers.

1. nonexpansive, Lipschitz, Hölder-Lipschitz and uniformly continuous mappings ([44]),

2. weakly quasi-nonexpansive mappings (first implicitly introduced - without a name - in [75] and - independently as 'J-type mappings' [38] and as 'weakly quasi nonexpansive mappings' in [44]; used in proof mining in [75]),

3. asymptotically nonexpansive mappings (introduced in [45] and used in proof mining and discussed from a logical point of view in [75, 78]),

4. uniformly contractive mappings (introduced in [111] and used in proof mining and discussed from a logical point of view in [84, 43]),

5. uniformly generalized p-contractive mappings (introduced - as a uniform strengthening of a notion from [56] - and applied in proof mining in [20]; logically studied in [21]),

6. asymptotic contractions in the sense of Kirk (introduced in [57] and used in proof mining in [42, 19]; logically studied in [21]),

7. firmly (quasi-)nonexpansive mappings in geodesic and normed spaces (introduced for Hilbert spaces in [23], for general Banach spaces in [25], for the Hilbert ball in [47] and for W-hyperbolic spaces in [1] and used in proof mining in [108, 1, 2, 71, 82, 72]),

8. strongly (quasi-)nonexpansive mappings in geodesic and normed spaces (introduced in [28] in the normed setting and in the 'quasi'-form and for the Hilbert ball in [112, 15] and in metric spaces in [27]; used in proof mining and discussed from a logical point of view in [71, 72]),

9. pseudocontractive mappings in normed spaces (introduced by Browder [22] and used in proof mining in [89, 90, 88]),

10. strict pseudocontractions in Hilbert space (introduced in [24] and used in proof mining in [53, 81, 119]),

11. mappings satisfying Wittmann's condition (W) in Hilbert space (introduced in [129] and used in proof mining in [116, 67, 71]),

12. mappings satisfying condition (E) in geodesic spaces (introduced in [39] and used in proof mining in [81]; see above for a discussion from the point of logic),

13. majorizable mappings in metric spaces ([44])

as well as combinations thereof (e.g. Lipschitzian pseudocontractions).
Uniformly continuous selfmappings of geodesic spaces and Lipschitz continuous mappings on general metric spaces are always majorizable where an a-majorant can be computed in terms of a modulus of uniform continuity and an upper bound $b \geq d(a,Ta)$ for $a \in X$ (see the proof of Corollary 4.20 in [44]). Hence majorizability (and extensionality) follows for such mappings including mappings which are asymptotically nonexpansive, uniformly contractive, firmly or strongly nonexpansive or strict pseudocontractions.

Due to the lack of continuity one does not have extensionality for free in the other cases but majorizability holds for weakly quasi-nonexpansive mappings (see the proof of Corollary 4.20(5) in [44]) and so also for firmly quasi-nonexpansive and strongly quasi-nonexpansive mappings if the mappings possess a fixed point. Majorants can then be given in just an upper bound $b \geq d(a,p)$ for some reference point a and a respective fixed point p. For mappings satisfying the condition (E), majorizability has been shown above. As mentioned already, mappings satisfying condition (W) are trivially majorized by the identity map.

Let us discuss the issue of majorizability for asymptotic contractions in the sense of Kirk (the logical discussion in [21] is only in the context of bounded metric spaces where the majorizability is trivial but the results in [19] are proven for general metric spaces):

a function $T: X \to X$ is an asymptotic contraction in the sense of Kirk if there are continuous mappings $\phi, \phi_n : [0,\infty) \to [0,\infty)$ with $\phi(s) < s$ for $s > 0$ such that for all $n \in \mathbb{N}, x, y \in X$

$$d(T^n x, T^n y) \leq \phi_n(d(x,y))$$

and $\phi_n \to \phi$ uniformly on the range of d. While the continuity of the functions ϕ_n is not part of Kirk's original definition it is added as an assumption in his main result on these mappings in [57] and [19] officially included this condition in his definition of 'asymptotic contractions in the sense of Kirk'. If ϕ_1 (additionally) is assumed to be continuous, then the majorizability is shown as follows: let $a \in X$ be a reference point and $x \in X, n \in \mathbb{N}$ with $d(x,a) \leq n$. Define $T^*(n) := b + \sup_{y \in [0,n]} \phi_1(y)$, where $\mathbb{N} \ni b \geq d(a,Ta)$. Then

$$d(a,Tx) \leq d(a,Ta) + d(Ta,Tx) \leq b + \phi_1(d(a,x)) \leq T^*(n).$$

Hence T^* is a majorant for T w.r.t. a.

General pseudocontractions T, even in Hilbert spaces other than $\mathbb{R}$, do not seem to be majorizable in general. However, in the applications in proof mining they have either been used under the additional assumption of being Lipschitzian ([89]) or as selfmaps $T : C \to C$ of bounded convex subsets $C \subseteq X$, where majorizability is trivial ([90, 88]). An exception is Theorem 2.3 in [90] where, however, the assumption of the existence of a fixed point of T implies the boundedness of the path considered.

For all the function classes above, except for the 'quasi'-classes, the conditions become purely universal once the appropriate data and moduli are added to the definition. If one generalizes the classes of '(firmly, strongly) quasi-nonexpansive' mappings to their 'weakly' versions, where the condition in question is not claimed to hold for all of their fixed points but only for some fixed point, then these conditions have the form Δ and so are also amenable in logical metatheorems. It has been for this reason why we considered the class of weakly quasi-nonexpansive functions in [75] and also the use of firmly quasi-nonexpansive and strongly quasi-nonexpansive mappings in [71] is of this form.

2.3 New proof principles treated by proof interpretations

In the applications of proof mining in nonlinear analysis (mainly fixed point theory) up to 2008, functional analytic tools for the space X in question were only needed to a rather limited extent. This changed in connection with applications to nonlinear ergodic theory ([68, 70, 79] and convex optimization (see chapter 8 in [86]), where one had to rely on new finitary quantitative versions of appropriate projection and weak compactness arguments as well as the elimination of Banach limits (see [66, 69, 79]).

With respect to metric projections P_C of X onto a nonempty convex closed subset C (in the case of uniformly convex spaces X) one has to distinguish between the cases where C is an abstract convex closed subset just axiomatized to be so (for the convexity of C this is an obvious universal axiom while the closedness can axiomatized in a purely universal way using the completion operator from [65], see [48]). Then a projection P_C can be added equally axiomatic by the then universal axiom

$$\forall x \in X \, \forall y \in C \, (\|x - P_C x\| \leq \|x - y\|).$$

Given a modulus η of uniform convexity for X one can compute a modulus of uniqueness for being a metric projection (see [65], Proposition 17.4) which in turn gives a modulus of uniform continuity for P_C and hence full extensionality can be derived. A majorant of P_C is given by the function $P_C^*(n) := 2n + m$, where $\mathbb{N} \ni m \geq \|c\|$ for some $c \in C$ since

$$\|P_C x\| \leq \|x - P_C x\| + \|x\| \leq \|x - c\| + \|x\| \leq 2\|x\| + \|c\|$$

(see also [48]).
The situation is different if C is given by some formula $\varphi(x)$ in the language of $\mathcal{A}^\omega[X,\ldots]$ (which - if it expresses an $=_X$-extensional property of points in C - will never be quantifier-free)
$$\forall x \in X \, (x \in C \leftrightarrow \varphi(x)).$$
Then the existence of the unique best approximation of $x \in X$ by an element in C
$$\forall x \in X \, \exists! y \in C \, \forall z \in C \, (\|x - y\| \leq \|x - z\|)$$
can be proven e.g. in $\mathcal{A}^\omega[X, \|\cdot\|, \eta, \mathcal{C}]$ (which adds an abstract η-uniformly convex Banach space X to $\mathcal{A}^\omega$) using countable choice $\mathrm{AC}^{0,X}$ for points in X (see [66], Proposition 3.2 and Remark 4). As a consequence of this, the monotone functional interpretation (of the negative translation) of this principle has a solution by a bar recursive functional in Spector's calculus $T + BR$ (obtained by majorizing the bar recursive solution of $\mathrm{AC}^{0,X}$, see [44]).

The existence of ε-best approximations
$$(+) \ \forall k \in \mathbb{N} \, \forall x \in X \, \exists y \in C \, \forall z \in C \, (\|x - y\| \leq \|x - z\| + 2^{-k})$$
can be shown even without the uniform convexity of X and the convexity of C and without $\mathrm{AC}^{0,X}$ using only induction (see [66] Proposition 3.1) which then has a monotone functional interpretation by terms in Gödel's T alone.

A typical application of this in fixed point theory is the following: let $C \subset H$ be a nonempty bounded convex closed subset of a Hilbert space H and $T : C \to C$ be a nonexpansive selfmap. By a classical result due to Browder, Göhde and Kirk (independently), T has a fixed point. Hence the fixed point set $Fix(T)$ is nonempty and easily shown to be again closed and convex (all this holds also in general uniformly convex Banach spaces). Hence for each $v_0 \in H$ there exists a unique fixed point $u \in Fix(T)$ whose distance is closest to v_0 among all fixed points. When this is used in a proof of some concrete statement, e.g. expressing that a certain iteration procedure converges to u, proof mining usually reveals that a quantitative ε-version of this projection statement is all that is needed for the quantitative analysis of the convergence proof. Functional interpretation leads to the following result used in [68][3] for the quantitative analysis of a nonlinear ergodic theorem due to Wittmann [130] and recently used again in [86] in the analysis of an algorithm due to Yamada [132] in convex optimization (for technical reasons it is convenient in Hilbert space to use the square of the norms in $(+)$):

---
[3]Correction to [68]: in lemma 3.1 '$4dn(8dn + 2)$' instead of '$4dn(4dn + 2)$'.

Proposition 2.1 ([68]). : *Let H, C, v_0, T as above and $\mathbb{N} \ni d \geq diam(C)$. Let $\varepsilon \in (0,1], \Delta : C \times (C \to (0,1]) \to (0,1]$ and $V : C \times (C \to (0,1]) \to C$. Then one can construct $u \in C$ and $\varphi : C \to (0,1]$ such that*

$$(1) \quad \|u - T(u)\| < \Delta(u, \varphi)$$

and

$$(2) \quad \begin{cases} \|T(V(u,\varphi)) - V(u,\varphi)\| < \varphi(V(u,\varphi)) \to \\ \|v_0 - u\|^2 \leq \|v_0 - V(u,\varphi)\|^2 + \varepsilon. \end{cases}$$

In fact, u, φ can be defined explicitly as functionals in ε, Δ, V (as well as in v_0, T and some fixed point $p \in C$ of U which we, however, do not mention as arguments as these are fixed parameters) as follows: for $i < n_\varepsilon := \left\lceil \frac{d^2}{\varepsilon} \right\rceil$ we define $\varphi_i : C \to (0,1]$ and $u_i \in C$ inductively by

$$\varphi_0(v) := 1, \ \varphi_{i+1}(v) := \Delta(v, \varphi_i),$$
$$u_0 := p \in Fix(U), \ u_{i+1} := V(u_i, \varphi_{n_\varepsilon - i - 1}).$$

Then for some $i < n_\varepsilon$ (that we may find by bounded search, see Remark 2.5 in [68]) we have that $u := u_i$, $\varphi := \varphi_{n_\varepsilon - i - 1}$ satisfy the claim.

Let us briefly discuss the monotone version of the above statement which translates majorants for V, Δ, p, v_0 into majorants for u, φ. Since C is assumed to be bounded, majorants for V, u, p, v_0 are trivial, namely given simply by a bound on the norm of the elements in C, in fact, since the whole argument only involves C and not H, only a bound d on $diam(C)$ is needed. So it suffices to majorize φ given a majorant $\Delta^* \gtrsim \Delta$ for Δ, i.e. a $\Delta^* : \mathbb{N} \to \mathbb{N}$ s.t.

$$\forall n \in \mathbb{N} \, v \in C \, (\forall w \in C(1/n \leq \varphi(w)) \to 1/\Delta^*(n) \leq \Delta(v, \varphi)).$$

It is now easy to see that for all $w \in C$

$$1/(\Delta^*)^{(i)}(1) \leq \varphi_i(w)$$

and so the solution φ in the above proposition is majorized by

$$\max\{\,(\Delta^*)^{(i)}(1) : i < n_\varepsilon\,\}.$$

In addition to the use of the metric projection onto the fixed point set of T, Wittmann's proof in [130] also makes use of a weak sequential compactness argument. However, at the very end of the complete logical analysis of Wittmann's proof as given in [68] what remains from this use is a trivial lemma (namely Lemma 2.13 in [68], see the discussion after Lemma 5.1 in the same paper). In particular, as the use of weak compactness is totally eliminated, there is no contribution of the enormous complexity of the functionals needed to interpret the monotone functional interpretation of sequential compactness. The latter functionals are computed in [69] and make use of two nested applications of bar recursion $B_{0,1}$ of lowest type (corresponding to the usual strong sequential compactness of an appropriate compact Polish space and the proof of the Riesz representation theorem resp. both of which are used in the proof of weak compactness). As follows from [95], this complexity is optimal. The solution to the weak convergence of bounded sequences in Hilbert spaces was then used in the logical analysis of a proof of the famous nonlinear ergodic theorem of Baillon ([70], see also section 3.2 below) where it is applied again twice in a row. Since this bounds on the weak Cauchy property for the nonlinear ergodic theorem is of type 2 and the uses of $B_{0,1}$ are in a context which otherwise is in T_0, i.e. the fragment of Gödel's T with primitive recursion of type $\mathbb{N}$ only, it follows from the detailed analysis of the type-2 functionals definable in $T_0 + B_{0,1}$ (using results due to W.A. Howard and H. Schwichtenberg as well as a normalization argument due to the present author [62]) that this bound can be restated as a functional in T. This time it has not been possible to eliminate the use of weak compactness. But note that Baillon's theorem itself is only a weak convergence result (and strong convergence is known to fail in general). In the famous special case of odd nonexpansive operators, where the convergence is again strong, one can indeed avoid the use of weak compactness (see [129]) and the proof-theoretic analysis of the corresponding proof then gives a primitive recursive (in the ordinary sense, i.e. in T_0,) rate of metastability ([116], see also section 3.2 below).

This leads to the question of whether one can isolate certain conditions that guarantee that a use of weak compactness can be eliminated from a proof of strong convergence from proofs that satisfy these conditions. Related to this is the interesting topic to give quantitative versions of so-called 'weak-to-strong' principles used in convex optimization to ensure strong convergence by suitable changes in only weakly convergent iterative algorithms (see e.g. [13]).

Proof-theoretic elimination techniques for the use of a non-principal ultrafilter $\mathcal{U}$ in favor of arithmetical comprehension have been developed in [94, 128]. Kreuzer [94] uses functional interpretation combined with a normalization argument to first reduce the use of $\mathcal{U}$ (in proofs of theorems of a suitable logical form) to the uniform arithmetical comprehension functional

$$(E^2): \exists E: \mathbb{N}^{\mathbb{N}} \to \mathbb{N}\, \forall f: \mathbb{N} \to \mathbb{N}\, (E(f) = 0 \leftrightarrow \exists x \in \mathbb{N}(f(x) = 0)).$$

By a method due to Feferman [36] (based again on functional interpretation and normalization), (E) is then further reduced to arithmetic comprehension. The latter has a functional interpretation by bar recursion of lowest type ([122], see [65]). Towsner [128] reduces $\mathcal{U}$ directly to arithmetic comprehension by a syntactic forcing translation.

As mentioned already above, the results in [49] suggest that many uses of ultrafilters made in connection with the definition of ultrapowers of metric and normed structures might be replaced by proof-theoretic techniques involving a suitable nonstandard uniform boundedness principle which in contrast to (non-principal) ultrafilters can never contribute to the complexity of extractable bounds.

Another use of ultrafilters is the construction of Banach limits as the limit along $\mathcal{U}$ of the Cesàro mean $x_1 + \ldots x_n/n$ of bounded sequences (x_n) in $\mathbb{R}$ (see [91], alternatively one can apply Hahn-Banach's theorem to l^∞). In [79, 80], we extracted rates of metastability (see section 2.6) for Halpern iterations in CAT(0) ([79]) and uniformly smooth Banach spaces ([80]) from given convergence proofs that made use of Banach limits. Here, however, special Banach limits are used that are shown to exist by the Hahn-Banach theorem applied to l^∞, namely - given a fixed sequence $(a_n) \in l^\infty$ - one uses a Banach limit $\mu : l^\infty \to \mathbb{R}$ that satisfies $\mu \leq q$ with $\mu((a_k)) = q((a_k))$, where

$$q : l^\infty \to \mathbb{R}, \quad q((a_k)) := \limsup_{p \to \infty} \sup_{n \geq 1} \frac{1}{p} \sum_{i=n}^{n+p-1} a_i$$

is a sublinear functional that can be defined by the comprehension functional (E^2). The proof is then modified so that instead of μ only q is used which then in turn is eliminated in terms of an elementary lemma on the finite averages $C_{n,p}((a_k)) = \frac{1}{p} \sum_{i=n}^{n+p-1} a_i$ with an at most simple polynomial contribution to the final extracted bound. In the addendum to [79], we observed that the analysis given in [79, 80] could be trivially seen to avoid this polynomial contribution altogether so that no trace of the use of Banach limits remains. Subsequently, [102] extended the results from [79] to the technically very involved case of CAT(κ)-spaces for $\kappa > 0$ (since a CAT(0) space is CAT(κ) for every $\kappa > 0$, this is an - in fact far reaching - generalization).

Very recently, we applied proof mining to Bauschke's solution ([12]) of the so-called 'minimal displacement conjecture' which is a very concrete asymptotic regularity statement for compositions of arbitrary metric projections onto closed and convex subsets in Hilbert space. The proof, however, uses a large arsenal of prima-facie very noneffective abstract operator theory: Minty's theorem (Zorn's lemma), Brézis-Haraux theorem, Rockafellar's maximal monotonicity and sum theorems, Bruck-Reich theory of strongly nonexpansive mappings, conjugate functions, normal cone operator etc. Nevertheless, the computational contribution of the use of these principles turned out to be of very low complexity resulting in a rate of convergence which is a simple polynomial in the data (see [72]). We, therefore, believe that the abstract theory of maximally monotone operators should be studied more systematically from the perspective of proof mining.

2.4 Hybrid interpretations of partially constructive proofs

The logical metatheorems developed in [63, 44, 65] apply to theories $\mathcal{A}^\omega[X,\ldots]$ that are based on classical logic. As a consequence of this, the formula $A_\exists$ in bound extractions from proofs of theorems

$$\forall T \in \mathcal{F} \, \exists n \in \mathbb{N} \, A_\exists(T, n)$$

must be purely existential. If, however, one uses intuitionistic logic instead then

1. A may be a formula of arbitrary logical complexity,

2. one may add the axiom of full extensionality to this intuitionistic system and

3. one may add highly noneffective principle such as the schema of comprehension in all types for arbitrary negated formulas (resulting in a theory that is proof-theoretically of the same strength as classical simple type theory but in contrast to the latter has he same provably recursive functions as intuitionistic arithmetic HA)

$$\mathrm{CA}_\neg : \exists \Phi \leq_{\rho \to \mathbb{N}} \lambda x^\rho . 1 \forall y^\rho \, (\Phi(y) =_\mathbb{N} 0 \leftrightarrow \neg A(y)).$$

Note that $\mathrm{CA}_\neg$ implies the law-of-excluded-middle schema for negated formulas.

The extraction technique is then based on a monotone modified realizability interpretation.

Remark 2.2. *Alternatively, one may use Markov's principle in all types and König's lemma KL (instead of CA_{neg}) but then - again - one has to replace full extensionality by the quantifier-free extensionality rule. Here one uses plain monotone functional interpretation, i.e. without any preceding negative translation.*

These and many related results are proved in [43] (for theories without abstract spaces X already in [61]) and have been adapted to UCW-spaces in [100]. In [100], Leuștean gives a very interesting proof analysis in the context of fixed point theory in UCW-spaces (namely the extraction of a so-called rate of asymptotic regularity for the Ishikawa iteration of nonexpansive mappings in such spaces) in which certain parts of the proof, that are basically constructive, are analyzed by the semi-constructive approach from [43] while other parts of the proof that use more heavily classical logic are interpreted by the methods for classical systems from [44]. In fact, it is shown that the analysis given in [99] can be logically understood as such a hybrid approach combining two different proof-theoretic methods.

Another application of such a hybrid approach in nonlinear analysis has recently been given in [120].

Together with general logical theorems on related hybrid proof interpretations due to [52, 109], this suggests that this approach has a large potential for further applications.

2.5 Alternative proof interpretation with potential in proof mining

There are several new forms of proof interpretations related to the Gödel functional interpretation which have been proposed in recent years for the use of analyzing proofs in analysis. Similarly to the monotone (see [60, 65]) and bounded (see [41] and, for the extension to abstract normed spaces, [35]) functional interpretations which only aim after bounds rather than exact realizers, these methods typically also extract only some weaker partial information that is relevant in the case at hand. E.g. in [6], a version of functional interpretation related to [5] is used to bound the ordinal level sufficient in the transfinite hierarchy of so-called distal factors used in their analysis of an ergodic-theoretic proof due to Furstenberg and Katznelson of a multidimensional Szemerédi theorem.

In [16], a functional interpretation of certain nonstandard extensions of systems of arithmetic and analysis in the language of functionals of all finite types is developed. The approach is inspired by Nelson's internal set-theory as the system is based on a unary predicate symbol for 'being standard'. Formulas that do not contain this symbol are called 'internal' and the interpretation acts trivially on those while it extracts finite lists of witnessing candidates for external quantifiers.

In [40], a bounded functional interpretation (in the sense of [41]) is developed for another nonstandard extension of finite type arithmetic in which the 'finite lists of witnessing candidates' extracted by [16] are replaced by extracting majorants for the external quantifiers.

Recent papers by Sanders (see e.g. [117]) suggest that the techniques from [16, 40] can be used to extract computational information from proofs in nonstandard analysis and that, in particular, [40] can be applied also to purely standard proofs by translating them appropriately into the nonstandard framework. It remains to be seen whether this approach may lead to new results when carried out in sufficiently nontrivial case studies.

2.6 Logical aspects of convergence statements

Let (x_n) be a Cauchy sequence in a complete metric space (X, d), i.e.

$$\forall k \in \mathbb{N} \, \exists n \in \mathbb{N} \, \forall i, j \geq n \, (d(x_i, x_j) \leq \frac{1}{k+1}) \in \forall \exists \forall.$$

Often a computable rate of convergence does not exist even for computable sequences (x_n). In fact, as shown in [121], there is a primitive recursive decreasing sequence $(x_n) \subset [0, 1] \cap \mathbb{Q}$ with no computable Cauchy rate. The noneffectivity of the Cauchy property of monotone bounded sequences of reals corresponds precisely to the law-of-excluded-middle-principle for Σ_1^0-formulas Σ_1^0-LEM (see [127]). In fact, if only weaker forms of LEM such as either LEM for arbitrary negated formulas or, alternatively, the Markov Principle plus the so-called LLPO-principle ('lesser-limited-principle-of-omniscience') are used (relative to a suitable intuitionistic framework) in a proof of a Cauchy statement, then effective rates of convergence can be extracted. For finite type systems over $\mathbb{N}$ this follows from Corollary 7.7 resp. Theorem 9.3 in [65]: note that the '$\exists$'-quantifier in the Cauchy property is monotone, i.e. any any upper bound is already a witness so that these bound extraction theorems are applicable (as already mentioned, LEM for negated formulas follows immediately from $CA_\neg$ and LLPO follows from WKL which can be written as an axiom Δ). For some extensions to theories with abstract spaces X see [43]. So at least Σ_1^0-LEM is needed to create a noneffective Cauchy statement (for a computable sequence). Conversely, it has been shown in [85] that over rather general intuitionistic frameworks $\mathcal{A}_i[X, \ldots]$, proofs of Cauchy statements can be transformed so that **only** Σ_1^0-LEM is needed: this follows by showing that $\mathcal{A}_i[X, \ldots] + \Sigma_1^0$-LEM is closed under the Σ_2^0-DNE rule, where

$$\Sigma_2^0\text{-DNE} : \neg\neg \exists x \in \mathbb{N} \, \forall y \in \mathbb{N} \, A_{qf}(x, y, \underline{a}) \to \exists x \in \mathbb{N} \, \forall y \in \mathbb{N} \, A_{qf}(x, y, \underline{a}),$$

(A_{qf} quantifier-free), suffices to prove that the negative translation of the Cauchy statement implies the original Cauchy statement.

So the intuitionistic reverse mathematics for Cauchy statements that do not admit a computable rate of convergence essentially is trivial as they will in all practical cases be equivalent to Σ_1^0-LEM. This also applies to the corresponding convergence statements which will follow from the Π_3^0-Cauchy property by applying countable choice for numbers (to create a fast converging sequence and hence a limit) which usually is included in intuitionistic frameworks. Classically, this need to apply Π_1^0-choice for numbers to create a limit will (in connection with the fact that the Π_3^0-property implies Σ_1^0-LEM) result in arithmetical comprehension being implied by the convergence theorem (and conversely, arithmetical comprehension suffices to prove the convergence from the Cauchy property). So to say something specific about the computational content of a concrete (noneffective) convergence statement one has to investigate the numerical content of the Cauchy statement once the latter is (classically equivalent) reformulated in such a way that it has a computational solution: the Cauchy property **noneffectively** is equivalent to

$$(*) \ \forall k \in \mathbb{N} \, \forall g \in \mathbb{N}^{\mathbb{N}} \, \exists n \in \mathbb{N} \, \forall i, j \in [n; n + g(n)] \ (d(x_i, x_j) < \frac{1}{k+1}) \in \forall \exists$$

which is the Herbrand normal form of the original Cauchy formulation and which has been - in the specific situation at hand - called metastability by Tao [125, 126].

We call a bound $\Phi(k, g)$ on '$\exists n$' in the latter formula a rate of metastability (this is essentially the Kreisel no-counterexample interpretation of the Cauchy statement in the sense of [92, 93]).

Since $(*)$ is (equivalent to a formula) of the form $\forall \exists$, the logical metatheorems discussed in section 2 above can be used to extract highly uniform such rates of metastability whose subrecursive complexity reflects the specific computational content of the proof.

Usually, convergence theorems not only state the plain convergence of some sequence (x_n) in X but also that the limit $x := \lim x_n$ satisfies some property which often can be written in the form $F(x) =_\mathbb{R} 0$, where $F : X \to \mathbb{R}$ is continuous. Then a rate of metastability should satisfy

$$(**) \ \forall k \in \mathbb{N} \, \forall g \in \mathbb{N}^{\mathbb{N}} \, \exists n \leq \Phi(k, g) \, \forall i, j \in [n; n + g(n)] \ \left(d(x_i, x_j), |F(x_i)| \leq \frac{1}{k+1} \right).$$

This formulation is significant for the following reasons:

- $(**)$ is purely universal and so is a **real statement** in the sense of Hilbert (the universal quantifier behind the bounded ones hidden in $\leq$ can be avoided by switching to appropriate rational approximations of $d(x_i, x_j), |F(x_i)|$).

- By negative translation, $(**)$ always has a **constructive** proof.

- By classical logic (and QF-AC$^{\mathbb{N},\mathbb{N}}$, i.e. closure under recursion), $(**)$ mathematically trivially implies (by a fixed piece of proof) that (x_n) is Cauchy.

- By arithmetical comprehension, $(**)$ mathematically trivially implies (by a fixed piece of proof) that (x_n) is convergent and $F(\lim x_n) = 0$.

- Under certain conditions (e.g. uniqueness or monotonicity properties), rates of metastability even yield rates of convergence.

- The structure of Φ yields information on the learnability of a convergence rate and sometimes oscillation bounds ([85]).

Let us discuss the last item in some more detail (the other items following already by the preceding discussion or being self-explanatory):

Definition 2.3 ([85], Definition 2.4). Consider a Σ_2^0 formula $\varphi \equiv \exists n \in \mathbb{N} \forall x \in \mathbb{N} \, \varphi_{qf}(x, n, \underline{a})$ (with quantifier-free φ_{qf} and all its free variables contained in $\underline{a}$) which is monotone in n, i.e.

$$\forall n \in \mathbb{N} \forall n' \geq n \forall x \in \mathbb{N} \, (\varphi_{qf}(x, n, \underline{a}) \to \varphi_{qf}(x, n', \underline{a})).$$

We call such a formula φ *(B,L)-learnable*, if there are function(al)s B and L such that the following holds:

$$\exists i \leq B(\underline{a}) \, \forall x \in \mathbb{N} \, \varphi_{qf}(x, c_i, \underline{a}),$$

where

$$c_0 := 0,$$

$$c_{i+1} := \begin{cases} L(x, \underline{a}), & \text{for the } x \text{ with } \neg \varphi_{qf}(x, c_i, \underline{a}) \land \forall y < x \, \varphi_{qf}(y, c_i, \underline{a}) \text{ if it exists} \\ c_i, & \text{otherwise.} \end{cases}$$

In [85](Theorem 2.11) it is shown that if the number of (parallel) instances of Σ_1^0-LEM used in a proof of a Cauchy statement is (implicitly) bounded then one can extract concrete terms (B, L) from the proof. Moreover, Proposition 2.16 (together with Proposition 2.5) in [85] shows that the (B,L) learnability of a Cauchy rate with majorizable B, L implies that the Cauchy statement has a rate of metastability given by (essentially)

$$\Phi(k, g, \underline{a}^*) = \left((\lambda x^{\mathbb{N}}.L^*(x, k, \underline{a}^*) \circ \tilde{g}\right)^{(B^*(n,\underline{a}^*))}(0),$$

where B^*, L^* are majorants of B, L, and $\underline{a}^*$ are majorants for the parameters $\underline{a}$ used to define the sequence in question and $\tilde{g}(n) := \max\{n, \max_{m \leq n}\{g(m)\}\}$. By Remark 2.17 in [85] also a kind of converse of this holds, i.e. given a rate of metastability of the above form, then (essentially) a Cauchy rate is (B^*, L^*) learnable. With the notable exceptions of the rates in [70, 116], all rates of metastability extracted so far have this simple structure and so give rise to explicit learnability information on the rate of convergence of the respective sequence. While this in general is not sufficient to infer an effective bound on the number of ε-fluctuations of the Cauchy statement (see Propositions 4.7 and 4.11 in [85] for a counterexample), the latter does follow if an additional gap condition is satisfied by L (Proposition 5.1 and Remark 5.2 in [85]). This e.g. is the case for the rate of metastability extracted for the von Neumann mean ergodic theorem in uniformly convex spaces in [77] and explains why this could be strengthened to a bound on the number of ε-fluctuations in [4].

3 Recent Applications to Nonlinear Analysis

3.1 Metric Fixed Point Theory

Metric fixed point theory has been the area to which the proof mining methodology has been applied most extensively since 2000. Even the results since 2010 are too many to mention all of them here. Instead we only focus on a few developments.

3.1.1 Rates of asymptotic regularity for families of mappings and strongly and firmly nonexpansive mappings

In [96], the following iteration schema is considered: Let C be a nonempty convex subset of a Banach space X and $\{T_i : 1 \leq i \leq k\}$ be a finite family of nonexpansive self-mappings $T_i : C \to C$. Let $U_0 = Id$ be the identity mapping and $0 < \lambda < 1$, then using the mappings

$$U_1 = (1-\lambda)Id + \lambda T_1 U_0$$
$$U_2 = (1-\lambda)Id + \lambda T_2 U_1$$
$$\vdots$$
$$U_k = (1-\lambda)Id + \lambda T_k U_{k-1},$$

one defines

$$x_0 \in C, \ x_{n+1} := (1-\lambda)x_n + \lambda T_k U_{k-1} x_n, \ n \geq 0. \quad (1.1)$$

The following result is implicit in [96]:

Theorem 3.1. *Let X be strictly convex, $C \subseteq X$ nonempty compact and closed, $T_1, \ldots, T_k : C \to C$ nonexpansive and $F := \bigcap_{i=1}^k Fix(T_i) \neq \emptyset$. Then for the sequence defined above one has the following asymptotic regularity result:*

$$\lim_{n \to \infty} \|x_n - T_i x_n\| = 0 \ \text{ for all } 1 \leq i \leq k.$$

In [54], a quantitative version of this theorem is obtained: in order to apply the logical metatheorems discussed in the introduction, strict convexity needs to be upgraded to uniform convexity. As shown in [54], the compactness assumption can then be dropped and one obtains a full rate of convergence in the above asymptotic regularity result even in the case of uniformly convex W-hyperbolic spaces (more precisely the UCW-spaces mentioned in section 2.1).

Theorem 3.2 ([54]). *Let C be a nonempty convex subset of a UCW-space with a monotone modulus of convexity η and let $\{T_i\}_{i=1}^k$ be a finite family of nonexpansive self-mappings $T_i : C \to C$ with $F = \bigcap_{i=1}^k F(T_i) \neq \emptyset$. Let $p \in F, x_0 \in C$ and $D > 0$ be such that $d(x_0, p) \leq D$. Then for the sequence (x_n) generated by Kuhfittig's schema above we have*

$$\forall \varepsilon \in (0,2] \, \forall n \geq \Phi_i(D, \varepsilon, \lambda, \eta) \ (d(x_n, T_i x_n) \leq \varepsilon) \ \text{ for } 1 \leq i \leq k,$$

where

$$\Phi_i := \theta\left(\widehat{\eta}^{(k-i+\min(1,k-1))}\left(\frac{\varepsilon}{2}\right)\right)$$

with

$$\theta(\varepsilon) := \left\lceil \frac{D}{\widehat{\eta}(\varepsilon)} \right\rceil, \ \text{ where } \widehat{\eta}(\varepsilon) := \lambda(1-\lambda)\eta\left(D, \frac{\varepsilon}{D+1}\right)\varepsilon.$$

If $\eta(r,\varepsilon)$ can be written as $\eta(r,\varepsilon) := \varepsilon\tilde{\eta}(r,\varepsilon)$, where $\tilde{\eta}$ increases with ε (i.e. $\forall \varepsilon_2 \geq \varepsilon_1 > 0$ $(\tilde{\eta}(r,\varepsilon_2) \geq \tilde{\eta}(r,\varepsilon_1))$, then we can replace η by $\tilde{\eta}$ in the bound: $\Phi_i(D,\varepsilon,\lambda,\tilde{\eta})$. Using $N \in \mathbb{N}$ with $1/N \leq \lambda(1-\lambda)$ one can replace in the above bound the dependency of the bound on λ by that on N.
In the special case of CAT(0)-spaces one may take $\tilde{\eta}(\varepsilon) := \varepsilon/8$.

Let us briefly discuss the logical reason why one obtains in this case a full rate of convergence. The key result in [96] is that any fixed point of the nonexpansive mapping $S := T_k U_{k-1}$ is a common fixed point of $T_1, \ldots, T_k$ which prenexes as follows

$$\forall q \in C \, \forall \varepsilon > 0 \, \exists \delta > 0 \, \underbrace{\left(d(Sq,q) \leq \delta \rightarrow \bigwedge_{i=1}^{k} (d(T_i q, q) < \varepsilon) \right)}_{\in \Sigma_1^0}.$$

By an appropriate logical metatheorem, one then extracts a uniform (positive lower) bound (and hence witness) $\Psi(D,\varepsilon,N,\eta)$ for δ which only depends on ε, some bound $D \geq d(x_0, p)$ for some $p \in F$, a modulus η of uniform convexity for X and λ (actually only an $N \in \mathbb{N}$ such that $1/N \leq \lambda(1-\lambda)$ is needed). Since Kuhfittig's iteration schema (x_n) is nothing else but a Krasnoselski-Mann iteration of S (with constant λ) one can use a previously extracted rate $\Theta(D,\varepsilon,N,\eta)$ of asymptotic regularity for the latter from [98] and simply put

$$\Phi(D,\varepsilon,N,\eta) := \Theta(D,\Psi(D,\varepsilon,N,\eta),N,\eta).$$

The extraction of a full rate Θ is possible since the sequence $(d(x_n, Sx_n))_{n \in \mathbb{N}}$ is non-increasing so that $d(x_n, Sx_n) \to 0$ is in Π_2^0.

For a different type of iteration, an explicit rate of asymptotic regularity for compositions for nonexpansive mappings in general classes of geodesic spaces is obtained using proof mining in [101].

For so-called strongly (quasi-)nonxpansive mappings (see [28] for the definition of 'strongly nonexpansive' and [27] for 'strongly quasi-nonexpansive') $T_1, \ldots, T_k : S \to S$ with $\bigcap_{i=1}^{k} Fix(T_i) \neq \emptyset$, where $S \subseteq X$ is an arbitrary subset of a metric space X it is known ([27]) that $Fix(T) = \bigcap_{i=1}^{k} Fix(T_i)$, where $T := T_k \circ \ldots \circ T_1$.
In [71], we extracted from the proof of this fact a uniform bound Ψ such that

$$\forall \varepsilon > 0 \, \left(d(Tx,x) \leq \Psi(\varepsilon) \rightarrow \bigwedge_{i=1}^{k} d(T_i x, x) < \varepsilon \right)$$

which, in addition to ε, only depends on k, a common modulus of strong quasi-nonexpansiveness, a bound $d \geq d(x,p)$ for some $p \in \bigcap_{i=1}^{k} Fix(T_i)$ and a common modulus of uniform continuity α_d for $T_1, \ldots, T_k$ on $S_d := \{y \in S : d(y,p) \leq d\}$ ([71], Proposition 4.15). Moreover, we extracted a rate of asymptotic regularity Φ for $(x_n := T^n x)$ in the case where the $T_1, \ldots, T_k$ are nonexpansive, in addition to being strongly quasi-nonexpansive, where Φ, in addition to ε, only depends on k, a common modulus of strong quasi nonexpansiveness and a bound $d \geq d(x,p)$ ([71], Theorems 4.6,4.7). Put together, this yields:

$$\forall \varepsilon > 0 \, \forall n \geq \Phi(\Psi(\varepsilon)) \left(\bigwedge_{i=1}^{k} d(T_i x_n, x_n) < \varepsilon \right).$$

In UCW-spaces, so-called firmly nonexpansive mappings due to [25] are strongly quasi-nonexpansive and nonexpansive (in uniformly convex Banach spaces even strongly nonexpansive) and so these results apply to the firmly nonexpansive mappings. The latter contain in the context of CAT(0)-spaces (and so in particular in Hilbert spaces) all metric projections onto closed convex subsets as well as resolvents of convex lower semicontinuous mappings (in Hilbert spaces even of general maximally monotone operators). Thus the most important mappings used in convex optimization are firmly nonexpansive. In [1], explicit rates of asymptotic regularity for the Picard iteration $T^n x$ of firmly nonexpansive mappings in UCW-space are extracted which become quadratic in the CAT(0)-case. In [71] we reproved these rates as instances of the more general results for strongly quasi-nonexpansive mappings discussed above. The latter class of mappings covers even metric projections in CAT(κ)-spaces for $\kappa > 0$ which no longer are firmly nonexpansive and, in fact, not even nonexpansive. One then, however, can still obtain metastable versions of the above results which suffices to obtain rates of metastability for $(T^n x)$ in the compact case, where T is the composition of metric projections onto closed convex sets that have a nonempty intersection, in the context of CAT(κ)-spaces for $\kappa > 0$ (see section 3.1.2 below). This situation is studied in convex optimization under the label of convex feasibility problems and we will comment further on this in section 3.3.1.

Whereas strong nonexpansivity in uniformly Banach spaces is implied by being firmly nonexpansive, this is false in general Banach spaces where these two concepts are independent. Asymptotic regularity of firmly nonexpansive mappings in general Banach spaces was first established in [113]. In [108], this is generalized to general W-hyperbolic spaces and an exponential rate of asymptotic regularity is extracted:

Theorem 3.3 ([108]). *Let X be a W-hyperbolic space and $C \subseteq X$ be a nonempty bounded subset and let b be an upper bound on the diameter of C. Let $T: C \to C$ be λ-firmly nonexpansive, i.e. (with $(1-\lambda)a \oplus \lambda b := W(a,b,\lambda)$)*

$$d(Tx, Ty) \leq d((1-\lambda)x \oplus \lambda Tx, (1-\lambda)y \oplus \lambda Ty), \quad \text{for all } x, y \in C,$$

for some $\lambda > 0$ and $N \in \mathbb{N}$ with $N \geq 1/\lambda$. Then

$$\forall x \in C \, \forall \varepsilon > 0 \, \forall n \geq \Phi(\varepsilon, N, b) \left(d(T^n x, T^{n+1} x) \leq \varepsilon \right),$$

where

$$\Phi(\varepsilon, N, b) := M \left\lceil \frac{2b(1+e^{NM})}{\varepsilon} \right\rceil \quad \text{with } M := \left\lceil \frac{4b}{\varepsilon} \right\rceil.$$

3.1.2 Rates of metastability for strong convergence theorems based on Fejér monotonicity

In this section we give a brief account of some of the results in [81]. In the following, (X, d) is a metric space.

Definition 3.4 ([81]). 1. *Let $F_k \subseteq X$ and define $F := \bigcap_{k \in \mathbb{N}} F_k$. Points of $AF_k := \bigcap_{i \leq k} F_i$ are called approximate F-points.*

2. *A sequence (x_n) in X has approximate F-points if*

$$\forall k \in \mathbb{N} \, \exists n \in \mathbb{N} \, (x_n \in AF_k).$$

3. *F is explicitly closed (w.r.t. (F_k)) if ($\overline{B}(p,\varepsilon)$ is the closed ε-ball with center p)*

$$\forall p \in X \, (\forall N, M \in \mathbb{N}(AF_M \cap \overline{B}(p, 1/(N+1)) \neq \emptyset) \to p \in F).$$

The canonical examples one has in mind here are the following:

- **Fixed point sets:** Let $C \subseteq X$ and $T: C \to C$ and define

$$F_k := \left\{ x \in C : d(x, Tx) \leq \frac{1}{k+1} \right\}.$$

Then the F- (resp. AF_k)-points are the T-fixed points (resp. $1/(k+1)$-approximate fixed points of T). Note that if T is continuous, then F is always explicitly closed.

- **Zero sets of maximally monotone operators:** Let X be a real Hilbert space and (for $\gamma > 0$) $J_\gamma A := (Id + \gamma A)^{-1}$ be the resolvent of γA for a maximally monotone operator $A : X \to 2^X$. Let (γ_n) be a sequence in $(0, \infty)$ and define
$$F_k := \bigcap_{i \leq k} \left\{ x \in H \ : \ \|x - J_{\gamma_i A} x\| \leq \frac{1}{k+1} \right\}.$$
Then $F = zer(A) := \{ x \in X \ : \ 0 \in Ax \}$ (see also section 3.3.3 below for the significance of this example).

Definition 3.5. $(x_n) \subset X$ is called Fejér monotone w.r.t. $F(\neq \emptyset)$ if
$$\forall n \in \mathbb{N} \, \forall p \in F \ (d(x_{n+1}, p) \leq d(x_n, p)).$$

Remark 3.6. [81] *actually considers a much more general form of Fejér monotonicity. In this definition one can also incorporate error terms $\delta_n \geq 0$ with $\sum \delta_n < \infty$: quasi-Fejér monotonicity.*

Proposition 3.7 ([81]). *Let X be a **compact** metric space and F be explicitly closed. If $(x_n) \subset X$ has approximate F-points and is Fejér monotone, then it converges to a point $x \in F$.*

The proof uses that sequences in X have convergent subsequences. Using results due to [107] it follows that for most of the usual iterations (x_n), the convergence (Cauchyness) above already in the case $X = [0, 1]$ implies the convergence (Cauchyness) of monotone sequences in (x_n) and hence ACA (Σ_1^0-LEM). So effective rates of convergence are largely ruled out.

The main result of [81] is a quantitative metastable version of the above theorem. In order to state this we first have to introduce the appropriate uniform quantitative versions of the concepts involved in the above proposition:

$$\begin{aligned}
\text{compactness} &\to \text{modulus } \gamma \text{ of total boundedness}\\
\text{explicit closedness} &\to \text{moduli } \omega, \delta \text{ of uniform closedness}\\
\text{approximate } F\text{-points} &\to \text{approximate } F\text{-point bound } \Phi\\
\text{Fejér monotonicity} &\to \text{modulus } \chi \text{ of uniform Fejér monotonicity.}
\end{aligned}$$

We now give the definition of these concepts:

Definition 3.8 ([81]). *1. $\gamma : \mathbb{N} \to \mathbb{N}$ is a modulus of total boundedness for X if for all $k \in \mathbb{N}$ and any sequence (x_n) in X*
$$\exists i < j \leq \gamma(k) \left(d(x_i, x_j) \leq \frac{1}{k+1} \right).$$

2. F (more precisely (F_k)) is uniformly closed with moduli $\delta_F, \omega_F : \mathbb{N} \to \mathbb{N}$ if

$$\forall k \in \mathbb{N} \, \forall p, q \in X \, \left(q \in AF_{\delta_F(k)} \,\wedge\, d(p,q) \leq \frac{1}{\omega_F(k)+1} \to p \in AF_k \right)$$

(compare $(*)$ in section 2.2).

3. Let (x_n) be a sequence with approximate F-points. $\Phi : \mathbb{N} \to \mathbb{N}$ is an approximate F-bound bound for (x_n) if it is nondecreasing and

$$\forall k \in \mathbb{N} \, \exists N \leq \Phi(k) \, (x_N \in AF_k).$$

(x_n) has the $\liminf$-property with bound $\Phi : \mathbb{N}^2 \to \mathbb{N}$ if $\Phi(k,n)$ is monotone in both k, n and

$$\forall k, n \in \mathbb{N} \, \exists N \leq \Phi(k,n) \, (N \geq n \wedge x_N \in AF_k).$$

4. (x_n) is uniformly Fejér monotone w.r.t. F (more precisely: w.r.t. (F_k)) with modulus $\chi : \mathbb{N}^3 \to \mathbb{N}$ if for all $m, n, r \in \mathbb{N}$

$$\forall p \in X \, \left(p \in AF_{\chi(n,m,r)} \to \forall i \leq m \, \left(d(x_{n+i}, p) < d(x_n, p) + \frac{1}{r+1} \right) \right).$$

Theorem 3.9 ([81]). *Let X be totally bounded with modulus γ, (x_n) be uniformly Fejér monotone w.r.t. F with modulus χ and let (x_n) have approximate F-points with bound Φ. Then (x_n) is Cauchy and for all $k \in \mathbb{N}$ and all $g : \mathbb{N} \to \mathbb{N}$:*

$$\exists N \leq \Psi \, \forall i, j \in [N, N + g(N)] \, \left(d(x_i, x_j) \leq \frac{1}{k+1} \right),$$

where $\Psi(k, g, \Phi, \chi, \gamma) := \Psi_0(P)$ with

$$\Psi_0(0) := 0,$$
$$\Psi_0(n+1) := \Phi\left(\chi_g^M(\Psi_0(n), 4k+3)\right),$$

$$\chi_g^M(n, k) := \max_{i \leq n}\{\chi(i, g(i), k),\ P := \gamma(4k+3).$$

Additional results:

- If F is additionally uniformly closed with moduli δ_F, ω_F then

$$\exists N \leq \tilde{\Psi} \,\forall i,j \in [N, N+g(N)] \left(d(x_i, x_j) \leq \frac{1}{k+1} \text{ and } x_i \in AF_k \right),$$

where $\tilde{\Psi}$ results from Ψ by replacing k and χ by

$$k' := \max\{k, \lceil ((\omega_F(k)-1)/2 \rceil\} \text{ and } \chi'(n,m,r) := \max\{\delta_F(k), \chi(n,m,r)\}.$$

- Theorem 3.9 can be adapted to uniformly quasi-Fejér monotone sequences if Φ is a lim inf-bound.

Using these quantitative results, rates of metastability have been obtained in the following situations (the first item was already obtained earlier and served as motivation for the general approach):

- Krasnoselski-Mann iterations of asymptotically nonexpansive mappings in uniformly convex spaces ([64]).

- Picard iterations of firmly nonexpansive mappings in uniformly convex W-hyperbolic ('UCW'-)spaces ([81, 1]).

- Ishikawa iterations of nonexpansive mappings in UCW-spaces ([81, 99]).

- Mann iterations of strict pseudo-contractions in Hilbert spaces ([81, 53], see also section 3.1.3 below).

- The proximal point algorithm for the zeroes of maximally monotone operators in Hilbert space ([81]; see also section 3.3.3 below).

- Mann iterations of mappings satisfying condition (E) ([81]).

- Convex feasibility problems in CAT(κ) spaces ([71]).

- Minimization problems for two maps ([82], see also section 3.3.2 below).

3.1.3 Rates of asymptotic regularity and metastability for pseudocontractive mappings

An important generalization of the class of nonexpansive mappings are the so-called pseudocontractive mappings that were introduced by Browder:

Definition 3.10 ([22]). *Let X be a normed linear space and $C \subseteq X$ be a nonempty convex subset of X. A mapping $T : C \to C$ is called pseudocontractive if it satisfies*

$$\forall u, v \in C \, \forall \lambda > 1 \, ((\lambda - 1)\|u - v\| \leq \|(\lambda Id - T)(u) - (\lambda Id - T)(v)\|),$$

where Id denotes the identity mapping.

In a real Hilbert space X this is equivalent to

$$\forall u, v \in C \, (\langle Tu - Tv, u - v \rangle \leq \|u - v\|^2)$$

which in turn is equivalent to

$$\forall u, v \in C \, (\|Tu - Tv\|^2 \leq \|u - v\|^2 + \|u - Tu - (v - Tv)\|^2).$$

In [89], a rate of asymptotic regularity is extracted from a convergence proof due to [30] for an iteration schema due to [26] for Lipschitzian pseudocontractive mappings in general Banach spaces. In the case where C is bounded, the rate is polynomial. In the situation where X is a real Hilbert space, a rate of metastability for the strong convergence of that iteration is obtained in [90]. Finally, analyzing a proof in [26], Körnlein [88] recently extracted a rate of metastability for pseudocontractions that are only assumed to be demicontinuous.

Whereas pseudocontractive mappings (already on $\mathbb{R}$) in general are not continuous, the smaller class of strict pseudocontractions (introduced by Browder and Petryshyn in [24]) is even Lipschitzian:

Definition 3.11. *Let X be a real Hilbert space and $C \subseteq X$ be nonempty and convex. Let $T : C \to C$ and $0 \leq \kappa < 1$. Then T is called a κ-strict pseudocontraction if*

$$\forall u, v \in C \, (\|Tu - Tv\|^2 \leq \|u - v\|^2 + \kappa \|u - Tu - (v - Tv)\|^2).$$

In [53], the following rate of asymptotic regularity is extracted for Krasnoselski-Mann iterations

$$x_0 := x, \quad x_{n+1} := (1 - \lambda_n)x_n + \lambda_n T x_n$$

of κ-strict pseudocontractions in real Hilbert spaces (see also [119] for related results):

Theorem 3.12 ([53]). *Let C be additionally bounded and $b \geq \text{diam}(C)$. If (λ_n) is a sequence in $(\kappa, 1)$ satisfying $\sum_{n=0}^{\infty}(\lambda_n - \kappa)(1 - \lambda_n) = \infty$ with rate of divergence $\theta : \mathbb{N} \to \mathbb{N}$, then*

$$\Phi(k, b, \theta) = \theta\left(\lceil b^2 \rceil (k+1)^2\right)$$

is a rate of asymptotic regularity for (x_n), i.e.

$$\forall k \in \mathbb{N} \, \forall n \geq \Phi(k,b,\theta) \left(\|x_n - Tx_n\| \leq \frac{1}{k+1} \right).$$

3.2 Nonlinear Ergodic Theory

The classical von Neumann mean ergodic theorem, in the formulation due to Riesz, states that the sequence $(x_n)_{n \geq 0}$ of Cesàro means

$$x_n := \frac{1}{n+1} \sum_{i=0}^{n} T^i x$$

of a linear nonexpansive selfmapping $T : X \to X$ of a Hilbert space X starting from $x \in X$ strongly converges. In [77], a rate of metastability is extracted for a generalization of this result to uniformly convex Banach spaces from a proof due to G. Birkhoff which - as mentioned already at the end of section 2.6 - was subsequently improved to a bound on the number of ε-fluctuations in [4].

If the linearity of T is dropped then one has (in the case of Hilbert spaces) weak convergence by the famous Baillon nonlinear ergodic theorem ([8], also for $T : C \to C$, where $C \subseteq X$ is a closed convex subset) but strong convergence is known to fail. In [70] we extracted from an alternative proof of Baillon's theorem in [18] an explicit rate φ of metastability for the weak Cauchy property of (x_n) (here $\mathbb{N} \ni b \geq \|x\|$)

$$\forall \varepsilon > 0 \, \forall g : \mathbb{N} \to \mathbb{N} \, \forall w \in B_1(0) \, \exists n \leq \varphi(\varepsilon, b, g) \, \forall i, j \in [n; n+g(n)] \, (|\langle x_i - x_j, w \rangle| < \varepsilon)$$

which, however, is too complicated to state here in detail but whose complexity was already discussed in section 2.3.

In order to get strongly convergent nonlinear ergodic theorems one either has

(i) to add some weak form of linearity of T, e.g. being odd, or

(ii) to change the form of the sequence (x_n), where in the presence of full linearity of T this new sequence, nevertheless, coincides with the Cesàro means.

That (for symmetric C) the assumption of T being odd (in addition to being nonexpansive) implies the strong convergence of the sequence of Cesàro means was again shown by Baillon ([9]) and much later generalized by Wittmann in [129] to the situation, where K is an arbitrary subset of X and $T : K \to K$ any (not even necessarily continuous) mapping satisfying the already mentioned condition

$$(W): \forall x, y \in K \ (\|Tx + Ty\| \leq \|x + y\|)$$

which is implied (for symmetric K) if T is nonexpansive and odd.
In [116], a rate of metastability is extracted from Wittmann's proof:

Theorem 3.13 ([116]). *Let $K \subseteq X$ be any subset of a Hilbert space and $T : K \to K$ any mapping satisfying the condition (W). Then the sequence (x_n) of Cesàro means of T starting from $x \in K$ with $\|x\| \leq b$ strongly converges with rate of metastability*

$$\forall k \in \mathbb{N} \, \forall g : \mathbb{N} \to \mathbb{N} \, \exists m \leq \Phi(k, b, g^M) \, \forall i, j \in [m, m + g(m)] \ (\|x_i - x_j\| \leq 2^{-k}),$$

where

$$\Phi(k, b, g) := (N(2k+7, g) + P(2k+7, g)) \cdot b \cdot 2^{2k+8} + 1,$$
$$P(k, g) := P_0(k, F(k, g, N(k, g))),$$
$$F(k, g, n)(p) := p + n + \tilde{g}((n+p) \cdot b \cdot 2^{k+1}),$$
$$L(k, g)(n) := n + P_0(k, F(k, g, n)) + \tilde{g}((n + P_0(k, F(k, g, n))) \cdot b \cdot 2^{k+1}),$$
$$N(k, g) := (L(k, g))^{(b^2 2^{k+2})}(0),$$
$$P_0(k, f) := \tilde{f}^{(b^2 2^k)}(0), \ \tilde{f}(n) := n + f(n), \ f^M(n) := \max_{i \leq n+1} f(i).$$

For other metastability results for iterations of odd operators see [67, 71].

Instead of adding a weak form of linearity one may also change the sequence (x_n) in the nonlinear case to achieve strong convergence: let us consider the so-called Halpern iteration of $T : C \to C$

$$x_0 := x, \quad x_{n+1} := \lambda_{n+1} u + (1 - \lambda_{n+1}) T x_n$$

with starting point $x \in C$ and the anchor $u \in C$, where $(\lambda_n)_{n \geq 1}$ is a sequence in $[0, 1]$ (Halpern [50] only considered the case $u := 0$). In a celebrated paper [130], Wittmann for first time proved the strong convergence of (x_n) (X Hilbert space, $C \subseteq X$ closed and convex, $T : C \to C$ nonexpansive with $Fix(T) \neq \emptyset$ and $u = x$) under conditions on (λ_n) that permit the case $\lambda_n := 1/(n+1)$. With this choice of (λ_n), the sequence of Halpern iterates coincides with the sequence of Cesàro means if T is assumed to be linear. In [68], we extracted a rate of metastability for this strong convergence from Wittmann's original proof making use of the quantitative form of the projection to the fixed point set of T discussed already in section 2.3.

[79] extracted such a rate from a rather different proof due to [115] for a generalization of Wittmann's result to CAT(0) spaces X (and general $u, x \in C$). For the aforementioned choice $\lambda_n := 1/(n+1)$ and bounded C the result is:

Theorem 3.14 ([79]). *Let (X, ρ) be a CAT(0) space, $C \subseteq X$ be convex, $diam(C) \leq M$, (x_n) as above and $\varepsilon \in (0, 1)$. Then (x_n) is strongly convergent with rate of metastability:*

$$\forall g : \mathbb{N} \to \mathbb{N} \, \exists k \leq \Sigma(\varepsilon, g, M) \, \forall i, j \in [k, k + g(k)] \, (\rho(x_i, x_j) \leq \varepsilon),$$

where[4]

$$\Sigma(\varepsilon, g, M) := \left\lceil \frac{12(M^2 \chi_L^*(\varepsilon^2/12) + 1)}{\varepsilon^2} \right\rceil - 1, \text{ with } L := \widetilde{h^*}^{(\lceil M^2/\varepsilon_0^2 \rceil)}(0) + \left\lceil \frac{1}{\varepsilon_0} \right\rceil,$$

$$\chi_k^*(\varepsilon) := \left\lceil \frac{12M^2(k+1)}{\varepsilon} + \frac{288M^4(k+1)^2}{\varepsilon^2} \right\rceil - 1, \quad \varepsilon_0 := \varepsilon^2/24(d+1)^2,$$

$$\Theta_k(\varepsilon) := \left\lceil \frac{3M^2(\chi_k^*(\varepsilon/3) + 1)}{\varepsilon} \right\rceil - 1, \quad \Delta_k^*(\varepsilon, g) := \frac{\varepsilon}{3g_{\varepsilon,k}(\Theta_k(\varepsilon) - \chi_k^*(\varepsilon/3))},$$

$$g_{\varepsilon,k}(n) := n + g\left(n + \chi_k^*\left(\frac{\varepsilon}{3}\right)\right), \quad h(k) := \max\left\{ \left\lceil \frac{M^2}{\Delta_k^*(\varepsilon^2/4, g)} \right\rceil, k \right\} - k,$$

$$h^*(k) := h\left(k + \left\lceil \frac{1}{\varepsilon_0} \right\rceil\right) + \left\lceil \frac{1}{\varepsilon_0} \right\rceil, \quad \widetilde{h^*}(k) := k + h^*(k).$$

Further consequences of the analysis:

1. A quadratic rate of convergence for the asymptotic regularity $\rho(x_n, T(x_n)) \to 0$:

$$\forall \varepsilon > 0 \, \forall n \geq \Psi(\varepsilon, M) := \frac{4M}{\varepsilon} + \frac{32M^2}{\varepsilon^2} \, (\rho(x_n, T(x_n)) \leq \varepsilon).$$

This rate can be easily combined with the rate of metastability Σ from the previous theorem: define for $g : \mathbb{N} \to \mathbb{N}$ and $N \in \mathbb{N}$ a new function $g_N(n) := g(n + N) + N$ and put

$$\Sigma'(\varepsilon, g, M) := \Sigma(\varepsilon, g_{\Psi(\varepsilon, M)}, M) + \Psi(\varepsilon, M),$$

then

$$\forall g \in \mathbb{N}^{\mathbb{N}} \, \exists k \leq \Sigma'(\varepsilon, g, M) \, \forall i, j \in [k, k + g(k)] \, \forall l \geq k \, (\rho(x_i, x_j), \rho(x_l, T(x_l)) \leq \varepsilon).$$

---
[4]Correction to [79]: on p.2534, line 6, 'ε' in the condition on D_k must be '$\varepsilon/2$' and hence the factor '2304' on line 19 (resp. '144' on line 1 of p.651 in the addendum to [79]) must be '4608' (resp. '288'). Also on p.2534, line 19 the factor '$48M$' must be '$48M^2$' and on p.2543, line 2 '$16M^2$' must be '$32M^2$'.

2. Let z_k^u be the unique fixed point of the contraction

$$T_k(x) := \frac{1}{k}u \oplus (1 - \frac{1}{k})T(x).$$

Then the analysis yields primitive recursively in a given rate α of convergence of the resolvent (z_k^u) a rate of convergence of (x_n).

Similar results for so-called modified Halpern iterations are obtained in [118]. A highly nontrivial extension of the analysis to the much more general case of CAT(κ) spaces (with $\kappa > 0$) is given in [102].

In [80], a rate of metastability of (x_n) for the case of uniformly smooth Banach spaces is given relative to rate of metastability for the resolvent sequence (z_k^u) which is assumed to exist. Whereas, such a rate for the latter has been computed in [68] for Hilbert spaces and in [79] for CAT(0)-spaces, it is subject of ongoing research to achieve this even for L^p-spaces $(1 < p < \infty)$ other than L^2. The uniformly smooth case is also treated in [87] under a somewhat different set of conditions on (λ_n) due to [131] which also include the case $\lambda_n := 1/(n+1)$.

3.3 Convex Optimization

In this section we discuss some recent applications of proof mining in the area of convex optimization.

3.3.1 Convex feasibility problems

Let X be a real Hilbert space and $P_i : X \to C_i$ be the metric projections onto the closed and convex subsets $C_1, \ldots, C_r \subseteq X$ with $C := \bigcap_{i=1}^r C_i \neq \emptyset$.

The so-called convex feasibility problem (also called 'image recovery problem') is to find a point $p \in C$ in the 'image set' C.

For $1 \leq i \leq r$, define $T_i := Id + \lambda_i(P_i - Id)$ for $0 < \lambda_i \leq 2$, $\lambda_1 < 2$ and put $T := \sum_{i=1}^r \alpha_i T_i$, where $\alpha_1, \ldots, \alpha_r \in (0,1)$ with $\sum_{i=1}^r \alpha_i = 1$.

By a result of Crombez [32, 33] one has $Fix(T) = C$ if C is nonempty. Moreover, T is asymptotically regular, i.e. $\|T^{n+1}x - T^n x\| \to 0$.

Define towards an ε-approximate version of the convex feasibility problem

$$C_{i,\varepsilon} := \bigcup_{x \in C_i} B_\varepsilon(x), \quad C_\varepsilon := \bigcap_{i=1}^r C_{i,\varepsilon}.$$

By analyzing the proof of the (nontrivial) inclusion $Fix(T) \subseteq C$ one can extract a bound (compare with the discussion after Theorem 3.2) $\delta(D,\varepsilon) > 0$ such that (for $x \in X, p \in C$ and $D \in \mathbb{N}$ with $D \geq ||x-p||$)

$$\forall \varepsilon \in (0,1) \ (||Tx - x|| \leq \delta(D,\varepsilon) \to x \in C_\varepsilon)$$

(see Theorem 3.1(i) in [55]).

This bound is then combined with a rate of asymptotic regularity (Theorem 2.3 in [55]) to finally obtain the following quantitative solution of the approximate convex feasibility problem:[5]

Theorem 3.15 ([55]). *Let $x_0 \in X$ and $D > ||x_0 - p||$ for some $p \in C$ and $N_1, N_2 \in \mathbb{N}$ s.t.*

$$\frac{1}{N_1} \leq \min\{\alpha_i \lambda_i : 1 \leq i \leq r\}, \ \frac{1}{N_2} \leq \min\{\alpha_1, 2 - \lambda_1\}.$$

Then for $x_n := T^n x_0$ one has:

$$\forall \varepsilon \in (0,1) \ \forall n \geq \Psi(D, N_1, N_2, \varepsilon) \ (x_n \in C_\varepsilon),$$

where

$$\Psi(D, N_1, N_2, \varepsilon) := \left\lceil \frac{1936 \cdot N_1^6 \cdot (D+1)^4 (4N_1 + 1)^2 \cdot (2N_2 + 1)^2}{\pi \cdot \varepsilon^4} \right\rceil.$$

A similar result also holds for uniformly convex Banach spaces X, where then, however, one has to restrict λ_i to the interval $(0,1)$.

For quantitative versions of convex feasibility problems obtained in the context of CAT(κ) spaces ($\kappa > 0$) via general proof mining results for iterations of compositions of strongly quasi-nonexpansive mappings (as discussed in section 3.1.1), see [71].

3.3.2 Quantitative results for the composition of two mappings

Whereas in the convex feasibility problems discussed in the previous section one assumes that the intersection of the convex sets is nonempty and so that a common fixed point of the respective projections exists, the significance of the next theorem is that it only assumes the existence of a fixed point of the composition of two mappings:

---

[5] Correction to [55]: in Corollary 4.3(i),(ii) replace '$F(T) \neq \emptyset$' by '$C_0 \neq \emptyset$' and drop the dummy argument 'd' in Ψ.

Theorem 3.16 ([2]). *Let X be a CAT(0)-space and $T_1, T_2 : X \to X$ satisfying the condition (which for Hilbert spaces X is equivalent to being firmly nonexpansive)*

$$(P): \quad 2d(T_i x, T_i y)^2 \leq d(x, T_i y)^2 + d(y, T_i x)^2 - d(x, T_i x)^2 - d(y, T_i y)^2.$$

Let $Fix(T_2 \circ T_1) \neq \emptyset$ and consider sequences $(x_n), (y_n)$ in X with

$$d(y_n, T_1 x_n) \leq \varepsilon_n \text{ and } d(x_{n+1}, T_2 y_n) \leq \delta_n, \text{ for all } n \in \mathbb{N},$$

where $\sum_{n=0}^{\infty} \varepsilon_n < \infty$ and $\sum_{n=0}^{\infty} \delta_n < \infty$. Then

$$\lim d(y_{n+1}, y_n) = \lim d(x_{n+1}, x_n) = 0.$$

The proof makes repeated use of the convergence of bounded monotone sequences in $\mathbb{R}$ and hence of arithmetical comprehension.

Motivation behind the theorem:

Consider two proper, convex and lower semi-continuous $f, g : X \to (-\infty, +\infty]$ and define (see [14] for the study of this problem in the context of Hilbert spaces and [11] for the generalization to CAT(0)-spaces)

$$\Phi(x, y) := f(x) + g(y) + \frac{1}{2\lambda} d(x, y)^2.$$

Then the resolvents $T_1 = J_\lambda^g, T_2 = J_\lambda^f$ of f, g satisfy (P), where

$$J_\lambda^g(x) := \underset{z \in X}{\mathrm{argmin}} \left[g(z) + \frac{1}{2\lambda} d(x, z)^2 \right]$$

(see [7] for the study of the resolvents in the context of CAT(0)-spaces).

Computing sequences $(x_n), (y_n)$ as above (which only requires to know the resolvents up to some error) provides ε-solutions for the minimization problem

$$\underset{(x,y) \in X \times X}{\mathrm{argmin}} \Phi(x, y).$$

For this particular case, a quadratic rate of asymptotic regularity is extracted in [2] in the absence of error terms (i.e. $\delta_n = \varepsilon_n = 0$) and extended to the situation with error terms in [82] (see Remark 3.4 in that paper).

In the general situation of Theorem 3.16 one has an exponential bound:

Theorem 3.17 ([82]). *Let α be a Cauchy-rate for $\sum_{n=0}^{\infty} \gamma_n$, where $\gamma_n := \varepsilon_n + \delta_n$ and let $B \in \mathbb{N}, b > 0$ be such that $\sum \gamma_n \leq B$ and $d(x_0, u) \leq b$ for some $u \in Fix(T_2 \circ T_1)$. Then*
$$\forall n \geq \Phi(\varepsilon, b, B, \alpha) \ (d(x_n, x_{n+1}) \leq \varepsilon),$$

where

$$\Phi(\varepsilon, b, B, \alpha) := \alpha(\varepsilon/3) + k \left\lceil \frac{12(1 + 2^k)(b + B)}{\varepsilon} \right\rceil + 1, \ k := \left\lceil \frac{12(b + B)}{\varepsilon} \right\rceil.$$

Similarly for (y_n) with $\Phi'(\varepsilon, b, B, \alpha) := \Phi(\varepsilon/2, b, B, \alpha)$.

The sequences $(x_n), (y_n)$ are easily be seen to be uniformly quasi-Fejér monotone. Hence using Theorem 3.9 (adapted to quasi-monotonicity) one obtains rates of metastability in the case where X is totally bounded:

Theorem 3.18 ([82]). *Let X additionally be totally bounded with a modulus γ. Then*

$$\forall k \in \mathbb{N} \forall g : \mathbb{N} \to \mathbb{N} \exists n \leq \Psi(k, g) \forall i, j \in [n, n + g(n)] \left(d(x_i, x_j) \leq \frac{1}{k+1} \right),$$

where

$$\Psi(k, g) := \Psi_0(P), \ P := \gamma(8k + 7) + 1, \ \xi(k) := \alpha(1/(k+1)),$$
$$\chi_g^M(n, k) := (\max_{i \leq n} g(i)) \cdot (k + 1),$$

and using $(\widehat{\Phi}(k, N) := \max\{N, \Phi(1/2(i+1)) : i \leq k\}$ with Φ from Theorem 3.17)

$$\Psi_0(0) := 0, \ \Psi_0(n+1) := \widehat{\Phi} \left(\chi_g^M \left(\Psi_0(n), 8k + 7 \right), \xi(8k + 7) \right).$$

A similar bound holds for (y_n).

Note that asymptotic regularity in the above situation is just the special case of metastability when $g(n) := 1$ for all $n \in \mathbb{N}$. The fact that the proof of asymptotic regularity does not use the total boundedness of X is reflected by the above rate of metastability: if $g(n)$ does not depend on n, then also $\chi_g^M(n, k)$ does not depend on n and so the recursive definition of $\Psi(n)$ becomes constant from $n = 1$ on. Hence the bound does not depend on the modulus of total boundedness γ (as this only enters into P) and the bound (essentially) collapses to the rate of metastability Φ from Theorem 3.17.

3.3.3 Proximal point algorithm

Let X be real Hilbert space and $A : X \to 2^X$ be a maximally monotone, $J_{\gamma A} = (Id + \gamma A)^{-1}$ be the resolvent of γA for $\gamma > 0$ and $(\gamma_n) \subset (0, \infty)$.
The famous proximal point algorithm (due in this setting to [114] but formulated in the important case of resolvents of convex lower semi-continuous functions already in [104]) is given by (see the 2nd example after Definition 3.4 above)

$$x_0 \in X, \quad x_{n+1} := J_{\gamma_n A} x_n.$$

Proposition 3.19 ([81]). *1. (x_n) is uniformly Fejér monotone (w.r.t.*
$F = \bigcap_{k \in \mathbb{N}} F_k$, *where* $F_k := \bigcap_{i \leq k} \left\{ x \in H : \|x - J_{\gamma_i A} x\| \leq \frac{1}{k+1} \right\}$) *with modulus*

$$\chi(n, m, r) := \max\{n + m - 1, m(r + 1)\}.$$

2. If $\sum \gamma_n^2 = \infty$ with rate of divergence θ, then

$$\Phi_{b,\theta}(k) := \theta(\lceil b^2(M_k + 1)^2 \rceil) \lceil b^2(M_k + 1)^2 \rceil - 1,$$

where $M_k := \lceil (k+1)(2 + \max_{0 \leq i \leq k} \gamma_i) \rceil - 1$ and $\|x_0 - p\| \leq b$ for some $p \in zer(A)$, is an approximate F-point bound for (x_n).

Hence, Theorem 3.9 can be applied to obtain a rate of metastability in the finite dimensional case.

3.3.4 The hybrid steepest descent method

Let X be real Hilbert space and consider a mapping $\Theta : X \to \mathbb{R}$. The goal is to solve $\min \Theta$ over a closed convex subset $S \subseteq X$.

Let the gradient $\mathcal{F} := \Theta'$ of Θ be κ-Lipschitzian and η-strongly monotone and $S = Fix(T)$ for some nonexpansive $T : X \to X$. Then the above goal is equivalent to solving the following variational inequality problem:

VIP: Find $u^* \in S$ s.t. $\langle v - u^*, \mathcal{F}(u^*) \rangle \geq 0$ for all $v \in S$.

In [132], Yamada showed that under suitable conditions on (λ_n) the scheme (with $\mu := \eta/\kappa^2$)

$$u_{n+1} := T(u_n) - \lambda_{n+1} \mu \mathcal{F} T(u_n)$$

converges strongly to a solution of VIP.

Very recently, Körnlein [86] extracted an explicit and highly uniform effective rate of metastability for (u_n). In fact, [86] does this also for a generalization to finite families of mappings $T_1, \ldots, T_N : C \to C$ (that is also considered by [132]) and for general τ-contractions $\mathcal{G} : C \to C$ (with $\tau \in (0,1)$ and C closed and convex, instead of $\mathcal{G} := Id - \mu \mathcal{F}$ only)

$$(*) \; u_{n+1} := (1 - \lambda_{n+1}) T_{[n+1]}(u_n) + \lambda_{n+1} \mathcal{G} T_{[n+1]}(u_n),$$

where $[n]$ is the 'modulo N' function (note that, for $N = 1$, this schema has also been known as Moudafi's viscosity method, see [106]). The conditions on (λ_n) are those considered in [130] which allow for the choice of $\lambda_{n+1} := 1/(n+1)$ (see the discussion before Theorem 3.14). [106] (and also Yamada [132] in his first proof for the case $N = 1$) used stronger conditions which do not permit this choice. However, in his proof for general N, which even for $N = 1$ is different from his first proof, Yamada needs only the Wittmann conditions (and also the viscosity method has later been studied under these conditions by various authors).

3.4 Nonlinear Semigroups and Abstract Cauchy Problems

3.4.1 Proof Mining in nonlinear semigroup theory

Let X be a Banach space, $C \subseteq X$ a nonempty convex subset and $\lambda \in (0,1)$.

Definition 3.20. *A family $\{T(t) : t \geq 0\}$ of nonexpansive mappings $T(t) : C \to C$ is a nonexpansive semigroup if*

$(i)\; T(s+t) = T(s) \circ T(t) \quad (s, t \geq 0),$
$(ii)\; \text{for each } x \in C, \text{ the mapping } t \mapsto T(t)x \text{ is continuous.}$

Theorem 3.21 ([123])**.** *Let $0 < \alpha < \beta$ be such that α/β is irrational. Then any fixed point $p \in C$ of*
$$S := \lambda T(\alpha) + (1-\lambda) T(\beta)$$
is a common fixed point of $T(t)$ for all $t \geq 0$, i.e. (note that trivially '$\supseteq$')
$$Fix(S) = \bigcap_{t \geq 0} Fix(T(t)).$$

Suzuki's proof uses weak König's lemma WKL in the form that a continuous function $[0, M] \to \mathbb{R}$ on a compact interval $[0, M]$ attains its maximum. General logical metatheorems for the extraction of uniform bounds become applicable once we assume additionally that $t \mapsto T(t)x$ is equicontinuous on norm-bounded subsets of C with a modulus $\omega : \mathbb{N}^3 \to \mathbb{N}$. This condition (which is usually satisfied in practice) is only needed to make the bound Φ (discussed below) independent of the point $p \in C$. Let $f_\gamma : \mathbb{N} \to \mathbb{N}$ be a modulus of irrationality (called effective irrationality measure in number theory) for $\gamma := \alpha/\beta$, $\Lambda, N, D \in \mathbb{N}$ be s.t. $1/\Lambda \leq \lambda, 1 - \lambda$ and $1/N \leq \beta \leq D$. Then one can extract a bound $\Phi(\varepsilon, M, d) := \Phi(\varepsilon, M, d, N, \Lambda, D, f_\gamma, \omega)$ s.t. for all $M, d \in \mathbb{N}$:

$$\forall p \in C \, \forall \varepsilon > 0 \, (\|p\| \leq d \wedge \|S(p) - p\| < \Phi(\varepsilon, M, d) \to \forall t \in [0, M] \, (\|T(t)p - p\| < \varepsilon)).$$

Let $x_{n+1} := \frac{1}{2}x_n + \frac{1}{2}Sx_n$ be a d-bounded Krasnoselski iteration of S with rate of asymptotic regularity $\Psi(\varepsilon, d)$, then

$$\forall n \geq \Psi(\Phi(\varepsilon, M, d), d) \, \forall t \in [0, M] \, (\|T(t)x_n - x_n\| < \varepsilon).$$

In the case at hand, the optimal rate Ψ is known. Combining this with the explicitly extracted Φ the following final rate is obtained:

Theorem 3.22 ([74]). *Under the previous assumptions:*

$$\forall M \in \mathbb{N} \, \forall m \in \mathbb{N} \, \forall n \geq \Omega(m, M, d) \, \forall t \in [0, M] \, (\|T(t)x_n - x_n\| < 2^{-m})$$

with

$$\Omega(m, M, d) = \frac{2^{2m+8} d^2 ((\sum_{i=1}^{\phi(k, f_\gamma) - 1} \Lambda^i + 1)(1 + MN))^2}{\pi},$$

where $d \geq \|x_0 - Sx_n\|, \|x_n\|$ for all n, $k := D2^{\omega_{D,b}(3 + [\log_2(1 + MN)] + m) + 1}$ and

$$\phi(k, f) := \max\{2f(i) + 6 \, : \, 0 < i \leq k\}.$$

Example: $\alpha = \sqrt{2}, \beta = 2, \lambda = 1/2$. Then we may take $\Lambda = 2, N = 1, D = 2, f_\gamma(p) = 4p^2$.

Remark 3.23. *A bound d on either of the sequences $(\|x_0 - Sx_n\|), (\|x_n\|)$ can be transformed into one of the other: let $b \geq \|x_n\|$ for all $n \in \mathbb{N}$. Then for $B \geq \|x_0 - Sx_0\|$ and using that $(\|x_n - Sx_n\|)$ is nonincreasing one gets that*

$$\|x_0 - Sx_n\| \leq \|x_0 - x_n\| + \|x_n - Sx_n\| \leq 2b + B.$$

Conversely, $b \geq \|x_0 - Sx_n\|$ for all $n \in \mathbb{N}$ implies for $B' \geq \|x_0\|$

$$\|x_n\| \leq \|x_0\| + \|x_0 - Sx_n\| + \|Sx_n - x_n\| \leq \|x_0\| + \|x_0 - Sx_n\| + \|Sx_0 - x_0\| \leq b + B + B'.$$

3.4.2 Cauchy problems and set-valued accretive operators

Let X be a real Banach space and let $D(A)$ denote the domain of a set-valued operator A. $A : D(A) \to 2^X$ is accretive if

$$\forall (x,u), (y,v) \in A\ (\langle u-v, x-y\rangle_+ \geq 0),$$

where $\langle y, x\rangle_+ := \max\{\langle y, j\rangle : j \in J(x)\}$ for the normalized duality map J of the Banach space X.

A with $0 \in Az$ is uniformly accretive at zero with modulus $\Theta : \mathbb{N}^2 \to \mathbb{N}$ if, moreover,

$$\forall k, K \in \mathbb{N} \forall (x,u) \in A\ (\|x-z\| \in [2^{-k}, K] \to \langle u, x-z\rangle_+ \geq 2^{-\Theta_K(k)})$$

([73]). E.g. this holds for m-ψ-strongly accretive operators or even for ϕ-accretive operators in the sense of García-Falset [38] if ϕ has some normal form (which is the case in many applications).

Consider the following homogeneous Cauchy problem for an accretive A (satisfying the so-called range condition):

$$(1) \quad \begin{cases} u'(t) + A(u(t)) \ni 0, & t \in [0, \infty) \\ u(0) = x_0, \end{cases}$$

which has a unique integral solution for $x_0 \in \overline{D(A)}$ given by the Crandall-Liggett formula

$$u(t) := S(t)(x_0) := \lim_{n\to\infty} (Id + \frac{t}{n}A)^{-n}(x_0).$$

A continuous $v : [0, \infty) \to \overline{D(A)}$ is an almost-orbit of the nonexpansive semigroup S if

$$\lim_{s\to\infty} (\sup_{t\in[0,\infty)} \|v(t+s) - S(t)v(s)\|) = 0.$$

Theorem 3.24 ([37]). *Let A be a ϕ-accretive at zero operator with the range condition s.t. (1) has a strong solution for each $x_0 \in D(A)$. Then every almost-orbit (for the semigroup S generated by $-A$) strongly converges to the zero z of A.*

Theorem 3.25 ([73]). *Same as above but A uniformly accretive at zero with modulus Θ. Then*

$$\forall k \in \mathbb{N} \forall g : \mathbb{N} \to \mathbb{N} \exists \overline{n} \leq \Psi \forall x \in [\overline{n}, \overline{n} + g(\overline{n})]\ (\|v(x) - z\| < 2^{-k}),$$

where

$$\Psi(k, \overline{g}, B, \Phi, \Theta) := \Phi(k+1, g) + h(\Phi(k+1, g)), \text{ with}$$
$$h(n) := (B(n) + 2) \cdot 2^{\Theta_{K(n)}(k+2)+1}, \ g(n) := \overline{g}(n + h(n)) + h(n),$$
$$K(n) := \left\lceil \sqrt{2(B(n)+1)} \right\rceil, \ B(n) \geq \tfrac{1}{2}\|v(n) - z\|^2 \ (B(n) \ nondecreasing),$$

and Φ is rate of metastability for v, i.e.

$$\forall k \in \mathbb{N} \, \forall g : \mathbb{N} \to \mathbb{N} \, \exists n \leq \Phi(k, g) \, \forall t \in [0, g(n)] \, (\|v(t+n) - S(t)v(n)\| \leq 2^{-k}).$$

Consider now the nonhomogeneous Cauchy problem (A as before):

$$(2) \quad \begin{cases} u'(t) + A(u(t)) \ni f(t), \ t \in [0, \infty) \\ u(0) = x, \end{cases}$$

where $f \in L^1(0, \infty, X)$.

Then for each $x \in \overline{D(A)}$ the integral solution $u(\cdot)$ of (2) is an almost-orbit ([105]).

Proposition 3.26 ([73]).

$$\Phi_M(k, g) := \tilde{g}^{(M \cdot 2^{k+1})}(0)$$

with

$$\tilde{g}(n) := n + g(n), \ M \geq \int_0^\infty \|f(\xi)\| d\xi$$

is a rate of metastability of u (and so can be used as Φ in the previous theorem).

References

[1] Ariza-Ruiz, D., Leuştean, L., López-Acedo, G., Firmly nonexpansive mappings in classes of geodesic spaces. Trans. Amer. Math. Soc. **366**, pp. 4299-4322 (2014).

[2] Ariza-Ruiz, D., López-Acedo, G., Nicolae, A., The asymptotic behavior of the composition of firmly nonexpansive mappings. J. Optim. Theory Appl. **167**, pp. 409-429 (2015).

[3] Avigad, J., Iovino, J., Ultraproducts and metastability. New York J. of Math. **19**, pp. 713-727 (2013).

[4] Avigad, J., Rute, J., Oscillation and the mean ergodic theorem for uniformly convex Banach spaces. Ergod. Th. & Dynam. Sys. **35**, pp. 1009-1027 (2015).

[5] Avigad, J., Towsner, H., Functional interpretation and inductive definitions. J. Symb. Logic **74**, pp. 1100-1120 (2009).

[6] Avigad, J., Towsner, H., Metastability in the Furstenberg-Zimmer tower. Fund. Math. **210**, pp. 243-268 (2010).

[7] Bačák, M.: Convex analysis and optimization in Hadamard spaces. De Gruyter Series in Nonlinear Analysis and Applications, vol. 22. Walter de Gruyter & Co., Berlin 2014.

[8] Baillon, J.B., Un théorème de type ergodique pour les contractions non linéaires dans un espace de Hilbert. C.R. Acad. Sci. Paris Sèr. A-B **280**, pp. 1511-1514 (1975).

[9] Baillon, J.-B., Quelques propriétés de convergence asymptotique pour les contractions impaires. C. R. Acad. Sei. Paris **283**, pp. 587-590 (1976).

[10] Baillon, J.-B., Bruck, R.E., Reich, S., On the asymptotic behavior of nonexpansive mappings and semigroups in Banach spaces. Houston J. Math. **4**, pp. 1-9 (1978).

[11] Banert, S., Backward-backward splitting in Hadamard spaces. J. Math. Anal. Appl. **414**, pp. 656-665 (2014).

[12] Bauschke, H.H., The composition of projections onto closed convex sets in Hilbert space is asymptotically regular. Proc. Amer. Math. Soc. **131**, pp. 141-146 (2003).

[13] Bauschke, H.H., Combettes, P.L., A weak-to-strong convergence principle for Fejér-monotone methods in Hilbert spaces. Mathematics of Operations Research **26**, pp. 248-264 (2001).

[14] Bauschke, H.H., Combettes, P.L., Reich, S., The asymptotic behavior of the composition of two resolvents. Nonlinear Anal. **60**, pp. 283-301 (2005).

[15] Bauschke, H., Matoušková E., Reich, S., Projections and proximal point methods: convergence results and counterexamples. Nonlinear Anal. **56**, pp. 715-738 (2004).

[16] van den Berg, B., Briseid, E., Safarik, P., A functional interpretation for nonstandard arithmetic. Ann. Pure Appl. Logic **163**, pp. 1962-1994 (2012).

[17] Bezem, M., Strongly majorizable functionals of finite type: a model for bar recursion containing discontinuous functionals. J. Symbolic Logic **50** pp. 652-660 (1985).

[18] Brézis, H., Browder, F.E., Nonlinear ergodic theorems. Bull. Amer. Math. Soc. **82**, pp. 959-961 (1976).

[19] Briseid, E.M., A rate of convergence for asymptotic contractions. J. Math. Anal. Appl. **330**, pp. 364-376 (2007).

[20] Briseid, E.M., Fixed points of generalized contractive mappings. J. of Nonlinear and Convex Anal. **9**, pp. 181-204 (2008).

[21] Briseid, E.M., Logical aspects of rates of convergence in metric spaces. J. Symb. Logic **74**, pp. 1401-1428 (2009).

[22] Browder, F.E., Nonlinear mappings of nonexpansive and accretive type in Banach spaces. Bull. Amer. Math. Soc. **73**, pp. 875-882 (1967).

[23] Browder, F.E., Convergence theorems for sequences of nonlinear operators in Banach spaces, Math. Z. **100**, pp. 201-225 (1967).

[24] Browder, F.E., Petryshyn, W.V., Construction of fixed points of nonlinear mappings in Hilbert spaces, J. Math. Anal. Appl. **20**, pp. 197-228 (1967).

[25] Bruck, R.E., Nonexpansive projections on subsets of Banach spaces. Pacific J. Math. **47**, pp. 341-355 (1973).

[26] Bruck, R.E., A strongly convergent iterative method for the solution of $0 \in U(x)$ for a maximal monotone operator U in Hilbert space. J. Math. Anal. Appl. **48**, pp. 114-126 (1974).

[27] Bruck, R.E., Random products of contractions in metric and Banach spaces. J. Math. Anal. Appl. **88**, pp. 319-332 (1982).

[28] Bruck, R.E., Reich, S., Nonexpansive projections and resolvents of accretive operators in Banach spaces. Houston J. Math. **3**, pp. 459-470 (1977).

[29] Chang, C., Keisler, H.J., Continuous model theory, Annals of Mathematics Studies **58**, Princeton Univ. Press, 1966.

[30] Chidume, C.E., Zegeye, H., Approximate fixed point sequences and convergence theorems for Lipschitz pseudocontractive maps. Proc. Amer. Math. Soc. **132**, pp. 831-840 (2004).

[31] Cho, S., A variant of continuous logic and applications to fixed point theory. Preprint 2016, arXiv:1610.05397.

[32] Crombez, G., Image recovery by convex combinations of projections. J. Math. Anal. Appl. **155**, pp. 413-419 (1991).

[33] Crombez, G., Parallel methods in image recovery by projections onto convex sets. Czechoslovak Math. J. **42**, pp. 445-450 (1992).

[34] Delzell, C.N., Kreisel's unwinding of Artin's proof-Part I. In: Odifreddi, P., Kreiseliana, 113-246, A K Peters, Wellesley, MA, 1996.

[35] Engrácia, P., Proof-theoretical studies on the bounded functional interpretation. PhD Thesis, Universidade de Lisboa, 2009.

[36] Feferman, S., Theories of finite type related to mathematical practice. In: Barwise, J. (ed.), Handbook of Mathematical Logic, pp. 913-972, North-Holland, Amsterdam (1977).

[37] García-Falset, J. The asymptotic behavior of the solutions of the Cauchy problem generated by ϕ-accretive operators, J. Math. Anal. Appl. **310**, pp. 594-608, 2005.

[38] García-Falset, J., Llorens-Fuster, E., Prus, S., The fixed point property for mappings admitting a center. Nonlinear Analysis **66**, pp. 1257-1274 (2007).

[39] García-Falset, J., Llorens-Fuster, E., Suzuki, T., Fixed point theory for a class of generalized nonexpansive mappings. J. Math. Anal. Appl. **375**, pp. 185-195 (2011).

[40] Ferreira, F., Gaspar, J., Nonstandardness and the bounded functional interpretation. Ann. Pure Appl. Logic **166**, pp. 701-712 (2015).

[41] Ferreira, F., Oliva, P., Bounded functional interpretation. Ann. Pure Appl. Logic **135**, pp. 73-112 (2005).

[42] Gerhardy, P., A quantitative version of Kirk's fixed point theorem for asymptotic contractions. J. Math. Anal. Appl. **316**, pp. 339-345 (2006).

[43] Gerhardy, P., Kohlenbach, U., Strongly uniform bounds from semi-constructive proofs. Ann. Pure Appl. Logic **141**, 89-107 (2006).

[44] Gerhardy, P., Kohlenbach, U., General logical metatheorems for functional analysis. Trans. Amer. Math. Soc. **360**, pp. 2615-2660 (2008).

[45] Goebel, K., Kirk, W.A., A fixed point theorem for asymptotically nonexpansive mappings. Proc. Amer. Math. Soc. **35**, pp. 171-174 (1972).

[46] Goebel, K., Kirk, W.A., Iteration processes for nonexpansive mappings. In: Singh, S.P., Thomeier, S., Watson, B., eds., Topological Methods in Nonlinear Functional Analysis. Contemporary Mathematics **21**, AMS, pp. 115-123 (1983).

[47] Goebel, K., Reich, S., Uniform convexity, hyperbolic geometry, and nonexpansive mappings. Monographs and Textbooks in Pure and Applied Mathematics, 83. Marcel Dekker, Inc., New York, ix+170 pp., 1984.

[48] Günzel, D., Logical Metatheorems in the Context of Families of Abstract Metric Structures. Master-Thesis, TU Darmstadt 2013.

[49] Günzel, D., Kohlenbach, U., Logical metatheorems for abstract spaces axiomatized in positive bounded logic. Adv. Math. **290**, 503-551 (2016).

[50] Halpern, B., Fixed points of nonexpanding maps. Bull. Amer. Math. Soc. **73**, pp., 957-961 (1967).

[51] Henson, C.W., Iovino, J., Ultraproducts in Analysis, In: Analysis and Logic, London Math. Soc. Lecture Note Ser., vol. 262, Cambridge Univ. Press, 2002, 1-115.

[52] Hernest, M.-D., Oliva, P., Hybrid functional interpretations. In: Beckmann, A., et al. eds., Proceedings of CiE'08. Springer LNCS **5028**, pp. 251-260 (2008).

[53] Ivan, D., Leuştean, L., A rate of asymptotic regularity for the Mann iteration of κ-strict pseudo-contractions, Numer. Funct. Anal. Optimiz. **36**, pp. 792-798 (2015).

[54] Khan, M.A.A., Kohlenbach, U., Bounds on Kuhfittig's iteration schema in uniformly

convex hyperbolic spaces. J. Math. Anal. Appl. **403**, pp. 633-642 (2013).

[55] Khan, M.A.A., Kohlenbach, U., Quantitative image recovery theorems. Nonlinear Anal. **106**, pp. 138-150 (2014).

[56] Kincses, J., Totik, V., Theorems and counterexamples on contractive mappings. Mathematica Balkanica, New Series **4**, pp. 69-90 (1990).

[57] Kirk, W.A., Fixed points of asymptotic contractions. J. Math. Anal. Appl. **277**, pp. 645-650 (2003).

[58] Kohlenbach, U., Theorie der majorisierbaren und stetigen Funktionale und ihre Anwendung bei der Extraktion von Schranken aus inkonstruktiven Beweisen: Effektive Eindeutigkeitsmodule bei besten Approximationen aus ineffektiven Beweisen. PhD Thesis, Frankfurt am Main, xxii+278pp. (1990).

[59] Kohlenbach, U., Effective bounds from ineffective proofs in analysis: an application of functional interpretation and majorization. J. Symbolic Logic **57**, pp. 1239-1273 (1992).

[60] Kohlenbach, U., Analysing proofs in analysis. In: W. Hodges, M. Hyland, C. Steinhorn, J. Truss, editors, *Logic: from Foundations to Applications. European Logic Colloquium* (Keele, 1993), pp. 225–260, Oxford University Press (1996).

[61] Kohlenbach, U., Relative constructivity. J. Symbolic Logic **63**, pp. 1218-1238 (1998).

[62] Kohlenbach, U., On the no-counterexample interpretation. J. Symbolic Logic **64**, pp. 1491-1511 (1999).

[63] Kohlenbach, U., Some logical metatheorems with applications in functional analysis. Trans. Amer. Math. Soc. **357**, no. 1, pp. 89-128 (2005).

[64] Kohlenbach, U., Some computational aspects of metric fixed point theory. Nonlinear Analysis **61**, pp. 823-837 (2005).

[65] Kohlenbach, U., Applied Proof Theory: Proof Interpretations and their Use in Mathematics. Springer Monographs in Mathematics. xx+536pp., Springer Heidelberg-Berlin, 2008.

[66] Kohlenbach, U., On the logical analysis of proofs based on nonseparable Hilbert space theory. In: Feferman, S., Sieg, W. (eds.), 'Proofs, Categories and Computations. Essays in Honor of Grigori Mints'. College Publications, pp. 131-143 (2010).

[67] Kohlenbach, U., On the asymptotic behavior of odd operators. J. Math. Anal. Appl. **382**, pp. 615-620 (2011).

[68] Kohlenbach, U.,On quantitative versions of theorems due to F.E. Browder and R. Wittmann. Advances in Mathematics **226**, pp. 2764-2795 (2011).

[69] Kohlenbach, U., Gödel functional interpretation and weak compactness. Ann. Pure Appl. Logic **163**, pp. 1560-1579 (2012).

[70] Kohlenbach, U., A uniform quantitative form of sequential weak compactness and Bail-

lon's nonlinear ergodic theorem. Comm. Contemp. Math. **14**, 20pp. (2012).

[71] Kohlenbach, U., On the quantitative asymptotic behavior of strongly nonexpansive mappings in Banach and geodesic spaces. Israel Journal of Mathematics **216**, pp. 215-246 (2016).

[72] Kohlenbach, U., A polynomial rate of asymptotic regularity for compositions of projections in Hilbert space. To appear in: FoCM.

[73] Kohlenbach, U., Koutsoukou-Argyraki, A., Rates of convergence and metastability for abstract Cauchy problems generated by accretive operators. J. Math. Anal. Appl. **423**, 1089-1112 (2015).

[74] Kohlenbach, U., Koutsoukou-Argyraki, A., Effective asymptotic regularity for one-parameter nonexpansive semigroups. J. Math. Anal. Appl. **433**, pp. 1883-1903 (2016).

[75] Kohlenbach, U., Lambov, B., Bounds on iterations of asymptotically quasi-nonexpansive mappings. In: G-Falset, J., L-Fuster, E., Sims, B. (eds.), Proc. International Conference on Fixed Point Theory, Valencia 2003, pp. 143-172, Yokohama Press, 2004.

[76] Kohlenbach, U., Leuştean, L., Mann iterates of directionally nonexpansive mappings in hyperbolic spaces. Abstract and Applied Analysis, vol. 2003, issue 8, pp. 449-477 (2003).

[77] Kohlenbach, U., Leuştean, L., A quantitative mean ergodic theorem for uniformly convex Banach spaces, Ergod. Th. & Dyn. Sys. **29**, pp. 1907-1915 (2009).

[78] Kohlenbach, U., Leuştean, L., Asymptotically nonexpansive mappings in uniformly convex hyperbolic spaces. J. European Math. Soc. **12**, pp. 71-92 (2010)

[79] Kohlenbach, U., Leuştean, L., Effective metastability of Halpern iterates in CAT(0) spaces. Adv. Math. **231**, pp. 2526-2556 (2012). Addendum in: Adv. Math. **250**, pp. 650-651, 2014)

[80] Kohlenbach, U., Leuştean, L., On the computational content of convergence proofs via Banach limits. Philosophical Transactions of the Royal Society A **370**, pp. 3449-3463 (2012).

[81] Kohlenbach, U., Leuştean, L., Nicolae, A., Quantitative results of Fejér monotone sequences. Communications in Contemporary Mathematics **20** (2018), 1750015 (42 pages). DOI: 10.1142/S0219199717500158.

[82] Kohlenbach, U., López-Acedo, G., Nicolae, A., Quantitative asymptotic regularity for the composition of two mappings. Optimization **66**, pp. 1291-1299 (2017).

[83] Kohlenbach, U., Nicolae, A., A proof-theoretic bound extraction theorem for CAT(κ)-spaces. Studia Logica **105**, pp. 611-624 (2017).

[84] Kohlenbach, U., Oliva, P., Proof mining: a systematic way of analysing proofs in math-

ematics. Proc. Steklov Inst. Math. **242**, pp. 136-164 (2003).

[85] Kohlenbach, U., Safarik, P., Fluctuations, effective learnability and metastability in analysis. Ann. Pure Appl. Log. **165**, pp. 266-304 (2014).

[86] Körnlein, D., Quantitative Analysis of Iterative Algorithms in Fixed Point Theory and Convex Optimization. PhD Thesis, TU Darmstadt 2016.

[87] Körnlein, D., Quantitative results for Halpern iterations of nonexpansive mappings. J. Math. Anal. Appl. **428**, pp. 1161-1172 (2015).

[88] Körnlein, D., Quantitative results for Bruck iterations of demicontinuous pseudocontractions. Preprint 2015, submitted.

[89] Körnlein, D., Kohlenbach, U., Effective rates of convergence for Lipschitzian pseudocontractive mappings in general Banach spaces. Nonlinear Analysis **74**, pp. 5253-5267 (2011).

[90] Körnlein, D., Kohlenbach, U., Rates of metastability for Bruck's iteration of pseudocontractive mappings in Hilbert space. Numer. Funct. Anal. Optimiz. **35**, pp. 20-31 (2014).

[91] Komjáth, P., Totik, V., Ultrafilters. Amer. Math. Monthly **115**, pp. 33-44 (2008).

[92] Kreisel, G., On the interpretation of non-finitist proofs, part I. J. Symbolic Logic **16**, pp.241-267 (1951).

[93] Kreisel, G., On the interpretation of non-finitist proofs, part II: Interpretation of number theory, applications. J. Symbolic Logic **17**, pp. 43-58 (1952).

[94] Kreuzer, A.P., Non-principal ultrafilters, program extraction and higher order reverse mathematics. J. of Mathematical Logic **12**. 16 pp. (2012).

[95] Kreuzer, A.P., On the strength of weak compactness. Computability **1**, pp. 171-179 (2012).

[96] Kuhfittig, P.K.F., Common fixed points of nonexpansive mappings by iteration. Pacific J. Math. **97**, pp. 137-139 (1981).

[97] Leuştean, L., Proof mining in ℝ-trees and hyperbolic spaces. Electronic Notes in Theoretical Computer Science (Proceedings of WoLLIC 2006) **165**, pp. 95-106 (2006).

[98] Leuştean, L., A quadratic rate of asymptotic regularity for CAT(0)-spaces, J. Math. Anal. Appl. **325**, pp. 386-399 (2007).

[99] Leuştean, L., Nonexpansive iterations in uniformly convex W-hyperbolic spaces. Nonlinear analysis and optimization I. Nonlinear analysis. Contemp. Math. **513**, Amer. Math. Soc., Providence, RI, pp. 193-210 (2010).

[100] Leuştean, L., An application of proof mining to nonlinear iterations. Ann. Pure Appl. Logic **165**, pp. 1484-1500 (2014).

[101] Leuştean, L., Nicolae, A., Effective results on compositions of nonexpansive mappings. J. Math. Anal. Appl. **410**, pp. 902-907 (2014).

[102] Leuştean, L., Nicolae, A., Effective results on nonlinear ergodic averages in CAT(k) spaces. Ergod. Th. & Dynam. Sys. **36**, pp. 2580–2601 (2016).

[103] Luckhardt, H., Herbrand-Analysen zweier Beweise des Satzes von Roth: Polynomiale Anzahlschranken. J. Symbolic Logic **54**, pp. 234-263 (1989).

[104] Martinet, B., Régularisation dinéquations variationnelles par approximations successives, Rev. Française Informat. Recherche Opérationnelle **4**, pp. 154-158 (1970).

[105] Miyadera, I., Kobayasi, K., On the asymptotic behaviour of almost-orbits of nonlinear contraction semigroups in Banach spaces, Nonlinear Analysis **6**, pp. 349-365 (1982).

[106] Moudafi, A., Viscosity approximation methods for fixed-point problems. J. Math. Anal. Appl. **241**, pp. 46-55 (2000).

[107] Neumann, E., Computational problems in metric fixed point theory and their Weihrauch degrees. Log. Method. Comput. Sci. **11**, 44pp. (2015).

[108] Nicolae, A., Asymptotic behavior of averaged and firmly nonexpansive mappings in geodesic spaces. Nonlinear Analysis **87**, pp. 102-115 (2013).

[109] Oliva, P., Hybrid functional interpretations of linear and intuitionistic logic. J. Logic Comput. **22**, pp. 305-328 (2012).

[110] Parsons, C., On n-quantifier induction. J. Symbolic Logic **37**, pp. 466-482 (1972).

[111] Rakotch, E., A note on contractive mappings. Proc. Amer. Math. Soc. **13**, pp. 459–465 (1962).

[112] Reich, S., The alternating algorithm of von Neumann in the Hilbert ball. Dynamical Systems Appl. **2**, pp. 21-25 (1993).

[113] Reich, S., Shafrir, I., The asymptotic behavior of firmly nonexpansive mappings. Proc. Amer. Math. Soc. **101**, pp. 246-250 (1987).

[114] Rockafellar, T., Monotone operators and the proximal point algorithm. SIAM J. Control Optim. **14**, pp. 877-898 (1976).

[115] Saejung, S., Halpern iterations in CAT(0) spaces. Fixed Point Theory and Applications, Volume 2010, Article ID 471781, 13pp.

[116] Safarik, P., A quantitative nonlinear strong ergodic theorem for Hilbert spaces. J. Math. Anal. Appl. **391**, pp. 26-37 (2012).

[117] Sanders, S., The unreasonable effectiveness of nonstandard analysis. Preprint 2016, arXiv:1508.07434v2.

[118] Schade, K., Kohlenbach, U., Effective metastability for modified Halpern iterations in CAT(0) spaces. Fixed Point Theory and Applications 2012:191, 19pp.

[119] Sipoş, A., A note on the Mann iteration for k-strict pseudocontractions in Banach spaces. Numer. Funct. Anal. Optimiz. **38**, pp. 80-90 (2017).

[120] Sipoş, A., Effective results on a fixed point algorithm for families of nonlinear mappings. Ann. Pure Appl. Log. **168**, 112-128 (2017).

[121] Specker, E., Nicht konstruktiv beweisbare Sätze der Analysis. J. Symb. Logic **14**, pp. 145-158 (1949).

[122] Spector, C., Provably recursive functionals of analysis: a consistency proof of analysis by an extension of principles formulated in current intuitionistic mathematics. In: Recursive function theory, Proceedings of Symposia in Pure Mathematics, vol. 5 (J.C.E. Dekker (ed.)), AMS, Providence, R.I., pp. 1-27 (1962).

[123] Suzuki, T., Common fixed points of one-parameter nonexpansive semigroups. Bull. London Math. Soc. **38**, pp. 1009-1018 (2006).

[124] Suzuki, T., Fixed point theorems and convergence theorems for some generalized nonexpansive mappings. J. Math. Anal. Appl. **340**, pp. 1088-1095 (2008).

[125] Tao, T., Soft analysis, hard analysis, and the finite convergence principle. Essay posted May 23, 2007. Appeared in: 'T. Tao, Structure and Randomness: Pages from Year One of a Mathematical Blog. AMS, 298pp., 2008'.

[126] Tao, T., Norm convergence of multiple ergodic averages for commuting transformations. Ergodic Theory and Dynamical Systems **28**, pp. 657-688 (2008).

[127] Toftdal, M., Calibration of Ineffective Theorems of Analysis in a Constructive Context. Master Thesis, Aarhus Universitet, 2004.

[128] Towsner, H., Ultrafilters in reverse mathematics. J. Math. Log. **14**, 11 pp. (2014).

[129] Wittmann, R., Mean ergodic theorems for nonlinear operators. Proc. Amer. Math. Soc. **108**, pp. 781-788 (1990).

[130] Wittmann, R., Approximation of fixed points of nonexpansive mappings. Arch. Math. **58**, pp. 486-491 (1992).

[131] Xu, H.-K., Iterative algorithms for nonlinear operators. J. Lond. Math. Soc. **66**, pp. 240-256 (2002).

[132] Yamada, I., The hybrid steepest descent method for the variational inequality problem over the intersection of fixed point sets of nonexpansive mappings. In: Y.C. Dan Butnariu and S. Reich, editors, Inherently Parallel Algorithms in Feasibility and Optimization and their Applications, volume 8 of Studies in Computational Mathematics, pp. 473-504, Elsevier 2001.

The Clique Covering Problem and Other Questions

Maryanthe Malliaris
Department of Mathematics, University of Chicago, Chicago, IL
`mem@math.uchicago.edu`

Abstract

We motivate several open questions about sequences of hypergraphs associated to first-order formulas, part of a developing constellation of ideas at the meeting point of model theory, graph theory, and the theory of ultrafilters.

1 Introduction

Ultraproducts are a kind of limit construction built to reflect the average behavior of a sequence of mathematical structures, such as graphs, groups, or fields. Ultraproducts average and, at the same time, amplify. Since the average is taken in the sense of an ultrafilter, as will be explained below, the average behavior visible in this limit structure is often not uniformly approximated in the index (factor) models. Sometimes it is approximated very obscurely. In fact, one of the most interesting things about the ultraproduct construction is precisely this distance between what appears 'on average', in the limit structure, and what appears in the factors.

Investigating this issue can be useful in two senses. On one hand, it may tell us something interesting about the structure of the ultrafilter – the object responsible for the averaging. On the other hand, it may tell us something about the objects being averaged: what in the nature of their basic combinatorial structure is amenable to the kind of amplification which the ultraproduct reflects, or is able to resonate with a particular known amplifier in the ultrafilter. This is especially useful in the case of an ultra*power*, when all the factor structures are isomorphic. In this case the ultraproduct may essentially function as scaffolding allowing us to identify inherent properties of the object itself which explain its amenability to amplification or, so to speak, its resonance.

Partially supported by NSF CAREER award 1553653 and by a Sloan fellowship.

The purpose of this note is to motivate and explain two open questions arising from the author's early papers. The problems have been discussed informally for several years following those papers, but prior to the present note the relevant definitions, theorems, elementary observations and intuitions were spread across several articles and in some cases not written down. The first is the (ultrapower) clique covering problem, which has a natural relation to the fundamental problem of Keisler's order. The second has to do with stability of the so-called characteristic sequence of a formula. Along the way, we define most terms and discuss interesting variants.

Thanks to R. Shore for some helpful comments.

It is a pleasure to thank here M. Baaz and V. Sabljakovic-Fritz.

Contents

1	**Introduction**	1
2	**Basic Definitions**	2
3	**The Clique Covering Problem**	8
4	**Interlude: Saturation of Ultrapowers**	11
5	**The Stable Sequences Problem**	15

2 Basic Definitions

In this section we define the basic objects: models, ultraproducts, and regular ultrafilters.

For a model theorist a structure (a graph, a group, a field...) is given in the following formal way. First we fix a language $\mathcal{L}$, which is a possibly infinite set of relation, function, and constant symbols (relation and function symbols are given with the data of their arity). We assume without further mention that all languages contain a binary relation = which will be interpreted to mean equality. For example, we might take the language $\{R\}$ containing a binary relation symbol, or the language $\{+, \times, 0, 1\}$ containing two binary functions and two constants. Now a model (i.e. a mathematical structure) in a given language $\mathcal{L}$ is given by the data of some set M – called the domain of the model – along with instructions for how to interpret each symbol of $\mathcal{L}$. That is, each constant of $\mathcal{L}$ is assigned to an element of the domain, each k-ary relation symbol is assigned to some (possibly empty) subset of M^k, and each k-ary function symbol is assigned to some subset of M^{k+1} which is the graph of a function with domain M^k. In slight abuse of notation, we will use M to refer both to the model and its domain. To see the use of this formalism, note that two models are defined to be isomorphic if there is a bijection between their domains which preserves relations, functions, and constants; in particular, this formalism is a way of pointing out exactly what structure we want to pay attention to while eliding what we do not. For a model theorist, if $\mathcal{L} = \{R\}$ is a single binary relation, then the model M with domain $\mathbb{Q}$ and $R^N = \{(p,q) \in \mathbb{Q} \times \mathbb{Q} : p < q$ in the usual ordering on the rationals$\}$ has many automorphisms which preserve the ordering, but not the arithmetic structure.[1]

Some notation: given a k-ary relation symbol R of $\mathcal{L}$, an $\mathcal{L}$-model M, and elements $a_1, \ldots, a_k \in M$, we say that R holds of $a_1, \ldots, a_k$ in N, in symbols $N \models R(a_1, \ldots, a_k)$, when $(a_1, \ldots, a_k) \in R^N$. Likewise, write $N \models f(a_1, \ldots, a_k) = a_{k+1}$ when $(a_1, \ldots, a_{k+1}) \in f^M$.

Given this setup, it is natural to fix a language $\mathcal{L}$ and consider averaging sequences of $\mathcal{L}$-models (for instance, a sequence of infinite graphs) to create a larger $\mathcal{L}$-structure where the interpretations of the symbols of $\mathcal{L}$ directly reflect the behavior of the factors. The definition of model has two steps – giving the domain and interpreting the symbols – so we will build an ultraproduct in two steps to reflect this. First we need to know which subsets of I are large. An ultrafilter is a coherent way of making this decision.[2]

Definition 2.1 (Ultrafilter). *An ultrafilter $\mathcal{D}$ on a set I is a collection $\mathcal{D} \subseteq \mathcal{P}(I)$ which does not contain the empty set, is upward closed ($A \in \mathcal{D}$ and $A \subseteq B \subseteq I$ implies $B \in \mathcal{D}$), is closed under finite intersection ($A, B \in \mathcal{D}$ implies $A \cap B \in \mathcal{D}$) and contains precisely one of A and $I \setminus A$ for all $A \subseteq I$.*

The simplest example of an ultrafilter is given by fixing some element $i_* \in I$ and declaring that a set is 'large' iff it contains i_*. This is indeed an ultrafilter, called the principal ultrafilter generated by $\{i_*\}$.

It follows from Zorn's lemma or by transfinite induction that so-called non-principal ultrafilters exist (necessarily, in these cases I is infinite and the ultrafilters in question contain all co-finite sets). In what follows, unless otherwise stated, all ultrafilters are non-principal.

---

[1] Note that the given choice of relation, function, or constant symbol makes no formal demands on how we interpret it, though in practice, when it is unlikely to cause confusion, we often use symbols such as $+, \times, <$ and assume they have the natural interpretation in the given context.

[2] Ultrafilters have a rich history in set theory and general topology, dating to early in the twentieth century. For a nice story of how they answer Bourbaki's question of whether the notion of limit can be liberated from the countable, see [20]. On ultrafilters and limits, see e.g. [8].

Suppose now that $\langle X_i : i \in I \rangle$ is a sequence of infinite sets and $\mathcal{D}$ is an ultrafilter. Consider the equivalence relation $\sim$ on the Cartesian product $\prod_{i \in I} X_i$ defined by:

$$\langle a_i : i \in I \rangle \sim \langle b_i : i \in I \rangle \text{ iff } \{ i \in I : a_i = b_i \} \in \mathcal{D}.$$

That $\sim$ is an equivalence relation follows from the fact that any ultrafilter is closed under finite intersection.

Definition 2.2 (Ultraproduct). *Fix a set I, and a language $\mathcal{L}$ and let $\langle M_i : i \in I \rangle$ be a sequence of $\mathcal{L}$-models, called the index models. Let $\mathcal{D}$ be an ultrafilter on I. The ultraproduct $N = M^I/\mathcal{D}$ is defined to be the following $\mathcal{L}$-model.*

- *The domain of N is the set of equivalence classes of M^I under the relation $\sim$. In what follows, we fix a representative for each equivalence class, so that given $a \in N$ and $i \in I$ it is well defined to write "$a[i]$" for the i-th coordinate.[3]*

- *For each k-ary relation symbol R of $\mathcal{L}$, and every $a_1, \ldots, a_k \in N$, we define*

$$(a_1, \ldots, a_k) \in R^N \iff \{ i \in I : (a_1[i], \ldots, a_k[i]) \in R^M \} \in \mathcal{D}.$$

- *Likewise, for each k-ary function symbol f of $\mathcal{L}$, and every $a_1, \ldots, a_k, b \in N$, we define*

$$f^N(a_1, \ldots, a_k) = b \iff \{ i \in I : f^M(a_1[i], \ldots, a_k[i]) = b[i] \} \in \mathcal{D}.$$

- *Likewise, for each constant symbol c of $\mathcal{L}$, and every $a \in N$, we define*

$$c^N = a \iff \{ i \in I : c^M = a[i] \} \in \mathcal{D}.$$

In other words, a relation, function, or constant holds of a tuple in N if and only if it holds of the projections to the index models $\mathcal{D}$-almost everywhere. We say that N is an ultrapower when the sequence $\langle M_i : i \in I \rangle$ is constant, and in this case often write simply $M^I/\mathcal{D}$.

For example, if $\mathcal{L} = \{R\}$ as above, $\langle M_i : i \in I \rangle$ is an infinite sequence of undirected graphs, and $N = \prod_{i \in I} M_i/\mathcal{D}$, then N is also an infinite undirected graph and elements $a, b \in N$ are connected by an edge in N precisely when on a $\mathcal{D}$-large set of i, a_i, b_i have an edge in M_i. The last item in the definition of ultrafilter guarantees that this is well defined.

---

[3]It is easy to see that the definition does not depend on the choice of representative.

In some sense, the ultraproduct amplifies and smooths.[4] For example, we will justify the following example presently.

Example 2.3. *If $\mathcal{L} = \{+, \times, 0, 1\}$, I is the set of primes, for each $p \in I$ $M_p = \bar{\mathbb{F}}_p$ is the algebraic closure of the finite field $\mathbb{F}_p$, and $\mathcal{D}$ is a nonprincipal ultrafilter on I, then $N = \prod_{p \in I} M_p / \mathcal{D}$ will be isomorphic to the field $\mathbb{C}$.*

Even when $\langle M_i : i \in I \rangle$ is a constant sequence there is a built in amplification. For example, if M is a countably infinite graph and $\mathcal{D}$ is a nonprincipal ultrafilter on a countable set then $M^I/\mathcal{D}$ will have cardinality continuum. We will return to the question of size in a moment.

By considering the definition of ultraproduct, it is easy to see that some kinds of structure immediately transfer from the factors M_i to N. For example, if $\mathcal{L} = \{R\}$ and the M_i are all undirected graphs with no loops and no multiple edges, the same is true of N (if some of the M_i are directed graphs, then whether or not N is directed depends on whether the set of i for which M_i is a directed graph is $\mathcal{D}$-large). Likewise, if $\mathcal{L} = \{+, \times, 0, 1\}$ and the M_i are all fields, then N is likewise a field. The right general statement is that we have a transfer principle for any mathematical statement expressible in first-order logic.

Theorem 2.4 (Łos' Theorem). *Fix a language $\mathcal{L}$. Suppose we are given an ultraproduct $N = \prod_{i \in I} M_i / \mathcal{D}$ of $\mathcal{L}$-structures, a formula $\varphi(x_1, \ldots, x_k)$ of first-order logic in the language $\mathcal{L}$, and elements $a_1, \ldots, a_k \in N$. Then*

$$N \models \varphi(a_1, \ldots, a_k) \iff \{i \in I : M \models \varphi(a_1[i], \ldots, a_k[i])\} \in \mathcal{D}.$$

In particular, for any sentence (=formula with no free variables) θ of $\mathcal{L}$,

$$N \models \theta \iff \{i \in I : M \models \theta\} \in \mathcal{D}.$$

Łos' theorem is often called the "fundamental theorem of ultraproducts." An easy but surprising consequence, the second fundamental theorem of ultraproducts, answers (among other things) the following question. Suppose we consider a model M consising of two disjoint parts: a copy of $(\mathbb{N}, <)$ and a copy of $(\mathbb{Q}, <)$. In the ultrapower $N = M^I/\mathcal{D}$ we will also have two parts: a large discrete linear order and a large dense linear order. Will these two parts have the same size?

---

[4] Although there were earlier precedents, the general ultraproduct construction was defined by Łos' in the 1950s, and subsequently developed by Tarski, by Frayne, Morel and Scott, by Keisler, by Kochen and others starting in the early 1960s. For more on the history, from one of the founders, see the recent survey of Keisler [7].

Theorem 2.5 (Ultrapowers commute with reducts). *Fix languages $\mathcal{L} \subseteq \mathcal{L}'$ and an $\mathcal{L}'$-model M. Let $\mathcal{D}$ be an ultrafilter on I. Write $\upharpoonright \mathcal{L}$ for the reduct to $\mathcal{L}$, that is, the model with the same domain obtained by forgetting the interpretations of symbols in $\mathcal{L}' \setminus \mathcal{L}$. Then*

$$(M \upharpoonright \mathcal{L})^I / \mathcal{D} = \left(M^I / \mathcal{D} \right) \upharpoonright \mathcal{L}.$$

In other words, if we forget the interpretation of some symbols before or after taking the ultrapower, it doesn't make a difference. This shows the answer to the question is yes. Why? Suppose we had considered M in a larger language $\mathcal{L}'$ in which there was a function symbol f interpreted as a set bijection between the two countable sets $\mathbb{N}$ and $\mathbb{Q}$. In the ultrapower, by Łos' theorem, f would remain a bijection so the two corresponding amplifications would have to have the same size. But the theorem says that we may forget f before or after taking the ultrapower without changing the result.[5]

The astute reader may now notice the following. By Łos' theorem, for any first-order formula $\varphi(\bar{x})$, and any first-order sentence ψ, ψ holds on the solution set of φ in N if and only if it holds on the solution set of φ in M_i for $\mathcal{D}$-almost all $i \in I$. In other words, a given first-order property is true of a given definable set in the ultraproduct if and only if it is true of its projections $\mathcal{D}$-a.e. So Theorem 2.5 entails that the statement just made remains true if we replace "definable set" by the more general "internal set":

Definition 2.6 (Internal set). *Let $N = \prod_{i \in I} M_i / \mathcal{D}$ and $X \subseteq N^k$. We say that X is* internal *if for some $\langle X_i : i \in I \rangle$ with $X_i \subseteq (M_i)^k$, if we were to expand the language by a new k-ary predicate P, interpret $P^{M_i} = X_i$ for each i, and let P^N be the interpretation induced on N by the definition of ultraproduct, then $P^N = X$.*

Internal functions are defined analogously. Internal sets and functions – those definable in any possible expansion of the index models to a larger language – faithfully reflect structure between the index models and the ultraproduct. As we will see, one way of studying the dissonance between what appears in the ultraproduct and what appears in the factor models has to do with how far ultraproduct phenomena are from being internal.

---

[5]The reader may modify this argument to show that if $\mathcal{L} = \{E\}$, a binary relation symbol, and M is a countable model in which E is an equivalence relation with infinitely many classes all of which are infinite, then in any ultrapower $N = M^I / \mathcal{D}$ all equivalence classes of E^N have the same size, and moreover this size is the same as the number of equivalence classes of E^N. In model-theoretic language, although this is not an uncountably categorical theory, its ultrapowers are effectively determined by their size. What if E has classes of arbitrarily large finite size? Here things get interesting; see Keisler's order, e.g. in the introduction to [18].

We need one more definition. For clarity in what follows we will focus on ultrafilters which have a finiteness property called regularity. When $\mathcal{D}$ is regular, any "small" set in any $\mathcal{D}$-ultraproduct may be thought of as being pseudofinite in the following sense. In the context of an ultraproduct $\prod_{i \in I} M_i/\mathcal{D}$, we will use "small" to mean "of size $\leq |I|$." Given an ultraproduct $N = \prod_{i \in I} M_i/\mathcal{D}$ and small $X \subseteq N$, we say that X is *covered by an ultraproduct of finite sets* if there exists a sequence $\langle X_i : i \in I \rangle$ with each X_i a finite subset of M_i, and such that $X \subseteq \prod_{i \in I} X_i/\mathcal{D}$.

Definition 2.7 (Regular ultrafilters). *Let I be an infinite set and let $\mathcal{D}$ be an ultrafilter on I. We say that $\mathcal{D}$ is regular if in every ultraproduct $N = \prod_{i \in I} M_i/\mathcal{D}$, every small $X \subseteq N$ is covered by an ultraproduct of finite sets.*

The reader may prefer one of the following equivalent definitions, expressing that there is a collection of large sets which are very spread out: (a) *Equivalently,* we say $\mathcal{D}$ is regular if there is a family of sets $\{ \mathcal{Y}_t : t < |I| \} \subseteq \mathcal{D}$, called a regularizing family, such that the intersection of any infinitely many elements of this family is empty. Note that the intersection of any finitely many is necessarily still in $\mathcal{D}$ by closure under intersection. (b) *Equivalently,* we say $\mathcal{D}$ is regular if there is a family of sets $\{ \mathcal{Y}_t : t < |I| \} \subseteq \mathcal{D}$ such that each $i \in I$ belongs to no more than finitely many elements of this family.

Any nonprincipal ultrafilter on a countable set will be regular: fix some enumeration of I as $\{ i_n : n < \omega \}$ and consider the family $\{ \mathcal{Y}_n := \{ i_m : m \geq n \} : n < \omega \}$. In general, regular ultrafilters on any infinite set always exist, see e.g. [3, Proposition 4.3.5]. (Donder proved that in the core model, all ultrafilters are regular [4].)

Convention 2.8. *For the rest of the paper, we assume all ultrafilters are regular, thus nonprincipal, unless otherwise stated.*

It is in general complicated to compute the size of an ultrapower, but when the ultrafilter is regular there is a nice answer, due to Frayne, Morel and Scott and to Keisler: it's the same size as the Cartesian power.

Theorem 2.9 (see [3] 4.3.7). *If $\mathcal{D}$ is a regular ultrafilter on I then for any infinite model M,*
$$|M^I/\mathcal{D}| = |M|^{|I|}.$$
Moreover, if $\langle M_i : i \in I \rangle$ is a sequence of models all of the same infinite size κ, then $|\prod_{i \in I} M_i/\mathcal{D}| = \kappa^{|I|}$.

Stating this theorem with separate cases for ultrapowers and ultraproducts is a red herring; size can be computed in the language with only equality, and on top of this one can build whatever index models one likes by quoting Theorem 2.5.

3417

As an exercise, let us apply the facts just explained to justify Example 2.3. Let N be the ultraproduct in that example. Any nonprincipal ultrafilter on a countable set will be regular, so N has cardinality continuum. In the given language we may write down the infinitely many axioms for an algebraically closed field, and observe for each such axiom that it is true in all of the index models, so by Łos' theorem it is true in N. On the other hand, for each prime p, let ψ_p say that $1+\cdots+1$ (p times) $=0$. Then each ψ_p is false $\mathcal{D}$-almost everywhere so by Łos' theorem false in N. So N is an algebraically closed field of characteristic zero and cardinality continuum, so it must be isomorphic to $\mathbb{C}$.

For more on ultraproducts, see [3] Chapters 4 and 6.

We may now give some precise versions of the questions raised in the introduction. To begin, we will focus on ultrapowers of graphs. In keeping with our model-theoretic notation, a graph $G = (V, E)$ is given by the data of a set of vertices V along with a symmetric irreflexive binary relation E on $V \times V$. We often identify G with its underlying vertex set, e.g. the size of G is the size of V, and "a subset X of G" means $X \subseteq V$ given along with the restriction of E to X, i.e. the subgraph induced on the set of vertices X. The expression $X \subseteq Y \subseteq V$ will always imply that X is an induced subgraph of Y, and that both X and Y are induced subgraphs of V. We call $X \subseteq G$ a *clique*, or complete graph, if every two distinct vertices of X are connected by an edge, and an *independent set*, or empty graph, if no two vertices of X are connected by an edge.

3 The Clique Covering Problem

Returning to the question of how average structure is reflected between the factor models and the ultraproduct or ultrapower, in this section, we fix $\mathcal{L} = \{R\}$ and all our models will be graphs. Here the clique covering problem is defined; its source and history are described in the next section.

Consider an ultrapower $N = M^I/\mathcal{D}$. If M is a clique, then necessarily so is N: we can either check the definition one pair at a time, or notice that $(\forall x)(\forall y)(R(x,y))$ is first order and quote Łos' theorem. Recall from the previous section that in the context of an ultrapower, *small* means $\leq |I|$. Suppose now that M is some arbitrary graph and $A \subseteq N = M^I/\mathcal{D}$ is a small infinite clique. There is a priori no reason that for a given $i \in I$, the projection $\{a[i] : a \in A\}$ will be a clique in the i-th copy of M. All that "A is a clique in N" means is that for each pair $a, b \in A$, there is *some* set $S_{a,b} \in \mathcal{D}$ such that $i \in S_{a,b}$ implies that $R(a[i], b[i])$ in M_i. The sets $S_{a,b}$ may vary quite a bit as a, b vary, and any given $i \in I$ many not belong to many of them. Moreover, even if we refine the projections in a reasonable way (by considering some function $d : A \to \mathcal{D}$ and considering projections of the form $\{a[i] : a \in A \wedge t \in d(a)\}$) it's not obvious we will get cliques everywhere or almost everywhere. This motivates the following definition.

Definition 3.1. *Let $\mathcal{D}$ be regular and suppose $X \subseteq G^I/\mathcal{D}$ is a clique. We say X is* covered by an ultraproduct of cliques *if there exists a sequence $\langle X_i : i \in I \rangle$ with each X_i a finite clique of G such that $X \subseteq \prod_{i \in I} X_i/\mathcal{D}$.*

We say the cliques of G are covered by $\mathcal{D}$ *if in the ultrapower $G^I/\mathcal{D}$, every small clique is covered by an ultraproduct of cliques.*

Put otherwise, 3.1 asks if a clique is "on average" a clique. Notice this is a two place relation, involving both the graph G and the ultrafilter $\mathcal{D}$. It has to do with the existence of a function $d : A \to \mathcal{D}$ as described above. Note that for small X, the condition that each X_i is finite comes for free from the definition of "$\mathcal{D}$ is regular". The question is whether we may find a possibly smaller pseudofinite set containing X which is internally a clique – whether the internal and external pictures are in some kind of alignment.

Definition 3.2. *Let G, G' be infinite graphs. Suppose that for every infinite cardinal λ, every set I of size λ, and every regular ultrafilter $\mathcal{D}$ on I, if the cliques of G' are covered by $\mathcal{D}$ then already the cliques of G are covered by $\mathcal{D}$. In this case we write*

$$G \trianglelefteq^g G'.$$

Notice that built into Definition 3.1 is a restriction on the size of the cliques we consider (they must all be small, i.e. of size $\leq |I|$). This definition gives a a pre-order on infinite graphs, which we will think of as a partial order on the $\trianglelefteq$-equivalence classes. The problem dates to [9], in a different language.

Problem 3.3 (The (Ultrapower) Clique Covering Problem).

1. *Determine the structure of the relation $\trianglelefteq^g$ on infinite graphs.*

2. *Give a graph-theoretic characterization of the relation $\trianglelefteq^g$ which makes no mention of ultrafilters.*

This problem is wide open. However, some things can be said. For example, by considerations which appear orthogonal to our discussion (explained in the next section), the following is true; the proof is deferred to the end of the next section.

Theorem 3.4. *There are at least three classes in $\trianglelefteq^g$.*

Discussion of the problem. A brief discussion of this problem is in order. There are some minor questions about the definition which may need to be resolved in order to take full advantage of the model theoretic connections described in the next section. It seems best to simply describe them here without making a final decision, as further work on the problem may make the choice clear.

The issue is how to handle the trivial cases, such as empty graphs and other graphs in which there are no non-trivial cliques to be covered. In the original context, described in the next section, the aim was not to classify all graphs but rather graphs arising as the incidence relations of first-order formulas, specifically those formulas which control saturation of ultrapowers. In these graphs there are generally many nontrivial cliques. What is possibly lost by allowing all graphs with only trivial cliques to go into the minimal class is that the complexity in the sense of classification theory of the $\trianglelefteq^g$-classes may no longer be so clean. For example, many properties of model theoretic interest (the order property, the independence property) are representable as bipartite graphs. If those graphs are truly bipartite in the graph theory sense of having no edges except between sides, then we won't have nontrivial cliques. This means, for instance, that as currently stated there are graphs in the minimal $\trianglelefteq^g$-class which are unstable in the sense of having the order property. It follows from earlier results that any graph containing the *complement* of a bipartite half-graph (that is, any graph containing distinct vertices $\{a_i : i < \omega\} \cup \{b_i : i < \omega\}$ with the a's forming a clique, the b's forming a clique, and $R(a_i, b_j)$ iff $i < j$) is $\trianglelefteq^g$-maximal.

One way of handling this is to redefine the $\trianglelefteq^g$-ordering so that the complexity of a graph is a function of the $\trianglelefteq^g$-complexity of the graph and the $\trianglelefteq^g$-complexity of its complement (same vertices, opposite edges). But this may introduce other kinds of noise.

A second way of handling this is to leave the definition as is, complete the classification, and then look for classification-theoretic interpretations only on the sub-class of graphs arising in the given special way as so-called characteristic sequences (defined below). It may well be that the the search for model-theoretic outside definitions of these classes should be left to a later stage.

There are many specific questions to ask about this order. For example, is it infinite? See the end of the next section for a comment on this.

4 Interlude: Saturation of Ultrapowers

The clique covering problem dovetails with, and arose from, a major open question and a corresponding research program around model-theoretic complexity, as we now briefly explain. In this section, for clarity, we assume all languages are countable.

As discussed above, given any infinite set X, a filter on X is a family of subsets of X which is nonempty, upward closed, and closed under finite intersection. The filter notion of limit has two major uses in model theory. One was already discussed: forming limits of sequences of models via the ultraproduct construction. The second is that the idea of a filter allows us to naturally define the "completeness" of any model. Let M be a model and let $\mathcal{S}$ be the boolean algebra of definable subsets of M (really, the boolean algebra of subsets identified up to logical equivalence). Filters or ultrafilters whose elements are definable sets ("filters of definable sets") describe the model-theoretic limit points of the structure, called types. Definable always means definable with parameters.[6]

Example 4.1. Let $M = (\mathbb{Q}; <)$. The types of M include all Dedekind cuts as well as $+\infty$, $-\infty$, and infinitesimals.

In the next definition, recall that we are assuming the language is countable.[7]

Definition 4.2. A model is κ-saturated if it contains the limit points for all filters generated by fewer than κ definable sets.

In model-theoretic language, a model is κ-saturated if it realizes all types over all sets $A \subseteq M$ with $|A| < \kappa$.

Part of the motivation for studying regular ultrafilters was Keisler's discovery that their saturation does not depend on the index model M chosen, only on its theory. Given a model M and an ultrafilter $\mathcal{D}$ on I, let us say that $\mathcal{D}$ *saturates* M if $M^I/\mathcal{D}$ is $|I|^+$-saturated (so it realizes all types over sets of size $\leq |I|$).

Theorem 4.3 (Keisler 1967). *If $\mathcal{D}$ is regular, then for any model M in a countable language and any $N \equiv M$, $\mathcal{D}$ saturates M if and only if $\mathcal{D}$ saturates N.*

So when $\mathcal{D}$ is regular, for any complete countable theory T, we may simply say "$\mathcal{D}$ saturates T" if $\mathcal{D}$ saturates some, equivalently every, model of T. Keisler [6] proposed the following pre-order on theories, usually considered as a partial order on the equivalence classes.

---

[6] Strictly speaking, a type is maximal consistent, so the analogue of an ultrafilter. The analogue of a filter is called a partial type.

[7] If it is uncountable, we should distinguish between 'κ-saturated,' which means we realize all types over parameter sets of size $< \kappa$, and 'κ-compact,' which means we realize all partial types consisting of $< \kappa$ formulas.

Definition 4.4 (Keisler's order). *Let T_1, T_2 be complete countable theories. Say $T_1 \triangleleft T_2$ if for every regular ultrafilter $\mathcal{D}$, if $\mathcal{D}$ saturates T_2 then $\mathcal{D}$ saturates T_1.*

Determining the structure of Keisler's order is a major open problem in model theory. Over the last fifty years, progress has been made, most recently in work of Malliaris starting around 2008 and in a very productive collaboration of Malliaris and Shelah starting from 2011 (see e.g. the introductions to [18] or [19], or the research announcement [16] for some of the history and connections of this problem to other areas of mathematics). We will not repeat that history here since it is well documented, but instead will explain how progress on the Clique Covering Problem would help to illuminate Keisler's order.[8]

The opening move is to notice that Keisler's order reduces to the study of φ-types (types in positive and negative instances of a single formula).

Theorem 4.5 (Malliaris [10]). *If T is countable, $\mathcal{D}$ is a regular ultrafilter on I, and $M \models T$, then $M^I/\mathcal{D}$ is $|I|^+$-saturated if and only if it is $|I|^+$-saturated for φ-types, for all formulas φ.*

This result suggests an approach to Keisler's order by investigating the patterns of consistency and inconsistency of instances of a given formula. Following a line of work in the author's [11], [12], [13], let us define a sequence of incidence hypergraphs on the parameter space of a given formula. In what follows, writing $\varphi(x;y)$ does not imply that either $\ell(x), \ell(y)$ are necessarily 1.

Definition 4.6 ([11]). *Given a theory T, $\varphi(x;y)$, and $n < \omega$, let*

$$P_n(y_1, \ldots y_n) = \exists x \bigwedge_{i \leq n} \varphi(x; y_i).$$

The characteristic sequence of φ (for the theory T) is $\langle P_n : n < \omega \rangle$.

Notice first that the characteristic sequence focuses on positive instances of the given formula, but this is not a real loss: we could always replace the formula ψ under consideration by one of the form

$$\varphi(x; y, z, w) = (\psi(x;y) \land z = w) \lor (\neg(\psi(x;y) \land z \neq w)$$

assuming there exist at least two elements in the model. Second, these P_n are definable in the original theory. It will be interesting to see what may be recovered from these hypergraphs alone by naming them and forgetting the ambient theory.

---

[8]As will be explained, the clique covering problem arose from the author's work on this order, but seems to be of independent interest.

Still, their definability means they have the usual compactness properties, and also that their classification-theoretic strength is no greater than that of the ambient theory T. Third, when do the P_n give new information for arbitrarily large n? It turns out this precisely characterizes the formula φ having the finite cover property, [11, 2.7]. If φ does not have f.c.p., there is some minimal k such that the k-hyperedges determine those of all higher arity, and we say φ has *support* k.

After fixing T and φ, in what follows we may (in slight abuse of notation) identify these predicates with their interpretation in some large, saturated model of the theory T. In this setup, call A a *positive base set* if $A \subseteq P_1$ such that $A^n \subseteq P_n$ for all $n < \omega$. So A is a positive base set if and only if $\{\varphi(x,a) : a \in A\}$ is a consistent partial type. We may also call a positive base set a P_∞-*complete graph*. The translation is:

Fact 4.7 ([13] Lemma 4.9). *Let $\mathcal{D}$ be a regular ultrafilter. The following are equivalent for a positive base set $A \subseteq N := M^\lambda/\mathcal{D}$ with $|A| \leq \lambda$.*

1. *The type $p = \{\varphi(x;a) : a \in A\}$ corresponding to A is realized in N.*

2. *There exists a map $d : A \to \mathcal{D}$ whose image is a regularizing family, and such that $\mathcal{D}$-almost everywhere, $A[t] := \{a[t] : t \in d(a)\}$ is a $P_{m[t]}$-complete graph, where $m[t] := |A[t]|$ is the cardinality of projection of the finite piece of A assigned by d to the index t.*

These conditions are not necessarily equivalent to finding an internal set containing A which is a P_∞-graph in N, but are equivalent to finding an internal set containing A which is a.e. a P_∞-complete graph.

In the case where φ is a formula with support 2, this amounts to clique covering:

Corollary 4.8. *Let $\mathcal{D}$ be a regular ultrafilter, $M \models T$, and let φ be a formula of T with support 2, so that the the characteristic sequence $\langle P_n : n < \omega \rangle$ of φ is determined by the graph edge relation P_2. Then the following are equivalent for the ultrapower $N := M^\lambda/\mathcal{D}$:*

1. *N is λ^+-saturated for φ-types.*

2. *The cliques of the graph (M, P_2) are covered by $\mathcal{D}$.*

This connection gives an independent motivation for studying the Clique Covering Problem (and explains its origin): among other things, it would settle Keisler's order for all theories whose saturation reduces to a formula with support 2. The last clause of Fact 4.7 addresses the natural question of whether it would be sufficient for Keisler's order to define the analogue of the clique covering problem for k-hypergraphs and solve this for each k. That is, when the ultrafilter is not so-called flexible and the formula has infinite support, these infinitely many approximations may not fit together. See, for example, [13, Conclusion 8.11].

Although clique covering asks about all graphs, not only those arising in the form of some P_2, already the information we have about the behavior of certain formulas of support 2 in Keisler's order is enough to justify Theorem 3.4 above, albeit in a rather indirect way:

Proof of Theorem 3.4. To prove there are at least three distinct classes in $\trianglelefteq^g$ it will suffice to show that there are pairs (T_0, φ_0), (T_1, φ_1), (T_2, φ_2) such that: (a) for $i = 0, 1, 2$, T_i is a complete countable theory with the property that its ultrapowers are saturated if and only if they are saturated for φ_i-types, (b) each φ_i has support 2, and (c) $T_0 \triangleleft T_1 \triangleleft T_2$ in Keisler's order. Let T_0 be the theory of an infinite set and let $\varphi_0 = \varphi_0(x, y) = x \neq y$; this is in the minimum class in Keisler's order. Let T_1 be the theory of the random graph and let $\varphi_1 = \varphi_1(x; y, z) = R(x, y) \wedge \neg R(x, z)$. This theory is not in the minimum Keisler class by Shelah's characterization of the first two Keisler classes as precisely the stable theories in [21, VI.4]. It is not in the maximal Keisler class by the main theorem of [17]. (In fact, this theory is minimum among the unstable theories in Keisler's order, see [13].) Let T_2 be the theory of $(\mathbb{Q}, <)$ and let $\varphi_2 = \varphi_2(x; y, z) = y > x > z$. This belongs to the maximum class in Keisler's order by [21, Theorem 2.6]. $\square$

As of writing, the currently known divisions in Keisler's order can be found in [19]. Several years older, but with much more detail on the material then known and on ultrafilter/theory correspondences, is the overview in [15, §4].

Embedded in the proof of Theorem 3.4 just given is the important fact that the maximal Keisler class contains a theory whose saturation depends on a formula of support 2. This has been recently used by Casey [2] to analyze the structure of certain regular ultrafilters, using the language of clique covering.

Although we do not even know if the order $\trianglelefteq^g$ is infinite, it is now known [19] that Keisler's order has infinitely many classes, and moreover is not a well order. The proof there, which overturned a long standing picture, used theories of hypergraphs of increasing arity. Can the arguments of that paper, or the ultrafilters built there, be used to reflect an infinite hierarchy into $\trianglelefteq^g$? It is not obvious which way this would go, especially because, in light of the discussion below, such an infinite hierarchy might reflect onto the so-called stable graphs.

5 The Stable Sequences Problem

In this section we explain the second problem, the Stable Sequences Problem. With the background aim of understanding saturation of ultrapowers, in [11]-[12]-[13] the investigations into the complexity of characteristic sequences were carried further with the aim of connecting graph-theoretic properties of these sequences to model-theoretic properties of the formulas. However, these questions soon took on a life of their own as it became apparent that graph theory could contribute to the picture of model theoretic complexity via the characteristic sequence (c.f. [12]). In this section we work in a large saturated model of the theory, not necessarily an ultrapower, and consider certain definable rather than internal sets.

It would be useful for the reader of this section to know a bit about the model theoretic classification of theories and in particular the definitions of stable, simple, NIP, SOP_2 and SOP_3.[9]

It is also useful to recall the local definition of stability: fixing a theory T, a formula $\varphi(x, y)$ is unstable if there are sequences $\langle a_i : i < \omega \rangle$, $\langle b_j : j < \omega \rangle$ such that $\varphi(a_i, b_j)$ holds iff $i < j$, i.e. if φ has the order property. Otherwise, there is a finite bound $n = n(\varphi)$ on the length of such a configuration in any model of T, and the formula is called stable. In graph theoretic language, the graph consisting of vertices $\{a_i : i < k\} \cup \{b_i : i < k\}$ with $R(a_i, b_j)$ iff $i < j$ is called a k-half graph. We will use half-graph to mean only this pattern of edges between a's and b's, with no assumptions about whether or not there are edges between the a's themselves, or between the b's themselves. We will use "stable graph" to mean a graph where there is a finite bound on the size of a half-graph, i.e., the edge relation is stable (or where the hypergraph relation is stable with respect to any partition of the variables). Note that this is a priori weaker than saying that the theory of the graph or hypergraph is stable, which would require stability for all formulas, not just the (hyper)edge relation.

Returning to the big picture, since the characteristic sequence is definable in the original theory, its graphs and hypergraphs are no more complex (in the sense of classification theory) than the original theory. In fact, they are often a lot less complex. For example, consider the formula $\varphi(x; y_1, y_2)$ in the random graph saying "$R(x, y_1) \land \neg R(x, y_2)$." It support 2 and no empty graphs of size greater than 4. So although it describes the pattern of incidence underlying saturation of the random graph, it is itself very far from random. It is possible to code the independence

---

[9] As a first approximation, one can look at the picture in the introduction to [19], which partially maps out the comparative complexity of theories by representing the known picture of Keisler's order. Some aspects of that picture may be misleading out of context: for example, it's not known whether all simple theories are ⊲-below all non-simple theories, see [18, Conclusion 8.4].

property in this graph, but only in an artificial way [11, Example 5.26]. Morally, it is a stable graph.

This observation leads to a theorem, described in item (3) below, and which requires a natural definition. Let A be a positive base set. An allowed definable set (for A) is a definable set of the form $P_m(x, c_1, \ldots, c_{m-1})$ which contains A, where $c_1, \ldots, c_{m-1}$ are elements of the (monster) model. A *localization* of A is a finite intersection of allowed definable sets.

The right setting for the theorem is of the form: around any positive base set in the monster model, there is always a localization on which the characteristic sequence is suitably well behaved.[10] Model-theoretically, in some sense, we can zoom in around any consistent partial type to find a region where things behave well. Three theorems of this kind proved in [11], stated in simplified form for readability, were the following.

(1) A formula φ is NIP only if for any positive base set A and every $n < \omega$ there is a localization Y such that $A \subset Y$ and Y is a P_n-complete graph [11, Corollary 5.16], see [11, Theorem 5.17] for iff. Informally, missing n-edges do not persist under localization around any type. This relates to the fact that missing edges represent inconsistent instances of φ, and so from persistence of missing edges one can extract a series of independent instances of inconsistency and so witness the independence property in the original formula.

(2) A formula is simple only if for any positive base set A and every $n < \omega$, there is a localization Y such that $A \subseteq Y$ and on Y there is a uniform finite bound on the size of a P_n-empty graph [11, Theorem 5.22] (see there for the wording of iff). Informally, P_n-empty graphs do not persist under localization around any type. This relates to the fact that a formula in a simple theory only divides finitely many times with respect to any given k.

The third theorem will motivate our question.

(3) If the formula is simple then we can always localize so that the characteristic sequence is stable (i.e. for each k, we can localize so that the predicates $P_2, \ldots, P_k$ do not have the order property on elements from the localization with respect to any partition of the variables). More precisely,

Theorem A (([11] Theorem 5.10, see also [12] Conclusion 2.9)). *Suppose T is simple, and fix φ and $\langle P_n \rangle$. Then for any positive base set A and for each $n < \omega$, there is a localization Y such that $A \subseteq Y$ and $P_2, \ldots, P_n$ are stable on Y.*

---

[10] The monster model allows for compactness arguments of the form: there must eventually be some suitable parameters allowing for the given localizations, else we would be able to form some forbidden configurations. The related question of how localizations work in specific, not too saturated models using only parameters from that model seems worth investigating: see below.

Definition 5.1. *Given a theory T, if a formula φ of T satisfies the conclusion of the previous theorem (i.e., if in the monster model of T, for any positive base set A and for each $n < \omega$, there is a localization Y such that $A \subseteq Y$ and $P_2, \ldots, P_n$ do not have the order property on Y with respect to any partition of the variables) then say that φ has* eventually stable sequences.

Problem 5.2 (The Stable Sequences Problem). *Give a useful model-theoretic characterization of those formulas whose characteristic sequence is eventually stable.*

One might want to call this the (*Eventually*) *Stable Sequences Problem for Formulas* and frame the (*Eventually*) *Stable Sequences Problem for Theories*, saying that a theory has eventually stable sequences if all its formulas do.

Parametrized versions of the problem may be especially interesting: characterizing when, for any μ-saturated model of T, any positive base set A of size $\leq \kappa$ and for each $n < \omega$, there is a localization Y such that $A \subseteq Y$ and $P_2, \ldots, P_n$ do not have the order property on Y with respect to any partition of the variables, i.e. when the characteristic sequence is eventually stable for (μ, κ). For applications to regular ultrapowers, $\mu = \aleph_1$ is a natural first case.

Note that a distinct version of this problem would be to replace the word "eventually" by "essentially," taken to mean that saturation (in the sense of saturation of ultrapowers) of a given theory can be reduced to saturation of φ-types for formulas φ with stable sequences. This second sense would likely require a careful analysis of how much ambient saturation is needed to find the given localization, as in the parametrized versions just given. Compare the result (1) just quoted to [12, Conclusion 6.15], which shows that any theory with SOP_3 has a strong version of the order property in some characteristic sequence; this may be the more accurate indicator of complexity. On stable sequences for the theory T_{feq}, the model completion of a parametrized family of crosscutting equivalence relations, see [11] Example 5.28 and Claim 5.29. Phenomena like this would need to be carefully sorted out.

The way instability appears may also be important. The section "Two kinds of order property," [12, §6], directs attention to the different forms the order property may take. Suppose for clarity that we restrict to formulas with support 2. Then by compactness instability in the characteristic sequence may show up in one of three ways: as an infinite half-graph where the a's form a clique and where the b's form a clique, where one side is a clique and one side an independent set, or where both are an independent set. The first, the so-called compatible order property, already implies that the theory is in the maximal $\trianglelefteq^g$-class, [12, 6.16]. Recalling that cliques in this context represent consistent partial types and independent sets indicate that (by compactness) the formula divides, it is natural that these three manifestations would have different effects: we're asking in the first case for a family of types fanned out along a linear order, and in the other cases for a series of possible tradeoffs between a type in a dividing sequence or two dividing sequences. Formally,

Definition 5.3 ([12] §6). *Let $\langle P_n \rangle$ be the characteristic sequence of φ.*

1. *φ has the n-compatible order property if there exists $\langle a_i, b_i : i < \omega \rangle$, a sequence of pairs of distinct elements of P_1, such that $P_2(a_i, b_j)$ holds iff $i < j$, and in addition, for all $m < \omega$, $P_{2m}(a_{i_1}, b_{j_1}, \ldots a_{i_m}, b_{j_m})$ holds iff $\max\{i_1, \ldots i_m\} < \min\{j_1, \ldots j_m\}$.*

2. *φ has the n-empty order property if there exists $\langle a_i, b_i : i < \omega \rangle$, a sequence of pairs of distinct elements of P_1, such that $P_2(a_i, b_j)$ holds iff $i < j$, and in addition, $\neg P_n(a_{i_1}, \ldots a_{i_n})$ and $\neg P_n(b_{i_1}, \ldots b_{i_n})$ hold for all $i_1, \ldots i_n < \omega$.*

3. *φ has the half-compatible order property if there exists $\langle a_i, b_i : i < \omega \rangle$, a sequence of pairs of distinct elements of P_1, such that $P_2(a_i, b_j)$ holds iff $i < j$, the a-s form a P_∞-complete graph, and the b-s form a P_2-empty graph.*

What makes 5.2 an interesting question?

First, we know relatively little about the region of non-simple $NSOP_3$ theories, and Problem 5.2 along with its variants just described suggest an approach. By [11] Claim 5.29, the class captured by Problem 5.2 will be strictly greater than simplicity.

Second, there is the possible relation to the problem of whether so-called SOP_2 implies SOP_3. From the definition of SOP_2, one can see that any formula with this property will have instances of the so-called half compatible order property (essentially, the half graph holding between a P_∞-complete graph and a P_2-independent set). On the other hand, the compatible order property implies SOP_3.

Third, there is the possibility of carrying further the project arising from [12]. There it was shown that combinatorial theorems about graphs and hypergraphs, such as Szemerédi regularity, may be applied to the characteristic sequence to give model theoretic information. In [14] we proved that both Szemerédi regularity and Ramsey's theorem work much better under the hypothesis of stability of the graph edge relation. Characterizing the formulas with essentially stable sequences could advance the program of applying these theorems on stable graphs and hypergraphs to extract model-theoretic information.

Recent work has improved our understanding and may well make the simplifying hypothesis of support 2 less necessary. For example, the improvement of Ramsey's theorem from [14] also works for hypergraphs. The hypergraph stable regularity lemma was recently announced by Ackerman, Freer, and Patel [1].

References

[1] N. Ackerman, C. Freer, and R. Patel. "Stable hypergraph regularity." In preparation. Announced in a talk of R. Patel at the ASL Annual Meeting, Storrs, May 2016.

[2] D. Casey, draft, 2016.

[3] C. C. Chang and H. J. Keisler, *Model Theory*. Third edition. North-Holland Publishing Co., Amsterdam, 1990. xvi+650 pp.

[4] H. D. Donder, "Regularity of ultrafilters and the core model." Israel Journal of Mathematics. Oct 1988. Volume 63, Issue 3, pp 289–322.

[5] H. J. Keisler, "Good ideals in fields of sets." Annals of Math. (2) 79 (1964), 338–359.

[6] H. J. Keisler, "Ultraproducts which are not saturated." J. Symb Logic 32 (1967) 23–46.

[7] H. J. Keisler. "The ultraproduct construction." In "Ultrafilters Across Mathematics", ed. by V. Bergelson et. al., *Contemporary Mathematics* 530 (2010), pp. 163-179.

[8] P. Komjáth and V. Totik. "Ultrafilters." A.M.M., 115, January 2008, 33–44.

[9] M. Malliaris, Ph. D. thesis, University of California, Berkeley (2009).

[10] M. Malliaris, "Realization of φ-types and Keisler's order." Ann. Pure Appl. Logic 157 (2009), no. 2-3, 220–224.

[11] M. Malliaris, "The characteristic sequence of a first-order formula." Journal of Symbolic Logic, 75, 4 (2010) 1415–1440.

[12] M. Malliaris, "Edge distribution and density in the characteristic sequence." Ann Pure Appl Logic 162, 1 (2010) 1–19.

[13] M. Malliaris, "Hypergraph sequences as a tool for saturation of ultrapowers." J Symbolic Logic, 77, 1 (2012) 195–223.

[14] M. Malliaris and S. Shelah, "Regularity lemmas for stable graphs." Trans. Amer. Math Soc, 366 (2014), 1551–1585.

[15] M. Malliaris and S. Shelah, "Constructing regular ultrafilters from a model-theoretic point of view." Trans. Amer. Math. Soc. 367 (2015), 8139–8173.

[16] M. Malliaris and S. Shelah, "General topology meets model theory, on $\mathfrak{p}$ and $\mathfrak{t}$." Proc Natl Acad Sci USA (2013) 110:33, 13300–13305.

[17] M. Malliaris and S. Shelah, "A dividing line within simple unstable theories." Advances in Math 249 (2013) 250–288.

[18] M. Malliaris and S. Shelah, "Existence of optimal ultrafilters and the fundamental complexity of simple theories." Advances in Math. 290 (2016) 614–618.

[19] M. Malliaris and S. Shelah, "Keisler's order has infinitely many classes." To appear, Israel J. Math.

[20] M. Mashaal, *Bourbaki: a secret society of mathematicians*. Amer. Math. Soc., 2006.

[21] S. Shelah, *Classification Theory and the number of non-isomorphic models*. North-Holland Publishing Co., first edition, 1978, rev. ed, 1990.

[22] E. Szemerédi, "On sets of integers containing no k elements in arithmetic progression," Acta Arith. 27 (1975), 199–245.

Useful axioms

Matteo Viale
Department of Mathematics, University of Torino, Italy
matteo.viale@unito.it

Abstract

We give a brief survey on the interplay between forcing axioms and various other non-constructive principles widely used in many fields of abstract mathematics, such as the axiom of choice and Baire's category theorem.

First of all we outline how, using basic partial order theory, it is possible to reformulate the axiom of choice, Baire's category theorem, and many large cardinal axioms as specific instances of forcing axioms. We then address forcing axioms with a model-theoretic perspective and outline a deep analogy existing between the standard Łoś Theorem for ultraproducts of first order structures and Shoenfield's absoluteness for Σ^1_2-properties. Finally we address the question of whether and to what extent forcing axioms can provide "complete" semantics for set theory. We argue that to a large extent this is possible for certain initial fragments of the universe of sets: The pioneering work of Woodin on generic absoluteness show that this is the case for the Chang model $L(\mathrm{Ord}^\omega)$ (where all of mathematics formalizable in second order number theory can be developed) in the presence of large cardinals, and recent works by the author with Asperó and with Audrito show that this can also be the case for the Chang model $L(\mathrm{Ord}^{\omega_1})$ (where one can develop most of mathematics formalizable in third order number theory) in the presence of large cardinals and maximal strengthenings of Martin's maximum or of the proper forcing axiom. A major open question we leave completely open is whether this situation is peculiar to these Chang models or can be lifted up also to $L(\mathrm{Ord}^\kappa)$ for cardinals $\kappa > \omega_1$.

Introduction

Since its introduction by Cohen in 1963 forcing has been the key and the most effective tool to obtain independence results in set theory. This method has found

---

This paper owes much of its clarity to the suggestions of Raphael Carroy, and takes advantage of several several fruitful discussions we shared on the material presented here.

The author acknowledges support from: Kurt Gödel Research Prize Fellowship 2010, PRIN grant 2012: Logica, modelli e insiemi, San Paolo Junior PI grant 2012.

applications in set theory and in virtually all fields of pure mathematics: in the last forty years natural problems of group theory, functional analysis, operator algebras, general topology, and many other subjects were shown to be undecidable by means of forcing. Starting from the early seventies and during the eighties it became transparent that many of these consistency results could all be derived by a short list of set theoretic principles, which are known in the literature as forcing axioms. These axioms gave set theorists and mathematicians a very powerful tool to obtain independence results: for any given mathematical problem we are most likely able to compute its (possibly different) solutions in the constructible universe L and in models of strong forcing axioms. These axioms settle basic problems in cardinal arithmetic like the size of the continuum and the singular cardinal problem (see among others the works of Foreman, Magidor, Shelah [10], Veličković [28], Todorčević [25], Moore [17], Caicedo and Veličković [5], and the author [29]), as well as combinatorially complicated ones like the basis problem for uncountable linear orders (see Moore's result [18] which extends previous work of Baumgartner [4], Shelah [23], Todorčević [24], and others). Interesting problems originating from other fields of mathematics and apparently unrelated to set theory have also been settled appealing to forcing axioms, as it is the case (to cite two of the most prominent examples) for Shelah's results [22] on Whitehead's problem in group theory and Farah's result [8] on the non-existence of outer automorphisms of the Calkin algebra in operator algebra. Forcing axioms assert that for a large class of compact topological spaces X Baire's category theorem can be strengthened to the statement that any family of $\aleph_1$-many dense open subsets of X has non empty intersection. In light of the success these axioms have met in solving problems a convinced platonist may start to argue that these principles may actually give a "complete" theory of a suitable fragment of the universe of sets. However it is not clear how one could formulate such a result. The aim of this paper is to explain in which sense we can show that forcing axioms can give such a "complete" theory and why they are so "useful".

Section 1 starts showing that two basic non-constructive principles which play a crucial role in the foundations of many mathematical theories, the axiom of choice and Baire's category theorem, can both be formulated as specific instances of forcing axioms. In section 2 we also argue that many large cardinal axioms can be reformulated in the language of partial orders as specific instances of a more general kind of forcing axioms. Sections 3 and 4 show that Shoenfield's absoluteness for Σ^1_2-properties and Łoś Theorem for ultraproducts of first order models are two sides of the same coins: recasted in the language of boolean valued models, Shoenfield's absoluteness shows that there is a more general notion of boolean ultrapower (of which the standard ultrapowers encompassed in Łoś Theorem are just special cases) and that in the specific case one takes a boolean ultrapower of a compact, second countable space X, the natural embedding of X in its boolean ultrapower is at least Σ_2-elementary. Section 5 embarks in a rough analysis of what is a maximal forcing axiom. We are led by two driving observations, one rooted in topological considerations and the other in model-theoretic arguments. First of all we outline how Woodin's generic absoluteness results for $L(\mathrm{Ord}^\omega)$ entail that in the presence of large cardinals the natural embeddings of a separable compact Hausdorff space X in its boolean ultrapowers are not only Σ_2-elementary but fully elementary. We then present other recent results by the author, with Aspero [1] and with Audrito [2] which show that, in the presence of natural strengthenings of Martin's maximum or of the proper forcing axiom, an exact analogue of Woodin's generic absoluteness result can be established also at the level of the Chang model $L(\mathrm{Ord}^{\omega_1})$ and/or for the first order theory of $H_{\aleph_2}$. The main open question left open is whether these generic absoluteness results are specific to the Chang models $L(\mathrm{Ord}^{\omega_i})$ for $i = 0, 1$ or can be replicated also for other cardinals. The paper is meant to be accessible to a wide audience of mathematicians, specifically the first two sections do not require any special familiarity with logic or set theory other than some basic cardinal arithmetic. The third section requires a certain familiarity with first order logic and the basic model theoretic constructions of ultraproducts. The fourth and fifth sections, on the other hand, presume the reader has some familiarity with the forcing method.

1 The axiom of choice and Baire's category theorem as forcing axioms

The axiom of choice AC and Baire's category theorem BCT are non-constuctive principles which play a prominent role in the development of many fields of abstract mathematics. Standard formulations of the axiom of choice and of Baire's category theorem are the following:

Definition 1.1. AC $\equiv \prod_{i \in I} A_i$ is non-empty for all families of non empty sets $\{A_i : i \in I\}$, i.e. there is a choice function $f : I \to \bigcup_{i \in I} A_i$ such that $f(i) \in A_i$ for all $i \in I$.

Theorem 1.2. $\mathsf{BCT}_0 \equiv$ *For all compact Hausdorff spaces* (X, τ) *and all countable families* $\{A_n : n \in \mathbb{N}\}$ *of dense open subsets of* X, $\bigcap_{n \in \mathbb{N}} A_n$ *is non-empty.*

There are large numbers of equivalent formulations of the axiom of choice and it may come as a surprise that one of these is a natural generalization of Baire's category theorem and naturally leads to the notion of forcing axiom.

Definition 1.3. $(P, \leq)$ is a *partial order* if $\leq$ is a reflexive and transitive relation on P.

Notation 1.4. Given a partial order $(P, \leq)$,
$$\uparrow A = \{p \in P : \exists q \in A : q \leq p\}$$
denotes the *upward closure* of A and similarly $\downarrow A$ will denote its *downward closure*.

- $A \subseteq P$ is *open* if it is a downward closed subset of P.

- The *order topology* τ_P on P is given by the downward closed subsets of P.

- D is *dense* if for all $p \in P$ there is some $q \in A$ refining p (q refines p if $q \leq p$),

- $G \subseteq P$ is a *filter* if it is upward closed and all $q, p \in G$ have a common refinement $r \in G$.

- p is *incompatible* with q ($p \perp q$) if no $r \in P$ refines both p and q.

- X is a *predense* subset of P if $\downarrow X$ is open dense in P.

- X is an *antichain* of P if it is composed of pairwise incompatible elements, and a maximal one if it is also predense.

- X is a *chain* of P if $\leq$ is a total order on X.

The terminology for open and dense subsets of P comes from the observation that the collection τ_P of downward closed subsets of P is a topology on the space of points P (though in general not a Hausdorff one), whose dense sets are exactly those satisfying the above property. Remark also that the downward closure of a dense set is open dense in this topology.

A simple proof of the Baire Category Theorem is given by a basic enumeration argument (which however needs some amount of the axiom of choice to be carried):

Lemma 1.5. $\mathsf{BCT}_1 \equiv$ *Let $(P, \leq)$ be a partial order and $\{D_n : n \in \mathbb{N}\}$ be a family of predense subsets of P. Then there is a filter $G \subseteq P$ meeting all the sets D_n.*

Proof. Build by induction a decreasing chain $\{p_n : n \in \mathbb{N}\}$ with $p_n \in \downarrow D_n$ and $p_{n+1} \leq p_n$ for all n. Let $G = \uparrow \{p_n : n \in \mathbb{N}\}$. Then G meets all the D_n. □

Baire's category theorem can be proved from the above Lemma (without any use of the axiom of choice) as follows:

Proof of BCT_0 *from* BCT_1. Given a compact Hausdorff space (X, τ) and a family of dense open sets $\{D_n : n \in \mathbb{N}\}$ of X, consider the partial order $(\tau \setminus \{\emptyset\}, \subseteq)$ and the family $E_n = \{A \in \tau : \mathrm{Cl}(A) \subseteq D_n\}$. Then it is easily checked that each E_n is dense open in the order topology induced by the partial order $(\tau \setminus \{\emptyset\}, \subseteq)$. By Lemma 1.5, we can find a filter $G \subseteq \tau \setminus \{\emptyset\}$ meeting all the sets E_n. This gives that for all $A_1, \ldots A_n \in G$

$$\mathrm{Cl}(A_1) \cap \ldots \cap \mathrm{Cl}(A_n) \supseteq A_1 \cap \ldots \cap A_n \supseteq B \neq \emptyset$$

for some $B \in G$ (where $\mathrm{Cl}(A)$ is the closure of $A \subseteq X$ in the topology τ.) By the compactness of (X, τ),

$$\bigcap \{\mathrm{Cl}(A) : A \in G\} \neq \emptyset.$$

Any point in this intersection belongs to the intersection of all the open sets D_n. □

Remark the interplay between the order topology on the partial order $(\tau \setminus \{\emptyset\}, \subseteq)$ and the compact topology τ on X. Modulo the prime ideal theorem (a weak form of the axiom of choice), BCT_1 can also be proved from BCT_0.

It is less well-known that the axiom of choice has also an equivalent formulation as the existence of filters on posets meeting sufficiently many dense sets. In order to proceed further, we need to introduce the standard notion of forcing axiom.

Definition 1.6. Let κ be a cardinal and $(P, \leq)$ be a partial order.

$\mathsf{FA}_\kappa(P) \equiv$ *For all families $\{D_\alpha : \alpha < \kappa\}$ of predense subsets of P, there is a filter G on P meeting all these predense sets.*

Given a class Γ of partial orders $\mathsf{FA}_\kappa(\Gamma)$ holds if $\mathsf{FA}_\kappa(P)$ holds for all $P \in \Gamma$.

Definition 1.7. Let λ be a cardinal. A partial order $(P, \leq)$ is $< \lambda$-*closed* if every decreasing chain $\{P_\alpha : \alpha < \gamma\}$ indexed by some $\gamma < \lambda$ has a lower bound in P.

Γ_λ denotes the class of $< \lambda$-closed posets. Ω_λ denotes the class of posets P for which $\mathsf{FA}_\lambda(P)$ holds.

It is almost immediate to check that $\Gamma_{\aleph_0}$ is the class of all posets, and that BCT_1 states that $\Omega_{\aleph_0} = \Gamma_{\aleph_0}$. The following formulation of the axiom of choice in terms of forcing axioms has been handed to me by Todorčević, I'm not aware of any published reference. In what follows, let ZF denote the standard first order axiomatization of set theory in the first order language $\{\in, =\}$ (excluding the axiom of choice) and ZFC denote $\mathsf{ZF}+$ the first order formalization of the axiom of choice.

Theorem 1.8. *The axiom of choice AC is equivalent (over the theory ZF) to the assertion that $\mathsf{FA}_\kappa(\Gamma_\kappa)$ holds for all regular cardinals κ.*

We sketch a proof of Theorem 1.8, the interested reader can find a full proof in [20, Chapter 3, Section 2] (see the following hyperlink: *Tesi-Parente*). First of all, it is convenient to prove 1.8 using a different equivalent formulation of the axiom of choice.

Definition 1.9. Let κ be an infinite cardinal. The *principle of dependent choices* DC_κ states the following:

For every non-empty set X and every function $F\colon X^{<\kappa} \to \mathcal{P}(X) \setminus \{\emptyset\}$, there exists
$$g\colon \kappa \to X \text{ such that } g(\alpha) \in F(g \restriction \alpha) \text{ for all } \alpha < \kappa.$$

Lemma 1.10. *AC is equivalent to* $\forall \kappa\, \mathsf{DC}_\kappa$ *modulo ZF.*

The reader can find a proof in [20, Theorem 3.2.3]. We prove the Theorem assuming the Lemma:

Proof of Theorem 1.8. We prove by induction on κ that DC_κ is equivalent to $\mathsf{FA}_\kappa(\Gamma_\kappa)$ over the theory $\mathsf{ZF} + \forall \lambda < \kappa\, \mathsf{DC}_\lambda$. We sketch the ideas for the case κ-regular[1]:

Assume DC_κ; we prove (in ZF) that $\mathsf{FA}_\kappa(\Gamma_\kappa)$ holds. Let $(P, \leq)$ be a $<\kappa$-closed partially ordered set, and $\{D_\alpha : \alpha < \kappa\} \subseteq \mathcal{P}(P)$ a family of predense subsets of P.

Given a sequence $\langle p_\beta : \beta < \alpha \rangle$ call $\xi_{\vec{p}}$ the least ξ such that $\langle p_\beta : \xi \leq \beta < \alpha \rangle$ is a decreasing chain if such a ξ exists, and fix $\xi_{\vec{p}} = \alpha$ otherwise. Notice that when the length α of $\vec{p}$ is successor then $\xi_{\vec{p}} < \alpha$.

We now define a function $F\colon P^{<\kappa} \to \mathcal{P}(P) \setminus \{\emptyset\}$ as follows: given $\alpha < \kappa$ and a sequence $\vec{p} \in P^{<\kappa}$,

$$F(\vec{p}) = \begin{cases} \{p_0\} & \text{if } \xi_{\vec{p}} = \alpha \\ \{d \in \downarrow D_\alpha : d \leq p_\beta \text{ for all } \xi_{\vec{p}} \leq \beta < \alpha\} & \text{otherwise.} \end{cases}$$

The latter set is non-empty since $(P, \leq)$ is $<\kappa$-closed, $\alpha < \kappa$, and D_α is predense. By DC_κ, we find $g\colon \kappa \to P$ such that $g(\alpha) \in F(g \restriction \alpha)$ for all $\alpha < \kappa$. An easy induction shows that for all α the sequence $g \restriction \alpha$ is decreasing, so $g(\alpha) \in \downarrow D_\alpha$ for all $\alpha < \kappa$. Then
$$G = \{p \in P : \text{there exists } \alpha < \kappa \text{ such that } g(\alpha) \leq p\}$$
is a filter on P, such that $G \cap D_\beta \neq \emptyset$ for all $\beta < \kappa$.

Conversely, assume $\mathsf{FA}_\kappa(\Gamma_\kappa)$, we prove (in ZF) that DC_κ holds.

Let X be a non-empty set and $F\colon X^{<\kappa} \to \mathcal{P}(X) \setminus \{\emptyset\}$. Define the partially ordered set
$$P = \{s \in X^{<\kappa} : \text{for all } \alpha \in \text{dom}(s),\ s(\alpha) \in F(s \restriction \alpha)\},$$

---

[1] In this case the assumption $\forall \lambda < \kappa\, \mathsf{DC}_\lambda$ is not needed, but all the relevant ideas in the proof of the equivalence are already present.

with $s \leq t$ if and only if $t \subseteq s$. Let $\lambda < \kappa$ and let $s_0 \geq s_1 \geq \cdots \geq s_\alpha \geq \ldots$, for $\alpha < \lambda$, be a chain in P. Then $\bigcup_{\alpha<\lambda} s_\alpha$ is clearly a lower bound for the chain. Since κ is regular, we have $\bigcup_{\alpha<\lambda} s_\alpha \in P$ and so P is $<\kappa$-closed. For every $\alpha < \kappa$, define
$$D_\alpha = \{s \in P : \alpha \in \mathrm{dom}(s)\},$$
and note that D_α is dense in P. Using $\mathsf{FA}_\kappa(\Gamma_\kappa)$, there exists a filter $G \subset P$ such that $G \cap D_\alpha \neq \emptyset$ for all $\alpha < \kappa$. Then $g = \bigcup G$ is a function $g \colon \kappa \to X$ such that $g(\alpha) \in F(g \restriction \alpha)$ for all $\alpha < \kappa$. □

2 Large cardinals as forcing axioms

From now on, we focus on boolean algebras rather than posets.

2.1 A fast briefing on boolean algebras

Definition 2.1. A *boolean algebra* B is a boolean ring i.e. a ring in which every element is idempotent. Equivalently a boolean algebra is a complemented distributive lattice $(\mathsf{B}, \wedge, \vee, \neg, 0, 1)$ (see [11]).

Notation 2.2. Given a boolean algebra $(\mathsf{B}, \wedge, \vee, \neg, 0, 1)$, the poset $(\mathsf{B}^+; \leq_\mathsf{B})$ is given by its non-zero elements, with order relation given by $b \leq_\mathsf{B} q$ iff $b \wedge q = b$ iff $b \vee q = q$.

A boolean ring $(\mathsf{B}, +, \cdot, 0, 1)$ has a natural structure of complemented distributive lattice $(\mathsf{B}, \wedge, \vee, \neg, 0, 1)$, for which the sum on the boolean ring becomes the operation Δ of symmetric difference ($a \Delta b = a \vee b \wedge (\neg(a \wedge b))$) on the complemented distributive lattice, and the multiplication of the ring the operation $\wedge$.

We refer to filters, antichains, dense sets, predense sets, open sets on B, meaning that these notions are declined for the corresponding partial order $(\mathsf{B}^+; \leq_\mathsf{B})$.

We also recall the following:

- An *ideal* I on B is a non-empty downward closed subset of B with respect to $\leq_\mathsf{B}$ which is also closed under $\vee$ (equivalently it is an ideal on the boolean ring $(\mathsf{B}, \delta, \wedge, 0, 1)$). Its *dual filter* $\check{I}$ is the set $\{\neg a : a \in I\}$. It is a filter on the poset $(\mathsf{B}^+; \leq_\mathsf{B})$ (equivalently I is an ideal in the boolean ring B).

- An ideal I on B is $<\delta$-*complete* (δ-complete) if all the subsets of I of size less than δ (of size δ) have an upper bound in I.

- A *maximal* ideal I is an ideal properly contained in B and maximal with respect to this property (equivalently it is a prime ideal on the boolean ring $(\mathsf{B}, \delta, \wedge, 0, 1)$). Its dual filter is an *ultrafilter*. An ideal I is maximal if and only if $a \in I$ or $\neg a \in I$ for all $a \in \mathsf{B}$.

- B is $< \delta$-complete (δ-complete) if all subsets of size less than δ (of size δ) have a supremum and an infimum.

- Given an ideal I on B, B$/I$ is the quotient boolean algebra given by equivalence classes $[a]_I$ obtained by $a =_I b$ iff $a \Delta b \in I$.

- B$/I$ is $< \kappa$-complete if I and B are both $< \kappa$-complete.

- B is *atomless* if there are no minimal elements in the partial order $(B^+; \leq_B)$.

- B is *atomic* if the set of minimal elements in the partial order $(B^+; \leq_B)$ is open dense.

Usually we insist in the formulation of forcing axioms on the requirement that for certain partial orders P any family of predense subsets of P of some fixed size κ can be met in a single filter. In order to obtain a greater variety of forcing axioms, we need to consider a much richer variety of properties which characterizes the families of predense sets of P which can be met in a single filter. Using boolean algebras, by considering partial orders of the form $(B^+; \leq_B)$ for some boolean algebra B, we can formulate (using the algebraic structure of B) a wide spectrum of properties each defining a distinct forcing axiom.

2.2 Measurable cardinals

A cardinal κ is measurable if and only if there is a uniform $< \kappa$-complete ultrafilter on the boolean algebra $\mathcal{P}(\kappa)$. The requirement that G is uniform amounts to say that G is disjoint from the ideal I on the boolean algebra $(\mathcal{P}(\kappa), \cap, \cup, \emptyset, \kappa)$ given by the bounded subsets of $\mathcal{P}(\kappa)$. This means that we are actually looking for an ultrafilter G on the boolean algebra $\mathcal{P}(\kappa)/I$. This is an atomless boolean algebra which is $< \kappa$-complete. The requirement that G is $< \kappa$-complete amounts to ask that G selects an unique member of any partition of κ in $< \kappa$-many pieces, moreover any maximal antichain $\{[A_i]_I : i < \gamma\}$ in the boolean algebra $\mathcal{P}(\kappa)/I$ of size γ less than κ is induced by a partition of κ in γ-many pairwise disjoint pieces.

All in all, we have the following characterization of measurability:

Definition 2.3. κ is a *measurable* cardinal if and only if there is a ultrafilter G on $\mathcal{P}(\kappa)/I$ (where I is the ideal of bounded subsets of κ) which meets all the maximal antichain on $\mathcal{P}(\kappa)/I$ of size less than κ.

In particular the measurability of κ holds if and only if $(\mathcal{P}(\kappa)/I)^+$ satisfies a certain forcing axiom stating that certain collections of predense subsets of $(\mathcal{P}(\kappa)/I)^+$ can be simultaneously met in a filter.

We are led to the following definitions:

Definition 2.4. Let $(P, \leq)$ be a partial order and $\mathcal{D}$ be a family of non-empty subsets of P. A filter G on P is $\mathcal{D}$-generic if $G \cap D$ is non-empty for all $D \in \mathcal{D}$.

Let $\phi(x, y)$ be a property and $(P, \leq)$ a partial order. $\mathsf{FA}_\phi(P)$ holds if for any family $\mathcal{D}$ of predense subsets of P such that $\phi(P, \mathcal{D})$ holds there is some $\mathcal{D}$-generic filter G on P.

For instance, $\mathsf{FA}_\kappa(P)$ says that $\mathsf{FA}_\phi(P)$ holds for $\phi(x, y)$ being the property:

"x is a partial order and y is a family of predense subsets of x of size κ"

The measurability of κ amounts to say that $\mathsf{FA}_\phi(P)$ holds for $\phi(x, y)$ being the property

"x is the partial order $(\mathcal{P}(\kappa)/I)^+$ and y is the (unique) family of predense subsets of x consisting of maximal antichains of $(\mathcal{P}(\kappa)/I)^+$ of size less than κ"

We do not want to expand further on this topic but many other large cardinal properties of a cardinal κ can be formulated as axioms of the form $\mathsf{FA}_\phi(P)$ for some property ϕ (for example this is the case for supercompactness, hugeness, almost hugeness, strongness, superstrongness, etc....).

In these first two sections we have already shown that the language of partial orders can accomodate three completely distinct and apparently unrelated families of non-constructive principles which are essential tools in the development of many mathematical theories (as it is the case for the axiom of choice and of Baire's category theorem) and of crucial importance in the current developments of set theory (as it is the case for large cardinal axioms).

3 Boolean valued models, Łoś theorem, and generic absoluteness

We address here the correlation between forcing axioms and generic absoluteness results. We show how Shoenfield's absoluteness for Σ^1_2-properties and Łoś Theorem are two sides of the same coin: more precisely they are distinct specific cases of a unique general theorem which follows from AC.

After recalling the basic formulation of Łoś Theorem for ultraproducts, we introduce boolean valued models, and we argue that Łoś Theorem for ultraproducts is the specific instance for complete atomic boolean algebras of a more general theorem which applies to a much larger class of boolean valued models. Then we introduce the concept of boolean ultrapower of a first order structure on a Polish space X

endowed with Borel predicates $R_1, \ldots, R_n$, and show that Shoenfield's absoluteness for Σ^1_2-properties amounts to say that the boolean ultrapower of $\langle X, R_1, \ldots, R_n \rangle$ by any complete boolean algebra is a Σ_2-elementary superstructure of $\langle X, R_1, \ldots, R_n \rangle$.

3.1 Łoś Theorem

Theorem 3.1. *Let $\{\mathfrak{M}_l : l \in L\}$ be models in a given first order signature*

$$\mathcal{L} = \{R_i : i \in I,\, f_j : j \in J,\, c_k : k \in K\},$$

i.e. each $\mathfrak{M}_l = (M_l, R_i^l : i \in I, f_j^l : j \in J, c_k^l : k \in K)$. Let G be a ultrafilter on L (i.e. its dual is a prime ideal on the boolean algebra $\mathcal{P}(L)$). Let

$$[f]_G = \left\{ g \in \prod_{l \in L} M_l : \{l \in L : g(l) = f(l)\} \in G \right\}$$

for each $f \in \prod_{l \in L} M_l$, and set

$$\prod_{l \in L} M_l / G = \left\{ [f]_G : f \in \prod_{l \in L} M_l \right\}.$$

For each $i \in I$ let $\bar{R}_i([f_1]_G, \ldots, [f_n]_G)$ hold on $\prod_{l \in L} M_l/G$ if and only if

$$\left\{ l \in L : \mathfrak{M}_l \models R_i^l(f_1(l), \ldots, f_n(l)) \right\} \in G.$$

Similarly interpret $\bar{f}_j : \prod_{l \in l}(M_l/G)^n \to \prod_{l \in L} M_l/G$ and $\bar{c}_k \in \prod_{l \in l} M_l^n/G$ for each $j \in J$ and $k \in K$.

Then:

1. *For all formulae $\phi(x_1, \ldots, x_n)$ in the signature $\mathcal{L}$*

$$(\prod_{l \in L} M_l/G, \bar{R}_i : i \in I,\, \bar{f}_j : j \in J,\, \bar{c}_k : k \in K) \models \phi([f_1]_G, \ldots, [f_n]_G)$$

if and only if

$$\{l \in L : \mathfrak{M}_l \models \phi(f_1(l), \ldots, f_n(l))\} \in G.$$

2. *Moreover if $\mathfrak{M}_l = \mathfrak{M}$ for all $l \in L$ (i.e. $\prod_{l \in L} M_j/G$ is the ultrapower of M by G), we have that the map $m \mapsto [c_m]_G$ (where $c_m : L \to M$ is constant with value m) defines an elementary embedding.*

It is a useful exercise to check that the axiom of choice is essentially used in the induction step for existential quantifiers in the proof of Łoś Theorem. Moreover Łoś Theorem is clearly a strenghtnening of the axiom of choice, for the very existence of an element in $\prod_{l \in L} M_l / G$ grants that $\prod_{l \in L} M_l$ is non-empty.

One peculiarity of the above formulation of Łoś theorem is that it applies just to ultrafilters on $\mathcal{P}(X)$. We aim to find a "most" general formulation of this Theorem, which makes sense also for other kind of "ultraproducts" and of ultrafilters on boolean algebras other than $\mathcal{P}(X)$. This forces us to introduce the boolean valued semantics.

3.2 A fast briefing on complete boolean algebras and Stone duality

Recall that for a given topological space (X, τ) the regular open sets are those $A \in \tau$ such that $A = \text{Reg}(A) = \text{Int}(\text{Cl}(A))$ (A coincides with the interior of its closure) and that $\text{RO}(X, \tau)$ is the complete boolean algebra whose elements are regular open sets and whose operations are given by $A \wedge B = A \cap B$, $\bigvee_{i \in I} A_i = \text{Reg}(\bigcup_{i \in I} A_i)$, $\neg A = X \setminus \text{Cl}(A)$.

For any partial order $(P, \leq)$ the map $i : P \to \text{RO}(P, \tau_P)$ given by $p \mapsto \text{Reg}(\downarrow \{p\})$ is order and incompatibility preserving and embeds P as a dense subset of the non-empty regular open sets in $\text{RO}(P, \tau_P)$.

Recall also that the Stone space $\text{St}(\mathsf{B})$ of a boolean algebra B is given by its ultrafilters G and it is endowed with a compact topology τ_B whose clopen sets are the sets $N_b = \{G \in \text{St}(\mathsf{B}) : b \in G\}$ so that the map $b \mapsto N_b$ defines a natural isomorphism of B with the boolean algebra $\text{CLOP}(\text{St}(\mathsf{B}))$ of clopen subset of $\text{St}(\mathsf{B})$. Moreover a boolean algebra B is complete if and only if $\text{CLOP}(\text{St}(\mathsf{B})) = \text{RO}(\text{St}(\mathsf{B}), \tau_\mathsf{B})$. Spaces X satisfying the property that its regular open sets are closed are extremally (or extremely) disconnected.

We refer the reader to [11] or [33, Chapter 1] (available at the following hyperlink: *Notes on Forcing*) for a detailed account on these matters.

3.3 Boolean valued models

In a first order model, a formula can be interpreted as true or false. Given a complete boolean algebra B, B-boolean valued models generalize Tarski semantics associating to each formula a value in B, so that propositions are not only true and false anymore (that is, only associated to 1_B and 0_B respectively), but take also other "intermediate values" of truth. A complete account of the theory of these boolean valued models can be found in [21]. We now recall some basic facts, an expanded version of the material of this section can be found in [26] (see also the following hyperlink: *Tesi-*

Vaccaro) and in [33, Chapter 3]. In order to avoid unnnecessary technicalities, we define boolean valued semantics just for relational first order languages (i.e. signatures with no function symobols).

Definition 3.2. Given a complete boolean algebra B and a first order relational language
$$\mathcal{L} = \{R_i : i \in I\} \cup \{c_j : j \in J\}$$
a B-*boolean valued model* (or B-valued model) $\mathcal{M}$ in the language $\mathcal{L}$ is a tuple
$$\langle M, =^{\mathcal{M}}, R_i^{\mathcal{M}} : i \in I, c_j^{\mathcal{M}} : j \in J \rangle$$
where:

1. M is a non-empty set, called *domain* of the B-boolean valued model, whose elements are called B-*names*;

2. $=^{\mathcal{M}}$ is the *boolean value* of the equality:
$$=^{\mathcal{M}}: M^2 \to \mathsf{B}$$
$$(\tau, \sigma) \mapsto [\![\tau = \sigma]\!]_{\mathsf{B}}^{\mathcal{M}}$$

3. The forcing relation $R_i^{\mathcal{M}}$ is the *boolean interpretation* of the n-ary relation symbol R_i:
$$R_i^{\mathcal{M}} : M^n \to \mathsf{B}$$
$$(\tau_1, \ldots, \tau_n) \mapsto [\![R_i(\tau_1, \ldots, \tau_n)]\!]_{\mathsf{B}}^{\mathcal{M}}$$

4. $c_j^{\mathcal{M}} \in M$ is the *boolean interpretation* of the constant symbol c_j.

We require that the following conditions hold:

- for $\tau, \sigma, \chi \in M$,
    1. $[\![\tau = \tau]\!]_{\mathsf{B}}^{\mathcal{M}} = 1_{\mathsf{B}}$;
    2. $[\![\tau = \sigma]\!]_{\mathsf{B}}^{\mathcal{M}} = [\![\sigma = \tau]\!]_{\mathsf{B}}^{\mathcal{M}}$;
    3. $[\![\tau = \sigma]\!]_{\mathsf{B}}^{\mathcal{M}} \wedge [\![\sigma = \chi]\!]_{\mathsf{B}}^{\mathcal{M}} \leq [\![\tau = \chi]\!]_{\mathsf{B}}^{\mathcal{M}}$;

- for $R \in \mathcal{L}$ with arity n, and $(\tau_1, \ldots, \tau_n), (\sigma_1, \ldots, \sigma_n) \in M^n$,
    1. $(\bigwedge_{h \in \{1,\ldots,n\}} [\![\tau_h = \sigma_h]\!]_{\mathsf{B}}^{\mathcal{M}}) \wedge [\![R(\tau_1, \ldots, \tau_n)]\!]_{\mathsf{B}}^{\mathcal{M}} \leq [\![R(\sigma_1, \ldots, \sigma_n)]\!]_{\mathsf{B}}^{\mathcal{M}}$;

Given a B-model $\langle M, =^M \rangle$ for equality, a forcing relation R on M is a map $R: M^n \to \mathsf{B}$ satisfying the above condition for boolean predicates.

The boolean valued semantics is defined as follows:

Definition 3.3. Let
$$\langle M, =^{\mathcal{M}}, R_i^{\mathcal{M}} : i \in I, c_j^{\mathcal{M}} : j \in J \rangle$$
be a B-valued model in a relational language
$$\mathcal{L} = \{R_i : i \in I\} \cup \{c_j : j \in J\},$$

ϕ a $\mathcal{L}$-formula whose free variables are in $\{x_1, \ldots, x_n\}$, and ν a valuation of the free variables in $\mathcal{M}$ whose domain contains $\{x_1, \ldots, x_n\}$. Since $\mathcal{L}$ is a relational languages, the terms of a formula are either free variable or constants, let us define $\nu(c_j) = c_j^M$ for c_j a constant of $\mathcal{L}$. We denote with $[\![\phi]\!]_{\mathsf{B}}^{\mathcal{M},\nu}$ the *boolean value of* ϕ with the assignment ν.

Given a formula ϕ, we define recursively $[\![\phi]\!]_{\mathsf{B}}^{\mathcal{M},\nu}$ as follows:

- for atomic formulae this is done letting
$$[\![t = s]\!]_{\mathsf{B}}^{\mathcal{M},\nu} = [\![\nu(t) = \nu(s)]\!]_{\mathsf{B}}^{\mathcal{M}},$$
and
$$[\![R(t_1, \ldots, t_n)]\!]_{\mathsf{B}}^{\mathcal{M},\nu} = [\![R(\nu(t_1), \ldots, \nu(t_n))]\!]_{\mathsf{B}}^{\mathcal{M}}$$

- if $\phi \equiv \neg \psi$, then
$$[\![\phi]\!]_{\mathsf{B}}^{\mathcal{M},\nu} = \neg [\![\psi]\!]_{\mathsf{B}}^{\mathcal{M},\nu};$$

- if $\phi \equiv \psi \wedge \theta$, then
$$[\![\phi]\!]_{\mathsf{B}}^{\mathcal{M},\nu} = [\![\psi]\!]_{\mathsf{B}}^{\mathcal{M},\nu} \wedge [\![\theta]\!]_{\mathsf{B}}^{\mathcal{M},\nu};$$

- if $\phi \equiv \exists y \psi(y)$, then
$$[\![\phi]\!]_{\mathsf{B}}^{\mathcal{M},\nu} = \bigvee_{\tau \in M} [\![\psi(y/\tau)]\!]_{\mathsf{B}}^{\mathcal{M},\nu};$$

If no confusion can arise, we omit the superscripts $\mathcal{M}, \nu$ and the subscript B, and we simply denote the boolean value of a formula ϕ with parameters in $\mathcal{M}$ by $[\![\phi]\!]$.

With elementary arguments it is possible prove the Soundness Theorem for boolean valued models.

Theorem 3.4 (Soundness Theorem). *Assume $\mathcal{L}$ is a relational language and ϕ is a $\mathcal{L}$-formula which is syntactically provable by a $\mathcal{L}$-theory T. Assume each formula in T has boolean value at least $b \in \mathsf{B}$ in a B-valued model $\mathcal{M}$ with valuation ν. Then $[\![\phi]\!]_\mathsf{B}^{\mathcal{M},\nu} \geq b$ as well.*

On the other hand the completeness theorem for the boolean valued semantics with respect to first order calculi is a triviality, given that 2 is complete boolean algebra.

We get a standard Tarski model from a B-valued model passing to a quotient by a ultrafilter $G \subseteq \mathsf{B}$.

Definition 3.5. Take B a complete boolean algebra, $\mathcal{M}$ a B-valued model in the language $\mathcal{L}$, and G a ultrafilter over B. Consider the following equivalence relation on M:
$$\tau \equiv_G \sigma \Leftrightarrow [\![\tau = \sigma]\!] \in G$$
The first order model $\mathcal{M}/G = \langle M/G, R_i^{\mathcal{M}/G} : i \in I, c_j^{\mathcal{M}/G} : j \in J \rangle$ is defined letting:

- For any n-ary relation symbol R in $\mathcal{L}$
$$R^{\mathcal{M}/G} = \{([\tau_1]_G, \ldots, [\tau_n]_G) \in (M/G)^n : [\![R(\tau_1, \ldots, \tau_n)]\!] \in G\}.$$

- For any constant symbol c in $\mathcal{L}$
$$c^{\mathcal{M}/G} = [c^{\mathcal{M}}]_G.$$

If we require $\mathcal{M}$ to satisfy a key additional condition, we get an easy way to control the truth value of formulas in $\mathcal{M}/G$.

Definition 3.6. A B-valued model $\mathcal{M}$ for the language $\mathcal{L}$ is *full* if for every $\mathcal{L}$-formula $\phi(x, \bar{y})$ and every $\bar{\tau} \in M^{|\bar{y}|}$ there is a $\sigma \in M$ such that
$$[\![\exists x \phi(x, \bar{\tau})]\!] = [\![\phi(\sigma, \bar{\tau})]\!]$$

Theorem 3.7 (Boolean Valued Models Łoś's Theorem). *Assume $\mathcal{M}$ is a full B-valued model for the relational language $\mathcal{L}$. Then for every formula $\phi(x_1, \ldots, x_n)$ in $\mathcal{L}$ and $(\tau_1, \ldots, \tau_n) \in M^n$:*

1. *For all ultrafilters G over B*
$$\mathcal{M}/G \models \phi([\tau_1]_G, \ldots, [\tau_n]_G) \text{ if and only if } [\![\phi(\tau_1, \ldots, \tau_n)]\!] \in G.$$

2. *For all $a \in \mathsf{B}$ the following are equivalent:*

(a) $[\![\phi(f_1,\ldots,f_n)]\!] \geq a$,

(b) $\mathcal{M}/G \models \phi([\tau_1]_G,\ldots,[\tau_n]_G)$ for all $G \in N_a$,

(c) $\mathcal{M}/G \models \phi([\tau_1]_G,\ldots,[\tau_n]_G)$ for densely many $G \in N_a$.

A key observation to relate standard ultraproducts to boolean valued models is the following:

Fact 3.8. Let $(M_x : x \in X)$ be a family of Tarski-models in the first order relational language $\mathcal{L}$. Then $N = \prod_{x \in X} M_x$ is a full $\mathcal{P}(X)$-model letting for each n-ary relation symbol $R \in \mathcal{L}$, $[\![R(f_1,\ldots,f_n)]\!]_{\mathcal{P}(X)} = \{x \in X : M_x \models R(f_1(x),\ldots,f_n(x))\}$.

Let G be any non-principal ultrafilter on X. Then, using the notation of the previous fact, N/G is the familiar ultraproduct of the family $(M_x : x \in X)$ by G, and the usual Łoś Theorem for ultraproducts of Tarski models is the specialization to the case of the full $\mathcal{P}(X)$-valued model N of Theorem 3.7. Notice that in this special case, if the ultraproduct is an ultrapower of a model M, the embedding $a \mapsto [c_a]_G$ (where $c_a(x) = a$ for all $x \in X$ and $a \in M$) is elementary.

3.4 Boolean ultrapowers of compact Hausdorff spaces and Shoenfield's absoluteness

Take X a set with the discrete topology, and let for any $a \in X$, $G_a \in \mathrm{St}(\mathcal{P}(X))$ denote the principal ultrafilter given by supersets of $\{a\}$. The map $a \mapsto G_a$ embeds X as an open, dense, discrete subspace of $\mathrm{St}(\mathcal{P}(X))$. In particular for any topological space (Y,τ), any function $f : X \to Y$ is continuous (since X has the discrete topology) and in the case Y is compact Hausdorff it induces a unique continuous $\bar{f} : \mathrm{St}(\mathcal{P}(X)) \to Y$ mapping $G \in \mathrm{St}(\mathcal{P}(X))$ to the unique point in Y which is in the intersection of $\{\mathrm{Cl}(A) : A \in \tau, f^{-1}[A] \in G\}$ (we are in the special situation in which $\mathrm{St}(\mathcal{P}(X))$ is also the Stone-Cech compactification of X).

This gives that for any compact Hausdorff space (Y,τ), the space $C(X,Y) = Y^X$ of (continuous) functions from X to Y is canonically isomorphic to the space $C(\mathrm{St}(\mathcal{P}(X)),Y)$ of continuous functions from $\mathrm{St}(\mathcal{P}(X))$ to Y.

What if we replace $\mathcal{P}(X)$ with an arbitrary (complete) boolean algebra? In view of the above remarks, it is a fair inference to state that $C(\mathrm{St}(\mathsf{B}),Y)$ is the B-ultrapower of Y for any compact Hausdorff space Y, since this is exactly what occurs for the case $\mathsf{B} = \mathcal{P}(X)$.

Let us examine closely this situation in the case $Y = 2^\omega$ with product topology. This will unfold the relation existing between the notion of Boolean ultrapowers of 2^ω and Shoenfield's absoluteness.

Let us fix B arbitrary (complete) boolean algebra, and set $M = C(St(\mathsf{B}), 2^\omega)$. Fix also R a Borel relation on $(2^\omega)^n$. The continuity of an n-tuple $f_1, \ldots, f_n \in M$ grants that the set
$$\{G : R(f_1(G) \ldots, f_n(G))\} = (f_1 \times \cdots \times f_n)^{-1}[R]$$
has the Baire property in $St(\mathsf{B})$ (i.e. it has symmetric difference with a unique regular open set — see [13, Lemma 11.15, Def. 32.21]), where $f_1 \times \cdots \times f_n(G) = (f_1(G), \ldots, f_n(G))$. So we can define
$$R^M : M^n \to \mathsf{B}$$
$$(f_1, \ldots, f_n) = \operatorname{Reg}\left(\{G : R(f_1(G), \ldots, f_n(G)\}\right).$$
Also, since the diagonal is closed in $(2^\omega)^2$,
$$=^M (f, g) = \operatorname{Reg}\left(\{G : f(G) = g(G)\}\right)$$
is well defined.

It is not hard to check that, for any Borel relation R on $(2^\omega)^n$, the structure $(M, =^M, R^M)$ is a full B-valued extension of $(2^\omega, =, R)$, where 2^ω is copied inside M as the set of constant functions. It is also not hard to check that whenever G is an ultrafilter on $St(\mathsf{B})$, the map $i_G : 2^\omega \to M/G$ given by $x \mapsto [c_x]_G$ (the constant function with value x) defines an injective morphism of the 2-valued structure $(2^\omega, R)$ into the 2-valued structure $(M/G, R^M/G)$. Nonetheless it is not clear whether this morphism is an elementary map or not. This is the case for $\mathsf{B} = \mathcal{P}(X)$, since in this case we are analyzing the standard embedding of the first order structure $(2^\omega, R)$ in its ultrapowers induced by ultrafilters on $\mathcal{P}(X)$. What are the properties of this map if B is some other complete boolean algebra?

We can relate the degree of elementarity of the map i_G with Shoenfield's absoluteness for Σ_2^1-properties. This can be done if one is eager to accept as a blackbox the identification of the B-valued model $C(St(\mathsf{B}), 2^\omega)$ with the B-valued model given by the family of B-names for elements of 2^ω in V^B (which is the canonical B-valued model for set theory), we will expand further on this identification in the next section. Modulo this identity, Shoenfield's absoluteness can be recasted as a statement about boolean valued models. We choose to name Cohen's absoluteness the following statement, which gives (as we will see) an equivalent reformulation of Shoenfield's absoluteness. Its proof (as we will see in the next section) ultimately relies on Cohen's forcing theorem, hence the name.

Theorem 3.9 (Cohen's absoluteness). *Assume B is a complete boolean algebra and $R \subseteq (2^\omega)^n$ is a Borel relation. Let $M = C(St(\mathsf{B}), 2^\omega)$ and $G \in St(\mathsf{B})$. Then*
$$(2^\omega, =, R) \prec_{\Sigma_2} (M/G, =^M /G, R^M/G).$$

4 Getting Cohen's absoluteness from Baire's category Theorem

Let us now show how Theorem 3.9 is once again a consequence of forcing axioms. To do so, we dwelve deeper into set theoretic techniques and assume the reader has some acquaintance with the forcing method. We give below a brief recall sufficient for our aims.

4.1 Forcing

Let V denote the standard universe of sets and ZFC the standard first order axiomatization of set theory by the Zermelo-Frankel axioms. For any complete boolean algebra $\mathsf{B} \in V$ let

$$V^{\mathsf{B}} = \left\{ f : V^{\mathsf{B}} \to \mathsf{B} \right\}$$

be the class of B-names with boolean relations $\in^{\mathsf{B}}, \subseteq^{\mathsf{B}}, =^{\mathsf{B}} \colon (V^{\mathsf{B}})^2 \to \mathsf{B}$ given by:

1.
$$\in^{\mathsf{B}} (\tau, \sigma) = [\![\tau \in \sigma]\!] = \bigvee_{\tau_0 \in \mathrm{dom}(\sigma)} ([\![\tau = \tau_0]\!] \wedge \sigma(\tau_0)).$$

2.
$$\subseteq^{\mathsf{B}} (\tau, \sigma) = \bigwedge_{\sigma_0 \in \mathrm{dom}(\tau)} (\neg \tau(\sigma_0) \vee [\![\sigma_0 \in \sigma]\!]).$$

3.
$$=^{\mathsf{B}} (\tau, \sigma) = [\![\tau = \sigma]\!] = [\![\tau \subseteq \sigma]\!] \wedge [\![\sigma \subseteq \tau]\!].$$

Theorem 4.1 (Cohen's forcing theorem I). $(V^{\mathsf{B}}, \in^{\mathsf{B}}, =^{\mathsf{B}})$ *is a full boolean valued model which assigns the boolean value* 1_{B} *to all axioms* $\phi \in$ ZFC.

V is copied inside V^{B} as the family of B-names $\check{a} = \left\{ \langle \check{b}, 1_{\mathsf{B}} \rangle : b \in a \right\}$ and has the property that for all Σ_0-formulae (i.e with quantifiers bounded to range over sets) $\phi(x_0, \ldots, x_n)$ and $a_0, \ldots, a_n \in V$

$$[\![\phi(\check{a}_0, \ldots, \check{a}_n)]\!] = 1_{\mathsf{B}} \text{ if and only if } V \models \phi(a_0, \ldots, a_n).$$

This procedure can be formalized in any first order model $(M, E, =)$ of ZFC for any $\mathsf{B} \in M$ such that $(M, E, =)$ models that B is a complete boolean algebra.

Two ingredients are still missing to prove Cohen's absoluteness (Theorem 3.9) from Baire's category theorem: the notion of M-generic filter and the duality between $C(\mathrm{St}(\mathsf{B}), 2^\omega)$ and the B-names in V^{B} for elements of 2^ω. We first deal with the duality.

4.2 $C(\mathrm{St}(\mathsf{B}), 2^\omega)$ is the family of B-names for elements of 2^ω

Definition 4.2. Let B be a complete boolean algebra. Let $\sigma \in V^\mathsf{B}$ be a B-name such that $[\![\sigma : \check{\omega} \to \check{2}]\!]_\mathsf{B} = 1_\mathsf{B}$. We define $f_\sigma : \mathrm{St}(\mathsf{B}) \to 2^\omega$ by

$$f_\sigma(G)(n) = i \iff [\![\sigma(\check{n}) = \check{i}]\!] \in G.$$

Conversely assume $g : \mathrm{St}(\mathsf{B}) \to 2^\omega$ is a continuous function, then define

$$\tau_g = \{\langle(\check{n,i}), \{G : g(G)(n) = i\}\rangle : n \in \omega, i < 2\} \in V^\mathsf{B}.$$

Observe indeed that

$$\{G \in \mathrm{St}(\mathsf{B}) : g(G)(n) = i\} = g^{-1}[N_{n,i}],$$

where $N_{n,i} = \{f \in 2^\omega : f(n) = i\}$. Since g is continuous, $g^{-1}[N_{n,i}]$ is clopen and so it is an element of B.

We can prove the following duality:

Proposition 4.3. *Assume that $[\![\sigma : \check{\omega} \to \check{2}]\!]_\mathsf{B} = 1_\mathsf{B}$ and $g : \mathrm{St}(\mathsf{B}) \to 2^\omega$ is continuous. Then*

1. *$\tau_g \in V^\mathsf{B}$;*
2. *$f_\sigma : \mathrm{St}(\mathsf{B}) \to 2^\omega$ is continuous;*
3. *$[\![\tau_{f_\sigma} = \sigma]\!]_\mathsf{B} = 1_\mathsf{B}$;*
4. *$f_{\tau_g} = g$.*

In particular letting

$$(2^\omega)^\mathsf{B} = \left\{\sigma \in V^\mathsf{B} : [\![\sigma : \check{\omega} \to \check{2}]\!]_\mathsf{B} = 1_\mathsf{B}\right\},$$

the 2-valued models $((2^\omega)^\mathsf{B}/G, =^\mathsf{B}/G)$ and $(C(\mathrm{St}(\mathsf{B}), 2^\omega), =^{\mathrm{St}(\mathsf{B})}/G)$ are isomorphic for all $G \in \mathrm{St}(\mathsf{B})$ via the map $[g]_G \mapsto [\tau_g]_G$.

This is just part of the duality, as the duality can lift the isomorphism also to all B-Baire relations on 2^ω, among which are all Borel relations. Recall that for any given topological space (X, τ) a subset Y of X is meager for τ if Y is contained in the countable union of closed nowhere dense (i.e. with complement dense open) subsets of X. Y has the Baire property if $Y \Delta A$ is meager for some unique regular open set $A \in \tau$.

Definition 4.4. $R \subseteq (2^\omega)^n$ is a B-Baire subset of $(2^\omega)^n$ if for all continuous functions $f_1, \ldots, f_n : \text{St}(\mathsf{B}) \to 2^\omega$ we have that

$$(f_1 \times \cdots \times f_n)^{-1}[A] = \{G : f_1 \times \cdots \times f_n(G) \in A\}$$

has the Baire property in $\text{St}(\mathsf{B})$.

$R \subseteq (2^\omega)^n$ is universally Baire if it is B-Baire for all complete boolean algebras B.

It can be shown in ZFC that Borel (and even analytic) subsets of $(2^\omega)^n$ are universally Baire (see [13, Def. 32.21]).

An important result of Feng, Magidor, and Woodin [9] can be restated as follows:

Theorem 4.5. $R \subseteq (2^\omega)^n$ is B-Baire if and only if there exist $\dot{R}^{\mathsf{B}} \in V^{\mathsf{B}}$ such that

$$\left[\!\!\left[\dot{R}^{\mathsf{B}} \subseteq (2^{\check{\omega}})^n \right]\!\!\right] = 1_{\mathsf{B}},$$

and for all $\tau_1, \ldots, \tau_n \in (2^\omega)^{\mathsf{B}}$

$$Reg(\{G : R(f_{\tau_1}(G), \ldots, f_{\tau_n}(G))\}) = \left[\!\!\left[(\tau_1, \ldots, \tau_n) \in \dot{R}^{\mathsf{B}} \right]\!\!\right].$$

In particular an easy Corollary is the following:

Theorem 4.6. Let $R \subseteq (2^\omega)^n$ be a B-baire relation. Then the map $[f]_G \mapsto [\tau_f]_G$ implements an isomorphism between

$$\langle C(\text{St}(\mathsf{B}))/G, R^{\text{St}(\mathsf{B})}/G \rangle \cong \langle (2^\omega)^{\mathsf{B}}/G, \dot{R}^{\mathsf{B}}/G \rangle$$

for any $G \in \text{St}(\mathsf{B})$.

These results can be suitably generalized to arbitrary Polish spaces. We refer the reader to [26] and [27]. [31] gives an application of this result to tackle a problem in number theory related to Schanuel's conjecture.

4.3 M-generic filters and Cohen's absoluteness

Definition 4.7. Let $(P, \leq)$ be a partial order and M be a set. A subset G of P is M-generic if $G \cap D$ is non-empty for all $D \in M$ predense subset of P.

By BCT_1 every countable set M admits M-generic filters for all partial orders P.

Theorem 4.8 (Cohen's forcing theorem II). *Assume $(N, \in)$ is a transitive model of* ZFC, $\mathsf{B} \in N$ *is a complete boolean algebra in N, and $G \in \mathrm{St}(\mathsf{B})$ is an N-generic filter for B^+.*

Let
$$\mathrm{val}_G : N^{\mathsf{B}} \to V$$
$$\sigma \mapsto \sigma_G = \{\tau_G : \exists b \in G \, \langle \tau, b \rangle \in \sigma\},$$

and $N[G] = \mathrm{val}_G[N^{\mathsf{B}}]$.

Then $N[G]$ is transitive, the map $[\sigma]_G \mapsto \sigma_G$ is the Mostowski collapse of the Tarski models $\langle N^{\mathsf{B}}/G, \in^{\mathsf{B}}/G \rangle$ and induces an isomorphism of this model with the model $\langle N[G], \in \rangle$.

In particular for all formulae $\phi(x_1, \ldots, x_n)$ and $\tau_1 \ldots, \tau_n \in N^{\mathsf{B}}$
$$\langle N[G], \in \rangle \models \phi((\tau_1)_G, \ldots, (\tau_n)_G)$$

if and only if $[\![\phi(\tau_1, \ldots, \tau_n)]\!] \in G$.

Recall that:

- For any infinite cardinal λ, H_λ is the set of all sets $a \in V$ such that $|\mathrm{trcl}(a)| < \lambda$ (where $\mathrm{trcl}(a)$ is the transitive closure of the set a).

- If κ is a strongly inaccessible cardinal (i.e. regular and strong limit), H_κ is a transitive model of ZFC.

- A property $R \subseteq (2^\omega)^n$ is Σ_2^1, if it is of the form
$$R = \{(a_1, \ldots, a_n) \in (2^\omega)^n : \exists y \in 2^\omega \, \forall x \in 2^\omega \, S(x, y, a_1, \ldots, a_n)\}$$
with $S \subseteq (2^\omega)^{n+2}$ a Borel relation.

- If $\phi(x_0, \ldots, x_n)$ is a Σ_0-formula and $M \subseteq N$ are transitive sets or classes, then for all $a_0, \ldots, a_n \in M$
$$M \models \phi(a_0, \ldots, a_n) \text{ if and only if } N \models \phi(a_0, \ldots, a_n).$$

Observe that for any theory $T \supseteq$ ZFC there is a recursive translation of Σ_2^1-properties (provably Σ_2^1 over T) into Σ_1-properties over H_{ω_1} (provably Σ_1 over the same theory T) [13, Lemma 25.25].

Lemma 4.9. *Assume $\phi(x, r)$ is a Σ_0-formula in the parameter $\vec{r} \in (2^\omega)^n$. Then the following are equivalent:*

1. $H_{\omega_1} \models \exists x \phi(x, r)$.

2. For all complete boolean algebra B $[\![\exists x \phi(x,r)]\!] = 1_\mathsf{B}$.

3. There is a complete boolean algebra B such that $[\![\exists x \phi(x,r)]\!] > 0_\mathsf{B}$.

Summing up we get: a Σ^1_2-statement holds in V iff the corresponding Σ_1-statement over H_{ω_1} holds in some model of the form V^B/G.

Combining the above Lemma with Proposition 4.3, we can easily infer the proof of Theorem 3.9.

Proof. We shall actually prove the following slightly stronger formulation of the non-trivial direction in the three equivalences above:

$H_{\omega_1} \models \exists x \phi(x, r)$ if $[\![\exists x \phi(x,r)]\!] > 0_\mathsf{B}$ for some complete boolean algebra $\mathsf{B} \in V$.

To simplify the exposition we prove it with the further assumption that that there exists an inaccessible cardinal $\kappa > \mathsf{B}$. With the obvious care in details the large cardinal assumption can be removed. So assume $\phi(x, \vec{y})$ is a Σ_0-formula and $[\![\exists x \phi(x, \check{\vec{r}})]\!] > 0_\mathsf{B}$ for some complete boolean algebra $\mathsf{B} \in V$ with parameters $\vec{r} \in (2^\omega)^n$. Pick a model $M \in V$ such that $M \prec (H_\kappa)^V$, M is countable in V, and $\mathsf{B}, \vec{r} \in M$. Let $\pi_M : M \to N$ be its transitive collapse (i.e. $\pi_M(a) = \pi_M[a \cap M]$ for all $a \in M$) and $\mathsf{Q} = \pi_M(\mathsf{B})$. Notice also that $\pi_M(\vec{r}) = \vec{r}$: since $\omega \in M$ is a definable ordinal *contained* in M, $\pi_M(\omega) = \pi_M[\omega] = \omega$, consequently π_M fixes also all the elements in $2^\omega \cap M$.

Since π_M is an isomorphism of M with N,

$$N \models \mathsf{ZFC} \land (b = [\![\exists x \phi(x, \check{\vec{r}})]\!] > 0_\mathsf{Q}).$$

Now let $G \in V$ be N-generic for Q with $b \in G$ (G exists since N is countable), then, by Cohen's theorem of forcing applied in V to N, we have that $N[G] \models \exists x \phi(x, \vec{r})$. So we can pick $a \in N[G]$ such that $N[G] \models \phi(a, \vec{r})$. Since $N, G \in (H_{\omega_1})^V$, we have that V models that $N[G] \in H_{\omega_1}^V$ and thus V models that a as well belongs to $H_{\omega_1}^V$. Since $\phi(x, \vec{y})$ is a Σ_0-formula, V models that $\phi(a, \vec{r})$ is absolute between the transitive sets $N[G] \subset H_{\omega_1}$ to which $a, \vec{r}$ belong. In particular a witnesses in V that $H_{\omega_1}^V \models \exists x \phi(x, \vec{r})$. □

5 Maximal forcing axioms

Guided by all the previous results we want to formulate maximal forcing axioms. We pursue two directions:

1. A direction led by topological considerations: we have seen that $\mathsf{FA}_{\aleph_0}(P)$ holds for any partial order P, and that AC is equivalent to the satisfaction of $\mathsf{FA}_\lambda(P)$ for all regular λ and all $< \lambda$-closed posets P.

   We want to isolate the largest possible class of partial orders Γ_λ for which $\mathsf{FA}_\lambda(P)$ holds for all $P \in \Gamma_\lambda$. The case $\lambda = \aleph_0$ is handled by Baire's category theorem, that shows that $\Gamma_{\aleph_0}$ is the class of all posets. We will outline that the case $\lambda = \aleph_1$ is settled by the work of Foreman, Magidor, and Shelah [10] and leads to Martin's maximum. On the other hand, the case $\lambda > \aleph_1$ is wide open and until recently only partial results have been obtained. New techniques to handle the case $\lambda = \aleph_2$ are being developed (notably by Neeman, and also by Asperò, Cox, Krueger, Mota, Velickovic, see among others [14, 15, 19]), however the full import of their possible applications is not clear yet.

2. A direction led by model-theoretic considerations: Baire's category theorem implies that the natural embedding of 2^ω into $C(\mathrm{St}(\mathsf{B}), 2^\omega)/G$ is Σ_2-elementary, whenever 2^ω is endowed with B-baire predicates (among which all the Borel predicates). We want to reinforce this theorem in two directions:

   (A) We want to be able to infer that (at least for Borel predicates) the natural embedding of 2^ω into $C(\mathrm{St}(\mathsf{B}), 2^\omega)/G$ yields a full elementary embedding of 2^ω into $C(\mathrm{St}(\mathsf{B}), 2^\omega)/G$.

   (B) We want to be able to define boolean ultrapowers M^B also for other first order structures M other than 2^ω and be able to infer that the natural embedding of M into M^B/G is elementary for these boolean ultrapowers.

Both directions (the topological and the model-theoretic) converge towards the isolation of certain natural forcing axioms. Moreover for each cardinal λ, the relevant stuctures for which we can define a natural notion of boolean ultrapower are either the structure H_{λ^+}, or the Chang model $L(\mathrm{Ord}^\lambda)$.

We believe that we have now a satisfactory understanding of the maximal forcing axioms one can get following both directions for the cases $\lambda = \aleph_0, \aleph_1$. The main open question remains how to isolate (if at all possible) the maximal forcing axioms for $\lambda > \aleph_1$.

5.1 Woodin's generic absoluteness for H_{ω_1} and $L(\mathrm{Ord}^\omega)$

We start by the model-theoretic direction, following Woodin's work in Ω-logic. Observe that a set theorist works either with first order calculus to justify some proofs over ZFC, or with forcing to obtain independence results over ZFC. However, in axiom systems extending ZFC there seems to be a gap between what we can achieve by ordinary proofs and the independence results that we can obtain by means of forcing. To close this gap we miss two desirable features of a "complete" first order theory T that contains ZFC, specifically with respect to the semantics given by the class of boolean valued models of T:

- T is complete with respect to its intended semantics, i.e for all statements ϕ only one among $T + \phi$ and $T + \neg\phi$ is forceable.

- Forceability over T should correspond to a notion of derivability with respect to some proof system, for instance derivability with respect to a standard first order calculus for T.

Both statements appear to be rather bold and have to be handled with care: Consider for example the statement $\omega = \omega_1$ in a theory T extending ZFC with the statements ω *is the first infinite cardinal* and ω_1 *is the first uncountable cardinal*. Then clearly T proves $|\omega| \neq |\omega_1|$, while if one forces with $\mathrm{Coll}(\omega, \omega_1)$ one produce a model of set theory where this equality holds (however the formula ω_1 *is the first uncountable cardinal* is now false in this model).

At first glance, this suggests that as we expand the language for T, forcing starts to act randomly on the formulae of T, switching the truth value of its formulae with parameters in ways which it does not seem simple to describe. However the above difficulties are raised essentially by our lack of attention to define the type of formulae for which we aim to have the completeness of T with respect to forceability. We can show that when the formulae are prescribed to talk only about a suitable initial segment of the set theoretic universe (i.e. H_{ω_1} or $L(\mathrm{Ord}^\omega)$), and we consider only forcings that preserve the intended meaning of the parameters by which we enriched the language of T (i.e. parameters in H_{ω_1}), this random behaviour of forcing does not show up anymore.

We take a platonist stance towards set theory, thus we have one canonical model V of ZFC of which we try to uncover the truths. To do this, we may use model theoretic techniques that produce new models of the part of $\mathsf{Th}(V)$ on which we are confident. This certainly includes ZFC, and (if we are platonists) all the axioms of large cardinals.

We may start our quest for uncovering the truth in V by first settling the theory of $H_{\omega_1}^V$ (the hereditarily countable sets), then the theory of $H_{\omega_2}^V$ (the sets of hereditarily cardinality $\aleph_1$) and so on and so forth, thus covering step by step all infinite cardinals. To proceed we need some definitions:

Definition 5.1. Given a theory $T \supseteq$ ZFC and a family Γ of partial orders definable in T, we say that ϕ is Γ-consistent for T if T proves that there exists a complete boolean algebra $\mathsf{B} \in \Gamma$ such that $[\![\phi]\!]_\mathsf{B} > 0_\mathsf{B}$.

Given a model V of ZFC we say that V models that ϕ is Γ-consistent if ϕ is Γ-consistent for $\mathsf{Th}(V)$.

Definition 5.2. Let
$$T \supseteq \mathsf{ZFC} + \{\lambda \text{ is an infinite cardinal}\}$$
Ω_λ is the definable (in T) class of partial orders P which satisfy $\mathsf{FA}_\lambda(P)$.

In particular Baire's category theorem amounts to say that $\Omega_{\aleph_0}$ is the class of all partial orders (denoted by Woodin as the class Ω). The following is a careful reformulation of Lemma 4.9 which do not require any commitment on the onthology of V.

Lemma 5.3 (Cohen's absoluteness Lemma). *Assume $T \supseteq \mathsf{ZFC} + \{p \subseteq \omega\}$ and $\phi(x,p)$ is a Σ_0-formula. Then the following are equivalent:*

- $T \vdash \exists x \phi(x,p)$,

- $T \vdash \exists x \phi(x,p)$ *is Ω-consistent.*

This shows that for Σ_1-formulae with real parameters the desired overlap between the ordinary notion of provability and the semantic notion of forceability is provable in ZFC. Now it is natural to ask if we can expand the above in at least two directions:

1. Increase the complexity of the formula,

2. Increase the language allowing parameters also for other infinite cardinals.

The second direction will be pursued in the next subsection. Concerning the first direction, the extent by which we can increase the complexity of the formula requires once again some attention to the semantical interpretation of its parameters and its quantifiers. We have already observed that the formula $\omega = \omega_1$ is inconsistent but Ω-consistent in a language with parameters for ω and ω_1. One of Woodin's main achievements[2] in Ω-logic show that if we restrict the semantic interpretation of ϕ to range over the structure $L([\mathrm{Ord}]^{\aleph_0})$ and we assume large cardinal axioms, we can get a full correctness and completeness result[3] [16, Corollary 3.1.7]:

---

[2] We follow Larson's presentation as in [16].
[3] The large cardinal assumptions on T of the present formulation can be significantly reduced. See [16, Corollary 3.1.7].

Theorem 5.4 (Woodin). *Assume T is a theory extending*

$$\mathsf{ZFC} + \{p \subset \omega\} + \text{there are class many supercompact cardinals,}$$

$\phi(x,y)$ is any formula in free variables x, y, $A \subseteq (2^\omega)^n$ is universally Baire. Then the following are equivalent (where $\dot{A}^\mathsf{B}$ is the B-name given by Theorem 4.5 lifting A to V^B):

- $T \vdash [L([\mathrm{Ord}]^{\aleph_0}, A) \models \phi(p, A)]$,
- $T \vdash \exists \mathsf{B} \; [\![L([\mathrm{Ord}]^{\aleph_0}, \dot{A}^\mathsf{B}) \models \phi(p, \dot{A}^\mathsf{B})]\!] > 0_\mathsf{B}$,
- $T \vdash \forall \mathsf{B} \; [\![L([\mathrm{Ord}]^{\aleph_0}, \dot{A}^\mathsf{B}) \models \phi(p, \dot{A}^\mathsf{B})]\!] = 1_\mathsf{B}$.

Remark that since $H_{\omega_1} \subseteq L([\mathrm{Ord}]^{\aleph_0})$, via Theorem 4.5 and natural generalizations of [13, Lemma 25.25] establishing a correspondence between Σ^1_{n+1}-properties and Σ_n-properties over H_{ω_1}, we obtain that for any complete boolean algebra B and any Σ^1_n-predicate $R \subseteq (2^\omega)^n$ the map $x \mapsto [c_x]_G$ of $(2^\omega, R)$ into $(C(\mathrm{St}(\mathsf{B}), 2^\omega), R^{\mathrm{St}(\mathsf{B})})$ is an elementary embedding. In particular the above theorem provides a first fully satisfactory answer to the question of whether the natural embeddings of 2^ω in its boolean ultrapowers can be elementary: the answer is yes if we accept the existence of large cardinal axioms!

The natural question to address now is whether we can step up this result also for uncountable λ. If so in which form?

5.2 Topological maximality: Martin's maximum MM

Let us now address the quest for maximal forcing axioms from the topological direction. Specifically: what is the largest class of partial orders Γ for which we can predicate $\mathsf{FA}_{\aleph_1}(\Gamma)$?

Shelah proved that $\mathsf{FA}_{\aleph_1}(P)$ fails for any P which does not preserve stationary subsets of ω_1. Nonetheless it cannot be decided in ZFC whether this is a necessary condition for a poset P in order to have the failure of $\mathsf{FA}_{\aleph_1}(P)$. For example let P be a forcing which shoots a club of ordertype ω_1 through a projectively stationary and costationary subset of $P_{\omega_1}(\omega_2)$ by selecting countable initial segments of this club: It is provable in ZFC that P preserve stationary subsets of ω_1 for all such P. However in L, $\mathsf{FA}_{\aleph_1}(P)$ fails for some such P while in a model of Martin's maximum MM, $\mathsf{FA}_{\aleph_1}(P)$ holds for all such P.

The remarkable result of Foreman, Magidor, and Shelah [10] is that the above necessary condition is consistently also a sufficient condition: it can be forced that $\mathsf{FA}_{\aleph_1}(P)$ holds if and only if P is a forcing notion preserving all stationary subsets of ω_1. This axiom is known in the literature as Martin's maximum MM. In view of Theorem 1.8, MM realizes a maximality property for forcing axioms: it can be seen as a maximal strengthening of the axiom of choice AC $\restriction_{\omega_2}$ for $\aleph_1$-sized families of non-empty sets. Can we strengthen this further? if so in which form? It turns out that stronger and stronger forms of forcing axioms can be expressed in the language of categories and provide means to extend Woodin's generic absoluteness results to third order arithmetic or more generally to larger and larger fragments of the set theoretic universe.

5.3 Category forcings and category forcing axioms

Assume Γ is a class of complete boolean algebras and $\to^\Theta$ is a family of complete homomorphisms between elements of Γ closed under composition and containing all identity maps. $(\Gamma, \to^\Theta)$ is the category whose objects are the complete boolean algebras in Γ and whose arrows are given by complete homomorphisms $i : \mathsf{B} \to \mathsf{Q}$ in $\to^\Theta$. We call embeddings in $\to^\Theta$, Θ-correct embeddings. Notice that these categories immediately give rise to natural class partial orders associated with them, partial orders whose elements are the complete boolean algebras in Γ and whose order relation is given by the arrows in $\to^\Theta$ (i.e. $\mathsf{B} \leq_\Theta \mathsf{C}$ if there exists $i : \mathsf{C} \to \mathsf{B}$ in $\to^\Theta$). We denote these class partial orders by $(\Gamma, \leq_\Theta)$.

Depending on the choice of Γ and $\to^\Theta$ these partial orders can be trivial (as forcing notions), for example:

Remark 5.5. Assume $\Omega = \Omega_{\aleph_0}$ is the class of all complete boolean algebras and $\to^\Omega$ is the class of all complete embeddings, then any two conditions in $(\Gamma, \leq_\Omega)$ are compatible, i.e. $(\Gamma, \leq_\Omega)$ is forcing equivalent to the trivial partial order. This is the case since for any pair of partial orders P, Q and X of size larger than $2^{|P|+|Q|}$ there are complete injective homomorphisms of $\mathsf{RO}(P)$ and $\mathsf{RO}(Q)$ into the boolean completion of $\mathrm{Coll}(\omega, X)$ (see [16, Thm A.0.7] and its following remark). These embeddings witness the compatibility of $\mathsf{RO}(P)$ with $\mathsf{RO}(Q)$.

On the other hand these class partial orders will in general be non-trivial: let SSP be the class of stationary set preserving forcings. Then the Namba forcing shooting a cofinal ω-sequence on ω_2 and $\mathrm{Coll}(\omega_1, \omega_2)$ are incompatible conditions in $(\mathsf{SSP}, \leq_\Omega)$: any forcing notion absorbing both of them makes the cofinality of ω_2^V at the same time of cofinality ω_1^V (using the generic filter for $\mathrm{Coll}(\omega_1, \omega_2)$) and countable (using the generic filter for Namba forcing); this means that this forcing must collapse ω_1^V to become a countable ordinal, hence cannot be stationary set preserving.

Forcing axioms as density properties of category forcings

The following results are one of the main reasons leading us to analyze in more details these type of class forcings:

Theorem 5.6 (Woodin, Thm. 2.53 [34]). *Assume there are class many supercompact cardinals. Then the following are equivalent for any complete cba B and cardinal κ:*

1. $\mathsf{FA}_\kappa(\mathsf{B})$;

2. *there is a complete homomorphism of B into a presaturated tower inducing a generic ultrapower embedding with critical point κ^+.*

Theorem 5.7 (V. Thm. 2.12 [32]). *Assume there are class many supercompact cardinals. Then the following are equivalent:*

1. MM^{++};

2. *the class of presaturated normal towers is dense in $(\mathsf{SSP}, \leq_\mathsf{SSP})$.*

It is not in the scope of this paper to dwelve into the definition and properties of presaturated tower forcings and of the axiom MM^{++}. Let us just remark the following two facts:

- MM^{++} is a natural strengthening of Martin's maximum whose consistency is proved by exactly the same methods producing a model of Martin's maximum.

- A presaturated tower $\mathcal{T}$ inducing a generic ultrapower embedding with critical point κ^+ is such that whenever G is V-generic for $\mathcal{T}$ we have that

$$H_{\kappa^+}^V \prec H_{\kappa^+}^{V[G]}. \tag{1}$$

In particolar the above theorems show that forcing axioms can be also stated as density properties of class partial orders. We will see that any $\mathsf{AX}(\Gamma, \kappa)$ yielding a dense class of forcings in $(\Gamma, \leq_\Gamma)$ whose generic extensions satisfy (1) produce generic absoluteness results. We refer the reader to [3, 2, 30] for details.

5.4 Iterated resurrection axioms and generic absoluteness for H_{κ^+}

The results and ideas of this subsection expand on the seminal work of Hamkins and Johnstone [12] on resurrection axioms.

Definition 5.8. Let Γ be a definable class of complete Boolean algebras closed under two step iterations. The *cardinal preservation degree* $\mathrm{cpd}(\Gamma)$ of Γ is the largest cardinal κ such that every $\mathsf{B} \in \Gamma$ forces that every cardinal $\nu \leq \kappa$ is still a cardinal in V^{B}. If all cardinals are preserved by Γ, we say that $\mathrm{cpd}(\Gamma) = \infty$.

The *distributivity degree* $\mathrm{dd}(\Gamma)$ of Γ is the largest cardinal κ such that every $\mathsf{B} \in \Gamma$ is $<\kappa$-distributive.

We remark that the supremum of the cardinals preserved by Γ is preserved by Γ, and the same holds for the property of being $<\kappa$ distributive. Furthermore, $\mathrm{dd}(\Gamma) \leq \mathrm{cpd}(\Gamma)$ and $\mathrm{dd}(\Gamma) \neq \infty$ whenever Γ is non trivial (i.e., it contains a Boolean algebra that is not forcing equivalent to the trivial Boolean algebra). Moreover $\mathrm{dd}(\Gamma) = \mathrm{cpd}(\Gamma)$ whenever Γ is closed under two steps iterations and contains the class of $<\mathrm{cpd}(\Gamma)$-closed posets.

Definition 5.9. Let Γ be a definable class of complete Boolean algebras. We let $\gamma = \gamma_\Gamma = \mathrm{cpd}(\Gamma)$.

For example, $\gamma = \omega$ if Γ is the class of all posets, while for axiom-A, proper, SP, SSP we have that $\gamma = \omega_1$, and for $<\kappa-$closed we have that $\gamma = \kappa$.

We aim to isolate for each cardinal γ classes of forcings Δ_γ and axioms $\mathsf{AX}(\Delta_\gamma)$ such that:

1. $\gamma = \mathrm{cpd}(\Delta_\gamma)$ and assuming certain large cardinal axioms, the family of $\mathsf{B} \in \Delta_\gamma$ which force $\mathsf{AX}(\Delta_\gamma)$ is dense in $(\Delta_\gamma, \leq_{\Delta_\gamma})$;

2. $\mathsf{AX}(\Delta_\gamma)$ gives generic absoluteness for the theory with parameters of H_{γ^+} with respect to all forcings in Δ_γ which preserve $\mathsf{AX}(\Delta_\gamma)$;

3. the axioms $\mathsf{AX}(\Delta_\gamma)$ are mutually compatible for the largest possible family of cardinals γ simultaneously;

4. the classes Δ_γ are the largest possible for which the axioms $\mathsf{AX}(\Delta_\gamma)$ can possibly be consistent.

Towards this aim remark the following:

- $\mathrm{dd}(\Gamma)$ is the least possible cardinal γ such that $\mathsf{AX}(\Gamma)$ is a non-trivial axiom asserting generic absoluteness for the theory of H_{γ^+} with parameters. In fact, $H_{\mathrm{dd}(\Gamma)}$ is never changed by forcings in Γ.

- $\mathrm{cpd}(\Gamma)$ is the largest possible cardinal γ for which an axiom $\mathsf{AX}(\Gamma)$ as above can grant generic absoluteness with respect to Γ for the theory of H_{γ^+} with parameters. To see this, let Γ be such that $\mathrm{cpd}(\Gamma) = \gamma$ and assume towards a contradiction that there is an axiom $\mathsf{AX}(\Gamma)$ yielding generic absoluteness with respect to Γ for the theory with parameters of H_λ with $\lambda > \gamma^+$.

Assume that $\mathsf{AX}(\Gamma)$ holds in V. Since $\mathrm{cpd}(\Gamma) = \gamma$, there exists a $\mathsf{B} \in \Gamma$ which collapses γ^+. Let $\mathsf{C} \leq_\Gamma \mathsf{B}$ be obtained by property (1) above for $\Gamma = \Delta_\gamma$, so that $\mathsf{AX}(\Gamma)$ holds in V^C, and remark that γ^+ cannot be a cardinal in V^C as well. Then γ^+ is a cardinal in H_λ and not in H_λ^C, witnessing failure of generic absoluteness and contradicting property (2) for $\mathsf{AX}(\Gamma)$.

We argue that there are axioms $\mathsf{RA}_\omega(\Gamma)$ satisfying the first two of the above requirements, and which are consistent for a variety of forcing classes Γ. These axioms also provide natural examples for the last two requirements. We will come back later on with philosophical considerations outlining why the last two requirements are also natural. We can prove the consistency of $\mathsf{RA}_\omega(\Gamma)$ for forcing classes which are definable in Gödel-Bernays set theory with classes NBG, closed under two steps iterations, weakly iterable (a technical definition asserting that most set sized descending sequences in $\leq_\Gamma$ have lower bounds in Γ, see [2] or [3] for details), and contain all the $< \mathrm{cpd}(\Gamma)$-closed forcings.

The axioms $\mathsf{RA}_\alpha(\Gamma)$ for α an ordinal can be formulated in the Morse Kelley axiomatization of set theory MK as follows:

Definition 5.10. Given an ordinal α and a definable[4] class of forcings Γ closed under two-steps iterations, the axiom $\mathsf{RA}_\alpha(\Gamma)$ holds if for all $\beta < \alpha$ the class

$$\left\{ \mathsf{B} \in \Gamma : \ H_{\gamma^+} \prec H_{\gamma^+}^\mathsf{B} \wedge V^\mathsf{B} \models \mathsf{RA}_\beta(\Gamma) \right\}$$

is dense in $(\Gamma, \leq_\Gamma)$ (where $\gamma = \gamma_\Gamma$).

$\mathsf{RA}_{\mathrm{Ord}}(\Gamma)$ holds if $\mathsf{RA}_\alpha(\Gamma)$ holds for all α.

Remark 5.11. The above definition can be properly formalized in MK (but most likely not in ZFC if α is infinite). The problem is the following: the axioms $\mathsf{RA}_\alpha(\Gamma)$ can be formulated only by means of a transfinite recursion over a well-founded relation which is not set-like. It is a delicate matter to argue that this transfinite recursion can be carried. [2] shows that this is the case if the base theory is MK.

The axiom $\mathsf{RA}_\omega(\Gamma)$ yields generic absoluteness by the following elementary argument:

---

[4] Γ must be definable by a formula with no class quantifier and no class parameter to be on the safe side for what concerns the definability issues regarding the iterated resurrection axioms raised by the remark right after this definition. All usual classes of forcings such as proper, semiproper, stationary set preserving, $< \kappa$-closed, etc.... are definable by formulae satisfying these restrictions.

Theorem 5.12. *Suppose $n \in \omega$, Γ is well behaved, $\mathsf{RA}_n(\Gamma)$ holds, and $\mathsf{B} \in \Gamma$ forces $\mathsf{RA}_n(\Gamma)$. Then $H_{\gamma^+} \prec_n H_{\gamma^+}^{\mathsf{B}}$ (where $\gamma = \gamma_\Gamma$).*

Proof. We proceed by induction on n. Since $\gamma^+ \leq (\gamma^+)^{V^{\mathsf{B}}}$, $H_{\gamma^+} \subseteq H_{\gamma^+}^{\mathsf{B}}$ and the thesis holds for $n = 0$ by the fact that for all transitive structures M, N, if $M \subset N$ then $M \prec_0 N$. Suppose now that $n > 0$, and fix G V-generic for B. By $\mathsf{RA}_n(\Gamma)$, let $\mathsf{C} \in V[G]$ be such that whenever H is $V[G]$-generic for C, $V[G*H] \models \mathsf{RA}_{n-1}(\Gamma)$ and $H_{\gamma^+}^V \prec H_{\gamma^+}^{V[G*H]}$. Hence we have the following diagram:

$$H_{\gamma^+}^V \xrightarrow{\Sigma_\omega} H_{\gamma^+}^{V[G*H]}$$
$$\Sigma_{n-1} \searrow \quad \nearrow \Sigma_{n-1}$$
$$H_{\gamma^+}^{V[G]}$$

obtained by inductive hypothesis applied both on V, $V[G]$ and on $V[G]$, $V[G*H]$ since in all those classes $\mathsf{RA}_{n-1}(\Gamma)$ holds.

Let $\phi \equiv \exists x \psi(x)$ be any Σ_n formula with parameters in $H_{\gamma^+}^V$. First suppose that ϕ holds in V, and fix $\bar{x} \in V$ such that $\psi(\bar{x})$ holds. Since $H_{\gamma^+}^V \prec_{n-1} H_{\gamma^+}^{V[G]}$ and ψ is Π_{n-1}, it follows that $\psi(\bar{x})$ holds in $V[G]$ hence so does ϕ. Now suppose that ϕ holds in $V[G]$ as witnessed by $\bar{x} \in V[G]$. Since $H_{\gamma^+}^{V[G]} \prec_{n-1} H_{\gamma^+}^{V[G*H]}$ it follows that $\psi(\bar{x})$ holds in $V[G*H]$, hence so does ϕ. Since $H_{\gamma^+}^V \prec H_{\gamma^+}^{V[G*H]}$, the formula ϕ holds also in V concluding the proof. □

Corollary 5.13. *Assume Γ is closed under two-steps iterations and contains the $<$ cpd(Γ)-closed forcings. If $\mathsf{RA}_\omega(\Gamma)$ holds, and $\mathsf{B} \in \Gamma$ forces $\mathsf{RA}_\omega(\Gamma)$, then $H_{\gamma^+} \prec H_{\gamma^+}^{\mathsf{B}}$ (where $\gamma = \gamma_\Gamma$).*

Regarding the consistency of the axioms $\mathsf{RA}_\omega(\Gamma)$ we have the following:

Proposition 5.14. *Assume there are class-many Woodin cardinals. Then $\mathsf{RA}_{\mathrm{Ord}}(\Omega)$ holds.*

Theorem 5.15. *$\mathsf{RA}_1(\Gamma)$ implies $H_{\gamma^+} \prec_1 V^{\mathsf{B}}$ for all $\mathsf{B} \in \Gamma$, hence it is a strenghtening of the bounded forcing axiom[5] $\mathsf{BFA}_\gamma(\Gamma)$ (where $\gamma = \gamma_\Gamma$).*

Theorem 5.16 ([2]). *Assume there is a super huge cardinal.[6]*

---
[5]The bounded forcing axiom $\mathsf{BFA}_\gamma(\Gamma)$ asserts that $H_{\gamma^+} \prec_1 V^{\mathsf{B}}$ for all $\mathsf{B} \in \Gamma$.

[6]A cardinal κ is *super huge* iff for every ordinal α there exists an elementary embedding $j : V \to M \subseteq V$ with $\mathrm{crit}(j) = \kappa$, $j(\kappa) > \alpha$ and $^{j(\kappa)}M \subseteq M$.

Then $\mathsf{RA}_{\mathrm{Ord}}(\mathsf{SSP}) + \mathsf{MM}^{++}$ *and* $\mathsf{RA}_{\mathrm{Ord}}(\mathsf{proper}) + \mathsf{PFA}^{++}$ *are consistent. For the consistency of* $\mathsf{RA}_{\mathrm{Ord}}(\mathsf{proper})$ *a Mahlo cardinal suffices.*

Moreover it is also consistent relative to a Mahlo cardinal that $\mathsf{RA}_{\mathrm{Ord}}(\Gamma_\kappa)$ *holds simultaneously for all cardinals* κ *(where* Γ_κ *is the class of* $<\kappa$*-closed forcings)*[7].

In this regard the axioms $\mathsf{RA}_\alpha(\Gamma)$ for $\Gamma \supseteq \Gamma_\kappa$ (Γ_κ being the class of $<\kappa$-closed forcings) appear to be natural companions of the axiom of choice, while the axioms $\mathsf{RA}_{\mathrm{Ord}}(\Omega)$ and $\mathsf{RA}_{\mathrm{Ord}}(\mathsf{SSP}) + \mathsf{MM}$ are natural maximal strengthenings of the axiom of choice at the levels ω and ω_1. Hence it is in our opinion natural to try to isolate classes of forcings Δ_κ as κ ranges among the cardinals such that:

1. $\kappa = \mathrm{cpd}(\Delta_\kappa)$ for all κ.

2. $\Delta_\kappa \supseteq \Gamma_\kappa$ for all κ.

3. $\mathsf{FA}_\kappa(\Delta_\kappa)$ and $\mathsf{RA}_\omega(\Delta_\kappa)$ are simultaneously consistent for all κ.

4. For all cardinals κ, Δ_κ is the largest possible Γ with $\mathrm{cpd}(\Gamma) = \kappa$ for which $\mathsf{FA}_\kappa(\Delta_\kappa)$ and $\mathsf{RA}_\omega(\Delta_\kappa)$ are simultaneously consistent (and if possible for all κ simultaneously).

Compare the above requests with requirements (3) and (4) in the discussion motivating the introduction of the iterated resurrection axioms on page 28. In this regard it appears that we have now a completely satisfactory answer on what are Δ_ω and Δ_{ω_1}: i.e., respectively the class of *all* forcing notions and the class of all SSP-forcing notions.

5.5 Boosting Woodin's absoluteness to $L(\mathrm{Ord}^\kappa)$: the axioms CFA(Γ)

We gave detailed arguments bringing us to axioms which can be stated as density properties of certain category forcings and yielding generic absoluteness results for the theory of H_{κ^+} for various cardinals κ. Carving in Woodin's proof for the generic absoluteness of the Chang model $L(\mathrm{Ord}^\omega)$ one can get an even stronger type of category forcing axioms yielding generic absoluteness results for the Chang models $L(\mathrm{Ord}^\kappa)$. The best result we can currently produce is the following (we refer the interested reader to [1, 3, 30] for details):

---

[7]It is also consistent the following:

$$\mathsf{RA}_{\mathrm{Ord}}(\Omega_{\aleph_0}) + \mathsf{RA}_{\mathrm{Ord}}(\mathsf{SSP}) + \forall \kappa > \omega_1\, \mathsf{RA}_{\mathrm{Ord}}(\Gamma_\kappa)$$

Theorem 5.17. *Let Γ be a κ-suitable class of forcings[8].*
Let MK* *stands for[9]*

> MK + *there are stationarily many inaccessible cardinals.*

There is an axiom[10] CFA(Γ) *which implies* FA$_\kappa(\Gamma)$ *as well as* RA$_{\mathrm{Ord}}(\Gamma)$ *and is such that for any T^* extending*

$$\mathrm{MK}^* + \mathrm{CFA}(\Gamma) + \kappa \text{ is a regular cardinal } + S \subset \kappa,$$

and for any formula $\phi(S)$, the following are equivalent:

1. $T^* \vdash [L(\mathrm{Ord}^\kappa) \models \phi(S)]$,

2. T^* *proves that for some forcing* $\mathsf{B} \in \Gamma$

$$[\![\mathrm{CFA}(\Gamma)]\!]_\mathsf{B} = [\![L(\mathrm{Ord}^\kappa) \models \phi(S)]\!]_\mathsf{B} = 1_\mathsf{B}.$$

We also have that

Theorem 5.18 ([1, 3]). *Assume Γ is κ-suitable. Then* CFA(Γ) *is consistent relative to the existence of a 2-superhuge cardinal[11].*

While the definition of κ-suitable Γ is rather delicate, it can be shown that many interesting classes are ω_1-suitable, among others: proper, semiproper, ω^ω-bounding and (semi)proper, preserving a suslin tree and (semi)proper. [1] contains a detailed list of classes which are ω_1-suitable. It is not known whether there can be κ-suitable classes Γ for some $\kappa > \omega_1$.

---

[8]This is a lenghty and technical definition; roughly it requires that:
- Γ is closed under two steps iterations, and contains all the $< \kappa$-closed posets (where $\kappa = \mathrm{cpd}(\Gamma)$),
- there is an iteration theorem granting that all set sized iterations of posets in Γ has a limit in Γ,
- Γ is defined by a syntactically simple formula (i.e. Σ_2 in the Levy hierarchy of formulae),
- Γ has a dense set of Γ-rigid elements (i.e. the $\mathsf{B} \in \Gamma$ admitting at most one $i : \mathsf{B} \to \mathsf{C}$ witnessing that $\mathsf{C} \leq_\Gamma \mathsf{B}$ for all $\mathsf{C} \in \Gamma$ form a dense subclass of Γ).

[9]In MK one can define the club filter on the class Ord, hence the notion of stationarity for classes of ordinals makes sense.

[10]CFA(Γ) can be formulated as a density property of the class forcing $(\Gamma, \leq_\Gamma)$.

[11]A cardinal κ is 2-superhuge if it is supercompact and this can be witnessed by 2-huge embeddings.

6 Some open questions

Here is a list of questions for which we do not have many clues.....

1. What are the Γ which are κ-suitable for a given cardinal $\kappa > \aleph_1$ (i.e. such that $\mathsf{CFA}(\Gamma)$ is consistent)?

2. Do they even exist for $\kappa > \aleph_1$?

3. In case they do exist for some $\kappa > \aleph_1$, do we always have a unique maximal Γ such that $\mathrm{cpd}(\Gamma) = \kappa$ as it is the case for $\kappa = \aleph_0$ or $\kappa = \aleph_1$?

Any interesting iteration theorem for a class $\Gamma \supseteq \Gamma_{\omega_2}$ closed under two steps iterations can be used to prove that $\mathsf{RA}_{\mathrm{Ord}}(\Gamma)$ is consistent relative to suitable large cardinal assumptions and freezes the theory of H_{ω_3} with respect to forcings in Γ preserving $\mathsf{RA}_\omega(\Gamma)$ (see [2]). It is nonetheless still a mystery which classes $\Gamma \supseteq \Gamma_{\omega_2}$ can give us a nice iteration theorem, even if the recent works, by Neeman, Asperò, Krueger, Mota, Velickovic and others are starting to shed some light on this problem (see among others [14, 15, 19]).

We can dare to be more ambitious and replicate the above type of issue at a much higher level of the set theoretic hierarchy. There is a growing set of results regarding the first-order theory of $L(V_{\lambda+1})$ assuming λ is a very large cardinal (i.e., for example admitting an elementary $j : L(V_{\lambda+1}) \to L(V_{\lambda+1})$ with critical point smaller than λ, see among others [6, 7, 35]). It appears that large fragments of this theory are generically invariant with respect to a great variety of forcings.

> Assume $j : L(V_{\lambda+1}) \to L(V_{\lambda+1})$ is elementary with critical point smaller than λ. Can any of the results presented in this paper be of any use in the study of which type of generic absoluteness results may hold at the level of $L(V_{\lambda+1})$?

The reader is referred to [1, 3, 2, 30, 32] for further examinations of these topics.

References

[1] David Asperó and Matteo Viale, *Category forcings*, In preparation, 2017.

[2] G. Audrito and M. Viale, *Absoluteness via resurrection*, J. Math. Log. (2017), On-line first.

[3] Giorgio Audrito, Raphaël Carroy, Silvia Steila, and Matteo Viale, *Iterated forcing, category forcings, generic ultrapowers, generic absoluteness*, Book in preparation, 2017.

[4] J. E. Baumgartner, *All $\aleph_1$-dense sets of reals can be isomorphic*, Fund. Math. **79** (1973), no. 2, 101–106. MR 317934

[5] A. E. Caicedo and B. Veličković, *The bounded proper forcing axiom and well orderings of the reals*, Math. Res. Lett. **13** (2006), no. 2-3, 393–408. MR 2231126 (2007d:03076)

[6] Vincenzo Dimonte and Sy-David Friedman, *Rank-into-rank hypotheses and the failure of GCH*, Arch. Math. Logic **53** (2014), no. 3-4, 351–366. MR 3194669

[7] Vincenzo Dimonte and Liuzhen Wu, *A general tool for consistency results related to I1*, Eur. J. Math. **2** (2016), no. 2, 474–492. MR 3498993

[8] I. Farah, *All automorphisms of the Calkin algebra are inner*, Ann. of Math. (2) **173** (2011), no. 2, 619–661. MR 2776359

[9] Qi Feng, Menachem Magidor, and Hugh Woodin, *Universally Baire sets of reals*, Set theory of the continuum (Berkeley, CA, 1989), Math. Sci. Res. Inst. Publ., vol. 26, Springer, New York, 1992, pp. 203–242. MR 1233821

[10] M. Foreman, M. Magidor, and S. Shelah, *Martin's maximum, saturated ideals, and nonregular ultrafilters. I*, Ann. of Math. (2) **127** (1988), no. 1, 1–47. MR 924672

[11] Steven Givant and Paul Halmos, *Introduction to Boolean algebras*, Undergraduate Texts in Mathematics, Springer, New York, 2009. MR 2466574

[12] Joel David Hamkins and Thomas A. Johnstone, *Resurrection axioms and uplifting cardinals*, Arch. Math. Logic **53** (2014), no. 3-4, 463–485. MR 3194674

[13] T. Jech, *Set theory*, Springer Monographs in Mathematics, Springer, Berlin, 2003, The third millennium edition, revised and expanded. MR 1940513

[14] John Krueger, *Adding a club with finite conditions, Part II*, Arch. Math. Logic **54** (2015), no. 1-2, 161–172. MR 3304741

[15] John Krueger and Miguel Angel Mota, *Coherent adequate forcing and preserving CH*, J. Math. Log. **15** (2015), no. 2, 1550005, 34. MR 3447934

[16] Paul B. Larson, *The stationary tower*, University Lecture Series, vol. 32, American Mathematical Society, Providence, RI, 2004, Notes on a course by W. Hugh Woodin. MR 2069032

[17] J. T. Moore, *Set mapping reflection*, J. Math. Log. **5** (2005), no. 1, 87–97. MR 2151584

[18] ———, *A five element basis for the uncountable linear orders*, Ann. of Math. (2) **163** (2006), no. 2, 669–688. MR 2199228

[19] Itay Neeman, *Forcing with sequences of models of two types*, Notre Dame J. Form. Log. **55** (2014), no. 2, 265–298. MR 3201836

[20] Francesco Parente, *Boolean valued models, saturation, forcing axioms*, (2015), Unpublished Master Thesis, University of Pisa 2015.

[21] Helena Rasiowa and Roman Sikorski, *The mathematics of metamathematics*, third ed., PWN—Polish Scientific Publishers, Warsaw, 1970, Monografie Matematyczne, Tom 41. MR 0344067

[22] S. Shelah, *Infinite abelian groups, Whitehead problem and some constructions*, Israel J. Math. **18** (1974), 243–256. MR 0357114 (50 #9582)

[23] _____, *Decomposing uncountable squares to countably many chains*, J. Combinatorial Theory Ser. A **21** (1976), no. 1, 110–114. MR 409196

[24] S. Todorcevic, *Basis problems in combinatorial set theory*, Proceedings of the International Congress of Mathematicians, Vol. II (Berlin, 1998), no. Extra Vol. II, 1998, pp. 43–52 (electronic). MR 1648055

[25] _____, *Generic absoluteness and the continuum*, Math. Res. Lett. **9** (2002), no. 4, 465–471. MR 1928866 (2003f:03067)

[26] A. Vaccaro and M. Viale, *Generic absoluteness and boolean names for elements of a Polish space*, Boll Unione Mat Ital (2017).

[27] Andrea Vaccaro, *C*-algebras and B-names for complex numbers*, (2015), Unpublished Master Thesis, University of Pisa 2015.

[28] B. Veličković, *Forcing axioms and stationary sets*, Adv. Math. **94** (1992), no. 2, 256–284. MR 1174395 (93k:03045)

[29] Matteo Viale, *A family of covering properties*, Math. Res. Lett. **15** (2008), no. 2, 221–238. MR 2385636

[30] _____, *Category forcings, MM^{+++}, and generic absoluteness for the theory of strong forcing axioms*, J. Amer. Math. Soc. **29** (2016), no. 3, 675–728. MR 3486170

[31] _____, *Forcing the truth of a weak form of schanuelÔs conjecture*, Confluentes Math. **8** (2016), no. 2, 59–83.

[32] _____, *Martin's maximum revisited*, Arch. Math. Logic **55** (2016), no. 1-2, 295–317. MR 3453587

[33] _____, *Notes on forcing*, 2017.

[34] W. H. Woodin, *The axiom of determinacy, forcing axioms, and the nonstationary ideal*, de Gruyter Series in Logic and its Applications, vol. 1, Walter de Gruyter & Co., Berlin, 1999. MR 1713438

[35] W. Hugh Woodin, *Suitable extender models II: beyond ω-huge*, J. Math. Log. **11** (2011), no. 2, 115–436. MR 2914848

The Strength of Abstraction with Predicative Comprehension

Sean Walsh

Department of Logic and Philosophy of Science, 5100 Social Science Plaza, University of California, Irvine, Irvine, CA 92697-5100, U.S.A., swalsh108@gmail.com or walsh108@uci.edu

Abstract

Frege's theorem says that second-order Peano arithmetic is interpretable in Hume's Principle and full impredicative comprehension. Hume's Principle is one example of an *abstraction principle*, while another paradigmatic example is Basic Law V from Frege's *Grundgesetze*. In this paper we study the strength of abstraction principles in the presence of predicative restrictions on the comprehension schema, and in particular we study a predicative Fregean theory which contains all the abstraction principles whose underlying equivalence relations can be proven to be equivalence relations in a weak background second-order logic. We show that this predicative Fregean theory interprets second-order Peano arithmetic (cf. Theorem 3.2).

Keywords: abstraction principles, logicism, Frege, predicativity

1 Introduction

The main result of this paper is a predicative analogue of Frege's Theorem (cf. Theorem 3.2). Roughly, Frege's theorem says that one can recover all of second-order Peano arithmetic using only the resources of Hume's Principle and second-order logic. This result was adumbrated in Frege's *Grundlagen* of 1884 ([9], [11]) and the contemporary interest in this result is due to Wright's 1983 book *Frege's Conception of Numbers as Objects* ([32]). For more on the history of this theorem, see the careful discussion and references in Heck [18] pp. 4-6 and Beth [1].

This paper was originally published as: The strength of abstraction with predicative comprehension. *The Bulletin of Symbolic Logic*, 22(1):105–120, 2016. The copyright is held by the Association for Symbolic Logic and is being reprinted with their permission.

More formally, Frege's theorem says that second-order Peano arithmetic is interpretable in second-order logic plus the following axiom, wherein the cardinality operator # is a type-lowering function from second-order entities to first-order entities:

$$\textit{Hume's Principle}: \forall\, X, Y\ (\#X = \#Y \leftrightarrow \exists\text{ bijection } f : X \to Y) \qquad (1.1)$$

Of course, one theory is said to be *interpretable* in another when the primitives of the interpreted theory can be defined in terms of the resources of the interpreting theory so that the translations of theorems of the interpreted theory are theorems of the interpreting theory (cf. [28] §2 or [20] pp. 96-97 or [15] pp. 148-149 or [25] §2.2). For a proof of Frege's Theorem, see Chapter 4 of Wright's book ([32]) or §2.2 pp. 1688 ff of [27].

The second-order logic used in the traditional proof of Frege's Theorem crucially includes impredicative instances of the comprehension schema. Intuitively, the comprehension schema says that every formula $\varphi(x)$ in one free first-order variable determines a second-order entity:

$$\exists\, F\, \forall\, x\, (Fx \leftrightarrow \varphi(x)) \qquad (1.2)$$

The traditional proof of Frege's Theorem uses instances of this comprehension schema in which some of the formulas in question contain higher-order quantifiers (cf. [27] p. 1690 equations (44)-(45)). However, there is a long tradition of *predicative mathematics*, in which one attempts to ascertain how much one can accomplish without directly appealing to such instances of the comprehension schema. This was the perspective of Weyl's great book *Das Kontinuum* ([31]) and has been further developed in the work of Feferman ([6], [7]). Many of us today learn and know of this tradition due to its close relation to the system ACA_0 of Friedman and Simpson's project of reverse mathematics ([13], [24]).

However, outside of the inherent interest in predicative mathematics, considerations related to Frege's philosophy of mathematics likewise suggest adopting the predicative perspective. For, Wright and Hale ([16], cf. [4]) have emphasized that Hume's Principle (1.1) is a special instance of the following:

$$A[E]: \quad \forall\, X, Y\ (\partial_E(X) = \partial_E(Y) \leftrightarrow E(X, Y)) \qquad (1.3)$$

wherein $E(X, Y)$ is a formula of second-order logic and ∂_E is a type-lowering operator taking second-order entities and returning first-order entities. These principles were called *abstraction principles* by Wright and Hale, who pointed out that the following crucial fifth axiom of Frege's *Grundgesetze* of 1893 and 1903 ([10], [12]) was also an abstraction principle:

$$\text{Basic Law } V : \quad \forall\, X, Y\, (\partial(X) = \partial(Y) \leftrightarrow X = Y) \tag{1.4}$$

The operator ∂ as governed by Basic Law V is called the *extension* operator and the first-order entities in its range are called *extensions*. Regrettably, there is no standard notation for the extension operator, and so some authors write $\S X$ in lieu of $\partial(X)$. In what follows, the symbol ∂ without any subscripts will be reserved for the extension operator, whereas the subscripted symbols ∂_E will serve as the notation for the type-lowering operators present in arbitrary abstraction principles (1.3).

While the Russell paradox shows that Basic Law V is inconsistent with the full comprehension schema (1.2) (cf. [27] p. 1682), nevertheless Basic Law V is consistent with predicative restrictions, as was shown by Parsons ([23]), Heck ([17]), and Ferreira-Wehmeier ([8]). This thus suggests the project of understanding whether there is a version of Frege's theorem centered around the consistent predicative fragments of the *Grundgesetze*. This project has been pursued in the last decades by many authors such as Heck ([17]), Ganea ([14]), and Visser ([26]). Their results concerned the restriction of the comprehension schema (1.2) to the case where no higher-order quantifiers are permitted. One result from this body of work says that Basic Law V (1.4) coupled with this restriction on the comprehension schema is mutually interpretable with Robinson's Q. Roughly, Robinson's Q is the fragment of first-order Peano arithmetic obtained by removing all the induction axioms. (For a precise definition of Robinson's Q, see [15] p. 28, [24] p. 4, [27] p. 1680, [28] p. 106). Additional work by Visser allows for further rounds of comprehension and results in systems mutually interpretable with Robinson's Q plus iterations of the consistency statement for this theory, which are likewise known to be interpretable in other weak arithmetics ([26] p. 147). In his 2005 book ([3]), Burgess surveys these kinds of developments, and writes:

> [...] I believe that no one working in the area seriously expects to get very much further in the sequence Q_m while working in predicative Fregean theories of whatever kind ([3] p. 145).

Here Q_m is the expansion of Robinson's Q by finitely many primitive recursive function symbols and their defining equations along with induction for bounded formulas ([3] pp. 60-63), so that Burgess records the prediction that predicative Fregean theories will be interpretable in weak arithmetics.

The main result of this paper suggests that this prediction was wrong, and that predicative Fregean theories can interpret strong theories of arithmetic (cf. Theorem 3.2). While we turn presently to developing the definitions needed to precisely state this result, let us say by way of anticipation that part of the idea is to work both with (i) an expanded notion of a "Fregean theory," so that it includes several abstraction principles, such as Basic Law V, in addition to Hume's Principle, and (ii) an expanded notion of "predicativity," in which one allows some controlled instances of higher-order quantifiers within the comprehension schema (1.2). Hence, of course, it might be that Burgess and others had merely conjectured that predicative Fregean theories in a more limited sense were comparatively weak.

This paper is part of a series of three papers, the other two being [29] and [30]. These papers collectively constitute a sequel to our paper [27], particularly as it concerns the methods and components related to Basic Law V. In that earlier paper, we showed that Hume's Principle (1.1) with predicative comprehension did not interpret second-order Peano arithmetic with predicative comprehension (cf. [27] p. 1704). Hence at the outset of that paper, we said that "in this specific sense there is no predicative version of Frege's Theorem" ([27] p. 1679). The main result of this present paper (cf. Theorem 3.2) is that when we enlarge the theory to a more inclusive class of abstraction principles containing Basic Law V, we do in fact succeed in recovering arithmetic.

This paper depends on [29] only in that the consistency of the predicative Fregean theory which we study here was established in that earlier paper (cf. discussion at close of next section). In the paper [30], we focus on embedding the system of the *Grundgesetze* into a system of intensional logic. The alternative perspective of [30] then suggests viewing the consistent fragments of the *Grundgesetze* as a species of intensional logic, as opposed to an instance of an abstraction principle.

This paper is organized as follows. In §2 we set out the definitions of the predicative Fregean theory. In §3 it is shown how this predicative Fregean theory can recover full second-order Peano arithmetic. In §4 it is noted that some theories which are conceptually proximate to the predicative Fregean theory are nonetheless inconsistent.

2 Defining a theory of abstraction with predicative comprehension

The predicative Fregean theory with which we work in this paper is developed within the framework of second-order logic. The language L_0 of the background second-order logic is an ω-sorted system with sorts for first-order entities, unary second-order entities, binary second-order entities etc. Further, following the Fregean tradition, the first-order entities are called *objects*, the unary second-order entities are called *concepts*, and the n-ary second-order entities for $n \geq 1$ are called *n-ary concepts*.

Rather than introduce any primitive notation for the different sorts, we rather employ the convention of using distinctive variables for each sort: objects are written with lower-case Roman letters $x, y, z, a, b, c \ldots$, concepts are written with upper-case Roman letters $X, Y, Z, A, B, C, F, G, H, U, \ldots$, n-ary concepts for $n > 1$ are written with the upper case Roman letters R, S, T, and n-ary concepts are written with the Roman letters f, g, h when they are graphs of functions.

Besides the sorts, the other basic primitive of the signature of the background second-order logic L_0 are the predication relations. One writes Xa to indicate that object a has property or concept X. Likewise, there are predication relations for n-ary concepts, which we write as $R(a_1, \ldots, a_n)$. The final element of the signature L_0 of the background second-order logic are the projection symbols. The basic idea is that one wants, primitive in the signature L_0, a way to move from the binary concept R and the object a to its projection $R[a] = \{b : R(a,b)\}$. We assume that the signature L_0 of the background second-order logic is equipped with symbols $(R, a_1, \ldots, a_m) \mapsto R[a_1, \ldots, a_m]$ from $(m+n)$-ary concepts R and an m-tuple of objects $(a_1, \ldots, a_m)$ to an n-ary concept $R[a_1, \ldots, a_m] = \{(b_1, \ldots, b_n) : R(a_1, \ldots, a_m, b_1, \ldots, b_n)\}$. Further, typically in what follows we avail ourselves of the tuple notation $\bar{a} = a_1, \ldots, a_n$ and thus write predication and projection more succinctly as $R(\bar{a})$ and $R[\bar{a}]$, respectively.

All this in place, we can then formally define the signature L_0 of the background second-order logic as follows:

Definition 2.1. *The signature L_0 of the background second-order logic is a many-sorted signature which contains (i) a sort for objects and for each $n \geq 1$ a sort for n-ary concepts, (ii) for each $n \geq 1$, an $(n+1)$-ary predication relation symbol $R(a_1, \ldots, a_n)$ which holds between an n-ary concept R and an n-tuple of objects $a_1, \ldots, a_n$, and (iii) for each $n, m \geq 1$, an $(m+1)$-ary projection function symbol $(R, a_1, \ldots, a_m) \mapsto R[a_1, \ldots, a_m]$ from an $(m+n)$-ary concept R and an m-tuple of objects $(a_1, \ldots, a_m)$ to an n-ary concept $R[a_1, \ldots, a_m]$.*

As is usual in many-sorted signatures, we adopt the convention that each sort has its own identity symbol, so that technically cross-sortal identities are not well-formed. But we continue to write all identities with the usual symbol "=" for the ease of readability.

The expansions of second-order logic with which we work are designed to handle abstraction principles (1.3). Hence, suppose that L is an expansion of L_0. Suppose that $E(R, S)$ is an L-formula with two free n-ary relation variables for some $n \geq 1$, with all free variables of $E(R, S)$ explicitly displayed. Then we may expand L to a signature $L[\partial_E]$ which contains a new function symbol ∂_E which takes n-ary concepts R and returns the object $\partial_E(R)$. Then the following axiom, called *the abstraction principle associated to* E, is an $L[\partial_E]$-sentence:

$$A[E]: \quad \forall\, R, S\ (\partial_E(R) = \partial_E(S) \leftrightarrow E(R, S)) \tag{2.1}$$

This generalizes the notion of an abstraction principle (1.3) described in the previous section in that the domain of the operator ∂_E can be n-ary concepts for any specific $n \geq 1$.

This generalization is warranted by several key examples, such as that of ordinals. Let R be a binary concept and let Field(R) be the unary concept F such that Fx iff there is a y such that Rxy or Ryx. Then consider the following formula $E(R, S)$ on binary concepts:

$$[(\mathrm{Field}(R), R) \models \mathrm{wo} \vee (\mathrm{Field}(S), S) \models \mathrm{wo}] \rightarrow \tag{2.2}$$
$$\exists\ \mathrm{isomorphism}\ f : (\mathrm{Field}(R), R) \rightarrow (\mathrm{Field}(S), S)$$

In this, "wo" denotes the natural sentence in the signature of second-order logic which says that a binary concept is a well-order, i.e. a linear order such that every non-empty subconcept of its domain has a least element. It's not too difficult to see that $E(R, S)$ is an equivalence relation on binary concepts, and that two well-orders will be E-equivalent if and only if they are order-isomorphic. Just as the Russell paradox shows that Basic Law V (1.4) is inconsistent with the full comprehension schema, so one can use the Burali-Forti paradox to show that $A[E]$ for this E in equation (2.2) is inconsistent with the full comprehension schema (cf. [19] p. 138 footnote, [2] pp. 214, 311). To handle these abstraction principles we need to adopt restrictions on the comprehension schema, to which we presently turn.

There are three traditional predicative varieties of the comprehension schema: the first-order comprehension schema, the Δ_1^1-comprehension schema, and the Σ_1^1-choice schema (cf. [24] VII.5-6, [27] Definition 5 p. 1683). However, to make the comparison with the full comprehension schema (1.2) precise, we should restate it to include not only concepts but n-ary concepts for all $n \geq 1$ and to indicate its explicit dependence on a signature:

Definition 2.2. *Suppose that L is an expansion of L_0. Then the* Full Comprehension Schema *for L-formulas consists of all axioms of the form* $\exists\, R\, \forall\, \overline{a}\ (R\overline{a} \leftrightarrow \varphi(\overline{a}))$, *wherein $\varphi(\overline{x})$ is an L-formula, perhaps with parameters, and $\overline{x}$ abbreviates $(x_1, \ldots, x_n)$ and R is an n-ary concept variable for $n \geq 1$ that does not appear free in $\varphi(\overline{x})$.*

The most restrictive predicative version of the comprehension schema is then the following, where the idea is that *no* higher-order quantifiers are allowed in the formulas:

Definition 2.3. *Suppose that L is an expansion of L_0. The* First-Order Comprehension Schema *for L-formulas consists of all axioms of the form $\exists R \, \forall \, \overline{a} \, (R\overline{a} \leftrightarrow \varphi(\overline{a}))$, wherein $\varphi(\overline{x})$ is an L-formula with no second-order quantifiers but perhaps with parameters, and $\overline{x}$ abbreviates $(x_1, \ldots, x_n)$ and R is an n-ary concept variable for $n \geq 1$ that does not appear free in $\varphi(\overline{x})$.*

A more liberal version of the comprehension schema is the Δ_1^1-comprehension schema. A Σ_1^1-formula (resp. Π_1^1-formula) is one which begins with a block of existential quantifiers (resp. universal quantifiers) over n-ary concepts for various $n \geq 1$ and which contains no further second-order quantifiers. One then defines:

Definition 2.4. *Suppose that L is an expansion of L_0. Then the Δ_1^1-Comprehension Schema for L-formulas consists of all axioms of the form*

$$(\forall \, \overline{x} \, \varphi(\overline{x}) \leftrightarrow \psi(\overline{x})) \rightarrow \exists R \, \forall \, \overline{a} \, (R\overline{a} \leftrightarrow \varphi(\overline{a})) \tag{2.3}$$

wherein $\varphi(\overline{x})$ is a Σ_1^1-formula in the signature of L and $\psi(\overline{x})$ is a Π_1^1-formula in the signature of L that may contain parameters, and $\overline{x}$ abbreviates $(x_1, \ldots, x_n)$, and R is an n-ary concept variable for $n \geq 1$ that does not appear free in $\varphi(\overline{x})$ or $\psi(\overline{x})$.

Finally, traditionally one also includes amongst the predicative systems the following choice principle:

Definition 2.5. *Suppose that L is an expansion of L_0. The Σ_1^1-Choice Schema for L-formulas consists of all axioms of the form*

$$[\forall \, \overline{x} \, \exists \, R' \, \varphi(R', \overline{x})] \rightarrow \exists R \, [\forall \, \overline{x} \, \varphi(R[\overline{x}], \overline{x})] \tag{2.4}$$

wherein the L-formula $\varphi(R', \overline{x})$ is Σ_1^1, perhaps with parameters, and $\overline{x}$ abbreviates $(x_1, \ldots, x_m)$ and R is an $(m+n)$-ary concept variable for $n, m \geq 1$ that does not appear free in $\varphi(R', \overline{x})$ where R' is an n-ary concept variable.

The Σ_1^1-Choice Schema and the First-Order Comprehension Schema together imply the Δ_1^1-Comprehension Schema (cf. [24] Theorem V.8.3 pp. 205-206, [27] Proposition 6 p. 1683). Hence, even if one's primary interest is in the latter schema, typically theories are axiomatized with the two former schemas since they are deductively stronger, and that is how we proceed in this paper.

To the signature L_0 of the weak background second-order logic, we want to associate a certain weak background L_0-theory. Some of the axioms of this background theory axiomatize the behavior of the predication symbols and the projection symbols. For each $m \geq 1$, one has the following *extensionality axiom*, wherein R, S are m-ary concept variables and $\overline{a} = a_1, \ldots, a_m$ are object variables:

$$\forall \, R, S \, [R = S \leftrightarrow (\forall \overline{a} \, (R(\overline{a}) \leftrightarrow S(\overline{a})))] \tag{2.5}$$

But it should be noted that some authors don't explicitly include the identity symbol for concepts or higher-order entities and simply take it as an abbreviation for coextensionality (cf. [24] pp. 2-3, [3] pp. 14-15). Second, for each $n, m \geq 1$, one has the following *projection axioms* governing the behavior of the projection symbols, wherein R is an $(m+n)$-ary concept variable and $\bar{a} = a_1, \ldots, a_m, \bar{b} = b_1, \ldots, b_n$ are object variables:

$$\forall R \, \forall \, \bar{a}, \bar{b} \, [(R[\bar{a}])(\bar{b}) \leftrightarrow R(\bar{a}, \bar{b})] \tag{2.6}$$

Finally, with all this in place, we can define the weak background theory of second-order logic:

Definition 2.6. *The weak background theory of second-order logic Σ_1^1-OS is L_0-theory consisting of (i) the extensionality axioms (2.5) and the projection axioms (2.6) and (ii) the Σ_1^1-Choice Schema for L_0-formulas (Definition 2.5) and (iii) the First-Order Comprehension Schema for L_0-formulas (Definition 2.3).*

In the theory Σ_1^1-OS and its extensions, we use standard abbreviations for various operations on concepts, for instance $X \cap Y = \{z : Xz \, \& \, Yz\}$ and $\{x\} = \{z : z = x\}$ and $X \times Y = \{(x, y) : Xx \, \& \, Yy\}$ and $\emptyset = \{x : x \neq x\}$. In general, we use $\{x : \Phi(x)\}$ as an abbreviation for the concept F such that Fx iff $\Phi(x)$, assuming that $\Phi(x)$ is a formula which falls under one of the comprehension principles available in the theory in which we are working.

This weak background theory Σ_1^1-OS of second-order logic is used to define the following Fregean theory at issue in this paper. If $E(R, S)$ is an L_0-formula with two free n_E-ary concept variables and no further free variables, then we let Equiv(E) abbreviate the L_0-sentence expressive of E being an equivalence relation on n_E-ary concepts, i.e. the universal closure of the following, wherein R, S, T are n_E-ary concept variables:

$$[E(R, R) \, \& \, (E(R, S) \rightarrow E(S, R)) \, \& \, ((E(R, S) \, \& \, E(S, T)) \rightarrow E(R, T))] \tag{2.7}$$

Then consider the following collection of L_0-formulas which consists of all the L_0-formulas $E(R, S)$ with two free n_E-ary concept variables and no further free variables such that Σ_1^1-OS proves Equiv(E):

$$\text{ProvEquiv}(L_0) = \{E(R, S) \text{ is an } L_0 \text{ formula} : \Sigma_1^1\text{-OS} \vdash \text{Equiv}(E)\} \tag{2.8}$$

Then define the following expansion of L_1 of L_0:

Definition 2.7. *Let L_1 consist of the expansion of the signature L_0 (2.1) by a new function symbol ∂_E from n_E-ary concepts to objects for each E from* ProvEquiv(L_0) *(2.8).*

Then we define the predicative theory as follows:

Definition 2.8. *The predicative Fregean theory, abbreviated* PFT*, is the L_1-theory consisting of (i) the extensionality axioms (2.5) and the projection axioms (2.6) and (ii) the Σ_1^1-Choice Schema for L_1-formulas (Definition 2.5) and (iii) the First-Order Comprehension Schema for L_1-formulas (Definition 2.3), and (iv) the abstraction principle $A[E]$ (2.1) for each E from* $\mathrm{ProvEquiv}(L_0)$ *(2.8).*

Hence, the theory PFT is a recursively enumerable theory in a recursively enumerable signature L_1. If one desired a recursive signature, one could alternatively define L_1 to consist of function symbols ∂_E from n_E-ary concepts to objects for *each* L_0-formula E, regardless of whether it was in $\mathrm{ProvEquiv}(L_0)$ (2.8). This is because clause (iv) in Definition 2.8 only includes the abstraction principle $A[E]$ (2.1) when the formula E is in fact in the set $\mathrm{ProvEquiv}(L_0)$ (2.8).

While this definition is technically precise, the niceties ought not obscure the intuitiveness of the motivating idea. For, the idea behind this predicative Fregean theory is that it conjoins traditional predicative constraints on comprehension together with the idea that abstraction principles associated to certain L_0-formulae are always available. More capaciously: if we start from weak background theory of second-order logic Σ_1^1-OS and if we can prove in this theory that an L_0-formula $E(R, S)$ in the signature of this weak background logic is an equivalence relation on n_E-ary concepts for some $n_E \geq 1$, then the predicative Fregean theory PFT includes the abstraction principle $A[E]$ (2.1) associated to E. Hence the theory PFT includes the abstraction principles associated to number, extension, and ordinal, namely Hume's Principle (1.1), Basic Law V (1.4) and the abstraction principle associated to ordinals (cf. (2.2) above).

One of the aims of the earlier paper [29] was to establish the following:

Theorem 2.9. *The theory* PFT *is consistent.*

Proof. Let $E_1, \ldots, E_n, \ldots$ enumerate the elements of the collection $\mathrm{ProvEquiv}(L_0)$ from equation (2.8). By compactness, it suffices to establish, for each $n \geq 1$, the consistency of the subsystem of PFT which is formed by restricting part (iv) of the Definition of PFT to the abstraction principles $A[E_1], \ldots, A[E_n]$. But then this theory is a subtheory of the theory which, in the paper [29], we called $\Sigma_1^1-[\mathrm{E}_1, \ldots, \mathrm{E_n}]\mathrm{A} + \mathrm{SO} + \mathrm{GC}$. The consistency of this theory was established in the Joint Consistency Theorem of that paper. □

3 Interpreting second-order arithmetic in the theory

While the predicative Fregean Theory only explicitly includes predicative instances of the comprehension schema for L_0-formulas, surprisingly it is able to deductively recover all instances of the Full Comprehension Schema for L_0-formulas.

Theorem 3.1. PFT *proves each instance of the Full Comprehension Schema for L_0-formulas.*

Proof. Let $\Phi(x, G)$ be an L_0-formula with all free variables displayed, wherein x is an object variable and G is a unary concept variable. Let us first show that PFT proves the following instance of the Full Comprehension Schema for L_0-formulas (Definition 2.2):

$$\forall\, G\, \exists\, F\, \forall\, x\, (Fx \leftrightarrow \Phi(x, G)) \tag{3.1}$$

After we finish the proof of this instance, we'll comment on how to establish the general case.

First consider the following L_0-formulas $\mu(R, S), \nu(R, S)$ with all free variables displayed, where R, S are binary concept variables:

$$\mu(R, S) \equiv [\exists\,!\, x, G \text{ with } R = \{x\} \times G]\, \&\, [\exists\,!\, y, H \text{ with } S = \{y\} \times H]$$
$$\&\, \forall\, x, G, y, H\, [(R = \{x\} \times G\, \&\, S = \{y\} \times H) \to (\Phi(x, G) \leftrightarrow \Phi(y, H))]$$
$$\nu(R, S) \equiv \neg[\exists\,!\, x, G \text{ with } R = \{x\} \times G]\, \&\, \neg[\exists\,!\, y, H \text{ with } S = \{y\} \times H]$$

In this, the identity $R = \{x\} \times G$ is an abbreviation for the claim that

$$\forall\, a, b\, (R(a, b) \leftrightarrow ((a = x)\, \&\, Gb)) \tag{3.2}$$

Hence, $\mu(R, S)$ expresses that R can be written uniquely as $\{x\} \times G$ for some x, G, while S can be written uniquely as $\{y\} \times H$ for some y, H, and that $\Phi(x, G) \leftrightarrow \Phi(y, H)$. The circumstance in which a binary relation R can be written as $\{x\} \times G$ but not *uniquely* so is when G is empty, since in this case $\{x\} \times G = \{x'\} \times G$ for any objects x, x'. Finally, consider the following L_0-formula $E(R, S)$ where again R, S are binary concept variables and all free variables are displayed:

$$E(R, S) \equiv (\mu(R, S) \vee \nu(R, S)) \tag{3.3}$$

The weak background theory Σ_1^1-OS proves that $E(R, S)$ is an equivalence relation on binary concepts. For reflexivity, either R can be written uniquely as $\{x\} \times G$ for some x, G, or not. If so, then one trivially has $\Phi(x, G) \leftrightarrow \Phi(x, G)$. This then implies $\mu(R, R)$ and so $E(R, R)$. If not, then of course $\nu(R, R)$ and so $E(R, R)$.

For symmetry, it simply suffices to note that both μ and ν are symmetric in that $\mu(R,S)$ implies $\mu(S,R)$ and likewise for ν. For transitivity, suppose that $E(R,S)$ and $E(S,T)$. Because of the disjunctive definition of E in (3.3), there are three cases to consider. First suppose that $\mu(R,S)$ and $\mu(S,T)$. Then we may uniquely write $R = \{x\} \times G, S = \{y\} \times H, T = \{z\} \times I$, and from $\Phi(x,G) \leftrightarrow \Phi(y,H)$ and $\Phi(y,H) \leftrightarrow \Phi(z,I)$ we may conclude that $\Phi(x,G) \leftrightarrow \Phi(z,I)$. Hence we then have $\mu(R,T)$ and thus $E(R,T)$. Second suppose that $\nu(R,S)$ and $\nu(S,T)$. These two assumptions imply that we can't write any of R,S,T uniquely as the product of a singleton and a unary concept, and hence that $\nu(R,T)$ and $E(R,T)$. Finally, suppose that $\mu(R,S)$ and $\nu(S,T)$ (or vice-versa). But this case leads to a contradiction, since $\mu(R,S)$ implies that we can write S uniquely as the product of a singleton and a unary concept, while $\nu(S,T)$ says that we can't. Hence $E(R,S)$ is indeed an equivalence relation on binary concepts, and provably so in the weak background theory Σ_1^1-OS.

Then the L_0-formula $E(R,S)$ is in the set $\text{ProvEquiv}(L_0)$ (2.8). Hence the theory PFT contains the abstraction principle $A[E]$ (2.1). Before we verify (3.1), let us introduce another abstraction principle. Consider the following L_0-formulas $\mu'(X,Y), \nu'(X,Y)$ with all free variables displayed, where X, Y are unary concept variables:

$$\mu'(X,Y) \equiv \exists\, x\, \exists\, y\; X = \{x\}\; \&\; Y = \{y\}\; \&\; (\Phi(x,\emptyset) \leftrightarrow \Phi(y,\emptyset))$$
$$\nu'(X,Y) \equiv \neg(\exists\, x\; X = \{x\})\; \&\; \neg(\exists\, y\; Y = \{y\})$$

Then consider the following L_0-formula $E'(X,Y)$ where again X,Y are unary concept variables and all free variables are displayed:

$$E'(X,Y) \equiv (\mu'(X,Y) \vee \nu'(X,Y)) \tag{3.4}$$

By the same argument as the previous paragraph, Σ_1^1-OS proves that $E'(X,Y)$ is an equivalence relation unary concepts. So the theory PFT contains the abstraction principle $A[E']$ (2.1)

Now, working in PFT, let us verify (3.1). There are three cases. First suppose that there is no x_0 with $\Phi(x_0, G)$. Then to establish (3.1) one can take $F = \emptyset$.

As a second case, suppose that there is a x_0 with $\Phi(x_0, G)$ and that G is non-empty. Then observe that the graph of the function $f(x) = \partial_E(\{x\} \times G)$ has both a Σ_1^1- and a Π_1^1-definition:

$$f(x) = y \leftrightarrow \exists\, R\; (\forall\, a,b\; R(a,b) \leftrightarrow (a = x\; \&\; Gb))\; \&\; \partial_E(R) = y$$
$$\leftrightarrow \forall\, R\; (\forall\, a,b\; R(a,b) \leftrightarrow (a = x\; \&\; Gb)) \rightarrow \partial_E(R) = y \tag{3.5}$$

These are equivalent because we can use the First-Order Comprehension Schema for L_1-formulas to secure that the binary relation $R = \{x\} \times G$ exists. Hence by the Δ_1^1-Comprehension Schema for L_1-formulas, the equivalence in (3.5) implies that the graph of f exists as a binary concept. Then by First-Order Comprehension Schema for L_1-formulas, the following unary concept exists:

$$F = \{x : f(x) = \partial_E(\{x_0\} \times G)\} \tag{3.6}$$

Now let's argue that $F = \{x : \Phi(x, G)\}$. First suppose that Fx. Then $f(x) = \partial_E(\{x_0\} \times G)$ and hence $\partial_E(\{x\} \times G) = \partial_E(\{x_0\} \times G)$. Then $E(\{x\} \times G, \{x_0\} \times G)$ and since G is non-empty we have $\mu(\{x\} \times G, \{x_0\} \times G)$. Then $\Phi(x, G) \leftrightarrow \Phi(x_0, G)$. Since we're assuming that $\Phi(x_0, G)$, we then conclude that $\Phi(x, G)$, which is what we wanted to show. For the converse, suppose that $\Phi(x, G)$. Since we're assuming that $\Phi(x_0, G)$ and that G is non-empty we may conclude that $\mu(\{x\} \times G, \{x_0\} \times G)$ and thus $E(\{x\} \times G, \{x_0\} \times G)$ and $\partial_E(\{x\} \times G) = \partial_E(\{x_0\} \times G)$. By the definition of f, we then have $f(x) = \partial_E(\{x_0\} \times G)$ which by the definition of F implies that Fx, which is what we wanted to show.

As a third case, suppose that there is an x_0 with $\Phi(x_0, G)$ but that G itself is empty. Then we argue as before that the graph of $g(x) = \partial_{E'}(\{x\})$ exists as a binary concept, that $F = \{x : g(x) = \partial_{E'}(\{x_0\})\}$ exists as a unary concept, and that $F = \{x : \Phi(x, G)\}$.

This finishes the proof of (3.1) in PFT. The proof of the general case of the Full Comprehension Schema for L_0-formulas (Definition 2.2) differs only in that unary concept variable F from (3.1) might instead be an n-ary concept variable and there may be more than one concept parameter G, as well as some additional object parameters. But the proof of this general case is directly analogous to the proof of (3.1). The only difference is that the number of abstraction principles used in the proof will increase with the number of concept parameters. In general if there are m-concept parameters $G_1, \ldots, G_m$, then there will be 2^m different abstraction principles used in the proof, since one must consider a case corresponding to the finite binary sequence $(i_1, \ldots, i_m)$, wherein $i_k = 0$ indicates that G_k is empty, and $i_k = 1$ indicates that G_k is non-empty. $\square$

Before turning to the proof that PFT interprets second-order Peano arithmetic, let's briefly note that in the consistency proof from [29] invoked in the proof of Theorem 2.9, we explicitly verified the Full Comprehension Schema for L_0-formulas. (In the language of that paper, these were part of the theory SO, and the interested reader may consult the proof of the Joint Consistency Theorem in that paper).

While the theory PFT only explicitly includes some instances of the Full Comprehension Schema for L_0-formulas in its definition (cf. Definition 2.8), the previous theorem says that it proves all of them. However, even in this predicative setting, the Russell paradox can be used to show that there is no concept consisting of the extensions, i.e. the range of the extension operator ∂ from Basic Law V (1.4). For a proof, see [27] Proposition 29 p. 1692. Now the formula $\mathrm{rng}(\partial)$ is definable by a Σ_1^1-formula of the signature $L_0[\partial]$. Further $L_0[\partial]$ is included in the signature L_1 of PFT. Hence, since the L_1-theory PFT is consistent by Theorem 2.9, it follows that PFT does not prove all instances of the Full Comprehension Schema for L_1-formulas.

This kind of situation is of course not entirely unfamiliar. For instance, Presburger arithmetic yields a complete axiomatization of the structure $(\mathbb{Z}, 0, 1+, <)$ (cf. Marker [22] pp. 82 ff). So this axiomatization proves each instance of the following induction schema in the signature $L = \{0, 1, +, <\}$:

$$[\varphi(0) \: \& \: \forall \: x \geq 0 \: (\varphi(x) \rightarrow \varphi(x+1)))] \rightarrow [\forall \: x \geq 0 \: \varphi(0)] \tag{3.7}$$

Consider a non-standard model $G = (G, 0, 1, +, <)$ of Presburger arithmetic, and extend L to L' by adding a new unary predicate Z which is interpreted on G as the integers $\mathbb{Z}$. Then of course the axioms of Presburger arithmetic do not imply all instances of the schema (3.7) in the expanded signature L'. So of course it's consistent for there to be a schema and an L'-theory and a subsignature L of L' such that the theory proves all instances of the L-schema but not every instance of the L'-schema.

Now let's show that PFT interprets second-order Peano arithmetic PA^2. These axioms are the natural set of axioms used to describe the standard model of second-order arithmetic; see [24] p. 4 or [27] p. 1680 or [28] p. 106 for an explicit list of these axioms.

Theorem 3.2. *The predicative Fregean theory* PFT *interprets second-order Peano arithmetic* PA^2.

Proof. First note that the predicative Fregean theory PFT proves the existence of the graph of the function $s(x) = \partial(\{x\})$ (cf. [27] Proposition 27 p. 1691), where this is the abstraction operator associated to Basic Law V (1.4). For, note that in PFT, for all objects x, y, one has that the following Σ_1^1-condition and Π_1^1-conditions are equivalent:

$$[\exists \: X \: (X = \{x\} \: \& \: \partial(X) = y)] \leftrightarrow [\forall \: X \: (X = \{x\} \rightarrow \partial X = y)] \tag{3.8}$$

By the Δ_1^1-Comprehension Schema for L_1-formulas, there is then a binary relation which holds of objects x, y iff either the Σ_1^1-condition holds or the Π_1^1-condition holds. And this binary relation is obviously the graph of the function $s(x) = \partial(\{x\})$.

Let M be $\{x : x = x\}$, which exists by Full Comprehension for L_0-formulas, and let $0 = \partial(\emptyset)$. Then one has that the triple $(M, 0, s)$ satisfies the first two axioms of Robinson's Q:

$$\forall\, x\ s(x) \neq 0, \qquad \forall\, x, y\ (s(x) = s(y) \to x = y) \tag{3.9}$$

For, suppose that $s(x) = 0$. Then $\partial(\{x\}) = \partial(\emptyset)$ and then by Basic Law V (1.4) one has that $\{x\} = \emptyset$, a contradiction. Similarly, suppose that $s(x) = s(y)$. Then $\partial(\{x\}) = \partial(\{y\})$ and so by Basic Law V (1.4) one has that $\{x\} = \{y\}$ and hence $x = y$. Thus (3.9) follows immediately from Basic Law V (1.4).

But then standard arguments allow one to interpret second-order Peano arithmetic PA^2 by taking the natural numbers N to be the sub-concept of M consisting of all those subconcepts of M which are "inductive," that is which contain zero and closed under successor. Here of course for the existence of N and the verification of the other axioms of arithmetic, one appeals to the Full Comprehension Schema for L_0-formulas, using $M, 0, s$ as parameters (cf. [27] Theorem 16 p. 1688). □

4 The fragility of abstraction with predicative comprehension

However, in spite of its technical strength, the conceptual basis of the predicative Fregean theory PFT is rather fragile. For, the L_1-theory PFT was formed by adding the abstraction principle $A[E]$ associated to the L_0-formulas $E(R, S)$ when this formula could be proven to be an equivalence relation in the background second-order logic $\Sigma_1^1\text{-}\mathsf{OS}$. But one cannot successively iterate this idea. For, suppose that in analogue to $\mathrm{ProvEquiv}(L_0)$ in equation (2.8), one defines:

$$\mathrm{ProvEquiv}(L_1) = \{E(R, S) \text{ is an } L_1 \text{ formula} : \mathsf{PFT} \vdash \mathrm{Equiv}(E)\} \tag{4.1}$$

And further suppose that one defines L_2 to be the expansion of L_1 by the addition of a function symbol ∂_E from n_E-ary concepts to objects for each L_1-formula $E(R, S)$ in $\mathrm{ProvEquiv}(L_1)$. Finally, suppose one defines the following iteration of PFT (cf. Definition 2.8):

Definition 4.1. *The theory PFT_2 is the L_2-theory consisting of (i) the extensionality axioms (2.5) and the projection axioms (2.6) and (ii) the Σ_1^1-Choice Schema for L_2-formulas (Definition 2.5) and (iii) the First-Order Comprehension Schema for L_2-formulas (Definition 2.3), and (iv) the abstraction principle $A[E]$ (2.1) for each E which is from $\mathrm{ProvEquiv}(L_0)$ (2.8) or from $\mathrm{ProvEquiv}(L_1)$ (4.1).*

Then the same argument as in the proof of Theorem 3.1 establishes that PFT$_2$ proves each instance of the Full Comprehension Schema for L_1-formulas. But then PFT$_2$ is inconsistent, since on pain of the Russell paradox there is no concept of all extensions (cf. [27] Proposition 29 p. 1692), where again the extensions are the range of the abstraction operator ∂ associated to Basic Law V (1.4). Hence, while the predicative Fregean theory PFT is consistent, when one tries to iterate its underlying idea of adding abstraction principles when their equivalence relations can be proven to be equivalence relations, one again runs up against the Russell paradox. This indicates that the resource of abstraction principles in the predicative setting is unlike that of typed theories of truth or second-order logic, which we may consistently add to any consistent theory.

This point is underscored when one observes that the same considerations show the inconsistency of an axiom-based analogue of the rule-based predicative Fregean theory PFT. In particular, suppose that we recursively defined a signature L^* extending L_0 so that if $E(R, S)$ is an L^*-formula in exactly two free n_E-ary concept variables then L^* also contains a function symbol ∂_E which takes n_E-ary concepts to objects and which does not occur in E. One could then define the following L^*-theory:

Definition 4.2. *The theory* PFT* *is the L^*-theory consisting of (i) the extensionality axioms (2.5) and the projection axioms (2.6) and (ii) the Σ_1^1-Choice Schema for L^*-formulas (Definition 2.5) and (iii) the First-Order Comprehension Schema for L^*-formulas (Definition 2.3), and (iv) the axiom* Equiv$(E) \to A[E]$ *for each L^*-formula E.*

In this, Equiv(E) is the sentence which says that E is an equivalence relation (cf. (2.7)) and $A[E]$ is the abstraction principle (2.1), so that the axiom Equiv$(E) \to A[E]$ says that if E is an equivalence relation, then $A[E]$ holds. The considerations of the previous paragraphs can be replicated in this theory PFT*, showing it to be inconsistent. However, the conceptual distance between the inconsistent L^*-theory PFT* and the consistent L_1-theory PFT is rather slim. The difference is merely a difference between a rule and an axiom: whereas the rule-based PFT only includes an abstraction principle when the underlying equivalence relation is expressible in the weak background logic and is *provably* an equivalence relation there, the axiom-based PFT* includes a commitment to either the truth of the abstraction principle or the falsity of its underlying formula being an equivalence relation.

In response to this, one might try to restrain the predicative Fregean theory PFT so that the analogously defined iterated version of it and the analogously defined axiom-based version of it were consistent. For instance, one might consider restricting the abstraction principles added to the theory PFT to those whose underlying equivalence relation was expressible both as a Σ_1^1-formula and a Π_1^1-formula in the background second-order logic. This, it might be suggested, would be a genuinely predicative theory of abstraction principles. Such a move would block the proof of Theorem 3.1. For, the equivalence relation $E(R, S)$ (3.3) used in that proof is not obviously expressible in such a way. However, it is unknown to us how much arithmetic this more austerely predicative theory could interpret, and it is not obvious to us whether the analogously defined iterated version of it (or axiom-based version of it) is consistent.

Another way forward might be to find some principled way to focus attention on abstraction principles which are somehow more like the paradigmatic Basic Law V (1.4) and Hume's Principle (1.1) and the abstraction principle associated to ordinals (2.2), and somehow less like the seemingly ad-hoc abstraction principles constructed in the proof of Theorem 3.1. But to do so would be to lose some of the original motivation for focusing on predicative abstraction principles. For, part of the attraction was supposed to be that more abstraction principles became consistent and jointly consistent. And indeed, as the predicative Fregean Theory PFT attests, a good deal of joint consistency is available in this setting. Hence in the earlier paper [29] we said that we had resolved an analogue of the joint consistency problem. But as we have seen in this section, when we try to iterate the underlying idea of abstraction principles in the predicative setting, we again run into inconsistency and seem back in the situation of trying to discern ways to weed out the acceptable from the unacceptable abstraction principles. For an overview of the various candidates for acceptable abstraction principles in the general impredicative setting, see [21] or [5].

Perhaps another way forward might be to give up on the idea of abstraction principles altogether and find principled reasons for studying systems centered around either Basic Law V (1.4) itself or Hume's Principle (1.1) itself or the abstraction principle associated to ordinals (2.2) all by itself. With respect to Basic Law V (1.4), this is the perspective of [30], where the idea is to work within an intensional logic and see the extension operator as selecting a sense for each concept, just like we might select a specific Turing machine index for each computable function. But much remains unknown about the individual abstraction principles at the predicative level. For instance, it is to our knowledge unknown whether Basic Law V (1.4) or the abstraction principle associated to ordinals (2.2), equipped with the Σ_1^1-choice schema and the First-Order Comprehension Schema, interprets the analogous predicative versions of arithmetic (cf. [27] p. 1707). In this paper, the idea for interpreting arithmetic was to collect together all the predicative abstraction principles so that they could effect the interpretation together, and it is in general unclear to us what happens when one focuses on the abstraction principles one by one.

Acknowledgements

I was lucky enough to be able to present parts of this work at a number of workshops and conferences, and I would like to thank the participants and organizers of these events for these opportunities. I would like to especially thank the following people for the comments and feedback: Robert Black, Roy Cook, Matthew Davidson, Walter Dean, Marie Duží, Kenny Easwaran, Fernando Ferreira, Martin Fischer, Rohan French, Salvatore Florio, Kentaro Fujimoto, Jeremy Heis, Joel David Hamkins, Volker Halbach, Ole Thomassen Hjortland, Luca Incurvati, Daniel Isaacson, Jönne Kriener, Graham Leach-Krouse, Hannes Leitgeb, Øystein Linnebo, Paolo Mancosu, Richard Mendelsohn, Tony Martin, Yiannis Moschovakis, John Mumma, Pavel Pudlák, Sam Roberts, Marcus Rossberg, Tony Roy, Gil Sagi, Florian Steinberger, Iulian Toader, Gabriel Uzquiano, Albert Visser, Kai Wehmeier, Philip Welch, Trevor Wilson, and Martin Zeman.

Finally, a special debt is owed to the editors and anonymous referees of this journal, to whom I express my gratitude. For, the proofs were greatly simplified by their suggestions and the previous reliance upon choice was removed by virtue of these suggestions. While composing this paper, I was supported by a Kurt Gödel Society Research Prize Fellowship and by Øystein Linnebo's European Research Council funded project "Plurals, Predicates, and Paradox."

References

[1] Evert W. Beth. Chapter 13: Logicism. In *The Foundations of Mathematics: A Study in the Philosophy of Science*, Studies in Logic and the Foundations of Mathematics, pages 353–364. North-Holland, Amsterdam, 1959.

[2] George Boolos. *Logic, logic, and logic*. Harvard University Press, Cambridge, MA, 1998. Edited by Richard Jeffrey.

[3] John P. Burgess. *Fixing Frege*. Princeton Monographs in Philosophy. Princeton University Press, Princeton, 2005.

[4] Roy T. Cook, editor. *The Arché papers on the mathematics of abstraction*, volume 71 of *The Western Ontario Series in Philosophy of Science*. Springer, Berlin, 2007.

[5] Roy T. Cook. Conservativeness, stability, and abstraction. *British Journal for the Philosophy of Science*, 63:673–696, 2012.

[6] Solomon Feferman. Systems of predicative analysis. *The Journal of Symbolic Logic*, 29:1–30, 1964.

[7] Solomon Feferman. Predicativity. In Stewart Shapiro, editor, *The Oxford Handbook of Philosophy of Mathematics and Logic*, pages 590–624. Oxford University Press, Oxford, 2005.

[8] Fernando Ferreira and Kai F. Wehmeier. On the consistency of the Δ^1_1-CA fragment of Frege's Grundgesetze. *Journal of Philosophical Logic*, 31(4):301–311, 2002.

[9] Gottlob Frege. *Die Grundlagen der Arithmetik*. Koebner, Breslau, 1884.

[10] Gottlob Frege. *Grundgesetze der Arithmetik: begriffsschriftlich abgeleitet*. Pohle, Jena, 1893, 1903. Two volumes. Reprinted in [?].

[11] Gottlob Frege. *The foundations of arithmetic: a logico-mathematical enquiry into the concept of number*. Northwestern University Press, Evanston, second edition, 1980. Translated by John Langshaw Austin.

[12] Gottlob Frege. *Basic laws of arithmetic*. Oxford University Press, Oxford, 2013. Translated by Philip A. Ebert and Marcus Rossberg.

[13] Harvey M. Friedman. Some systems of second-order arithmetic and their use. In *Proceedings of the International Congress of Mathematicians, Vancouver 1974*, volume 1, pages 235–242. 1975.

[14] Mihai Ganea. Burgess' PV is Robinson's Q. *Journal of Symbolic Logic*, 72(2):618–624, 2007.

[15] Petr Hájek and Pavel Pudlák. *Metamathematics of first-order arithmetic*. Perspectives in Mathematical Logic. Springer, Berlin, 1998.

[16] Bob Hale and Crispin Wright. *The reason's proper study*. Oxford University Press, Oxford, 2001.

[17] Richard G. Heck, Jr. The consistency of predicative fragments of Frege's Grundgesetze der Arithmetik. *History and Philosophy of Logic*, 17(4):209–220, 1996.

[18] Richard G. Heck, Jr. *Frege's theorem*. Oxford University Press, Oxford, 2011.

[19] Harold Hodes. Logicism and the Ontological Commitments of Arithmetic. *The Journal of Philosophy*, 81(3):123–149, 1984.

[20] Per Lindström. *Aspects of incompleteness*, volume 10 of *Lecture Notes in Logic*. Association for Symbolic Logic, Urbana, IL, second edition, 2003.

[21] Øystein Linnebo. Some criteria for acceptable abstraction. *Notre Dame Journal of Formal Logic*, 52(3):331–338, 2010.

[22] David Marker. *Model Theory: An Introduction*, volume 217 of *Graduate Texts in Mathematics*. Springer-Verlag, New York, 2002.

[23] Terence Parsons. On the consistency of the first-order portion of Frege's logical system. *Notre Dame Journal of Formal Logic*, 28(1):161–168, 1987. Reprinted in [?].

[24] Stephen G. Simpson. *Subsystems of second order arithmetic*. Cambridge University Press, Cambridge, second edition, 2009.

[25] Albert Visser. Categories of theories and interpretations. In Ali Enayat, Iraj Kalantari, and Mojtaba Moniri, editors, *Logic in Tehran*, volume 26 of *Lecture Notes in Logic*, pages 284–341. Association for Symbolic Logic, La Jolla, 2006.

[26] Albert Visser. The predicative Frege hierarchy. *Annals of Pure and Applied Logic*, 160(2):129–153, 2009.

[27] Sean Walsh. Comparing Hume's principle, Basic Law V and Peano Arithmetic. *Annals*

of *Pure and Applied Logic*, 163:1679–1709, 2012.

[28] Sean Walsh. Logicism, interpretability, and knowledge of arithmetic. *The Review of Symbolic Logic*, 7(1):84–119, 2014.

[29] Sean Walsh. Fragments of Frege's Grundgesetze and Gödel's constructible universe. *The Journal of Symbolic Logic*, 81(2):605 – 628, 2016.

[30] Sean Walsh. Predicativity, the Russell-Myhill paradox, and Church's intensional logic. *The Journal of Philosophical Logic*, 45(3):277–326, 2016.

[31] Hermann Weyl. *Das Kontinuum. Kritische Untersuchungen über die Grundlagen der Analysis*. Veit, Leipzig, 1918.

[32] Crispin Wright. *Frege's conception of numbers as objects*, volume 2 of *Scots Philosophical Monographs*. Aberdeen University Press, Aberdeen, 1983.

Perspectives for Proof Unwinding by Programming Languages Techniques

Danko Ilik*
Inria & LIX, Ecole Polytechnique, Palaiseau, France
ilik.danko@orange.fr

Abstract

In this chapter, we propose some future directions of work, potentially beneficial to Mathematics and its foundations, based on the recent import of methodology from the theory of programming languages into proof theory. This scientific essay, written for the audience of proof theorists as well as the working mathematician, is not a survey of the field, but rather a personal view of the author who hopes that it may inspire future and fellow researchers.

1 Introduction

> *We cannot hope to prove that every definition, every symbol, every abbreviation that we introduce is free from potential ambiguities, that it does not bring about the possibility of a contradiction that might not otherwise have been present.*
>
> N. Bourbaki [4]

> *There is an error, I can confess now. Some 40 years after the paper was published, the logician Robert M. Solovay from the University of California sent me a communication pointing out the error. I thought: "How could it be?" I started to look at it and finally I realized [...]*
>
> John F. Nash Jr. [56]

---

*This chapter was prepared while the author was financed by the ERC Advanced Grant ProofCert.

> *Mathematics arises from all sorts of application or insights but in the end must always consist of proofs*, [but] *although a real proof is not simply a formalized document but a sequence of ideas and insights*, [a] *real proof is not something just probably correct.*
>
> Saunders Mac Lane [51]

What constitutes a *real proof* is a question at the origin of mathematical logic. In effect, a real proof is one that can be reduced to the use of only a few accepted 'ideal' principles such as the axioms for a set theory like ZFC. And yet certain ideal principles are far from self-evident as Euclid's axiomatic method would require them to be. Proof theory was conceived by Hilbert with the program to further "recognize the non-contradictory character of all the usual [ideal] mathematical methods without exceptions". Around 1960, these concerns were addressed for the theory of arithmetic and analysis in the so called *modified* Hilbert program using the early models of *computation*—proof theory was also pivotal for the development of computer science (Hilbert's Entscheidungsproblem).

Applying mathematical rigor to formal proofs as the object of study brought an answer to the question of what a real proof is: a formal proof can be given semantics in terms of Gödel's system T and Spector's bar recursion, thereby eliminating logic in favor of pure computation. However, these early models of computation that were used to provide the answer to the consistency question, although satisfying in terms of precision, are cumbersome to use in practice.

Firstly, it is far from clear why the old computational interpretations are the right ones, for it is often hard to distinguish them (bar recursion) from brute force search procedures. We would like to understand the computational answer to the main consistency questions in terms of modern and more finely grained computing abstractions, such as the ones developed over the course of the past four decades in the theory of programming languages—for research on (natural) models of computation surely did not end with the invention of the Turing machine and recursive function theory.

Secondly, the cumbersome machinery, although ingenious, makes it difficult to address the next level of research questions. Once that we have the answer to what a real proof *is*, we need to know what constitutes the *essential data* of a proof—curiously, this question of finding criteria of greatest simplicity for proofs was already listed as 24th in Hilbert's famous list of open problems, but being premature was not included among the ones finally published [63].

The title of this chapter refers to *proof unwinding*, the pioneering research program from the 1950's of Georg Kreisel [50], who started to use the computational approach, not for foundational purposes, but to extract numerical content from actual mathematical arguments. We aim at the *working mathematician*, a term used by Bourbaki [4] who meant by it a researcher with a pragmatic attitude toward foundations. The time is ripe for a leap forward, both in foundations and unwinding applications. The present chapter has as goal to propose bringing proof unwinding on a par with the latest computing abstractions from the theory of programming languages, with the ambition to turn such streamlined proof theoretic methods into a toolbox readily used by the working mathematician, rather than the rare specialist in proof theory as it has been the case up to now.

2 New Unwinding Toolbox

Conducted with the goal:

> *"To determine the constructive (recursive) content or the constructive equivalent of the non-constructive concepts and theorems used in mathematics"* [44],

Kreisel's research program applied the proof theory of the day, namely Hilbert's ϵ-substitution method, Herbrand's theorem, and the no-counterexample interpretation, combined with then brand new theory of recursive functions, to extract new bounds and algorithms from prima facie ineffective proofs. But, even in the hands of masters, the early unwinding methodology was apparently difficult to apply, if one is to judge from the lapses of time in between applications: Littlewood's theorem by Kreisel in 1951 [43], Artin's proof of Hilbert's 17th problem by Kreisel first in 1957 and again in 1978 [14]), the Thue-Siegel-Roth theorem by Kreisel and Macintyre in 1982 and Luckhardt in 1989 [46, 49], Van der Waerden's theorem by Girard in 1987 [20]. The unwinding methods are so complex that there are even doubts cast on some of the results by authorities in proof theory [16].

However, there is a more recent application of unwinding to functional analysis in the *proof mining* program of Kohlenbach [42]. This very successful unwinding program has at its methodological core the classic unwinding approach using Kolmogorov's double-negation translation (1929) and Gödel's functional 'Dialectica' interpretation (1941).

In parallel, in constructive mathematics, there have been equally significant results in the program of *constructive analysis* [2] and *constructive algebra* [54, 48], although these are primarily theory reconstruction programs and rely little on direct application of proof theoretic methods to unwind ineffective proofs.

But, both mining and research in constructive mathematics have not sought to reap the benefits of notable proof theoretic advances directly inspired by the theory of programming languages. This theory, a continuation of the work on the early models of computation, has arrived at highly abstract notions for structuring programs. We shall now describe the proof theoretic state-of-the-art for three such proof unwinding techniques. This new methodology will be referred to as the *unwinding toolbox*.

2.1 Computational Side-Effects

The first of these methods concerns *computational side-effect*. Namely, since the work of Griffin [25], it has been known that the principle of proof by contradiction can be interpreted by a programming language mechanism (a computational side-effect) for *control operators*. Although, in absence of mathematical axioms additional to the reductio-ad-absurdum principle, control operators provide not much more than a very elegant way to obtain Herbrand's theorem (an important very early result on classical first-order logic from 1930), in presence of additional axioms like induction or choice, when Herbrand's theorem no longer holds, one begins to get new results. For instance, by the use of computational side-effects, in set theory, Krivine has managed to extend Cohen's forcing method from the usual sets of conditions to realizability algebras [47].

However, the promise that control operators can turn every proof by contradiction into an effective one is a mirage: there are classically provable formulas whose effective proof would allow to decide the Halting problem. Whether an ineffectively proved formula can be unwound, in general needs to be considered on a case-by-case basis. Nevertheless, there *are* whole *classes* of formulas which we know can be unwound upfront, like the class of Π_2^0-formulas. *Delimiting* control operators only to formulas in these classes allows to get a new constructive logic. This logic still respects the existence property, characteristic of intuitionistic logic that is at the bases of current constructive mathematics, but the obtained new constructive logic manages to prove intuitionistically non-provable principles.

For instance, Herbelin [29] showed that Markov's principle (MP),

$$\neg\neg \exists x A_0(x) \to \exists x A_0(x),$$

where $x \in \mathbb{N}$ and A_0 is quantifier-free, an axiom crucial for constructive proofs of completeness of first-order logic [31], can be interpreted with the help of a computational side-effect known as *(delimited) exceptions*. The author further showed [32] that the double negation shift principle (DNS),

$$\forall x \neg\neg A(x) \to \neg\neg \forall x A(x),$$

where $x \in \mathbb{N}$, a principle crucial for the interpretation of the classical axiom of choice—and that has only been interpreted before by the *generally*-recursive schema of bar recursion—can be interpreted computationally by a generalization of the exceptions effect to so called delimited continuations, or *delimited control operators*. The key observation from these results is that—when delimited—computational side-effects like control operators can be used to unwind ineffective proofs and at the same time not run into non-decidability problems. The newly obtained logics are intermediate logics, in between classical and intuitionistic logic, and take the best of both worlds.

These results are controversial from the point of view of the orthodox constructive mathematician who is used to intuitionistic logic as first codified by Heyting's analysis of Brouwer's work in intuitionistic mathematics. Namely, the only previous computational interpretation of MP were either trivial (as given by Gödel's functional interpretation) or proceeded by unbounded search. As for DNS, the computational interpretation was only given by the generally-recursive bar recursion schema, whose termination must be ensured by Brouwer's bar induction or continuity principle. As unbounded/general recursion can lead to an inconsistent formal system, intuitionists have been understandably wary of accepting these principle. By replacing the mentioned computational interpretations by computationally meaningful realizers, we not only propose to intuitionists to reconsider the constructivity of principles like MP and DNS, but we are re-establishing the link between modern proof theory and one of the offspring of Hilbert's proof theory, the theory of programming languages.

A further principle interpreted in this way, in a joint work of Nakata and the author [38], was the open induction principle,

$$\forall \alpha (\forall \beta < \alpha (\beta \in U) \to \alpha \in U) \to \forall \alpha (\alpha \in U),$$

where α, β range over Cantor space and U is open. This principle is the only known equivalent form of the axiom of choice that is stable under double-negation translation (even if we replace Cantor space by Baire space). The principle is also interesting for combinatorics, where it leads to a direct version of Nash-Williams' proof of Kruskal's tree theorem [65], as well as in algebra where it is used to replace Zorn's lemma [59].

We finally mention a last result [35], still in review, on the nature of the programming-language inspired proof rules. It concerns higher-type primitive recursion—Gödel's system T—versus general recursion—Spector's bar recursion. Namely, already in 1979, Schwichtenberg has shown that bar recursion of type 0 and 1 does not allow to define functions beyond system T [60]. Since a previous analysis of Kreisel [45] shows that these types are sufficient for all practical purposes (realizing Σ_2^0-theorems), it follows that we know for a long time that we should not need bar recursion for the computational interpretation of ideal proof principles. What we proposed is how to circumvent bar recursion and generate System T terms directly, using delimited control operators as an intermediary step. This also shows that the extensions of system T with computational side-effects are in fact conservative extensions. In order to establish this fact, we relied on *partial evaluation*, the second set of techniques of our unwinding toolbox that we explain in the following subsection.

2.2 Partial Evaluation and Formalization

The second programming languages method that we intend to employ for proof unwinding concerns formalization of proofs in proof assistant software and, more specifically, the use of formalized *(type-directed) partial evaluators*.

The topic of partial evaluators came up in our previous research on constructive versions of completeness theorems [31]. These logical theorems establish the adequacy of a given formal system to encode actual mathematical proofs. As it turns out, and thanks to initial work on the link between normalization proofs and completeness of intuitionistic logic for Kripke models [12], the computational content of proofs of intuitionistic completeness can be expressed by type-directed partial evaluation algorithms [13]. Having a rich theory of such algorithms in the theory of programming languages, allowed to cover cases of constructive completeness proofs that were beyond the previous state-of-the art in proof theory. More precisely, we now know how to partially evaluate (i.e. show constructively completeness for) not only classical logic [32], intuitionistic logic with disjunction [33], but also simultaneous presence of delimited control operators and higher-type primitive recursion [34, 35] (the second citation is still in review).

The development of these logical meta-theorems was conducted formally, in the Coq and Agda proof assistants. Since the formalized proofs are constructive, they can be used to *compute a proof transformation* for every actual formalized argument. What this allows is to perform unwinding of actual mathematical proofs more directly, by pushing the complexity of doing a manual double-negation transformation (like done in the classic unwinding approach and used, for instance, by Kohlenbach in his program of proof mining in analysis) into the realizability model, that is, into the reduction mechanism of the proof assistant used.

Proof assistants are most well known for their use in the full formalization of complex proofs, such as the four-color theorem [21], the Kepler conjecture [27], or the Feit-Thompson theorem [22]. However, as far as proof unwinding is concerned, one can in general *avoid* needing a *fully* formalized version of an actual mathematical proof. It suffices to notice that lemmas that have a computationally irrelevant form, such as Π_1^0, can be simply assumed without proof. A more refined analysis of computational relevance of formulas can be found in [61] where the classes of so called definite and goal formulas are isolated. This allows to greatly decrease the burden of formalizing i.e. we are only dealing with *partial formalization* which nonetheless contains as much of algorithmic content as a full formalization.

The important lesson that we learned from partial evaluation is that proofs need not be interpreted uniformly, by 'oracles' such as bar recursion that work uniformly (for example the realization of DNS by bar recursion is agnostic of the concrete formula A in the instance of DNS). Rather, it is possible to specialize (i.e. partially evaluate) proofs, even if they are highly ineffective, and, when one in addition uses a proof assistant like Coq, the specialization of the (partially) formalized theorem can become automatic. This is one of the principal advantages of our toolbox over the old toolbox built on Herbrand's theorem, ϵ-substitution, double-negation- and A-translation, and functional interpretation: while unwinding, the mathematician can concentrate on the essential parts of a proof rather than get lost in manual proof transformations.

2.3 Type Isomorphisms

The final third method of our unwinding toolbox concerns *type isomorphisms*. Mathematically, this notion is the same as the one of *constructive cardinality* of sets [54], saying not only that sets are of the same size, but moreover that they have indistinguishable structure. In programming languages theory, the notion allows to generalize the notion of type assigned to a program, which allows to test more easily if a programs conforms to a formal specification [58].

The link that brings us to the study of type isomorphisms is Tarski's high-school algebra problem [9]. This basic question, asking whether the system of eleven arithmetic equation taught in high-school suffices to derive all the true equations between *exponential* polynomials, had taken some time to be answered in mathematical logic. It turned out that the high-school system is not complete, as shown by a counter-example of Wilkie in 1981 [68], a true statement which is not derivable by only using the eleven equations. Gurevič further showed that the system cannot be completed by any additional finite set of axioms [26].

Now, by the Curry-Howard correspondence, formulas of intuitionistic logic can be seen as types (conjunction correspond to products, disjunction to coproducts, and implication to exponentials) and proofs of formulas can be seen as computer programs of the corresponding type. Following the correspondence, one gets a notion of strong equivalence, or formula isomorphism, from isomorphism of types. A new correspondence is thus obtained: the language of formulas is the same as the language of exponential polynomials—and, moreover—formula isomorphism generalizes equality of exponential polynomials in the standard model of positive natural numbers, that is

$$A \cong B \text{ implies } \mathbb{N}^+ \vDash A = B.$$

The link that one establishes in this way allows to use the rich theory on exponential polynomials to obtain proof theoretic results. For instance, Fiore, Di Cosmo, and Balat, showed that the non-finite-axiomatizability result of Gurevič also hold for the theory of type isomorphism [17]. Using results of Richardson, Martin, Levitz, Wilkie, Macintyre, Henson and Gurevič, the author proved that although not finitely axiomatizable, type i.e. formula isomorphism is recursively axiomatizable and moreover decidable [36]. The value of this unexpected positive result is still somewhat limited because the existence of a *practical* decision algorithm for type isomorphism is open.

Nevertheless, even if the meta-theory of type isomorphism has remaining open questions to be resolved, applications to proof theory are well under way. Recently, the author proposed a pseudo-normal form of types [37], inspired by the decomposition of the exponential function in exponential fields [28], called the exp-log normal form, that allows to decompose the axioms of the notoriously non-local theory of $\beta\eta$-equality for the lambda calculus with coproduct type. This equality can be seen as the essence of *identity of proofs* for intuitionistic propositional logic with disjunction. An extension to the first-order case has also been proposed in a joint work with Brock-Nannestad [5], where the normal form appears to produce the first arithmetical hierarchy for formulas of intuitionistic logic that copes with both quantifiers equally well; the only previously known hierarchy, the one of Burr [8], covers well only the universal quantifier. This has been a long standing open problem for constructive logic, although for classical logic an arithmetical hierarchy exists since the 1930s.

3 Perspectives

Today, a paradigm change in proof unwinding is possible, thanks to the notions from contemporary programming languages theory comprising our New Unwinding Toolbox. These long-evolved techniques provide proof-theoretic simplifications of the order that makes them more accessible even to non-specialists in proof theory.

The overall goal of this undertaking would be to exploit the full potential of the novel toolbox and apply it, beyond logic itself, to proofs of landmark results in number theory, combinatorics, and homotopy theory. In parallel, it would be necessary to address the foundations of unwinding i.e. tackle long-standing open questions in the foundations of constructive mathematics such as identity of proofs and simplified computational interpretations of semi-intuitionistic principles. We have thus two sets of objectives, work on applications and work on foundations.

Objective I — Applications of Proof Unwinding

The first set of objectives concerns applications to areas that are important for the 'working mathematician', that is, analysis, number theory, and combinatorics, as well as an application to unwinding incompleteness theorems in logic. Objective I would be achieved by tackling three more specific objectives, called *perspectives*: Perspective 1: Unwinding in Analysis Revisited, Perspective 2: Unwinding in Number Theory and Combinatorics, and Perspective 3: Unwinding Incompleteness Theorems.

Objective II — Foundations of Proof Unwinding

The second set of objectives concerns work on foundations of constructive mathematics that are both necessary to guarantee the soundness of applying unwinding and as an update to the current foundational theories. The two more specific objectives, or perspectives, to be tackled are: Perspective 4: Identity of Proofs and Homotopy Type Theory and Perspective 5: A Next Generation of Constructive Foundations.

The immediate effects of the project would be, on the one hand, to show that our new proof theoretic methods can be used by the working mathematician to extract numerical bounds and algorithms from prima facie ineffective proofs in analysis, number theory, combinatorics, homotopy theory, and logic, and, on the other hand, to update the current foundational theories of constructive mathematics with the powerful computing abstractions that computational side-effects, partial evaluators and type isomorphisms represent.

In the longer term, we can hope to see the streamlined proof unwinding methodology becoming an important toolbox across mathematics. We can also expect to see a synergy of the objectives. For instance, not only would unwinding efforts across mathematics become possible (Objective I), but, as the new constructive foundations (Objective II) get adopted in the community working on proof assistant systems, proof analysis and development would eventually be carried out even more efficiently with the help of a proof assistant.

The approach to fulfilling the two objectives would be through carrying out the five perspectives described in this section. We shall explain each one of the tasks in the context of its proper state-of-the-art, objectives, methodology, and feasibility.

3.1 Perspective 1: Unwinding in Analysis Revisited

Analysis is essentially the only area of mainstream mathematics to have benefited from direct application of proof unwinding techniques. In approximation theory, by using proof theory, Kohlenbach and his collaborators have managed to obtain explicit moduli of uniqueness, significantly better than previous ones, for best Chebyshev approximation, as well as to obtain a first effective rate of strong unicity in the case of best approximation for the L_1-norm [40]. How this works is that first logical meta-theorems are established [42], which are on one hand general enough to be applicable as analytic theorems, and on the other hand specific enough to enter in a class of statements, such as the Π_2^0-class of the arithmetical hierarchy, for which we know by proof theory that explicit functions or existence witnesses can be extracted. Moreover, such general logical meta-theorems are not only good for extracting numerical data from concrete proofs, but also for analyzing whether a known analytic theorem has optimal form. For instance, in the fixed point theory for functions of contractive type, one does not only get effective quantitative forms of theorems, but one can often also relax the compactness assumption for the metric space.

Why, then, when Kohlenbach's proof mining approach is already successful, do we propose a perspective on proof unwinding of analysis? There are two reasons. The first one is methodological: our form of unwinding has not been applied outside of logic, and proof mining provides the perfect test bed to make it grow up in the 'real world'. Second, even if we cannot pretend to analytic skills of the level of the ones present in mining, we do believe that the general logical meta-theorems can be unwound in a simpler way; this could lead to better extracted bounds even if we use the exact same analytic machinery as in mining.

To explain the difference and simplification mentioned, we briefly explain how the meta-theorems are established right now. The core is to show that in classical logic, and in presence of additional axioms for induction and choice, like the weak König's lemma, one can turn the $\forall\exists$ quantifier combination from $\forall x \exists y A_0(x, y)$, where A_0 is a quantifier-free formula, into an explicit recursive function f such that $\forall x A_0(x, f(x))$. One can further extend this to formulas beyond the strict class Π_2^0 and allow for instance any number of additional hypotheses of form Π_1^0. But, to obtain the recursive functions f, which, as explained before in the section Computational Side-Effects, needs an a priori generally recursive definition schema, one first has to transform by the double-negation translation all proofs of the original proof system (Peano arithmetic + axiom of choice) into proofs of a (semi-)intuitionistic system. This is a *non-local* transformation of proofs, and in particular the meaning of formulas can be changed by the transformation (hence the restriction to the Π_2^0-class of formulas). Once a (semi-)intuitionistic proof is obtained, one can use Gödel's functional interpretation to obtain a higher-type primitive recursive function, possibly also needing Spector's generally-recursive schema of bar recursion. Actually, more redefined versions of the Dialectica interpretation (monotone and bounded variants) and of bar recursion are used in practice.

Now, what our approach offers is first to push the technical complexity of the double negation translation into the realizability model based on computational side-effects (ex. control operators). Since these notions have a well-studied operational semantics, one can perform a more direct reduction of a proof to a program or a more direct reading off of witnesses (numeric bounds). With the additional help of a proof assistant like Coq, this can be further automated.

In addition, thanks to the reasons already explained in section Computational Side-Effects, our unwinding method makes it likely that in fact a pure system T witnessing terms can be extracted from any concrete proof, circumventing bar-recursion-like schemata altogether.

A third, orthogonal, improvement to the extraction process will be offered by use of richer data-types for extracted programs and bounds. Traditionally, one only uses the 'negative' function and product types. Although these can encode 'positive' types (for instance, sum types $\rho+\sigma$ can be encoded by $(\mathbb{N} \to \rho) \times (\mathbb{N} \to \sigma)$), encoding leads to an increase of the *degree* of the type. Simpler and more natural realizers can thus be extracted from disjunction and other inductively defined positive predicates.

Feasibility for Perspective 1

We will need to cope with semi-intuitionistic principles that we have not treated before, notably the weak Kőnig's lemma and the independence of premise schema. For these, we plan to use Computational Side-Effects, like we have done previously for the open induction principle: the fan theorem, a positive version of the weak Kőnig lemma, is implied by open induction. At the level of realizers, it will be necessary to use Type Isomorphisms to handle extensionality.

The risk for handling the logical part (meta-theorems) is moderate, hence it is possible to propose this for a subject of a PhD thesis. As for obtaining better bounds that the ones already obtained in proof mining, the risk is higher; in fact, it would be a success even if we manage to obtain the same bounds, since this would mean that our toolbox is ready to be used in the following, Perspective 2.

3.2 Perspective 2: Unwinding in Number Theory and Combinatorics

In this task, we should bring in our New Unwinding Toolbox to bear on highly non-effective proofs from number theory and combinatorics. The concrete goals will be to unwind landmark proofs in these areas, but what we see as equally important is to arrive at a situation where a sufficiently interesting intersection of proof theory and the application domain area is recognized. This kind of objective is only possible through a combination of expertise, and for its carrying out, it would be appropriate to engage two post-doctoral researchers, one in each application domain.

In number theory, we would intend to unwind Thue-Siegel-Roth's theorem on Diophantine approximations. Saying that an algebraic irrational number has only finitely many exceptionally good rational number approximations, this Σ_2 statement has first been tackled upon by Kreisel and Macintyre [46] using technology for obtaining Herbrand terms. However the combinatorial explosion arising from use of Herbrand's theorem apparently did not allow to obtain useful bounds on the number of rational approximations, and only Luckhardt [49] managed to limit the growth of Herbrand terms in order to obtain such a bound. This bound is essentially the same as the one obtained by Bombieri and van der Poorten [3].

In this case, even more advanced existing technology like Gödel's functional interpretation has not been applied. We suspect this is the case because, in order to apply it, one would first need to perform a double-negation translation of an actual proof of Thue-Siegel-Roth into a semi-intuitionistic theory—something possible to do in principle, but given the sophistication of the original proof, its translation would be an order of magnitude more complex. We propose thus to treat it directly using our approach with computational side-effects, i.e. without a preliminary double-negation translation followed by a functional interpretation. Technically, our approach can be seen as a version of the so called modified realizability technique but where the language of realizers is enriched to contain delimited control operators.

In combinatorics, we would intend to unwind Szemerédi's theorem saying that every subset of the natural numbers with positive upper density contains arithmetic progressions of arbitrary length. Conjectured by Erdős and Turán in 1936, this statement was only proved by Szemerédi in 1975 by an ingenious and complex combinatorial argument. In 1977, Furstenberg provided a proof using ergodic theory. The interest in giving a better proof of this theorem is still ongoing, and applications include for instance the recent work of Green and Tao on arbitrary long arithmetic progressions in the prime numbers [24].

We first intend to address an important special case of the theorem, the van der Waerden theorem, saying that if we use a finite number of colors to color the natural numbers, then there is at least one color containing arbitrarily long arithmetic progressions. The current upper bound for van der Waerden's number $W(k,r)$, where r is the number of colors and k is the requested length of an arithmetic progression, was obtained via Szemerédi's theorem and is due to Gowers [23]. What is intriguing in this subject is that the upper bounds appear to be heavy overestimates: for instance, the bound for $W(3,3)$ is of the order of 10^{14616}, while the exact value is 27.

Girard has previously analyzed Furstenberg and Weiss's proof of van der Waerden's theorem using cut elimination [20]. But, the bound that he arrived at was essentially the same upper bound obtained by Furstenberg and Weiss [53]. We could attack the problem by using our modified realizability based on computational side-effects and attempt to partially evaluate the latest available proofs for Szemerédi's and van der Waerden's theorem—that would avoid the need for having a fully formalized proof on hand.

Feasibility for Perspective 2

Although the *statements* of the mentioned theorems in number theory and combinatorics are arithmetical, their *proofs* are not arithmetical. The risk for the objectives of this task is to cope with the highly non-effective nature of proofs, as well as their considerable complexity (see Szemerédi's diagram of lemmas from his proof in [62]). After all, proofs of the corresponding theorems have brought Fields medals to both Roth and Szemerédi. The main proof theoretic question is which kind of ideal principles are at the core of arguments and can we provide a direct constructive justification for them. Sometimes, as in the case of Kruskal's theorem, another statement of Ramsey theory, the link to the open induction principle (analyzed previously in joint work with Nakata [38]) turns out to be direct [65].

We intend to use proof assistant technology and partial formalization to cope with the complexity of proofs. Concerning mathematical risk, given a choice of motivated post-doctoral researchers to work on this topics, I would say that it is medium. Work on ergodic theory done in the previous Perspective 1: Unwinding in Analysis Revisited would serve as preparatory work and would help to further mitigate the risk. This task demands more resources than the other ones.

3.3 Perspective 3: Unwinding Incompleteness Theorems

A statement is said to be independent from a theory if it can neither be proved nor disproved from the axioms of the theory. The *incompleteness* phenomenon is the fact that for *any* theory, under the assumption that it is consistent, there exist statements that are independent of the theory. One might wonder what is the nature of these statements, and whether they are relevant in practice. Indeed, the first such statement discovered by Gödel in 1931 has an 'artificial' flavor since it encodes the Epimenides' liar paradox. But, later, natural examples from Ramsey theory have been found, first by Paris and Harrington [55], and include important results like Kruskal's tree theorem. Finding concrete mathematical incompleteness statements is nowadays a fruitful field of research led by Friedman [18].

However, what we find especially interesting is the *limit* at which a statement starts to become independent from a theory. This phenomenon, called *phase transition* by analogy to thermodynamics, happens when the *provability* of a theorem, taking a rational number as parameter, depends on the *value* of this rational parameter. For instance, a parametrized version of Kruskal's theorem can be provable in Peano arithmetic (PA) below a certain value of the parameter, and becomes independent above that value—this is in fact a real number, often found by use of analytic combinatorics, and provides a measure of the strength of the axiom system. Phase transitions are a research program led by Weiermann [67].

In this task, we propose to develop a method for unwinding incompleteness theorems and phase transition phenomena based on programming language theory. The idea is that an incompleteness theorem, $PA \nvdash \bot \to PA \nvdash \text{Con}(PA)$, saying that no consistent formal system (in this case, Peano arithmetic (PA)) can prove its own consistency, can be rephrased positively as $PA \vdash \text{Con}(PA) \to PA \vdash \bot$. Translated in programming languages terms, $PA \vdash \text{Con}(PA)$ expresses the possibility of writing an interpreter for Gödel's system T inside system T itself—that is, a *self-interpreter*. Self-interpreters have not only been studied in programming languages theory, but they are a standard way to bootstrap a compiler for a programming language.

Nevertheless, self-interpreters are usually written for a Turing-complete languages like Scheme and ones without a strong typing discipline. If one adds a type system on top of Scheme one can retrieve system T in its λ-calculus formulation. There are recent intriguing results on typed self-interpreters. Brown and Palsberg have recently constructed the first *typed* self-interpreter [7]; their target was Girard's system U, and this is still 'acceptable', since system U is known to be inconsistent as a logical system. But, their latest result concerns Girard's system F_ω [6], which is a higher-order logic and considered to be consistent.

In this task, we would first investigate whether it is possible to construct a self-interpreter for system T. For the purpose of the paper [35], we have already developed a formally verified interpreter for system T^+ inside Martin-Löf type theory. Since this type theory has a realizability model based on system T, one comes close to having a self-interpreter. We would also have to study the recent results of Brown and Palsberg, and attempt to retrieve their result for system F_ω in system T.

Feasibility for Perspective 3

The proposed methodology involves a frontal attack on consistency of PA. Although the risk is high, the fact that the prior works of Brown and Palsberg, and the author, all involve formally checked proofs, gives us some confidence. If our effort succeeds, the gain one may have will be equally high as the risk. But, even if it turns out to be impossible to write a typed self-interpreter for T, we can aim to obtain solid results on interpreting Weiermann's phase transition, and hence characterizing the strength of a formal system, in terms of notions that are equally natural from the point of view of computation as analytic combinatorics are.

3.4 Perspective 4: Identity of Proofs and Homotopy Type Theory

Formal proofs are combinatorial objects meant to encode a fully correct mathematical argument, going down to the smallest details, but that makes it difficult to spot the most essential parts of an arguments. Curiously, finding "criteria of simplicity, or proof of the greatest simplicity of certain proofs" was already part of Hilbert's program, who even planned to include it as the 24th in his famous list of open problems [63]. In particular, Hilbert asked for a procedure to decide when two given proof are essentially the same. This problem known as *identity of proofs* is still open [15].

In this task, we would start by tackling the identity of proofs for constructive logic, before proceeding to a vast generalization of it, the computational interpretation of Voevodsky's univalence axiom in homotopy type theory [11] in the case when the underlying definitional equality has been strengthened to decide identity of proofs i.e. to convertibility modulo isomorphism for dependent types.

For intuitionistic *propositional* logic, the difficulty of deciding identity of proofs comes from the simultaneous presence of disjunction and implication. Nevertheless, if we follow the analogy between formulas, types, and exponential polynomials, explained in section Type Isomorphisms, we can re-express the problem precisely as that of the effective decidability of the $\beta\eta$-equational theory for the lambda calculus with coproducts. We have recently proposed a first step in this direction by showing how to decompose the equational theory for terms, by the use of the exp-log normal form for types in order to enlarge the $\beta\eta$-congruence classes of terms [37] (in review).

This exp-log normal form of types is extensible to the *first-order* case, when the quantifiers $\forall$ and $\exists$ are also present. Namely, recent work with Brock-Nannestad [5] shows that it leads to an intuitionistic arithmetical hierarchy, a classification of formulas that was elusive for intuitionistic logic, even though it has existed for classical logic since the 1930's where it is at the basis of results like the completeness theorem.

A further question is whether we can make the technique work for dependent types, an extension of the first-order case. Martin-Löf Type Theory has dependent types which allow it to have special treatment of equality. Basic equality between elements a, a' of a type A is encoded by the identity type for A, $\text{Id}_A(a, a')$. Identity of proofs in this context means extending the notion of definitional (computational) equality to cope with η-equality for coproducts (and other inductive types).

Pursuing generalization even further, we can talk about *identity between proofs of identity*, $\text{Id}_{\text{Id}_A}(p, p')$, that, in turn, endows ever type A with the structure of a groupoid. Iterating this construction, $\text{Id}_{\text{Id}_{\text{Id}\ldots \text{Id}_A}}$, allows to show that every type A is in fact endowed with the structure of an ∞-groupoid [30]. Using Grothendieck's correspondence between ∞-groupoids and homotopy types has led Voevodsky to give a homotopy theoretic interpretation of type theory in his model based on simplicial sets [39]. This model satisfies Voevodsky's *univalence axiom*, generalizing identity of proofs, and specializing to: equality at the level of propositions, bijection at the level of sets, categorical equivalence at the level of groupoids, etc. Adding this axiom on top of Martin-Löf's type theory produces homotopy type theory, which is a logical system formalizing the *univalent foundations* of mathematics [64].

What we propose to do is to build the convertibility of proof terms modulo type isomorphism into the definitional equality of Martin-Löf and homotopy type theory. An identity type then gets to cover equality between terms of a whole class of isomorphic types instead of only one type. We hope that in this way it will be possible to strengthen the notion of *transport of structures* and to show that important special cases of the univalence axioms satisfy a simple computational interpretation. The only existing computational interpretation of homotopy type theory appears in the effective version of the simplicial set model [1] and works for the standard (restricted) notion of identity type.

Feasibility for Perspective 4

The univalence axiom is known to imply a form of full functional extensionality in type theory. Given that extensionality of functions in general is undecidable, the risk for extending the computational interpretation for the univalence axiom defined over the notion of identity types strengthened to work modulo isomorphism is high. Nevertheless, by strengthening the underlying definitional equality of the type theory, we hope to diminish the need for resorting to full functional extensionality and even address important special cases of univalence more simply than before.

As concerns the identity of proofs for the propositional and first-order case, based on our preliminary investigations of this area, we would say that the risk is moderate.

3.5 Perspective 5: A Next Generation of Constructive Foundations

This task would serve as an umbrella for more specific but important problems that need to be tackled in the foundations of constructive mathematics, as well as an umbrella collecting the foundational implications of the previous four tasks of this chapter.

For instance, we already know that axioms which are independent of intuitionistic logic like double negation shift can be safely added to intuitionistic systems, but we have to establish the outer limits of the potential given by Computational Side-Effects. We need to develop direct computational interpretations of principles arising from the work in constructive reverse mathematics, such as the equivalent forms of the open induction principle [66], the extension of our work [38] to Baire space, and novel versions of Markov's principle [19].

Another important topic will be to provide a direct constructive proof of Goodman's theorem. This theorem says that the axiom of choice presents a conservative extension of higher type Heyting arithmetic concerning arithmetical formulas; for the meta-theory of constructive mathematics, it plays the role that Hilbert's ϵ-elimination theorems play for the proof theory of classical logic. There has recently been renewed interest about this old result of Goodman by other researchers as well [41, 10].

A third important topic will be to find practical decision algorithms for type isomorphism. As explain in the section Type Isomorphisms, although a decidability result holds for type isomorphisms [36], thanks to prior work of Richardson [57] and Macintyre [52], it is not clear at the moment whether a (practical) decision algorithm can be constructed. Arriving at such an algorithm would not only be useful for proof theory, but also for symbolic computation.

Finally, we would like to interact with the researchers working on proof assistant systems like Coq. The logical cores of proof assistants are lagging behind contemporary proof theory. For instance, program extraction from proofs in a state-of-the-art proof assistants such as Coq relies on the simplest possible realizability interpretation, the so called modified realizability interpretation of Kreisel. Integrating the techniques from the New Unwinding Toolbox would be beneficial for users of proof assistants because it would allow for easier formalization of many apparently ineffective proofs.

3.5.1 Feasibility for Perspective 5

The main challenge for this task is that, when we are interpreting semi-intuitionistic principles, we are working at the limit of computability: our realizability models for the classical axiom of choice refute the internal (formal) version of Church's thesis, but the external weak Church's rule still holds [35] (in review). It is thus hard to predict upfront how far the outer limits of constructive foundations can be extended. As concerns Goodman's theorem, we think the risk involved is not very high, since after all this result has been establish by non-direct methods. Finally, the risk on finding a practical algorithm deciding type isomorphism is hard to estimate; but even if we manage to find ones that only work for special cases, the benefits could spread also beyond proof theory.

References

[1] Marc Bezem, Thierry Coquand, and Simon Huber. A model of type theory in cubical sets. In *19th International Conference on Types for Proofs and Programs (TYPES 2013)*, volume 26, pages 107–128, 2014.

[2] Errett Bishop and Douglas S. Bridges. *Constructive Analysis*, volume 279 of *Grundlehren der mathematischen Wissenschaften*. Springer-Verlag Berlin Heidelberg, 1985.

[3] E Bombieri and AJ Van der Poorten. Some quantitative results related to roth's theorem. *Journal of the Australian Mathematical Society (Series A)*, 45(02):233–248, 1988.

[4] N. Bourbaki. Foundations of mathematics for the working mathematician. *The Journal of Symbolic Logic*, 14(1):1–8, 1949.

[5] Taus Brock-Nannestad and Danko Ilik. An intuitionistic formula hierarchy based on high-school identities. *arXiv:1601.04876*, 2016. Submitted.

[6] Matt Brown and Jens Palsberg. Breaking through the normalization barrier: A self-interpreter for F-omega. To appear in Proceedings of the 43rd Annual ACM SIGPLAN-SIGACT Symposium on Principles of Programming Languages.

[7] Matt Brown and Jens Palsberg. Self-representation in Girard's system U. In *Proceedings of the 42nd Annual ACM SIGPLAN-SIGACT Symposium on Principles of Programming Languages*, pages 471–484. ACM, 2015.

[8] Wolfgang Burr. Fragments of Heyting arithmetic. *The Journal of Symbolic Logic*, 65(3):1223–1240, 2000.

[9] Stanley N. Burris and Karen A. Yeats. The saga of the high school identities. *Algebra Universalis*, 52:325–342, 2004.

[10] Thierry Coquand. About Goodman's theorem. *Annals of Pure and Applied Logic*, 164(4):437–442, 2013.

[11] Thierry Coquand. Théorie des types dépendants et axiome d'univalence. *Séminaire BOURBAKI*, 66(1085), 2014.

[12] Thierry Coquand and Peter Dybjer. Intuitionistic model constructions and normalization proofs. *Mathematical Structures in Computer Science*, 7(1):75–94, 1997.

[13] Olivier Danvy. Type-directed partial evaluation. In *Proceedings of the Twenty-Third Annual ACM SIGPLAN SIGACT Symposium on Principles of Programming Languages (POPL'96)*, pages 242–257, 1996.

[14] Charles N. Delzell. Kreisel's unwinding of Artin's proof. In Piergiorgio Odifreddi, editor, *Kreiseliana. About and Around Georg Kreisel*, pages 113–246. A K Peters, 1996.

[15] Kosta Došen. Identity of proofs based on normalization and generality. *Bulletin of Symbolic Logic*, 9(4):477–503, 2003.

[16] Solomon Feferman. Kreisel's "Unwinding" Program. In Piergiorgio Odifreddi, editor, *Kreiseliana. About and Around Georg Kreisel*, pages 247–273. A K Peters, 1996.

[17] Marcelo Fiore, Roberto Di Cosmo, and Vincent Balat. Remarks on isomorphisms in typed lambda calculi with empty and sum types. *Annals of Pure and Applied Logic*, 141:35–50, 2006.

[18] Harvey Friedman. *Boolean Relation Theory and Incompleteness*. Lecture Notes in Logic. ASL Publications, 2015.

[19] Makoto Fujiwara, Hajime Ishihara, and Takako Nemoto. Some principles weaker than Markov's principle. *Archive for Mathematical Logic*, 54(7-8):861–870, 2015.

[20] Jean-Yves Girard. *Proof theory and logical complexity*, volume 1. Bibliopolis, Naples, 1987.

[21] Georges Gonthier. Formal proof—the four-color theorem. *Notices of the AMS*, 55(11):1382–1393, December 2008.

[22] Georges Gonthier, Andrea Asperti, Jeremy Avigad, Yves Bertot, Cyril Cohen, François Garillot, Stéphane Le Roux, Assia Mahboubi, Russell O'Connor, Sidi Ould Biha, Ioana Pasca, Laurence Rideau, Alexey Solovyev, Enrico Tassi, and Laurent Théry. A Machine-Checked Proof of the Odd Order Theorem. In Sandrine Blazy, Christine Paulin, and David Pichardie, editors, *ITP 2013, 4th Conference on Interactive Theorem Proving*, volume 7998 of *LNCS*, pages 163–179, Rennes, France, July 2013. Springer.

[23] W.T. Gowers. A new proof of Szemerédi's theorem. *Geometric & Functional Analysis GAFA*, 11(3):465–588, 2001.

[24] Ben Green and Terence Tao. The primes contain arbitrarily long arithmetic progressions. *Annals of Mathematics*, pages 481–547, 2008.

[25] Timothy G. Griffin. A formula-as-types notion of control. In *Conf. Record 17th Annual ACM Symp. on Principles of Programming Languages, POPL'90, San Francisco, CA, USA, 17-19 Jan 1990*, pages 47–58, 1990.

[26] R. H. Gurevič. Equational theory of positive numbers with exponentiation is not finitely axiomatizable. *Annals of Pure and Applied Logic*, 49:1–30, 1990.

[27] Thomas Hales. *Dense sphere packings: A blueprint for formal proofs*, volume 400 of *London Mathematical Society Lecture Note Series*. Cambridge University Press, 2012.

[28] Godfrey Harold Hardy. *Orders of Infinity. The 'Infinitärcalcül' of Paul Du Bois-Reymond*. Cambridge Tracts in Mathematic and Mathematical Physics. Cambridge University Press, 1910.

[29] Hugo Herbelin. An intuitionistic logic that proves Markov's principle. In *Proceedings*

of the 25th Annual IEEE Symposium on Logic in Computer Science, LICS 2010, 11-14 July 2010, Edinburgh, United Kingdom*, pages 50–56. IEEE Computer Society, 2010.

[30] Martin Hofmann and Thomas Streicher. The groupoid interpretation of type theory. In *Twenty-five years of constructive type theory (Venice, 1995)*, volume 36 of *Oxford Logic Guides*, pages 83–111. Oxford Univ. Press, New York, 1998.

[31] Danko Ilik. *Constructive Completeness Proofs and Delimited Control*. PhD thesis, École Polytechnique, Palaiseau, France, October 2010.

[32] Danko Ilik. Delimited control operators prove double-negation shift. *Annals of Pure and Applied Logic*, 163(11):1549 – 1559, 2012.

[33] Danko Ilik. Continuation-passing style models complete for intuitionistic logic. *Annals of Pure and Applied Logic*, 164(6):651 – 662, 2013.

[34] Danko Ilik. Type directed partial evaluation for level-1 shift and reset. In Ugo de'Liguoro and Alexis Saurin, editors, Proceedings First Workshop on *Control Operators and their Semantics*, Eindhoven, The Netherlands, June 24-25, 2013 , volume 127 of *Electronic Proceedings in Theoretical Computer Science*, pages 86–100. Open Publishing Association, 2013.

[35] Danko Ilik. An interpretation of the Sigma-2 fragment of classical Analysis in System T. *arXiv:1301.5089*, 2014. Submitted.

[36] Danko Ilik. Axioms and decidability for type isomorphism in the presence of sums. In *Proceedings of the Joint Meeting of the Twenty-Third EACSL Annual Conference on Computer Science Logic (CSL) and the Twenty-Ninth Annual ACM/IEEE Symposium on Logic in Computer Science (LICS)*, pages 53:1–53:7. ACM, 2014.

[37] Danko Ilik. The exp-log normal form of types: Decomposing extensional equality and representing terms compactly. In *Proceedings of the 44th ACM SIGPLAN Symposium on Principles of Programming Languages*, POPL 2017, pages 387–399, New York, NY, USA, 2017. ACM.

[38] Danko Ilik and Keiko Nakata. A direct version of Veldman's proof of open induction on Cantor space via delimited control operators. *Leibniz International Proceedings in Informatics (LIPIcs)*, 26:188–201, 2014.

[39] Chris Kapulkin, Peter LeFanu Lumsdaine, and Vladimir Voevodsky. The simplicial model of univalent foundations. *arXiv preprint arXiv:1211.2851*, 2012.

[40] Ulrich Kohlenbach. New effective moduli of uniqueness and uniform a priori estimates for constants of strong unicity by logical analysis of known proofs in best approximation theory. *Numerical Functional Analysis and Optimization*, 14(5-6):581–606, 1993.

[41] Ulrich Kohlenbach. A note on Goodman's theorem. *Studia Logica*, 63(1):1–5, 1999.

[42] Ulrich Kohlenbach. Some logical metatheorems with applications in functional analysis. *Transactions of the American Mathematical Society*, 357(1):89–128, 2005.

[43] Georg Kreisel. On the interpretation of non-finitist proofs—Part I. *The Journal of Symbolic Logic*, 16(04):241–267, 1951.

[44] Georg Kreisel. Mathematical significance of consistency proofs. *The Journal of Symbolic Logic*, 23(02):155–182, 1958.

[45] Georg Kreisel. Interpretation of analysis by means of constructive functionals of finite types. In Arend Heyting, editor, *Constructivity in Mathematics, Proceedings of the colloqium held at Amsterdam, 1957*, Studies in Logic and The Foundations of Mathematics, pages 101–127. North-Holland Publishing Company Amsterdam, 1959.

[46] Georg Kreisel and Angus MacIntyre. Constructive logic versus algebraization I. In A.S. Troelstra and D. van Dalen, editors, *The L.E.J. Brouwer Centenary Symposium*, pages 217–260. North-Holland Publishing Company, 1982.

[47] Jean-Louis Krivine. On the structure of classical realizability models of ZF. To appear.

[48] Henri Lombardi and Claude Quitté. *Algèbre commutative – Méthodes constructives*. Calvage & Mounet, Paris, 2011.

[49] Horst Luckhardt. Herbrand-Analysen zweier Beweise des Satzes von Roth: Polynomiale Anzahlschranken. *The Journal of Symbolic Logic*, 54(01):234–263, 1989.

[50] Horst Luckhardt. Bounds Extracted by Kreisel From Ineffective Proofs. In Piergiorgio Odifreddi, editor, *Kreiseliana. About and Around Georg Kreisel*, pages 289–300. A K Peters, 1996.

[51] Saunders Mac Lane. Despite physicists, proof is essential in mathematics. *Synthese*, 111(2):147–154, 1997.

[52] Angus Macintyre. *Model Theory and Arithmetic*, volume 890 of *Lecture Notes in Mathematics*, chapter The laws of exponentiation, pages 185–197. Springer Berlin Heidelberg, 1981.

[53] Angus Macintyre. The mathematical significance of proof theory. *Philosophical Transactions of the Royal Society of London A: Mathematical, Physical and Engineering Sciences*, 363(1835):2419–2435, 2005.

[54] Ray Mines and Fred Richman. *A course in constructive algebra*. Springer, 1988.

[55] Jeff Paris and Leo Harrington. A mathematical incompleteness in Peano arithmetic. *Handbook of mathematical logic*, 90:1133–1142, 1977.

[56] Martin Raussen and Christian Skau. Interview with Abel laureate John F. Nash Jr. *European Mathematical Society. Newsletter*, 97:26–31, September 2015.

[57] Daniel Richardson. Solution of the identity problem for integral exponential functions. *Zeitschrift für mathematische Logik und Grundlagen der Mathematik*, 15:333–340, 1969.

[58] Mikael Rittri. Using types as search keys in function libraries. *Journal of Functional Programming*, 1:71–89, 1991.

[59] Peter Schuster. Induction in algebra: A first case study. *Logical Methods in Computer Science*, 9(3):1–19, 2013.

[60] Helmut Schwichtenberg. On bar recursion of types 0 and 1. *The Journal of Symbolic Logic*, 44(3), 1979.

[61] Helmut Schwichtenberg and Stanley S. Wainer. *Proofs and Computations*. Perspectives in Logic. Cambridge University Press, 2012.

[62] Endre Szemerédi. On sets of integers containing no k elements in arithmetic progression. *Acta Arith*, 27(199-245):2, 1975.

[63] Rüdinger Thiele. Hilbert's twenty-fourth problem. *American Mathematical Monthly*,

2003.

[64] The Univalent Foundations Program. *Homotopy Type Theory: Univalent Foundations of Mathematics.* http://homotopytypetheory.org/book, Institute for Advanced Study, 2013.

[65] Wim Veldman. An intuitionistic proof of Kruskal's theorem. *Archive for Mathematical Logic*, 43:215–264, 2001.

[66] Wim Veldman. The principle of open induction on Cantor space and the approximate-fan theorem. *arXiv preprint 1408.2493*, 2014.

[67] Andreas Weiermann. Analytic combinatorics, proof-theoretic ordinals, and phase transitions for independence results. *Annals of Pure and Applied Logic*, 136, 2005.

[68] Alex Wilkie. On exponentiation – a solution to Tarski's high school algebra problem. *Quaderni di Matematica*, 6, 2000.

Regular Languages of Infinite Trees and Probability

Matteo Mio
École Normale Supérieure de Lyon (ENS-Lyon), France
miomatteo@gmail.com

1 Introduction

In Computer Science, *formal verification* is the process of proving or disproving the correctness of computing systems with respect to a certain property, expressed in some specification language, using rigorous mathematical methods. Examples of computing systems include digital circuits, communication protocols, software expressed as source code, *etc*.

A well established way of representing mathematically the semantics of many systems is in terms of (possibly infinite) trees: each node represents a *state* or *configuration* of the computation, with the root being the initial state, and the parent-children relation of the tree represents the *state-to-state transitions* allowed by the program which might by triggered by the reception of external inputs or other nondeterministic events.

The main gain of having an interpretation of computing systems in terms of some mathematical objects, like trees, is of course that it becomes possible to formally *express* interesting properties of systems and to formally *verify* if a certain property is fulfilled by a given system. For example, the property "the program will eventually halt on every input" might be formalized as the property "the tree does not have any infinite branch". The formal specification language adopted to express properties of trees is called a *tree (program) logic*.

Well known tree logics include *Computation Tree Logic* (CTL) [9][2, §6], its many variants such as CTL* [10][2, §6.8], the modal μ-calculus [17, 6] and the Monadic

---

The author would like to thank the Kurt Gödel Society and the John Templeton Foundation for supporting his research during the period July 2014–July 2016. The research described in this document has also been supported by grant "Projet Émergent" *PMSO* of the École Normale Supérieure de Lyon, by the grant "PEPS JCJC" *EMSO* of the CNRS (France) and by Polish National Science Centre grant no. 2014-13/B/ST6/03595.

Second Order Logic of the full binary tree (MSO, for short) [23][27]. Among these, the logic MSO plays a fundamental role because it is the most expressive as it can formulate the properties expressible in all other logics.

In his fundamental work [23] Michael Rabin proved that the theory of MSO is decidable: there is an algorithm that recognizes the set of valid MSO formulas, i.e., those satisfied by every tree. This theorem is widely regarded among the deepest decidability results in theoretical computer science. Firstly, it implies the decidability of the theory of all other tree logics mentioned above. Secondly, it implies the decidability of two fundamental problems in formal verification:

- Model Checking: given a program represented by a regular[1] tree T and a MSO formula ϕ, verify if T satisfies the property ϕ.

- Synthesis: given a satisfiable MSO formula ϕ, construct a tree T such that T satisfies ϕ.

After the seminal result of Rabin, the theory of the logic MSO has been further investigated and its development is still an active subject of research: connections found with automata theory (tree automata), algorithmic game theory (two-player parity games), descriptive set theory (topological properties of sets of trees definable by MSO formulas), among other mathematical fields, contribute to a rather rich theory.

Contribution The present document presents a high level overview of some of the research carried out by the author during the tenure of the "Gödel Research Prize Fellowship 2014", organized by the Kurt Gödel Society with support from the John Templeton Foundation, in the period July 2014–July 2016.

Three research papers [12, 19, 20] have been selected to illustrate same aspects of this research. The common goal in three these works has been to further advance the theory of the MSO logic by studying measure theoretic (probabilistic) properties of *regular languages of trees*, i.e., of those sets of infinite trees definable by MSO formulas. These three topics will be outlined in sections 3, 4 and 5 and Section 2 will provide some required technical background.

We conclude this section with a brief description of each topic.

Topic 1: Measurability of regular sets

---

[1] An infinite tree is *regular* if it is finitely representable by a finite directed graph.

A formula $\phi(X_1, \ldots, X_n)$ of the MSO logic, having n free variables, defines a collection of tuples $\langle t_1, \ldots, t_n \rangle$ of infinite trees that satisfies ϕ. This collection, denoted by $[\![\phi]\!]$, is called the *regular language of trees* defined by ϕ. The collection of all possible n-tuples is an uncountable set and, once equipped with a natural topology, is homeomorphic to $\mathcal{T}_{0,1}^n$ where $\mathcal{T}_{0,1}$ denotes the *Cantor space*. See Section 2 for more details. A natural question is then the following:

Question: Is $[\![\phi]\!] \subseteq \mathcal{T}_{0,1}^n$ a μ-measurable set, for all Borel measures μ on $\mathcal{T}_{0,1}$?

The question was first raised in the author's PhD thesis ([21], see also [22]) and does not have a straightforward answer because the sets $[\![\phi]\!]$ can have high topologically complexity: for example, they are generally not Borel and they are not even contained in the σ-algebra generated by the analytics sets.

The main theorem proved in [12] is that, indeed, the sets $[\![\phi]\!]$ are always measurable. This, informally, means that it makes mathematical sense to talk about the probability that a randomly generated tree belongs to $[\![\phi]\!]$.

Topic 2: Computing the probability of regular sets

Having proved that regular sets of infinite trees are measurable, the following question arises naturally:

Question: does there exist an algorithm which for a given regular language of trees $[\![\phi]\!] \subseteq \mathcal{T}_{0,1}^n$ computes the probability $\mu_c([\![\phi]\!])$?

where μ_c denotes the natural coin-flipping measure on $\mathcal{T}_{0,1}^n$ (see Section 2). While a complete answer to this question is still missing at the time of writing this report, in [19] we proved that, for a subclass of MSO formulas definable by *game automata*, the problem admits a positive answer.

In the study of [19] we have also been able to establish some basic facts about the nature of this problem. For example, there are formulas ϕ such that $\mu_c([\![\phi]\!])$ is irrational. Also, there are formulas ϕ such that $\mu_c([\![\phi]\!]) = 0$ but the regular language $[\![\phi]\!]$ is large topologically (it is the complement of a meager set). These facts contrast with, e.g., the theory of regular sets of infinite words where all languages have rational (coin-flipping) probability and a language has probability 0 if and only if it is small topologically (it is a meager set). Hence, the probabilistic properties of regular languages of trees seem to be significantly more refined than in the case of languages of infinite words.

Topic 3: MSO extended with the measure quantifier

Beside studying the properties of set of trees definable by the MSO logic (cf. Topic 1) an interesting topic of research is to look at variants or *extensions* of the MSO logic capable of expressing properties of trees not expressible in standard MSO. In the context of the research exposed in this document, it is of particular interests to study extensions of MSO capable of expressing *probabilistic properties* of trees. Indeed, many of the properties that can be formulated by logics for expressing properties of probabilistic programs, such as pCTL and its variants (see, e.g., [2] for an overview) are not expressible in the MSO logic.

In [20] we investigated an extension of the MSO logic, denoted by MSO + $\forall^{=1}$, with the so-called *measure quantifier* $\forall^{=1}$. The intuitive meaning of the formula $\forall^{=1} X.\phi$ is that the formula ϕ holds true for *almost all* trees t in the sense that the set $\{t \mid t \text{ satisfies } \phi\} \subseteq \mathcal{T}_{0,1}$ has coin-flipping measure one. The measure quantifier $\forall^{=1}$ has been studied, in the more general context of first-order logic, by Harvey Friedman in a series of unpublished manuscripts in 1978-1979.

The logic MSO + $\forall^{=1}$ is interesting because it strictly increases the expressive power of standard MSO and is capable of expressing the properties definable in logic for expressing properties of probabilistic programs, such as the qualitative fragment of pCTL (see, e.g., [2]). Our main result regarding MSO + $\forall^{=1}$ is, however, negative: the logic MSO + $\forall^{=1}$ has an undecidable theory. Nevertheless, the decidability of fragments sufficiently expressive to formulate most properties of interests in (probabilistic) program verification is an interesting problem for future research.

2 Technical Background

In this section we give the basic definitions of concepts from descriptive set theory, measure theory and the MSO logic required to read this document. We refer to [15] and [27] as references to these topics.

The set of natural numbers is either denoted by $\mathbb{N}$ or ω, the choice primarily depending on typographical constraints. Given two sets X and Y we denote with X^Y the space of functions $X \to Y$. We can view elements of X^Y as Y-indexed sequences $\{x_i\}_{i \in Y}$ of elements of X. We refer to X^ω as the collection of ω-*words* over X. The collection of *finite* sequences of elements in X is denoted by X^*. As usual we denote with ϵ the empty sequence and with ww' the concatenation of $w, w' \in X^*$.

Given a finite set Σ, the collection Σ^ω of ω-words over Σ, endowed with the product topology (where Σ is given the discrete topology) is called the *Cantor space*.

The Cantor space is *zero-dimensional*, i.e., it has a basis of *clopen* (both open and closed) sets.

The smallest σ-algebra of subsets of Σ^ω containing all open sets is denoted by $\mathcal{B}$ and its elements are called *Borel sets*. A *Borel probability measure* on Σ^ω is a function $\mu : \mathcal{B} \to [0, 1]$ such that: $\mu(\emptyset) = 0$, $\mu(\Sigma^\omega) = 1$ and, if $\{B_n\}_{n \in \omega}$ is a sequence of disjoint Borel sets, $\mu(\bigcup_n B_n) = \sum_n \mu(B_n)$. A subset $A \subseteq \Sigma^\omega$ is called μ-null if there exists a Borel set B such that $A \subseteq B$ and $\mu(B) = 0$. A subset $C \subseteq \{0,1\}^\omega$ is called μ-measurable if $C = A \cup B$ for some Borel set B and some μ-null set A.

We will be mostly interested in one specific Borel measure on the Cantor space which we refer to as *coin-flipping measure*. If $\Sigma = \{a_1, \ldots, a_n\}$, this is the unique Borel measure satisfying the equality $\mu(B_{m=a_1}) = \mu(B_{m=a_2}) = \ldots \mu(B_{m=a_n}) = \frac{1}{n}$ where, for $m \in \mathbb{N}$, we define $B_{m=a_i} = \{(b_i)_{i \in \mathbb{N}} \mid b_m = a_i\}$. Intuitively, the coin-flipping measure on Σ^ω generates an infinite sequence $(b_0, b_1, \ldots)$ by randomly choosing to fix $b_m = a_i$ with the uniform distribution on Σ, for every $m \in \omega$.

2.1 Syntax and Semantics of Monadic Second Order Logic

In this section we define the syntax and the semantics of the MSO logic interpreted over the full binary tree.

Definition 2.1 (Full Binary Tree). *The collection $\{L, R\}^*$ of finite words over the alphabet $\{L, R\}$ can be seen as the set of vertices of the infinite binary tree. We refer to $\{L, R\}^*$ as the full binary tree. We use the letters v and w to range over elements of the full binary tree.*

Definition 2.2 (Syntax). *The set of formulas of the logic MSO on the full binary tree is generated by the following grammar:*

$$\phi ::= \mathrm{Sing}(X) \mid \mathtt{succ}_L(X, Y) \mid \mathtt{succ}_R(X, Y) \mid X \subseteq Y \mid \neg\phi \mid \phi_1 \vee \phi_2 \mid \forall X.\phi$$

where X, Y range over a countable set of variables.

Hence MSO formulas are conventional first-order formulas over the signature $\mathcal{S}$ consisting of one unary symbol $Sing$ and three binary symbols $\mathtt{succ}L, \mathtt{succ}R, \subseteq$. We interpret MSO formulas over the collection $\{L, R\}^* \to \{0, 1\}$ of subsets of the full binary. To improve the notation, we denote with $\mathcal{T}_{0,1}$ the space (homeomorphic to the Cantor space) $\{L, R\}^* \to \{0, 1\}$. Thus MSO formulas are interpreted over the universe $\mathcal{T}_{0,1}$ with the following interpretations of the symbols in $\mathcal{S}$:

- $Sing^I(X) \Leftrightarrow X = \{v\}$, for some $v \in \{L, R\}^*$, i.e., if $X \in \mathcal{T}_{0,1}$ is a singleton.

- $\mathtt{succ}L^I(X, Y) \Leftrightarrow$ "$X = \{v\}$, $Y = \{w\}$ and $w = vL$.

- $\text{succ}R^I(X,Y) \Leftrightarrow$ "$X=\{v\}$, $Y=\{w\}$ and $w=vR$.

- $\subseteq^I (X,Y) \Leftrightarrow X \subseteq Y$, i.e., if X is a subset of Y.

Definition 2.3 (Semantics). *Let $\mathbb{T} = \langle \mathcal{T}_{0,1}, \text{Sing}^I, \text{succ}L^I, \text{succ}R^I, \subseteq^I \rangle$ be the relational structure as above. The truth of a MSO formula ϕ is given by the standard Tarski's satisfiability relation $\mathbb{T} \models \phi$ for first order logic. Given parameters $\vec{A} \in \mathcal{T}_{0,1}$, we write $\vec{A} \in \phi(\vec{X})$ to indicate that $\mathbb{T} \models \phi(A_1, \ldots, A_n)$, i.e., that $\mathbb{T}$ satisfies the formula ϕ with parameters $\vec{A}$.*

Lastly, we can introduce the notion of regular languages of trees.

Definition 2.4 (Regular Languages). *Given a MSO formula $\phi(X_1, \ldots, X_n)$, the set $[\![\phi]\!] \subseteq \mathcal{T}_{0,1}^n$ defined as:*

$$[\![\phi]\!] = \{\langle A_1, \ldots, A_n\rangle \mid \mathbb{T} \models \phi(A_1, \ldots, A_n)\}$$

is called the regular language *of trees defined by ϕ.*

3 Measurability of regular sets

In his PhD thesis [21], the author asked the following question (we adhere here to the notation introduced in Section 2):

Question: Given an arbitrary MSO formula $\phi(X_1, \ldots, X_n)$, is the regular set $[\![\phi]\!] \subseteq \mathcal{T}_{0,1}^n$ a μ-measurable set, for every Borel measure μ on $\mathcal{T}_{0,1}^n$?

This question does not admit a straightforward positive answer because regular sets generally belong to the $\mathbf{\Delta}_2^1$-class of sets in the projective hierarchy of Polish spaces. This high topological complexity is a concern due to a celebrated result of Kurt Gödel (see [13, §25]) which states that it is consistent with Zermelo-Fraenkel Set Theory with the Axiom of Choice (ZFC) that there exists a $\mathbf{\Delta}_2^1$ set which is not measurable. This means that it is not possible to prove (in ZFC) that all $\mathbf{\Delta}_2^1$-sets are measurable.

Measure theoretic problems such as the one formulated in the above Question have been investigated since the first developments of measure theory, in late 19th century, as the existence of non-measurable sets (e.g. Vitali sets [13]) was already known. The measure-theoretic foundations of probability theory are based around the concept of a σ-algebra of measurable events on a space of potential outcomes. Typically, the σ-algebra is assumed to contain all open sets. Hence the minimal σ-algebra under consideration consists of all Borel sets whereas the maximal consists, by definition, of the collection of all measurable sets. The Borel σ-algebra, while simple to work with, lacks important classes of measurable sets such as the *analytic* ($\mathbf{\Sigma}_1^1$) sets. On the other hand, the full σ-algebra of measurable sets may be difficult to work with since there is no constructive methodology for establishing its membership relation, i.e. for proving that a given set belongs to this σ-algebra. This picture led to a number of attempts to find larger σ-algebras, extending the Borel σ-algebra and including as many measurable sets as possible and, at the same time, providing practical techniques for establishing the membership relation.

A classical methodology for constructing such σ-algebras is to identify a family $\mathcal{F}$ of "safe" operations on sets which, when applied to measurable sets are guaranteed to produce measurable sets. When the operations considered have countable arity (e.g. countable union), the σ-algebra generated by the open sets closed under the operations in $\mathcal{F}$ admits a transfinite decomposition into ω_1 levels, and this allows the membership relation to be established inductively. The simplest case is given by the σ-algebra of Borel sets, with $\mathcal{F}$ consisting of the operations of complementation and countable union. Other less familiar examples include $\mathcal{C}$-sets studied by E. Selivanovski [24], Borel programmable sets proposed by D. Blackwell [5] and $\mathcal{R}$-sets proposed by A. Kolmogorov [16].

Most measurable sets arising in ordinary mathematics are $\mathcal{R}$-sets belonging to the lower levels of the transfinite hierarchy of $\mathcal{R}$-sets. For example, all Borel sets, analytic sets, co-analytic sets and Selivanovski's $\mathcal{C}$-sets lie in the first two levels [8]. Furthermore, the inductive proof method for establishing membership in the class of $\mathcal{R}$-sets has allowed the development of a rich theory of $\mathcal{R}$-sets. Beside the original work of Kolmogorov [16], fundamental results were obtained by Lyapunov [18] and, more recently, by Burgess [8]. Further progress can be found in the work of Barua [3, 4]. We refer to [14] for a modern introduction to the subject.

Our main result is the following (cf. Theorem 1 in [12]).

Theorem 3.1. *Every regular set $[\![\phi]\!] \subseteq \mathcal{T}_{0,1}^n$ is a $\mathcal{R}$-set belonging to a finite level of the hierarchy of $\mathcal{R}$-sets.*

By applying the fact, from Kolmogorov, that every $\mathcal{R}$-set is measurable (with respect to any Borel measure), we get a positive answer to our original question:

Corollary 3.1. *Every regular set $[\![\phi]\!] \subseteq \mathcal{T}_{0,1}^n$ is measurable.*

As already mentioned earlier, most sets appearing in ordinary mathematics belong to very low levels of the hierarchy of $\mathcal{R}$-sets, generally at the second level. This is arguably why the theory of $\mathcal{R}$-sets is a rather exotic and very specialistic topic in descriptive set theory.

To our surprise, when proving Theorem 3.1 above, we discovered that regular sets can (strictly) belong to arbitrarily high levels of the hierarchy of $\mathcal{R}$-sets (cf. Theorem 1 in [12]).

Theorem 3.2. *For every $n \in \mathbb{N}$ there is a regular language $[\![\phi]\!] \subseteq \mathcal{T}_{0,1}^n$ such that $[\![\phi]\!]$ belongs to the $(n+1)$-th level of the hierarchy of $\mathcal{R}$-sets but not to the n-th level.*

Hence the theory of MSO, and of regular languages of trees, provides concrete examples of sets belonging to high levels of the hierarchy of $\mathcal{R}$-sets.

4 Computing the probability of regular sets

As discussed in the previous section (cf. Theorem 3.1), regular languages of infinite trees are μ-measurable for all Borel measures μ on the n-dimensional space $\mathcal{T}_{0,1}^n$ of all infinite trees.

A particularly natural Borel measure on $\mathcal{T}_{0,1}^n$ is the *coin-flipping* measure (see Section 2) denoted, in what follows, by μ_c. The following question is then quite natural:

Question: does there exist an algorithm which for a given regular language of trees $[\![\phi]\!] \subseteq \mathcal{T}_{0,1}^n$ computes the probability $\mu_c([\![\phi]\!])$?

As a standard notion in recursion theory, a real number $r \in [0, 1]$ is computable if there exists an effective procedure P_r that, for each input number $n \in \mathbb{N}$, produces as output the first n digits of the decimal representation of r. Examples of computable numbers include the rational numbers, the algebraic numbers and π. Hence the question above asks for the existence of an algorithm taking as input a regular language, represented by a formula ϕ, and returning (the code of) a procedure P_r computing the real $r = \mu_c([\![\phi]\!])$.

At the time of writing this document, a full answer to the question above is still missing. In what follows we report some preliminary results from [19] on this topic.

The main result from [19] is that, for a special class of MSO formulas ϕ, the question above has a positive answer, i.e., it is possible to compute the probability $\mu_c([\![\phi]\!])$. This class of formulas is defined indirectly as those formulas ϕ such that the language $[\![\phi]\!]$ can be defined by a so-called *game automaton*. The definition of game automata is rather technical and we omit it here (see Section 2.2 of [19] for definitions and further references). An important point, proved in [11], is that given a formula ϕ it is decidable to determine if $[\![\phi]\!]$ can be recognized by a game automaton or not. So the class of formulas definable by game automata is effective. The main result from [19] is the following (cf. Theorem 1 in [19]):

Theorem 4.1. *Let ϕ by a MSO formula definable by a game automaton. Then the probability $\mu_c(\llbracket\phi\rrbracket)$ is computable and is an algebraic number.*

The procedure for computing the probability of a game-automata definable formula is the following. First, from the formula ϕ, a game automaton $\mathcal{A}_\phi$ defining the language $\llbracket\phi\rrbracket$ is constructed. This automaton $\mathcal{A}_\phi$ is then itself transformed into a *two-player stochastic meta-game* G_ϕ, a class of games introduced by the author in his PhD thesis [21]. These transformations are done in such a way that the value of G_ϕ coincide with $\mu_c(\llbracket\phi\rrbracket)$. Finally, the value v of G_ϕ is expressed as the unique real number satisfying a formula $F(x)$, with only one free variable x, in the first order theory of real closed fields. By Tarski's celebrated *quantifier elimination* procedure, the value $v \in [0,1]$ is a computable algebraic number.

This result, while partial, is interesting because it sheds some light on the full problem. For example, it has been possible to exhibit a formula ϕ, definable by game automata such that (cf. Proposition 2 in [19]):

Proposition 4.1. *There exists a game automata definable formula ϕ such that $\llbracket\phi\rrbracket$ is irrational.*

This contrasts with the well known fact that the (coin-flipping) probability of a regular set of infinite words has always rational probability.

As another example, we have shown (cf. Proposition 3 in [19]) that there exists a game automata definable formula ϕ such that the regular set $\llbracket\phi\rrbracket$ has probability 0 but is *comeager*, i.e., it is the complement of a meager set.

Proposition 4.2. *There exists a game automata definable formula ϕ such that $\llbracket\phi\rrbracket$ is a comeager set having probability 0.*

Once again, this contrasts with the theory of regular languages of infinite words where, as proved by Staiger [25], a regular set has probability 0 if and only if it is meager.

Hence, the probabilistic properties of regular languages of trees seem to be significantly more refined than in the case of languages of infinite words.

5 MSO extended with the measure quantifier

In this section we give an overview of the work presented in [20] where an extension of the MSO logic, denoted by $\text{MSO} + \forall^{=1}$, with the so-called *measure quantifier* $\forall^{=1}$, is investigated.

This logic is formally introduced by extending the definitions regarding the MSO logic of Section 2.1 as follows.

Definition 5.1 (Syntax). *The syntax of* $\text{MSO} + \forall^{=1}$ *is obtained by extending that of MSO (Definition 2.2) with the quantifier* $\forall^{=1} X.\phi$ *as follows:*

$$\phi ::= \text{Sing}(X) \mid X < Y \mid X \subseteq Y \mid \neg \phi \mid \phi_1 \vee \phi_2 \mid \forall X.\phi \mid \forall^{=1} X.\phi$$

Definition 5.2 (Semantics). *Each formula* $\phi(X_1, \ldots, X_n)$ *of* $\text{MSO} + \forall^{=1}$ *is interpreted as a subset of* $\mathcal{T}_{0,1}^n$ *by extending Definition 2.3 with the following clause:*

$$\langle A_1, \ldots, A_n \rangle \in [\![\forall^{=1} X.\phi(X, Y_1, \ldots, Y_n)]\!]$$
$$\Leftrightarrow$$
$$\mu_c(\{B \mid \langle B, A_1, \ldots, A_n \rangle \in [\![\phi(X, Y_1, \ldots, Y_n)]\!]\}) = 1$$

where A_i, B *range over the space of trees* $\mathcal{T}_{0,1}$ *and* μ_c *is the coin-flipping measure on* $\mathcal{T}_{0,1}$.

The set denoted by $\forall^{=1} X.\phi(X, \vec{Y})$ can be illustrated as in Figure 1, as the collection of tuples $\vec{A}$ having a *large* section $\phi(X, \vec{A})$, that is a section having coin-flipping measure 1. Informally, $\langle A_1, \ldots, A_n \rangle$ satisfies $\forall^{=1} X.\phi(X, \vec{Y})$ if "for almost all" $B \in \mathcal{T}_{0,1}$, the tuple $(B, A_1, \ldots, A_n)$ satisfies ϕ.

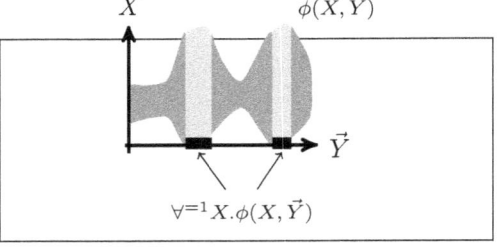

Figure 1: *The large sections selected by the quantifier* $\forall^{=1}$ *are marked in grey.*

The quantifier $\forall^{=1}$ has been originally investigated, in the general context of first order logic, by Harvey Friedman in unpublished manuscripts in 1978–79. See [26] for an overview on Friedman's research.

An interesting fact about $\text{MSO} + \forall^{=1}$ (cf. Theorem 5 in [20]) is that this logic is capable of expressing the properties definable in qualitative pCTL (see, e.g., [2]) and other similar logics for expressing properties of probabilistic programs. Remarkably, the decidability of the theory of such logics, is a long-standing open problem (see, e.g., [7] for references on the problem).

Our main result about $\text{MSO} + \forall^{=1}$ is, however, a negative one (cf. Theorem 1 in [20]).

Theorem 5.1. *The logic* $\text{MSO} + \forall^{=1}$ *has an undecidable theory.*

The proof of this theorem is by a reduction to a problem, in the theory of probabilistic automata, recently proved undecidable in [1].

In an attempt to circumvent this negative results, in [20, §6] a fragment of $\text{MSO} + \forall^{=1}$ has been identified and named $\text{MSO} + \forall^{=1}_\pi$. Interestingly, this fragment is still sufficiently expressive to interpret most logics of probabilistic programs, such as pCTL and its (un)decidability has not yet been determined. In particular, the proof method adopted to prove undecidability of MSO + $\forall^{=1}$ does not seem to be adaptable to the case of MSO + $\forall^{=1}_\pi$. The further study of the logic MSO + $\forall^{=1}_\pi$ appears to be an interesting subject of future research.

References

[1] Christel Baier, Marcus Grösser, and Nathalie Bertrand. Probabilistic ω-automata. *Journal of the ACM*, 59(1), 2012.

[2] Christel Baier and Joost Pieter Katoen. *Principles of Model Checking*. The MIT Press, 2008.

[3] Rana Barua. R-sets and category. *Transactions of the American Math. Society*, 286, 1984.

[4] Rana Barua. *Studies in Set-Theoretic Hierarchies: From Borel sets to R-sets*. PhD thesis, Indian Statistical Institute, Calcutta, 1986.

[5] D. Blackwell. Borel–programmable functions. *Ann. of Probability*, 6:321–324, 1978.

[6] Julian C. Bradfield and Colin Stirling. Modal logics and mu-calculi: an introduction. In *Handbook of Process Algebra*. Elsevier, 2001.

[7] Tomáš Brázdil, Vojtech Forejt, Jan Kretínský, and Antonín Kucera. The satisfiability problem for probabilistic ctl. In *Proceedings of the 2008 23rd Annual IEEE Symposium on Logic in Computer Science*, pages 391–402, 2008.

[8] John P. Burgess. Classical hierarchies from a modern standpoint. II. R-sets. *Fund. Math.*, 115(2):97–105, 1983.

[9] E. M. Clarke and E. A. Emerson. Design and synthesis of synchronisation skeletons using branching time temporal logic. In *Logic of Programs, Proceedings of Workshop, Lecture Notes in Computer Science*, volume 131, pages 52–71, 1981.

[10] E. A. Emerson and Joseph Y. Halpern. "sometimes" and "not never" revisited: on branching versus linear time temporal logic. *Journal of the ACM (JACM)*, 33, 1986.

[11] Alessandro Facchini, Filip Murlak, and Michal Skrzypczak. Rabin-Mostowski index problem: A step beyond deterministic automata. In *Proc. of LICS*, pages 499–508, 2013.

[12] Tomasz Gogacz, Henryk Michalewski, Matteo Mio, and Michał Skrzypczak. Measure Properties of Game Tree Languages. In *Proc. of MFCS*, 2014.

[13] Thomas Jech. *Set Thery*. Springer Monographs in Mathematics. Springer, 2002.

[14] Vladimir G. Kanovei. Kolmogorov's ideas in the theory of operations on sets. *Russian Math. Surveys*, 43(6):111–155, 1988.

[15] A. S. Kechris. *Classical Descriptive Set Theory*. Springer Verlag, 1994.

[16] A.N. Kolmogorov. Operations sur des ensembles (in Russian, summary in French). *Mat. Sb.*, 35:415–422, 1928.

[17] D. Kozen. Results on the propositional mu-calculus. In *Theoretical Computer Science*, pages 333–354, 1983.

[18] A. A. Lyapunov. $\mathcal{R}$-sets. *Trudy Mat. Inst. Steklov.*, 40:3–67, 1953.

[19] Henryk Michalewski and Matteo Mio. On the problem of computing the probability of regular sets of trees. In *In Proceedings of FSTTCS*, pages 489–502, 2015.

[20] Henryk Michalewski and Matteo Mio. Measure quantifier in monadic second order logic. In *In Proceedings of LFCS*, pages 267–282, 2016.

[21] Matteo Mio. *Game Semantics for Probabilistic μ-Calculi*. PhD thesis, School of Informatics, University of Edinburgh, 2012.

[22] Matteo Mio. Probabilistic Modal μ-Calculus with Independent product. *Logical Methods in Computer Science*, 8(4), 2012.

[23] Michael O. Rabin. Decidability of second-order theories and automata on infinite trees. *Transactions of American Mathematical Society*, 141:1–35, 1969.

[24] E. Selivanowski. Sur une classe d'ensembles effectifs (ensembles C) (in Russian, summary in French). *Mat. Sb.*, 35:379–413, 1928.

[25] Ludwig Staiger. Rich omega-words and monadic second-order arithmetic. In *Proc. of CSL*, pages 478–490, 1997.

[26] C. I. Steinhorn. *Borel Structures and Measure and Category Logics*, volume 8 of *Perspectives in Mathematical Logic*, chapter 14, pages 579–596. Springer-Verlag, 1985.

[27] Wolfgang Thomas. Languages, automata, and logic. In *Handbook of Formal Languages*, pages 389–455. Springer, 1996.

Orbit Equivalence Relations

Marcin Sabok
Department of Mathematics and Statistics, McGill University, Canada and Institute of Mathematics, Polish Academy of Sciences, Poland
`marcin.sabok@mcgill.ca`

Abstract

We survey recent results obtained in the area of orbit equivalence relations and the complexity of classification problems.

1 Introduction

During the last two years there have been dramatic breakthroughs in the theory of orbit equivalence relations leading to solutions of long standing open problems and applications in other areas such as the classification program of nuclear simple separable C*-algebras. The purpose of this article is to give an overview of these results and the proof techniques that stand behind them.

Many classification problems appearing in mathematics can be understood in terms of analytic equivalence relations. A working assumption here is a version of the Church–Turing thesis saying that the classes of (usually separable) objects that appear in mathematical practice admit natural and unique Borel structures. This is the case in all examples considered in this article: separable metric spaces, separable Banach spaces, separable C*-algebras, metrizable compact spaces and metrizable Choquet simplices. While there may be more than one natural Polish topology on the classes of objects, all such topologies lead to the same standard Borel structure. From this point of view, isomorphism relation considered on such classes of objects is always an analytic equivalence relation as isomorphism of two objects is expressed using one existential quantifier.

One situation when the isomorphism relation can be perfectly understood is when we are able to assign simple invariants, such as numbers (natural, or even reals), to the objects that are being classified and obtain a complete classification:

The author acknowledges support of the John Tempelton Foundation through the Kurt Gödel Research Prize Fellowship in the category Logical Foundations of Mathematics.

two objects are isomorphic if and only if the corresponding numbers are the same. While this does happen is some cases, such as the classification of closed orientable surfaces via their genus or the classification of Bernoulli shifts via their entropy, such results are extremely rare. One should be careful here, however, as the assignment of numbers should be definable in a suitable sense. Without this assumption, one can always use the axiom of choice to assign such complete invariants. Therefore, the map computing the real invariants should be a Borel map from the standard Borel space of objects to the reals. Such equivalence relations (or the corresponding classification problems) are called *smooth*.

The complexity theory for classification classification problems began with the observation that there exist Borel equivalence relations which are not smooth. The simplest such example is the so-called *Vitali equivalence relation* and arises as the coset equivalence relation of the rationals in the additive group of real numbers. The first example appearing as a classification problem (in representation theory and operator algebras) came with the work of Mackey and Glimm on the classification of unitary irreducible group representations. Glimm proved that this classification is not smooth if and only if the group is not of type I. This led Glimm to a general result saying that for a continuous action of a locally compact group, the equivalence relation arising from the orbits is not smooth if and only if the Vitali equivalence relation is Borel-reducible to this equivalence relation (for definitions, see Section 3). The latter was later generalized by Effros and finally by Harrington, Kechris and Louveau to what is called today the Glimm–Effros dichotomy and provides the first dividing line in the complexity theory of analytic equivalence relations.

The complexity theory of analytic equivalence relations has been developed extensively over the last forty years and is now a rich and powerful theory having connections to model theory, ergodic theory and operator algebras. The goal of this article is to survey some very recent developments which, although originating within the modern framework of descriptive set theory, still have connections to classification problems in operator algebras, in the spirit of Mackey and Glimm.

2 Polish group actions

We consider various classes of structures: these include some metric structures, as defined in [2] but also topological (compact) spaces. A structure X is *universal* for a class C if any structure in C can be embedded into X. Particular examples of universal structures that are interesting from our point of view are:

- the Urysohn space, universal for separable metric spaces,
- the Hilbert cube, universal for compact metrizable spaces,

- the Cuntz algebra $\mathcal{O}_2$, universal for separable nuclear C*-algebras,
- the Gurarij space, universal for separable Banach spaces,
- the Poulsen simplex, universal for metrizable Choquet simplices.

Automorphism groups of structures as above have natural Polish topologies. Given a separable metric structure X, the group of its automorphisms $\mathrm{Aut}(X)$, i.e. isometries that preserve the additional structure is endowed with the pointwise convergence topology and is a Polish group. The group of homeomorphisms of the Hilbert cube and the group of affine homeomorphisms of the Poulsen simplex are also Polish groups with the compact-open topologies. The existence of well-behaved metrics on the automorphism group of a metric structure X often depend on the structure X but there is always a complete (usually not left-invariant, though) metric on $\mathrm{Aut}(X)$.

A Polish group G is a *universal Polish group* if every Polish group H is isomorphic as a topological group to a (necessarily closed) subgroup of G. Interestingly, there are many (nonisomorphic) examples of universal Polish groups that almost always arise as automorphism or homeomorphism groups of structures that are universal in their respective categories. The following groups are universal Polish groups:

- the isometry group of the Urysohn space (Uspenskij [35]),
- the linear isometry group of the Gurarij space (Ben Yaacov [1]),
- the homeomorphism group of the Hilbert cube (Uspenskij [34]),
- the affine homeomorphism group of the Poulsen simplex.

For the Cuntz algebra, the structure of its automorphism group is less understood and it is an open problem whether this group is a universal Polish group:

Question 1. ([31]) Is the automorphism group of the Cuntz algebra $\mathcal{O}_2$ a universal Polish group?

We are interested in particular actions of the automorphism groups of universal structures that induce equivalence relations of isomorphism on these classes of structures. Given a class Z of closed substructures of a structure X we say that Z has the *extension property* if every isomorphism between substructures in Z can be extended to an automorphism of X. Examples of classes of closed substructures with the extension property include:

- Z-subsets of the Hilbert cube,

- separable metric spaces embedded in the Urysohn space via the Katětov construction,

- separable nuclear C*-algebras embedded in the Cuntz algebra $\mathcal{O}_2$ via the tensor product with $\mathcal{O}_2$ (by the Kirchberg embedding theorem),

- proper faces of the Poulsen simplex.

In all the above examples, given a universal structure X for a class of objects C, the family Z of its closed substructures with the extension property is invariant under homeomorphisms of X and is already universal, i.e. and any structure in C can be embedded into X in such a way that the image of this embedding lies in Z. In such case, the automorphism group group of X induces the action on Z, and the orbits of this action can be identified with the isomorphism classes of objects in C.

3 The hierarchy of Borel reducibility

The theory of analytic equivalence relations was developed partly in the hope of providing means of attack to the Vaught conjecture but it quickly became clear that it can be useful in much broader context and describe structure of many classification problems appearing in mathematics.

Note at this point that if a Polish group $G \curvearrowright Z$ acts in a Borel way on a standard Borel space Z, then the induced orbit equivalence relation E_Z^G is analytic: for two points $z_1, z_2 \in Z$ we have $z_1 \, E_Z^G \, z_2$ if and only if $\exists g \in G \quad g \cdot z_1 = z_2$. Such analytic equivalence relations are called *orbit equivalence relation*.

The relative structure of orbit equivalence relations is measured in terms of the Borel reducibility order. Although there are equivalence relations that are not orbit equivalence relations (nor even Borel reducible to orbit equivalence relations), within this article we restrict attention to orbit equivalence relations. Suppose G_1 acts on Z_1 and G_2 acts on Z_2 in a Borel way. We say that the relation $E_{Z_1}^{G_1}$ is *Borel reducible to* the relation $E_{Z_2}^{G_2}$ if there exists a Borel map $f : Z_1 \to Z_2$ such that for every $z, z' \in Z_1$ we have

$$z \, E_{Z_1}^{G_1} \, z' \quad \text{if and only if} \quad f(z) \, E_{Z_2}^{G_2} \, f(z').$$

Note that even in the case $G_1 = G_2$, the Borel reduction need not be equivariant, i.e. may not preserve the group action in any way. From this point of view, smooth equivalence relations are those which are Borel reducible to the equality relation on the reals.

Within orbit equivalence relations, certain classes have been extensively studied, for instance those *classifiable by countable structures*, i.e. Borel reducible to Borel actions of the group S_∞ of all permutations of the natural numbers. The theory or turbulence developed by Hjorth [22] provides the dividing line for those orbit equivanence relations which are not classifiable by countable structures.

The following definition is the central notion discussed in this article.

Definition 2. A group action $G \curvearrowright Z$ induces a *complete orbit equivalence relation* if for any Polish group action $H \curvearrowright Y$ the relation E_Y^H is Borel reducible to E_Z^G.

One can construct a complete orbit equivalence relation purely abstractly using the Mackey–Hjorth extension theorem for group actions [18, Theorem 3.5.2] and the result of Becker and Kechris [18, Theorem 3.3.4], that for any Polish group G there exists a universal G-action X_G, i.e. such that any other G-action can be embedded into X_G via a G-equivariant Borel map.

The first natural example of a complete orbit equivalence relation was identified by Gao and Kechris and, independently, Clemens [19, 4]. It is the action of the isometry group of the Urysohn space on the space of its closed subsets. Subsequently, Melleray [28] showed that the action of the linear group of the Gurarij space on its closed linear subspaces induces a complete orbit equivalence relation. In both cases, the restrictions of the actions to subsets with the extension property (subspaces embedded via the Katětov construction) are also complete orbit equivalence relations.

In all of the above cases, the group that induces a complete orbit equivalence relation is a universal Polish group. This is the case for the group of isometries of the Urysohn space and the linear group of the Gurarij space. Interestingly, however, also non-universal Polish groups can induce complete orbit equivalence relations because there exist surjectively universal but not universal Polish groups [6].

An important open problem is whether the unitary group of the infinite-dimensional separable Hilbert space can induce a complete orbit equivalence relation. This group is not universal, as there are exotic L_0 Polish groups that do not admit any notrivial unitary representations and thus cannot embed into the unitary group [21].

Among the universal Polish groups we have, however, the homeomorphism group of the Hilbert cube [24]. The action of the homeomorphism group of the Hilbert cube on compact subsets of the cube has been long conjectured to be a complete orbit equivalence relation but this remained open until last year. Suprisingly, the first proof relied heavily on the classification problem for separable simple nuclear C*-algebras, as decribed below.

4 Separable simple nuclear C*-algebras

In the applications of operator algebras, nuclear C*-algebras play a crucial role. A C*-algebra A is *nuclear* if for every C*-algebra B, the maximal and minimal norms on the tensor product $A \otimes B$ agree. Equivalently, the identity map $i : A \to A$ can be approximately factored through completely positive maps $i_1 : A \to F$ and $i_2 : F \to A$ with F finite algebras.

The isomorphism problem for various classes of separable simple nuclear C*-algebras has been studied since the work of Glimm in the 1960's and evolved into the Elliott program that classifies C*-algebras via their K-theoretic invariants. Glimm's result [20], restated in modern language, implies that the isomorphism relation for UHF algebras is smooth (see [18, Chapter 5.4]). In the 1970's the classification has been pushed forward to AF algebras via the K_0 group [8]. The Elliott invariant, which consists of the groups K_0 and K_1 together with the tracial simplex and the pairing map, was conjectured (see [10, 15]) to completely classify all infinite-dimensional, separable, simple nuclear C*-algebras. The conjecture has been verified for various classes of C*-algebras, e.g certain classes of real rank zero algebras, AH algebras of slow dimension growth or separable, simple, purely infinite, nuclear algebras (modulo the Universal coefficient theorem) [9, 12, 13, 26, 14, 29, 27] and there have been dramatic breakthroughs in the program, including the counterexamples to the general classification conjecture constructed by Rørdam [30] and Toms [33].

The classification program of separable C*-algebras can be studied from the point of view of descriptive set-theoretic complexity theory (cf [11]). The framework here has been set up in 1996 by Kechris [25] and more recently by Farah, Toms and Törnquist [16, 17].

The standard Borel space of separable simple nuclear C*-algebras can be identified with a Borel subset of the space of all subalgebras of the Cuntz algebra O_2 and Farah, Toms an Törnquist [17] showed that there is a Borel subset Z of the space of subalgebras the Cuntz algebra which has the extension property and thus the isomorphism problem for separable simple nuclear C*-algebras is an orbit equivalence relation induced by the group $\text{Aut}(O_2)$. Farah, Toms and Törnquist showed then [17] that the classification of separable simple nuclear C*-algebras is not classifiable by countable structures.

The standard Borel space of all separable C*-algebras is constructed in a slightly different way than that of nuclear ones, due to the fact that there does not exist a universal separable C*-algebra [23]. One can, however, parametrize all separable C*-algebras via their generating sequences in $B(H)$ (see [17]). In [7] the authors showed that the isomorphism relation for all separable C*-algebras (and, in fact, the isometry relation for any class of Polish metric structures) is Borel reducible to an action of the isometry group of the Urysohn space.

The following result pinned down the complexity for both separable simple nuclear and all separable C*-algebras:

Theorem 3 (Sabok, [31]). *The isomorphism relation of separable simple nuclear C*-algebras is a complete orbit equivalence relation.*

The proof of the above theorem is based on earlier result of Farah, Tomas and Törnquist, who, using the results of Thomsen [32], showed that the relation of affine homeomorphism of Choquet simplices is Borel reducible to the isomorphism relation of separable, simple nuclear C*-algebras. Theorem 3 thus follows from the following.

Theorem 4 (Sabok, [31]). *The relation of affine homeomorphism of Choquet simplices is a complete orbit equivalence relation.*

The proof of Theorem 4 goes in several steps, which ultimately construct a Borel reduction from the isometry of separable metric spaces to the affine homeomorphism of Choquet simplices. First, it is shown how to realize a given Polish metric space as a dense subset of the extreme boundary of some compact convex set in the Hilbert cube. Moreover, if a metric space is a special subset of the Urysohn space, then the compact convex set is actually a simplex. Embedding the space in the special way into the Urysohn space involves a Katětov construction. This step is done in an invariant way, i.e. starting with two isometric spaces we end up with two affinely homeomorphic simplices.

The second step builds on the first step and ensures that the construction is actually a reduction, i.e. starting with two non-isometric spaces we will get two simplices which are not affinely homeomorphic. This is done by encoding the metric structure of the space into the affine structure of the simplex. The key observation is that the metric is always encoded in countable amount of data: the distances between points in a countable dense subset. The encoding of the distances between points x, y (now in the extreme boundary) is done by distinguishing points of the form $\lambda x' + (1-\lambda)y'$ where λ is the distance that is being encoded. The distinguishing of these points is done by ensuring that they are the only limit points of isolated extreme points (so-called *cone points*) in the interval between x' and y'. This is done only for special pairs of points x' and y'. Without going into the details of the construction, let us only mention that the coding makes use only of the topological and affine structure, namely of isolated extreme points and of the affine combinations that can encode real numbers.

5 Compact metric spaces and abelian C*-algebaras

The complexity of homeomorphism of compact metric spaces was studied already in the 1990. It was observed, for example by Clemens, Gao and Kechris [5] that this relation is Borel bi-reducible with the isometry relation for Banach spaces of the form $C(K)$, as follows from the classical Banach–Stone duality, as well as with the isomorphism relation for abelian unital C*-algebras, as follows from the Gelfand–Najmark theorem. Kechris and Solecki observed that the homeomorphism relation for compact metric spaces is an orbit equivalence relation induced by the group of homeomorphisms of the Hilbert cube acting on Z-sets in the Hilbert cube.

The question whether the homeomorphism relation for compact metric spaces is a complete orbit equivalence relation has been a notorious open problem.

Recently, Zielinski [36] used Theorem 4 to answer this question in the affirmative.

Theorem 5 (Zielinski [36]). *The homeomorphism relation for compact metric spaces is a complete orbit equivalence relation.*

Zielinski shows that the affine homeomorphism relation for Choquet simplices can be reduced to the homeomorphism relation of compact metrizable spaces. The key observation is that a homeomorphism $f : K \to L$ from a simplex K to a simplex L is affine if and only if f preserves the set of triples $(z, y, \frac{1}{2}(x+y))$, i.e. $f^3(K_{\frac{1}{2}}) = L_{\frac{1}{2}}$, where

$$K_{\frac{1}{2}} = \{(x, y, z) \in K^3 : z = \frac{1}{2}(x + y)\},$$

$$L_{\frac{1}{2}} = \{(x, y, z) \in L^3 : z = \frac{1}{2}(x + y)\}.$$

The proof of Theorem 5 proceeds in several steps. First, the author considers the set of pairs (K, K_3) where K_3 is a compact subset of $K \times K \times K$ and K is a compact metric space, and the equivalence relation $\equiv_3$ of homeomorphism that preserves the second component. By the above observation and Theorem 4, this relation is a complete orbit equivalence relation.

In the second step, Zielinski shows that one can reduce the above relation to the homeomorphism relation of compact metric spaces. This is done first by considering the set of infinite sequences $(K, K_0, K_1, \ldots)$, where $K_i \subseteq K$ are compact sets and considering the relation $\equiv_1$ on the above set which is the homeomorphism on the first coordinate that permutes the sets in the remaining coordinates. Zielinski then shows that $\equiv_3$ is Borel reducible to $\equiv_1$. Finally, in the last step, he shows that homeomorphism of compact metric spaces can be Borel reduced to the relation $\equiv_1$.

6 Further directions

One of the main further problems that are left open is to determine for which classes of compact metrizable spaces the homeomorphism relation is a complete orbit equivalence.

A very recent paper of Chang and Gao [3] shows that the homeomorphism relation for compact connected spaces (continua) is complete. This is an important improvement to Theorem 5 as the proof of Zielinski makes essential use of isolated points in the compact spaces.

It is worth noting that the first result on the complexity of homeomorphism relation of compact spaces goes back to Hjorth [22] who showed that the homeomorphism relation for subsets of square $[0,1]^2$ is not classifiable by countable structures. This was one of the first applications of the theory of turbulence. On the other hand, homeomorphism relation for subsets of the interval $[0,1]$ is classifiable by countable structures.

This leads to the following question:

Question 6. Is the homeomorphism relation for subsets of $[0,1]^2$ a complete orbit equivalence relation?

The question is also open in the following version

Question 7. What are the complexities of the homeomorphism relations of n-dimensional compact spaces?

The proof of Theorem 5 makes essential use of infinite-dimensional spaces and the above questions pose a serious challenge.

Finally, as most of the coding in the compact spaces is usually done with the use of special points, the following question is especially interesting.

Question 8. What is the complexity of homeomorphism of homogeneous compact spaces?

References

[1] Itaï Ben Yaacov. The linear isometry group of the Gurarij space is universal. *Proc. Amer. Math. Soc.*, 142(7):2459–2467, 2014.

[2] Itaï Ben Yaacov, Alexander Berenstein, C. Ward Henson, and Alexander Usvyatsov. Model theory for metric structures. In *Model theory with applications to algebra and analysis. Vol. 2*, volume 350 of *London Math. Soc. Lecture Note Ser.*, pages 315–427. Cambridge Univ. Press, Cambridge, 2008.

[3] C. Chang and S. Gao. The complexity of the classification problem of continua. *Proc. Amer. Math. Soc.* to appear.

[4] John D. Clemens. Isometry of Polish metric spaces. *Ann. Pure Appl. Logic*, 163(9):1196–1209, 2012.

[5] John D. Clemens, Su Gao, and Alexander S. Kechris. Polish metric spaces: their classification and isometry groups. *Bull. Symbolic Logic*, 7(3):361–375, 2001.

[6] Longyun Ding. On surjectively universal Polish groups. *Adv. Math.*, 231(5):2557–2572, 2012.

[7] G.A. Elliott, I. Farah, V. Paulsen, C. Rosendal, A. Toms, and A. Törnquist. The isomorphism relation for separable C*-algebras. *Mathematical Research Letters*. to appear.

[8] George A. Elliott. On the classification of inductive limits of sequences of semisimple finite-dimensional algebras. *J. Algebra*, 38(1):29–44, 1976.

[9] George A. Elliott. On the classification of C^*-algebras of real rank zero. *J. Reine Angew. Math.*, 443:179–219, 1993.

[10] George A. Elliott. The classification problem for amenable C^*-algebras. In *Proceedings of the International Congress of Mathematicians, Vol. 1, 2 (Zürich, 1994)*, pages 922–932, Basel, 1995. Birkhäuser.

[11] George A. Elliott. Towards a theory of classification. *Adv. Math.*, 223(1):30–48, 2010.

[12] George A. Elliott and Guihua Gong. On the classification of C^*-algebras of real rank zero. II. *Ann. of Math. (2)*, 144(3):497–610, 1996.

[13] George A. Elliott, Guihua Gong, and Liangqing Li. On the classification of simple inductive limit C^*-algebras. II. The isomorphism theorem. *Invent. Math.*, 168(2):249–320, 2007.

[14] George A. Elliott and Mikael Rørdam. Classification of certain infinite simple C^*-algebras. II. *Comment. Math. Helv.*, 70(4):615–638, 1995.

[15] George A. Elliott and Andrew S. Toms. Regularity properties in the classification program for separable amenable C^*-algebras. *Bull. Amer. Math. Soc. (N.S.)*, 45(2):229–245, 2008.

[16] Ilijas Farah, Andrew Toms, and Asger Törnquist. The descriptive set theory of C*-algebra invariants. *Int. Math. Res. Not. IMRN*, (22):5196–5226, 2013. Appendix with Caleb Eckhardt.

[17] Ilijas Farah, Andrew S. Toms, and Asger Törnquist. Turbulence, orbit equivalence, and the classification of nuclear C^*-algebras. *J. Reine Angew. Math.*, (688):101–146, 2014.

[18] Su Gao. *Invariant descriptive set theory*, volume 293 of *Pure and Applied Mathematics (Boca Raton)*. CRC Press, Boca Raton, FL, 2009.

[19] Su Gao and Alexander S. Kechris. On the classification of Polish metric spaces up to isometry. *Mem. Amer. Math. Soc.*, 161(766), 2003.

[20] James G. Glimm. On a certain class of operator algebras. *Trans. Amer. Math. Soc.*, 95:318–340, 1960.

[21] Wojchiech Herer and Jens Peter Reus Christensen. On the existence of pathological

submeasures and the construction of exotic topological groups. *Math. Ann.*, 213:203–210, 1975.

[22] Greg Hjorth. *Classification and orbit equivalence relations*, volume 75 of *Mathematical Surveys and Monographs*. American Mathematical Society, Providence, RI, 2000.

[23] M. Junge and G. Pisier. Bilinear forms on exact operator spaces and $B(H) \otimes B(H)$. *Geom. Funct. Anal.*, 5(2):329–363, 1995.

[24] Alexander S. Kechris. *Classical descriptive set theory*, volume 156 of *Graduate Texts in Mathematics*. Springer-Verlag, New York, 1995.

[25] Alexander S. Kechris. The descriptive classification of some classes of C^*-algebras. In *Proceedings of the Sixth Asian Logic Conference (Beijing, 1996)*, pages 121–149. World Sci. Publ., River Edge, NJ, 1998.

[26] Eberhard Kirchberg and N. Christopher Phillips. Embedding of exact C^*-algebras in the Cuntz algebra O_2. *J. Reine Angew. Math.*, 525:17–53, 2000.

[27] Eberhard Kirchberg and Mikael Rørdam. Non-simple purely infinite C^*-algebras. *Amer. J. Math.*, 122(3):637–666, 2000.

[28] Julien Melleray. Computing the complexity of the relation of isometry between separable Banach spaces. *MLQ Math. Log. Q.*, 53(2):128–131, 2007.

[29] Mikael Rørdam. Classification of certain infinite simple C^*-algebras. *J. Funct. Anal.*, 131(2):415–458, 1995.

[30] Mikael Rørdam. A simple C^*-algebra with a finite and an infinite projection. *Acta Math.*, 191(1):109–142, 2003.

[31] Marcin Sabok. Completeness of the isomorphism problem for separable C*-algebras. *Invent. Math.*, 204(3):833–868, 2016.

[32] K. Thomsen. Inductive limits of interval algebras: the tracial state space. *Amer. J. Math.*, 116(3):605–620, 1994.

[33] Andrew S. Toms. On the classification problem for nuclear C^*-algebras. *Ann. of Math. (2)*, 167(3):1029–1044, 2008.

[34] V. V. Uspenskij. A universal topological group with a countable basis. *Functional Analysis and Its Applications*, 20:86–87, 1986.

[35] V. V. Uspenskij. On the group of isometries of the Urysohn universal metric space. *Commentationes Mathematicae Universitatis Carolinae*, 31(1):181–182, 1990.

[36] Joseph Zielinski. The complexity of the homeomorphism relation between compact metric spaces. *Adv. Math.*, 291:635–645, 2016.

Reasoning about Coalition Structures in Social Environments via Weighted Propositional Logic

Gianuigi Greco
University of Calabria, 87036, Rende, Italy
`ggreco@mat.unical.it`

Abstract

Decision-making is studied in a setting where agents' preferences are expressed via weighted propositional logic and where the goal is to compute social desirable solutions w.r.t. both the utilitarian social welfare and the egalitarian social welfare. Differently from the classical perspective studied in the literature where all agents are required to jointly take a decision (e.g., select a belief or a course of action among several alternative possibilities), it is assumed that agents can form coalitions, each of them possibly taking a different decision. In particular, it is assumed that agents belong to a social environment, so that their utilities depend not only of their absolute preferences but also on the number of "neighbors" occurring with them in the coalition that emerged.

The proposed setting is formalized and analyzed from the computational complexity viewpoint, in particular, by focusing on the problem of assessing how agents can partition themselves with the aim of guaranteeing some desired level of social welfare. A number of intractability results have been pointed out, and efforts have been spent to identify tractable scenarios based on qualitative restrictions as well as on structural restrictions on the underlying environments.

1 Introduction

Whenever a group of agents is involved in some decision process, their preferences over the alternatives have to be suitably taken into account in order to end up with a socially desirable outcome [6]. Prominent examples include allocation of indivisible goods and voting on combinatorial domains, just to name a few (see, e.g., [3, 29]).

In fact, a drastic division exists between *ordinal* settings, where agents express preference relations over alternatives, and *cardinal* settings, where they express utility functions mapping the alternatives to some suitable numerical scale. The latter settings are considered in the paper, and the question of how to aggregate such cardinal preferences into a collective utility function is investigated.

Now, in group decision-making, the set of possible solutions has often a combinatorial structure: possible allocations of items to agents, binary vectors in multiple referenda, subsets of k candidates in committee elections, etc. The exponential size of the set of solutions implies a tension between expressivity (allowing the agents to express any possible utility function) and elicitation and computation complexity (avoiding the agents to spend hours specifying their preferences, and the computer to spend hours computing the optimal solution). A common way that sacrifices expressivity but makes elicitation (and often computation) easy consists in assuming that utility functions are *additive*, that is, described only by their values on singletons, the utility of a tuple of values being then the sum of utilities of the individual values. For instance, when expressing utilities over sets of goods, the utility value given by an agent to a set of goods is the sum of all values she gives to the individual goods in the set. However, assuming additivity implies a huge loss of expressivity, because it does not allow the agents to express *preferential dependencies*. On the other hand, allowing agents to express arbitrary utility functions over a combinatorial set of solutions by listing all solutions together with their utility is clearly unpractical, because it would amount to ask each agent to provide an exponentially large list of values. An approach to reconciliate expressivity and complexity is to use a *compact* representation language for representing utility functions (see, e.g., [18, 19, 28, 35]).

Weighted propositional logic is a language of this kind that attracted much attention in the literature: Each individual expresses her preferences as a set of propositional formulas associated with numerical values. Given an interpretation σ assigning a truth value to each variable, the utility of the individual is defined as the *sum*[1] of the values associated with the formulas satisfied by σ. Moreover, in order to preserve the semantics of the application, interpretations might be restricted to those satisfying some given constraints. In fact, expressing utilities by weighted formulas is more succinct than expressing them directly, and is fully expressive, in the sense that every utility function can be expressed by some set of weighted formulas. Detailed results on the expressivity, succinctness and complexity of various fragments of this language are in [34]. The succinctness of weighted formulas with respect to other logical representation languages is also discussed in [10]. Weighted formulas have also been used to express values of coalitions in cooperative games and hedonic games, in so-called *marginal contribution nets* [26, 14, 15], as well as in fair division [5]. Moreover, related languages have been designed for supporting

---

[1] See [35] for an alternative approach where the utility is defined as the maximum over the values of the satisfied formulas, and [28] for a discussion on (further) possible aggregation functions.

bidding mechanisms in combinatorial auctions [4, 30, 20].

The language of weighted formulas to express individual preferences is the one adopted in the paper, and the question of how these preferences can be aggregated into a collective utility function is considered within the resulting framework.

The analysis is conducted by considering two different perspectives. First, the *utilitarian* perspective is considered, where the collective utility is the sum of the utilities of the individuals, called the *utilitarian social welfare*. While this is the classical approach in social choice theory, it is not desirable in a number of application domains where a "fair" approach would be more appropriate, with the goal being to maximize the *egalitarian social welfare*, that is, the satisfaction of the least satisfied agent (see, e.g., [2, 5]). For instance, under egalitarianism, finding an optimal allocation of indivisible goods to agents is the so-called *Santa Claus problem* [3, 2].

In addition to the kinds of social welfare considered in the specific scenario, the paper proposes a further classification of decision-making in two main categories. On the one hand, it considers the setting where all agents are required to take a decision involving all of them. This setting has been already analyzed in the literature [24] and relevant computational results derived for it are here recalled and discussed.

On the other hand, the paper studies a setting where agents might want to form *coalitions* in order to obtain higher worth by staying all together. In particular, the paper assumes that agents are part of a *social environment* so that their utility function depends not only of their absolute preferences but also on the number of neighbors occurring with them in the coalition that emerged. Note that the study of coalition formation processes for logic-based agents, formalized via weighted propositional logic, has recently attracted attention in the literature [25], where some reasoning problems emerging therein have been put under the computational lens.

That setting is reconsidered in this paper, by complementing known results with novel elaborations and with the analysis of different computational problems arising when reasoning about how agents can partition themselves with the aim of guaranteeing some desired level of social welfare. As the main novel technical contributions of this paper, a number of intractability results are pointed out and efforts are spent to identify tractable classes of instances based on qualitative restrictions and on structural restrictions over the underlying social environments.

Organization. The rest of the paper is organized as follows. Section 2 introduces basic concepts about weighted propositional logic and computational complexity. The classical setting where all agents are required to end up with a joint utility function is studied in Section 3, whereas its extension for social environments is discussed in Section 4. The computational complexity of the latter setting is studied in Section 5 and Section 6, with the latter section being in particular devoted to identify islands of structural tractability. A few final remarks and directions for further work are illustrated in Section 7.

2 Preliminaries

Weighted Propositional Logics. Throughout the paper, a universe $\mathcal{V}$ of variables is assumed to be given and the propositional language $\mathcal{L}$ is considered, which consists of all formulas built over $\mathcal{V}$ by using the Boolean connectives $\wedge$, $\vee$, and $\neg$, plus the constants $\top$ (true) and $\bot$ (false). For any propositional formula $\varphi \in \mathcal{L}$, $\text{dom}(\varphi)$ denotes the *domain* of φ, i.e., the set of all the variables in it. An *interpretation* $\sigma : \mathcal{W} \to \{\top, \bot\}$ over $\mathcal{W} \subseteq \mathcal{V}$ is a function assigning a Boolean value to each variable in $\mathcal{W}$. The set of all interpretations that are defined over $\mathcal{W}$ is denoted by $\mathcal{I}(\mathcal{W})$.

An interpretation $\sigma \in \mathcal{I}(\mathcal{W})$ associates a Boolean value to any formula $\varphi \in \mathcal{L}$ with $\mathcal{W} \supseteq \text{dom}(\varphi)$, by means of the inductive application of the following rules: $\sigma(\top) = \top$; $\sigma(\bot) = \bot$; $\sigma(\neg \phi) = \top$ iff $\sigma(\phi) = \bot$; $\sigma(\phi_1 \wedge \phi_2) = \top$ iff $\sigma(\phi_1) = \sigma(\phi_2) = \top$; and $\sigma(\phi_1 \vee \phi_2) = \bot$ iff $\sigma(\phi_1) = \sigma(\phi_2) = \bot$.

An interpretation $\sigma \in \mathcal{I}(\mathcal{W})$ such that $\mathcal{W} \supseteq \text{dom}(\varphi)$ and $\sigma(\varphi) = \top$ is a *model* of φ, shortly denoted as $\sigma \models \varphi$. A formula φ is *satisfiable* if it has a model.

A *weighted formula* is a pair $\langle \varphi, w \rangle$, where $\varphi \in \mathcal{L}$ is a propositional formula and where $w \in \mathbb{Q}$ is a rational number. A *goalbase* G is a finite set of weighted formulas, whose domain is $\text{dom}(G) = \bigcup_{\langle \varphi, w \rangle \in G} \text{dom}(\varphi)$. For any interpretation σ, the number $G(\sigma) = \sum_{\langle \varphi, w \rangle \in G} \text{ such that } \sigma \models \varphi} w$ is the *value* of σ w.r.t. G.

A *utility function* over $\mathcal{W}$ is a mapping $u : \mathcal{I}(\mathcal{W}) \to \mathbb{Q}$. Given the function u, we can always build a goalbase G_u with $\text{dom}(G_u) = \mathcal{W}$ and such that $G_u(\sigma) = u(\sigma)$, for each $\sigma \in \mathcal{I}(\mathcal{W})$ [10, 34].

Computational Complexity. Some basic definitions about complexity theory are recalled next. The reader is referred to [31] for more on this.

Decision problems are maps from strings (encoding the input instance over a suitable alphabet) to the set $\{\text{"yes"}, \text{"no"}\}$. A (possibly nondeterministic) Turing machine M answers a decision problem if on a given input x, (*i*) a branch of M halts in an accepting state iff x is a "yes" instance, and (*ii*) all the branches of M halt in some rejecting state iff x is a "no" instance.

The paper deals with three complexity classes, which are now discussed. The class **P** is the set of decision problems that can be answered by a deterministic Turing machine in polynomial time. The class of decision problems that can be solved by a nondeterministic Turing machine in polynomial time is denoted by **NP**, while the class of decision problems whose complementary problem is in **NP** is denoted by co-**NP**. The class **DP** is the class of problems defined as a conjunction of two independent problems, one from **NP** and one from co-**NP**, respectively.

In conclusion, it is useful to illustrate the notion of reduction for decision problems. A decision problem A_1 is *polynomially reducible* to a decision problem A_2 if there is a polynomial time computable function h such that for every x, $h(x)$ is defined and A_1 output "yes" on input x iff A_2 outputs "yes" on input $h(x)$. A decision problem A is *complete* for the class $\mathcal{C} \in \{\mathbf{NP}, \text{co-}\mathbf{NP}, \mathbf{DP}\}$ if A belongs to $\mathcal{C}$ and every problem in $\mathcal{C}$ is polynomially reducible to A.

3 Group Decision-Making

In this section, the language of weighted formulas is adopted to express individual preferences and a setting where such preferences have to be combined is studied, by focusing on the maximization either of the utilitarian or of the egalitarian social welfare. The setting will be also put under the lens of a complexity analysis, by summarizing some recent results that have been pointed out in the literature.

3.1 Utilitarian and Egalitarian Social Welfare

Let $\mathcal{A} = \{A_1, ..., A_n\}$ be a set of *agents*, with (the reasoning capabilities of) each agent being modeled as a goalbase, and let $\text{dom}(\mathcal{A}) = \bigcup_{i=1}^{n} \text{dom}(A_i)$ denote the *domain* of the set $\mathcal{A}$. Note that each agent $A_i \in \mathcal{A}$ is implicitly associated with the utility function u_i mapping any interpretation $\sigma \in \mathcal{I}(\mathcal{W})$ to the rational number $u_i(\sigma) = \sum_{\langle \varphi, w \rangle \in A_i, \sigma \models \varphi} w$.

An important task over sets of agents is to define appropriate ways to aggregate all their utilities into a collective utility function. While doing so, the aggregation process might be subject to constraints emerging from the application, which can be naturally modeled (again) as a formula in $\mathcal{L}$ that have to be satisfied by the candidate interpretations. Accordingly, we define a *decision-making* scenario as a pair $(\Gamma, \mathcal{A})$, where Γ is a satisfiable propositional formula and $\mathcal{A}$ is the set of agents.

An interpretation σ is *feasible* (w.r.t. Γ) if $\sigma \models \Gamma$.

Example 1 (cf. [24]). Consider an allocation problem with agents A_1 and A_2, and three indivisible goods g_1, g_2, g_3. Below, it is shown how this problem can be modeled as a decision-making scenario $(\Gamma, \{A_1, A_2\})$.

The scenario is defined over the set of Boolean variables $\mathcal{V} = \{X_{i,j} \mid i \in \{1, 2\}, j \in \{1, 2, 3\}\}$. An interpretation σ over $\mathcal{V}$ is therefore naturally associated with an allocation, where $X_{i,j}$ being true in σ means that A_i receives good g_j. Moreover, the focus is on those interpretations satisfying the formula $\Gamma = \bigwedge_j \bigwedge_{i \neq i'} \neg(X_{i,j} \wedge X_{i',j})$, which constrains each good to be allocated at most to one individual. Finally, goalbases associated with A_1 and A_2 are meant to encode their preferences over the allocations. For instance, assume that A_1 chooses to express her utility function by the set of weighted formulas $\{\langle X_{1,1} \vee (X_{1,2} \wedge X_{1,3}), \frac{3}{5}\rangle, \langle X_{1,1} \wedge X_{1,2}, \frac{2}{5}\rangle\}$, while A_2 has additive preferences, expressed by the set $\{\langle X_{2,1}, \frac{2}{5}\rangle, \langle X_{2,2}, \frac{1}{5}\rangle\}, \langle X_{2,3}, \frac{2}{5}\rangle\}$. Let π be the allocation giving $\{g_1, g_2\}$ to A_1 and $\{g_3\}$ to A_2. The interpretation corresponding to π is σ_π where the variables evaluating to true are those in $\{X_{1,1}, X_{1,2}, X_{2,3}\}$. The utility of A_1 (resp., A_2) in π is given by $u_1(\sigma_\pi) = \frac{3}{5} + \frac{2}{5} = 1$ (resp., $u_2(\sigma_\pi) = \frac{2}{5}$.). ◁

Assume that a decision-making scenario $(\Gamma, \mathcal{A})$ is given. According to the classical approach in social choice theory, while combining the utilities of the agents, one has to focus on maximizing their overall satisfaction, i.e., the utilitarian social welfare. Formally, the *utilitarian social welfare* of an interpretation $\sigma \in \mathcal{I}(\text{dom}(\Gamma) \cup \text{dom}(\mathcal{A}))$ is the value $\text{UT}(\sigma) = \sum_{A_i \in \mathcal{A}} A_i(\sigma)$. The interpretation σ is UT-*optimal* if it has the maximum utilitarian social welfare over all feasible interpretations (taken from $\mathcal{I}(\text{dom}(\Gamma) \cup \text{dom}(\mathcal{A}))$). The set of all UT-optimal interpretations is denoted by UT-OPT$(\Gamma, \mathcal{A})$, and their utilitarian social welfare is denoted by UT-OPT$(\Gamma, \mathcal{A})$.

In addition to the utilitarian social welfare, the paper considers also the egalitarian social welfare. The intuition in this case is to look for interpretations that are not "too far" from the optimum values that can be achieved when optimizing the preferences of the agents independently on the others. To express more clearly the requirement, it is convenient to assume hereinafter that $\max_{\sigma \in \mathcal{I}(\text{dom}(A_i))} u_i(\sigma) = 1$ and $\min_{\sigma \in \mathcal{I}(\text{dom}(A_i))} u_i(\sigma) = 0$, for each agent $A_i \in \mathcal{A}$. An agent enjoying this property will be said *normalized*. W.l.o.g, any agent can be normalized by rescaling the associated weighted formulas (cf. [24]). Dealing with normalized agents is a typical assumption in the literature (see, e.g., [18]).

Let $(\Gamma, \mathcal{A})$ be a decision-making scenario where $\mathcal{A}$ is a set of normalized agents, and let σ be an interpretation. The *egalitarian social welfare* of σ is the value $\text{EG}(\sigma) = \min_{A_i \in \mathcal{A}} A_i(\sigma)$, which evaluates the satisfaction of the least satisfied agent. The interpretation σ is EG-*optimal* if it has the maximum egalitarian social welfare over all feasible interpretations. The set of all EG-optimal interpretations is denoted by EG-OPT$(\Gamma, \mathcal{A})$, and their egalitarian social welfare is denoted by EG-OPT$(\Gamma, \mathcal{A})$.

Example 2. In the setting of Example 1, each agent can get all objects if she were alone. Hence, $\max_{\sigma \in \mathcal{I}(\text{dom}(A_i))} u_i(\sigma) = 1$, for each $i \in \{1, 2\}$. In particular, agents are already normalized, and one can directly apply the above definitions. For instance, consider the interpretation $\sigma_{\pi'}$, where the variables evaluating to true are precisely those in the set $\{X_{1,1}, X_{2,2}, X_{2,3}\}$. Note that $\text{EG}(\sigma_{\pi'}) = \min\{\frac{3}{5}, \frac{3}{5}\} = \frac{3}{5}$. In fact, it can be checked that $\sigma_{\pi'}$ is EG-optimal, hence EG-OPT$(\Gamma, \{A_1, A_2\}) = \frac{3}{5}$. ◁

3.2 Complexity Results

The careful reader might have already noticed that most of the interesting reasonings problems about decision-making scenarios will likely be intractable, since even the very basic problem of deciding whether a given propositional formula admits a model is **NP**-hard. This motivates a finer grained complexity analysis to identify the tractability boarder between polynomial and **NP**-hard settings, along several parameters such as the syntax of formulas, the allowed weights, as well as the number of agents, propositional symbols, and formulas per agent.

In order to embark on this analysis, consider the language $\mathcal{L}_{\{\wedge,\vee,\neg\}}$ consisting of all propositional formulas φ built according to the following grammar:

$$\varphi ::= X \mid \neg X \mid (\varphi \wedge \cdots \wedge \varphi) \mid (\varphi \vee \cdots \vee \varphi),$$

where X is any variable in $\mathcal{V}$. If $B \subseteq \{\wedge, \vee, \neg\}$ is a set of Boolean connectives, then $\mathcal{L}_C$ denotes the set of all the formulas in $\mathcal{L}_{\{\wedge,\vee,\neg\}}$ that do not contain symbols in $C \setminus \{\wedge, \vee, \neg\}$. Note that all formulas are assumed to be given in Negation Normal Form, that is, negation applies only over variables (if it could apply over general subformulas, then $\wedge$ would be expressible from $\vee$ and vice versa).

Let $h_1, h_2, h_3 \in \{1, c, \infty\}$, let $S \subseteq \{+, -\}$ with $|S| \geq 1$, and let $\mathcal{S}_{B,S}[h_1, h_2, h_3]$ be the set of all decision-making scenarios $(\Gamma, \mathcal{A})$ where:

- Γ as well as all formulas in $\mathcal{A}$ are taken from $\mathcal{L}_B$;

- if $S = \{+\}$ (resp., $S = \{-\}$), then every weighted formula is associated with a positive (resp., negative) weight; if $S = \{+, -\}$ then no restriction is imposed on the weights associated with the formulas;

- if $h_1 = 1$ (resp., $h_1 = c$), then $|\mathcal{A}| = 1$ (resp., $|\mathcal{A}|$ is bounded by a fixed constant); if $h_1 = \infty$, then no bound is required over $|\mathcal{A}|$;

- if $h_2 = 1$ (resp., $h_2 = c$), then $|A_i| = 1$ (resp., $|A_i|$ is bounded by a fixed constant), for each $A_i \in \mathcal{A}$; if $h_2 = \infty$, then no bound is required over any $|A_i|$; and

- if $h_3 = 1$ (resp., $h_3 = c$), then $|\text{dom}(\varphi_i)| = |\text{dom}(\Gamma)|$ are equal to 1 (resp., are bounded by a fixed constant) for each $A_i \in \mathcal{A}$ and for $\langle \varphi_i, w_i \rangle \in A_i$; if $h_3 = \infty$, then no bound is required over $|\text{dom}(\varphi_i)|$ and $|\text{dom}(\Gamma)|$.

Let $\text{X} \in \{\text{UT}, \text{EG}\}$ denote the specific social welfare under analysis, and consider the problem of computing an X-optimal interpretation restricted on $\mathcal{S}_{C,S}[h_1, h_2, h_3]$, which is hereinafter denoted as X-$\text{FIND}_{C,S}[h_1, h_2, h_3]$.

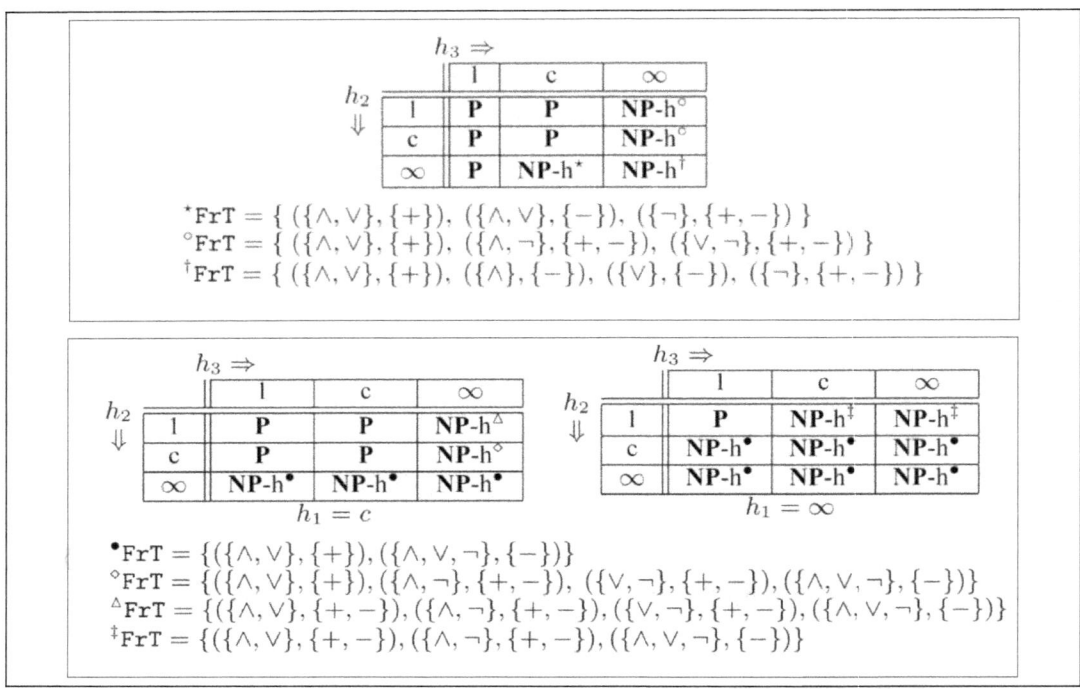

Figure 1: Summary of complexity results in Section 3.2. Individual decision-making (top) and egalitarian social welfare (bottom). Entries marked as "**P**" correspond to tractable settings without any restriction while, for the **NP**-hard scenarios, the tractability frontier is reported.

To analyze its complexity, it is useful to introduce a concept of *frontier of tractability*, which allows us to express more succinctly complexity results, by avoiding to get lost in a large number of different combinations. Formally, the *frontier of tractability* of $\text{X} \in \{\text{UT}, \text{EG}\}$ w.r.t. $h_1, h_2, h_3 \in \{1, c, \infty\}$, denoted by $\text{X-FrT}[h_1, h_2, h_3]$, is the *minimal* set of pairs (B, S) with $B \subseteq \{\vee, \wedge, \neg\}$ and $\emptyset \subset S \subseteq \{+, -\}$ such that:

- $\forall (B, S) \in \text{X-FrT}[h_1, h_2, h_3]$, $\text{X-FIND}_{B,S}[h_1, h_2, h_3]$ is in **P**;

- $\forall (B', S')$ with $B' \subseteq \{\wedge, \vee, \neg\}$ and $S' \subseteq \{+, -\}$ such that there is no pair $(B, S) \in \text{X-FrT}[h_1, h_2, h_3]$ with $B \cup S \supseteq C' \cup S'$, we have that the problem $\text{X-FIND}_{B,S}[h_1, h_2, h_3]$ is **NP**-hard.

Note that the notion precisely captures the intuition that the given pairs mark the boundary between tractable and intractable settings.

With these notions at hand, the reader can now have a look at Figure 1, where all complexity results for the given fragments are summarized—these results have been recently established by [24]. Note that the case where $h_1 = 1$ is the classical setting of individual decision-making (see, e.g., [34, 10]). Moving from $h_1 = 1$ to $h_1 \in \{c, \infty\}$, one can instead observe that the optimization of a set $\mathcal{G}$ of goalbases, from the viewpoint of the utilitarian social welfare, is equivalent to the optimization of $G = \{\langle \varphi, w \rangle \mid \exists G_i \in \mathcal{G}$ such that $\langle \varphi, w \rangle \in G_i\}$. Thus, for each $h_3 \in \{1, c, \infty\}$, the following picture easily emerge:

- UT-FrT$[c, 1, h_3]$ = UT-FrT$[c, c, h_3]$ = UT-FrT$[1, c, h_3]$;
- UT-FrT$[c, \infty, h_3]$ = UT-FrT$[1, \infty, h_3]$; and
- UT-FrT$[\infty, h_2, h_3]$ = UT-FrT$[1, \infty, h_3]$, $h_2 \in \{1, c, \infty\}$.

Finally, concerning the egalitarian social welfare, results are reported in the bottom of Figure 1. In this case, in order to make the analysis meaningful, scenarios are assumed to be normalized and *non-trivial*, i.e., where $h_1 > 1$ holds.

4 Decision-Making in Social Environments

In the previous section, decision-making has been considered in a setting where all agents have to simultaneously take part to the reasoning task. However, this hardly happens in practice. Indeed, in a number of scenarios, there are agents whose utility functions strongly contrast with those of the others, so that they will likely leave the group and possibly form a *coalition* with the agents on which some agreement can be find more easily. Actually, in these contexts, for reasons that might range from physical limitations and constraints to legal banishments, certain agents might not be allowed to form coalitions with certain others.

Sensor networks, communication networks, or transportation networks, within which units are connected through bilateral links, provide natural settings for such classes of games. In many multiagent coordination settings, agents might be restricted to communicate or interact with only a subset of other agents in the environment, due to limited resources or existing physical barriers. Another example is provided by hierarchies of employees within an enterprise, where employees at the same level work together under the supervision of a boss, i.e., of an employee at the immediately higher level in the hierarchy. In all these settings, the environment can be seen to possess some *structure* that forbids the formation of certain coalitions (see, e.g., [8], and the references therein). In particular, in a decision-making scenario, this structure can be viewed as being naturally induced by the goalbases associated with the various agents, by taking into account the constraint formula Γ.

Formally, every scenario $(\Gamma, \mathcal{A})$ can be associated with an *interaction graph* $\text{IG}(\Gamma, \mathcal{A}) = (\mathcal{A}, E)$ where the agents in $\mathcal{A}$ are transparently viewed as the nodes, and where an edge $\{A_i, A_j\}$ (resp., $\{A_i, \Gamma\}$) is in E if, and only if, at least one of the following two conditions hold:

- $\text{dom}(A_i) \cap \text{dom}(A_j) \neq \emptyset$, that is, the two agents directly share some variable in their domains;

- $\text{dom}(A_i) \cap \text{dom}(\Gamma) \neq \emptyset$ and $\text{dom}(A_j) \cap \text{dom}(\Gamma) \neq \emptyset$, that is, they both reason about variables that also occur in the constraint Γ (in this case, the interaction is indirect via Γ).

A coalition $C \subseteq \mathcal{A}$ of agents is *legal* if the subgraph of $\text{IG}(\Gamma, \mathcal{A})$ induced over the nodes in C is connected. It is assumed, w.l.o.g., that $\mathcal{A}$ is legal.

Example 3. Consider again the two agents in Example 1. Note that $\text{dom}(A_1) \cap \text{dom}(A_2) = \emptyset$. However, $\text{dom}(A_1) \cap \text{dom}(\Gamma) \neq \emptyset$ and $\text{dom}(A_2) \cap \text{dom}(\Gamma) \neq \emptyset$. Therefore, the interaction graph $\text{IG}(\Gamma, \{A_1, A_2\})$ is the graph built over the two agents with an edge connecting them. ◁

Now, while taking part to the coalition formation process, agents can be influenced by their "social" relationships, in that their own utility functions can be affected by the number of neighbors belonging to the coalitions they belong to. To define this influence, we first define the concept of neighborhood w.r.t. the underlying interaction graph. So, for each legal coalition $C \subseteq \mathcal{A}$ and agent $A_i \in C$, we define the neighbors of A_i that occur in C as the set $\text{neigh}(A_i, C)$ of all agents $A_j \in C$ such that there is an edge connecting A_i and A_j in $\text{IG}(\Gamma, \mathcal{A})$. With this definition in place, note that two agents A_i and A_j can be neighbors of each other even if $\text{dom}(A_i) \cap \text{dom}(A_j) = \emptyset$. This happens when $\text{dom}(A_i) \cap \text{dom}(\Gamma) \neq \emptyset$ and $\text{dom}(A_j) \cap \text{dom}(\Gamma) \neq \emptyset$. For instance, in Example 1, it can be easily checked that $\text{neigh}(A_1, \{A_1, A_2\}) = \{A_2\}$ and $\text{neigh}(A_2, \{A_1, A_2\}) = \{A_1\}$.

As a final ingredient of the formalization, by belonging to a social environment, it makes sense to assume that the utilities of the agents depend not only of their absolute preferences but also on the number of neighbors occurring with them in the coalition that emerged. In order to balance between their own utility functions and the social influence, we introduce a *social factor* α, which is a rational number with $0 \leq \alpha \leq 1$. Then, for each interpretation σ over a superset of $\text{dom}(C)$, we define the following "adjusted" utility (indeed depending on C):

$$u_i^\alpha(\sigma, C) = \alpha \times \frac{|\text{neigh}(A_i, C)|}{|\text{neigh}(A_i, \mathcal{A})|} + (1 - \alpha) \times u_i(\sigma).$$

Note that for $\alpha = 0$ the settings comes back to the classical setting discussed in the above section where the utility of the agents does not depend on the coalitions they belong to. Instead, for $\alpha = 1$ all agents would like to stay together, with each of them getting rid of the original utility function. Moreover, note that since $u_i^\alpha(\sigma, C)$ is defined as an affine combination of two independent factors, it is convenient to assume, as we already did in the context of the definition of the egalitarian social welfare that agents are normalized.

Example 4. Assume in the setting of Example 1 that $\alpha = \frac{1}{2}$, so each agent perfectly balances her social attitude with the goal of optimizing her own utility function.

Consider the interpretation σ_π where the variables evaluating true are those in $\{X_{1,1}, X_{1,2}, X_{1,3}\}$. Recall from Example 1 that $u_2(\sigma_\pi) = \frac{2}{5}$, and consider the adjusted utility (w.r.t. α) in the two cases when (i) A_2 is alone and when (ii) A_2 and A_1 belong to the same coalition, respectively:

(i) $u_2^\alpha(\sigma_\pi, \{A_2\}) = (1 - \alpha) \times u_2(\sigma_\pi) = \frac{1}{2} \times \frac{2}{5} = \frac{1}{5}$;

(ii) $u_2^\alpha(\sigma_\pi, \{A_1, A_2\}) = \alpha + (1 - \alpha) \times u_2(\sigma_\pi) = \frac{1}{2} + \frac{1}{5} = \frac{7}{10}$;

Therefore, in the former case, i.e., when A_2 is alone, she gets less than the baseline value $\frac{2}{5}$ associated with the case where she does not care of the other agents (i.e., for $\alpha = 0$). However, in the latter, she obtains an additional utility just by the fact of being together with A_1 in the same coalition. Concerning agent A_1, recall instead that $u_1(\sigma_\pi) = 1$, so that $u_1^\alpha(\sigma_\pi, \{A_1\}) = \frac{1}{2}$ and $u_1^\alpha(\sigma_\pi, \{A_1, A_2\}) = 1$. Again, the best choice is to form a coalition together with A_2. ◁

With the above concepts in place, the utilitarian and the egalitarian social welfare can be now re-defined, this time parametrically w.r.t. the social factor α. Formally, for any interpretation σ and coalition C, one can consider:

- the *utilitarian social welfare*, denoted by UT^α, where the utilities of the various agents are summed; hence, $\text{UT}^\alpha(\sigma, C) = \sum_{A_i \in C} u_i^\alpha(\sigma, C)$;

- the *egalitarian social welfare*, denoted by EG^α, where we take care of the least satisfied agent; hence, $\text{EG}^\alpha(\sigma, C) = \min_{A_i \in C} u_i^\alpha(\sigma, C)$.

Example 5. By considering the interpretation σ_π, the value $\alpha = \frac{1}{2}$, and the coalition formed by all agents, it can be derived that $\text{UT}^\alpha(\sigma_\pi, \mathcal{A}) = u_1^\alpha(\sigma_\pi, \{A_1, A_2\}) + u_2^\alpha(\sigma_\pi, \{A_1, A_2\}) = 1 + \frac{7}{10}$, whereas $\text{EG}^\alpha(\sigma_\pi, \mathcal{A}) = \frac{7}{10}$. ◁

Eventually, the concepts of optimal interpretations can be now adapted in order to fit the presence of a social environment. In particular, a feasible interpretation σ is said to be UT^α-*optimal for* C if it has the maximum utilitarian social welfare $\text{UT}^\alpha(\sigma, C)$ over all interpretations. The set of all UT^α-optimal interpretations is denoted by $\text{UT}^\alpha\text{-OPT}(\Gamma, C)$, and their utilitarian social welfare is denoted by $\text{UT}^\alpha\text{-VAL}(\Gamma, C)$. Similarly, σ is EG^α-*optimal for* C if it has the maximum egalitarian social welfare $\text{EG}^\alpha(\sigma, C)$ over all interpretations. The set of all EG^α-optimal interpretations is denoted by $\text{EG}^\alpha\text{-OPT}(\Gamma, C)$, and their utilitarian social welfare is denoted by $\text{EG}^\alpha\text{-VAL}(\Gamma, C)$.

Before leaving the section, it is worthwhile observing that the setting formalized above is an extension of a similar one recently proposed in the literature [25]. In fact, that setting [25] is rather limited from the knowledge representation viewpoint, as it does support the definition of constraints (i.e., Γ) on the possible interpretations and as it has been analyzed by considering logic formulas built without disjunctions (and where negation is atomic).

5 Reasoning about Coalitions

In the previous section, the formal framework for decision-making in social environments has been introduced. Now, some complexity issues arising therein are analyzed by focusing on a number of relevant decision problems. In the exposition below, let $\text{X} \in \{\text{UT}, \text{EG}\}$ denote the specific social welfare one is going to consider in the problem formulation.

5.1 Problems and Overview of the Results

Problems that have been classically considered in the literature about decision-making involve checking whether the utilitarian/egalitarial social welfare of the whole set of agents exceeds some given desired threshold. In the following, such problems are naturally reconsidered by focusing on arbitrary coalitions, rather than on the whole set $\mathcal{A}$ only. In addition, specific problems are formalized that exploit the peculiarity of the proposed setting. These problems are defined moving from the observation that low values for $\text{X}^\alpha(\sigma, \mathcal{A})$ are determined by the utility functions of some agents which contrast with those of others. Accordingly, it might be helpful to identify the (subset-)maximal coalitions over which good "agreements" can be found. Formally, let γ be a rational number, and let $C \subseteq \mathcal{A}$ be a legal coalition. Then, we say that C is X^α-*maximal* for γ (w.r.t. Γ) if $\text{X}^\alpha\text{-VAL}(\Gamma, C) \geq \gamma$ and there is no legal coalition $C' \supset C$ such that $\text{X}^\alpha\text{-VAL}(\Gamma, C') \geq \gamma$.

By summing up these observations, for each given social factor α, one can consider the following problems, all of them receiving as input a set $\mathcal{A}$ of *normalized* agents, a formula Γ in $\mathcal{L}$, and a rational number γ:

	arbitrary		positive	
	UT	EG	UT	EG
UT$^\alpha$/EG$^\alpha$-CHECK	**NP**-c	**NP**-c	**NP**-c	**NP**-c
UT$^\alpha$/EG$^\alpha$-EXISTENCE	**NP**-c	**NP**-c	**NP**-c	**NP**-c
UT$^\alpha$/EG$^\alpha$-MAXIMAL	**DP**-c	**DP**-c	**NP**-c	**DP**-c

Figure 2: Summary of complexity results in Section 5. All the entries in the table are shown to hold *for each social factor* α such that $0 \leq \alpha < 1$, and even for environments consisting of a fixed number of agents. In fact, the three problems are tractable when $\alpha = 1$.

X$^\alpha$-CHECK: Given a legal coalition $C \subseteq \mathcal{A}$, does X$^\alpha$-VAL$(\Gamma, C) \geq \gamma$ holds?

X$^\alpha$-EXISTENCE: Is there any legal coalition $C \subseteq \mathcal{A}$ such that X$^\alpha$-VAL$(\Gamma, C) \geq \gamma$ holds?

X$^\alpha$-MAXIMAL: Given a legal coalition $C \subseteq \mathcal{A}$, is C X$^\alpha$-maximal for γ?

A summary of the complexity results is in Figure 2. Note that, in addition to the case where arbitrary agents are considered, *positive* ones are considered too, i.e., such that every weighted formula occurring in them is associated with a positive weight (as in Section 3.2). Note that the dual setting where all weights are negative is not possible, since agents are assumed to be normalized. In the following, proofs are reported by distinguishing the case where $\alpha = 1$ from the case where $0 \leq \alpha < 1$.

Note that the decision problems under analysis have been also considered in the literature [25], where computational results have been derived for *social environments without any bound on the number of agents* and by considering reductions exhibited for some specific social factor. The main technical contribution of this section is to complement known hardness results *(i)* by focusing on settings where the number of agents is a given fixed constant and *(ii)* by exhibiting reductions that apply to *each* social factor α such that $0 \leq \alpha < 1$.

5.2 Analysis for $\alpha = 1$

The case where $\alpha = 1$ corresponds to scenarios where agents just care about the presence of their neighbors, without considering their weighted formulas at all. Indeed, for any given coalition $C \subseteq \mathcal{A}$ and formula Γ, we have that

- UT^1-VAL$(\Gamma, C) = \sum_{A_i \in C} \frac{|\text{neigh}(A_i, C)|}{|\text{neigh}(A_i, \mathcal{A})|}$; and

- $\text{EG}^1\text{-VAL}(\Gamma, C) = \min_{A_i \in C} \frac{|\texttt{neigh}(A_i, C)|}{|\texttt{neigh}(A_i, \mathcal{A})|}$.

Therefore, for each pair of coalitions C and C' with $C \subseteq C'$, we have that $\text{UT}^1\text{-VAL}(\Gamma, C) \leq \text{UT}^1\text{-VAL}(\Gamma, C')$ and $\text{EG}^1\text{-VAL}(\Gamma, C) \leq \text{EG}^1\text{-VAL}(\Gamma, C')$. Then, because of this monotonic behavior, the following is immediate.

Theorem 6. X^1-CHECK, X^1-EXISTENCE, and X^1-MAXIMAL are feasible in polynomial time, $\forall \text{X} \in \{\text{UT}, \text{EG}\}$.

Proof. For $\alpha = 1$, the expressions $\text{EG}^\alpha\text{-VAL}(\Gamma, C)$ and $\text{EG}^\alpha\text{-VAL}(\Gamma, C)$ depend only on the neighbors of the agents in C, and they can be trivially evaluated in polynomial time. Now, observe that due to the monotonicity of $\text{UT}^1\text{-VAL}$ and $\text{EG}^1\text{-VAL}$, there is a coalition C such that $\text{X}^1\text{-VAL}(\Gamma, C) \geq \gamma$ if, and only if, $\text{X}^1\text{-VAL}(\Gamma, \mathcal{A}) \geq \gamma$ holds. Therefore, X^1-EXISTENCE reduces to X^1-CHECK on input $\mathcal{A}$ and it is, hence, feasible in polynomial time. Similarly, X^1-MAXIMAL reduces to checking whether $C = \mathcal{A}$ and whether there is a positive answer of X-CHECK on input $\mathcal{A}$. Thus, it is feasible in polynomial time, too. □

5.3 Proofs of Results for $0 \leq \alpha < 1$

Consider now the case where $0 \leq \alpha < 1$, for which results are hereinafter provided *for each* given social factor. In particular, all reductions are designed in a way that their salient properties hold independently on the specific value being considered. Elaborations start with the results for the problems CHECK and EXISTENCE.

Theorem 7. Let α be any fixed social factor with $0 \leq \alpha < 1$. Then, X^α-CHECK is **NP**-complete, for each $\text{X} \in \{\text{UT}, \text{EG}\}$. Hardness results hold even over scenarios consisting of a fixed number of positive agents.

Proof. (Membership) The problem UT^α-CHECK (resp., EG^α-CHECK) can be solved in polynomial time by a nondeterministic Turing machine that guesses an interpretation σ and then checks that $\text{UT}^\alpha(\sigma, C) \geq \gamma$ (resp., $\text{EG}^\alpha(\sigma, C) \geq \gamma$).

(Hardness) Recall that deciding whether a Boolean formula is satisfiable is a well-known **NP**-hard problem. Given a Boolean formula φ, consider a variable X not in $\texttt{dom}(\varphi)$ and define the agents $A_\varphi = \{\langle X \wedge \varphi, 1 \rangle, \langle \neg X, 1 \rangle\}$ and $\bar{A}_\varphi = \{\langle X, 1 \rangle\}$. Note that $\text{UT}^\alpha(\sigma, \{A_\varphi, \bar{A}_\varphi\}) = 2 \times \alpha + (1 - \alpha) \times (A_\varphi(\sigma) + \bar{A}_\varphi(\sigma))$. Therefore, $\text{UT}^\alpha\text{-VAL}(\top, \{A_\varphi, \bar{A}_\varphi\}) \geq 2$ holds if, and only if, φ is satisfiable. So, UT^α-CHECK is **NP**-hard. Similarly, one has $\text{EG}^\alpha(\sigma, \{A_\varphi, \bar{A}_\varphi\}) = \alpha + (1 - \alpha) \times \min\{A_\varphi(\sigma), \bar{A}_\varphi(\sigma)\}$. Again, $\text{EG}^\alpha\text{-VAL}(\top, \{A_\varphi, \bar{A}_\varphi\}) \geq 1$ holds if, and only if, φ is satisfiable. So, EG^α-CHECK is **NP**-hard, too. □

Theorem 8. *Let α be any fixed social factor with $0 \leq \alpha < 1$. Then, $\textsc{x}^\alpha$-Existence is **NP**-complete, for each $\textsc{x} \in \{\textsc{ut},\textsc{eg}\}$. Hardness results hold even over scenarios consisting of two positive agents.*

Proof. $\textsc{ut}^\alpha$-Existence (resp., $\textsc{eg}^\alpha$-Existence) can be solved by first guessing an interpretation σ and a legal coalition C, and by subsequently checking that $\sigma \models \Gamma$ and $\textsc{ut}^\alpha(\sigma, C) \geq \gamma$ (resp., $\textsc{eg}^\alpha(\sigma, C) \geq \gamma$). So, the problems belong to **NP**. For the hardness, given a Boolean formula φ, consider again the reduction in the proof of Theorem 7. By inspecting that proof, note that if $\textsc{ut}^\alpha\text{-Val}(\top, C) \geq 2$, then it is necessarily the case that $C = \{A_\varphi, \bar{A}_\varphi\}$. Therefore, $\textsc{ut}^\alpha$-Existence is equivalent to $\textsc{ut}^\alpha$-Check on input $\{A_\varphi, \bar{A}_\varphi\}$. Hence, it is **NP**-hard, too. Similarly, if $\textsc{eg}^\alpha\text{-Val}(\top, \{A_\varphi, \bar{A}_\varphi\}) \geq 1$, then it is necessarily the case that $C = \{A_\varphi, \bar{A}_\varphi\}$. So, $\textsc{eg}^\alpha$-Existence is equivalent to $\textsc{eg}^\alpha$-Check on input $\{A_\varphi, \bar{A}_\varphi\}$, and **NP**-hard. □

Let us now analyze the complexity of Maximal, from the utilitarian perspective, by considering first the case where all agents are positive.

Theorem 9. *Let α be any fixed social factor with $0 \leq \alpha < 1$. Then, $\textsc{ut}^\alpha$-Maximal is **NP**-complete when restricted over positive agents.*

Proof. When agents in $\mathcal{A}$ are positive, it is immediate to check that a coalition $C \subseteq \mathcal{A}$ is $\textsc{ut}^\alpha$-maximal for γ (w.r.t. Γ) only if $C = \mathcal{A}$. Therefore, $\textsc{ut}^\alpha$-Maximal reduces to $\textsc{ut}^\alpha$-Check on input $\mathcal{A}$. Then, **NP**-completeness follows by inspection in the proof of Theorem 7. □

When agents are not necessarily positive, a complexity increase occurs.

Theorem 10. *Let α be any fixed social factor with $0 \leq \alpha < 1$. Then, $\textsc{ut}^\alpha$-Maximal is **DP**-complete (without any restriction on the agents). Hardness result holds even over scenarios consisting of three agents.*

Proof. (Membership) Let $C \subseteq \mathcal{A}$ be the legal coalition provided as input, and recall that C is is $\textsc{ut}^\alpha$-maximal for γ (w.r.t. Γ) if the following conditions hold:

(C1) $\textsc{ut}^\alpha\text{-Val}(\Gamma, C) \geq \gamma$, and

(C2) there is no legal coalition $C' \supset C$ such that $\textsc{ut}^\alpha\text{-Val}(\Gamma, C') \geq \gamma$.

Observe that (C1) can be checked in **NP**, as we shown in Theorem 7. Consider, then, the condition complementary to (C2), that is, that there is some legal coalition $C' \supset C$ such that $\textsc{ut}^\alpha\text{-Val}(\Gamma, C') \geq \gamma$. To verify whether the condition holds, one can just guess a legal coalition $C' \supset C$ and an interpretation σ, by subsequently checking that $\sigma \models \Gamma$ and $\text{UT}^\alpha(\sigma, C') \geq \gamma$ hold. This is feasible in **NP**, and hence (C2) can be checked in co-**NP**. Overall, the problem is in **DP**.

(Hardness) Recall that the problem receiving as input a pair (φ_1, φ_2) such that $\text{dom}(\varphi_1) \cap \text{dom}(\varphi_2) = \emptyset$ and asking of deciding whether φ_1 is satisfiable and φ_2 is not satisfiable is **DP**-hard. Let $\beta = \frac{1-3\times\alpha}{1-\alpha}$ and note that $\beta \leq 1$ holds.

Let X and Y be two variables not in $\text{dom}(\varphi_1) \cup \text{dom}(\varphi_2)$, and define the agents $A_{1,2} = \{\langle X \wedge \varphi_1, 1\rangle, \langle \neg X \wedge Y, 1\rangle\}$, $\bar{A}_{1,2} = \{\langle X \wedge \varphi_2, \beta\rangle, \langle \neg X \wedge \neg Y, 1\rangle, \langle X \wedge Y, -1\rangle\}$, and $A_0 = \{\langle X \wedge Y, 1\rangle\}$. Note that these agents are normalized. Moreover, $\text{dom}(A_0) \cap \text{dom}(A_{1,2}) \cap \text{dom}(\bar{A}_{1,2}) \neq \emptyset$. Therefore, for each interpretation σ, the following expressions can be derived

$$\text{UT}^\alpha(\sigma, \{A_0, A_{1,2}\}) = \alpha + (1-\alpha) \times (A_0(\sigma) + A_{1,2}(\sigma)); \text{ and} \tag{1}$$

$$\text{UT}^\alpha(\sigma, \{A_0, A_{1,2}, \bar{A}_{1,2}\}) = 3 \times \alpha + (1-\alpha) \times (A_0(\sigma) + A_{1,2}(\sigma) + \bar{A}_{1,2}(\sigma)). \tag{2}$$

Now, it can be claimed that: $\{A_0, A_{1,2}\}$ *is* UT^α-*maximal for* $2 - \alpha$ *(w.r.t.* $\top$*) if, and only if, φ_1 is satisfiable and φ_2 is not satisfiable.*

(only-if) Assume that $\{A_0, A_{1,2}\}$ is UT^α-maximal for $2 - \alpha$. First, this implies that there is an interpretation σ such that $\text{UT}^\alpha(\sigma, \{A_0, A_{1,2}\}) \geq 2 - \alpha$. By Equation 1, we get that $A_0(\sigma) + A_{1,2}(\sigma) = 2$. As agents are normalized, this entails that $A_0(\sigma) = 1$ and $A_{1,2}(\sigma) = 1$. Therefore, $\sigma \models X \wedge Y$ and $\sigma \models \varphi_1$. That is, φ_1 is satisfiable.

Second, from the fact that $\{A_0, A_{1,2}\}$ is UT^α-maximal for $2-\alpha$, one also derives that $\text{UT}^\alpha\text{-VAL}(\top, \{A_0, A_{1,2}, \bar{A}_{1,2}\}) < 2 - \alpha$. For the sake of contradiction assume that φ_2 is satisfiable, too. Then, there is an interpretation σ^* such that $\sigma^* \models X \wedge Y \wedge \varphi_1 \wedge \varphi_2$. For this interpretation, we get $A_0(\sigma^*) + A_{1,2}(\sigma^*) + \bar{A}_{1,2}(\sigma^*) = 1 + \beta$. By Equation 2, one gets $\text{UT}^\alpha(\sigma^*, \{A_0, A_{1,2}, \bar{A}_{1,2}\}) = 3 \times \alpha + (1-\alpha) \times (1+\beta) < 2 - \alpha$. By algebraic manipulations, it can be derived $\beta < \frac{1-3\times\alpha}{1-\alpha}$, which is impossible.

(if) Assume that φ_1 is satisfiable and φ_2 is not satisfiable. Since φ_1 is satisfiable, by Equation 1, we immediately have that $\text{UT}^\alpha\text{-VAL}(\top, \{A_0, A_{1,2}\}) = 2 - \alpha$. Assume now, for the sake of contradiction, that there is an interpretation σ^* such that $\text{UT}^\alpha(\sigma^*, \{A_0, A_{1,2}, \bar{A}_{1,2}\}) \geq 2 - \alpha$. By Equation 2, one derives $A_0(\sigma^*) + A_{1,2}(\sigma^*) + \bar{A}_{1,2}(\sigma^*) \geq 1 + \frac{1-3\times\alpha}{1-\alpha} = 1 + \beta$. This entails that $\sigma^* \models X \wedge Y \wedge \varphi_1 \wedge \varphi_2$. Indeed, just check that for each interpretation σ such that $\sigma \not\models X \wedge Y \wedge \varphi_1 \wedge \varphi_2$, one has $A_0(\sigma) + A_{1,2}(\sigma) + \bar{A}_{1,2}(\sigma) < 1 + \beta$. Then, the proof concludes by noticing that φ_2 is satisfied by σ^*, which is impossible.

The claim above implies that EG$^\alpha$-CHECK is **DP**-hard. $\square$

The counterpart under the egalitarian social welfare is next proven. In this case, **DP**-hardness can be established even on positive agents.

Theorem 11. *Let α be any fixed social factor with $0 \leq \alpha < 1$. Then, EG$^\alpha$-MAXIMAL is **DP**-complete. Hardness holds even over scenarios consisting of positive agents.*

Proof. Membership in **DP** can be established with the same argument as in the proof of Theorem 10, by replacing the notions related to the utilitarian social welfare with the corresponding ones for the egalitarian social welfare.

For the hardness, it can be considered again the problem receiving as input a pair (φ_1, φ_2) such that $\mathsf{dom}(\varphi_1) \cap \mathsf{dom}(\varphi_2) = \emptyset$ and asking of deciding whether φ_1 is satisfiable and φ_2 is not satisfiable. Let X and Y be two variables not in $\mathsf{dom}(\varphi_1) \cup \mathsf{dom}(\varphi_2)$, and define the agents $A_{1,2} = \{\langle X \wedge \varphi_1, 1\rangle, \langle \neg X, 1\rangle\}$, $\bar{A}_{1,2} = \{\langle X \wedge \varphi_2, 1\rangle, \langle \neg X, 1\rangle\}$, and $A_0 = \{\langle X, 1\rangle\}$. Moreover, let $n = \max\{4, \lceil \frac{\alpha}{1-\alpha} \rceil + 1\}$ and define the set of agents $\mathcal{B} = \{B_1, ..., B_{n-2}\}$, each one being identical to A_0. Note that agents are normalized (and that n is a constant).

Let β be any rational number such that $1 \geq \alpha \times \frac{n-1}{n} + (1-\alpha) > \beta > \alpha$. Note that β is well defined, since $\alpha < 1$ and given the choice of n.

It can be claimed that: $\{A_0, A_{1,2}\} \cup \mathcal{B}$ is EG$^\alpha$-maximal for β (w.r.t. $\top$) if, and only if, φ_1 is satisfiable and φ_2 is not satisfiable.

(**only-if**) Assume that $\{A_0, A_{1,2}\} \cup \mathcal{B}$ is EG$^\alpha$-maximal for β. Then, there is an interpretation σ such that EG$^\alpha(\sigma, \{A_0, A_{1,2}\}) \geq \beta$. This means that $\alpha \times \frac{n-1}{n} + (1-\alpha) \times A_0(\sigma) \geq \beta$ and $\alpha \times \frac{n-1}{n} + (1-\alpha) \times A_{1,2}(\sigma) \geq \beta$. Since $\beta > \alpha > \alpha \times \frac{n-1}{n}$, one derives $A_0(\sigma) > 0$ and $A_{1,2}(\sigma) > 0$. Given the form of the two agents, it can be concluded that $\sigma \models X \wedge \varphi_1$. Hence, φ_1 is satisfiable. Moreover, it must hold that for each interpretation σ, EG$^\alpha(\sigma, \{A_0, A_{1,2}, \bar{A}_{1,2}\}) < \beta$. Thus, $\alpha + (1-\alpha) \times \bar{A}_{1,2}(\sigma) < \beta$. Given that $\beta \leq 1$, it is impossible that $\bar{A}_{1,2}(\sigma) = 1$. That is, φ_2 is not satisfiable.

(**if**) Assume that φ_1 is satisfiable and φ_2 is not satisfiable. Since φ_1 is satisfiable, one immediately has that there is an interpretation σ such that $\sigma \models X \wedge \varphi_1$, and EG$^\alpha(\sigma, \{A_0, A_{1,2}\}) = \alpha \times \frac{n-1}{n} + (1-\alpha)$. Thus, it can be concluded that EG$^\alpha(\sigma, \{A_0, A_{1,2}\}) > \beta$. Assume now, for the sake of contradiction, that $\{A_0, A_{1,2}\}$ is not EG$^\alpha$-maximal for β. Hence, there is an interpretation σ such that EG$^\alpha(\sigma, \{A_0, A_{1,2}, \bar{A}_{1,2}\}) \geq \beta$. Thus, $\alpha + (1-\alpha) \times \bar{A}_{1,2}(\sigma) \geq \beta$. Since $\beta > \alpha$, it is impossible that $\bar{A}_{1,2}(\sigma) = 0$. So, φ_2 is satisfiable, which is impossible.

The claim above implies that EG$^\alpha$-MAXIMAL is **DP**-hard. $\square$

6 Reasoning about Coalition Structures

In this section, rather than focusing on the coalition formation processes by considering one coalition at time, a more ambitious goal is considered by studying how agents can form a *coalition structure*, i.e., how they can partition themselves with the aim of guaranteeing some desired level of social welfare. This study and the related complexity results have no counterpart in the literature about logic-based agents formalized via weighted propositional logic.

A crucial notion in the subsequent analysis is that of a coalition structure Π for a set $\mathcal{A}$ of agents, which is defined as any set of disjoint legal coalitions such that $\bigcup_{C \in \Pi} C = \mathcal{A}$. Assessing which coalition structure might emerge in a given scenario is a fundamental problem in the study of multi-agent systems, which attracted much research in earlier literature (see, e.g., [33, 16]). In fact, while the utilitarian social welfare viewpoint is classically adopted in this context, given the perspective of the paper, it make sense to consider the egalitarian viewpoint, too. Accordingly, given a social factor α, one can focus on studying the following two problems, receiving as input a set $\mathcal{A}$ of normalized agents, a formula Γ, and a threshold γ:

UT$^\alpha$-CSG-EXISTENCE: Decide whether there is a coalition structure Π for $\mathcal{A}$ such that $\sum_{C \in \Pi}$ UT$^\alpha$-VAL$(\Gamma, C) \geq \gamma$;

EG$^\alpha$-CSG-EXISTENCE: Decide whether there is a coalition structure Π for $\mathcal{A}$ such that $\min_{C \in \Pi}$ EG$^\alpha$-VAL$(\Gamma, C) \geq \gamma$.

In the following, the complexity of these problems is analyzed by stressing their intractability and by identifying restrictions that lead to polynomial time scenarios.

6.1 Basic Results

The first observation is that when one focuses on the extreme values for α, X$^\alpha$-CSG-EXISTENCE is tractable for each X $\in \{$UT, EG$\}$.

Theorem 12. *For each* X $\in \{$UT, EG$\}$ *and* $\alpha \in \{0, 1\}$, X$^\alpha$-CSG-EXISTENCE *is feasible in polynomial time.*

Proof. Tractability of X^1-CSG-EXISTENCE immediately follows by the arguments in the proof of Theorem 6. Let us therefore focus on problems UT^0-CSG-EXISTENCE and EG^0-CSG-EXISTENCE. Observe that, for each $A_i \in \mathcal{A}$, UT^0-VAL$(\Gamma, \{A_i\})$ = EG^0-VAL$(\Gamma, \{A_i\}) = 1$ since normalized agents are considered. Therefore, if $\Pi^* = \{\{A_i\} \mid A_i \in \mathcal{A}\}$ is the partition where all agents form singleton coalitions, then

- $\sum_{\{A_i\} \in \Pi^*}$ UT$^\alpha$-VAL$(\Gamma, \{A_i\}) = |\mathcal{A}|$, and

- $\min_{\{A_i\}\in\Pi^*} \text{UT}^\alpha\text{-VAL}(\Gamma,\{A_i\}) = 1$.

In both cases, the values are clearly the best possible ones that can be achieved over all possible coalition structures. □

For intermediate values of α, the problem becomes intractable.

Theorem 13. *Let α be any fixed social factor with $0 < \alpha < 1$. Then, X^α-CSG-EXISTENCE is **NP**-complete, for each $\text{X} \in \{\text{UT}, \text{EG}\}$. Hardness holds even over scenarios with two positive agents.*

Proof. Membership in **NP** is immediate. For the hardness, given a Boolean formula φ, consider again the reduction in the proof of Theorem 7. In particular, recall that we have the two agents $A_\varphi = \{\langle X \wedge \varphi, 1\rangle, \langle \neg X, 1\rangle\}$ and $\bar{A}_\varphi = \{\langle X, 1\rangle\}$. Let Π be any coalition structure. Note that $\sum_{C\in\Pi} \text{UT}^\alpha\text{-VAL}(\Gamma,C) = 2$ holds if, and only if, φ is satisfiable. Similarly, $\min_{C\in\Pi} \text{UT}^\alpha\text{-VAL}(\Gamma,\{A_i\}) = 1$ holds if, and only if, φ is satisfiable. Thus, X^α-CSG-EXISTENCE is **NP**-hard. □

By inspecting the simple proof of the above result, it clearly emerges that the source of the intractability straightforwardly follows from the fact that one agent has to reason on the satisfiability of an entire Boolean formula. Therefore, for a finer grained analysis, it make sense to focus on cases where the reasoning capabilities of each agent are "bounded". To this end, let $h > 0$ be a fixed natural number, and assume that $|\text{dom}(\Gamma)| \leq h$ and that for each legal coalition C and each $A_i \in C$, $|\text{dom}(A_i)| \leq h$ holds. A scenario enjoying these properties will be said h-*bounded*.

Now, the crucial observation is that over h-bounded scenarios the set of all feasible interpretations for each agent is polynomially bounded, so that we can efficiently reason about them. However, the subtle interplay that can emerge among the agents still suffices to keep X^α-GSC-EXISTENCE **NP**-hard.

Theorem 14. *Let α be any fixed social factor with $0 \leq \alpha < 1$. Then, X^α-EXISTENCE and X^α-CSG-EXISTENCE are **NP**-complete, for each $\text{X} \in \{\text{UT}, \text{EG}\}$. Hardness results hold even when over positive agents and 3-bounded scenarios.*

Proof. Consider a Boolean formula $\varphi = c_1 \wedge ... \wedge c_n$ in conjunctive normal form such that $|\text{dom}(c_i)| \leq 3$, for each $i \in \{1,...,n\}$. Deciding the satisfiability of these formulas is a well-known **NP**-complete problem. W.l.o.g., assume that each variable occurs in at least two clauses. Based on φ, consider the agents $A_i = \{\langle c_i, 1\rangle\}$, for each $i \in \{1,...,n\}$. Then, observe that there is a coalition structure Π such that $\sum_{C\in\Pi} \text{UT}^\alpha\text{-VAL}(\top,C) = n$ (resp., $\min_{C\in\Pi} \text{EG}^\alpha\text{-VAL}(\top,C) = 1$) if, and only if, φ is satisfiable. That is, X^α-CSG-EXISTENCE is **NP**-hard. □

3551

Motivated by the above bad news, the question of whether h-bounded scenarios become tractable when further restrictions are considered is explored in the following, by looking at the structure of the possible interactions among the agents.

6.2 Structural Tractability

Many **NP**-hard problems in different areas such as AI, Database Systems, Game theory, and Network Design, are known to be efficiently solvable when restricted to instances whose underlying structures can be modeled via acyclic graphs. Indeed, for such restricted classes of instances, solutions can usually be computed via dynamic programming. However, as a matter of fact, (graphical) structures arising from real applications are in most relevant cases not properly acyclic. Yet, they are often not very intricate and exhibit some rather limited degree of cyclicity, which suffices to retain most of the nice properties of acyclic instances. Therefore, many efforts have been spent to investigate graph properties that are best suited to identify nearly-acyclic graph, leading to the definition of a number of *structural decomposition methods* (see, e.g., [22]).

In this section, the question of whether these methods can be used to identify islands of tractability for the reasoning problems addressed in the paper is analyzed, by considering the underlying interaction graph as the reference graphical structure. Moreover, the concept of treewidth [32], based on tree decompositions of graphs, is considered to identify such structural restrictions. Indeed, there are different possible notions to measure how far a graph is from a tree, that is, to measure its degree of cyclicity or, dually, its tree-likeness (see, e.g., [23]). Among them, treewidth is a powerful one, in that it is able to extend the nice computational properties of trees to the largest possible classes of graphs, in many applications from different fields.

Definition 15 ([32]). A *tree decomposition* of a graph $G = (N, E)$ is a pair $\langle T, \chi \rangle$, where $T = (V, F)$ is a tree, and χ is a labeling function assigning to each vertex $p \in V$ a set of vertices $\chi(p) \subseteq N$, such that the following three conditions are satisfied: (1) for each node b of G, there exists $p \in V$ such that $b \in \chi(p)$; (2) for each edge $(b, d) \in E$, there exists $p \in V$ such that $\{b, d\} \subseteq \chi(p)$; and (3) for each node b of G, the set $\{p \in V \mid b \in \chi(p)\}$ induces a connected subtree of T.

The *width* of $\langle T, \chi \rangle$ is the number $\max_{p \in V}(|\chi(p)| - 1)$. The *treewidth* of G, denoted by $tw(G)$, is the minimum width over all its tree decompositions. □

Note that treewidth is a true generalization of graph acyclicity. Indeed, a graph G is acyclic if and only if $tw(G) = 1$. For example, the graph reported in Figure 3 is cyclic and its treewidth is 2, as it is witnessed by the width-2 tree decomposition depicted in the same figure.

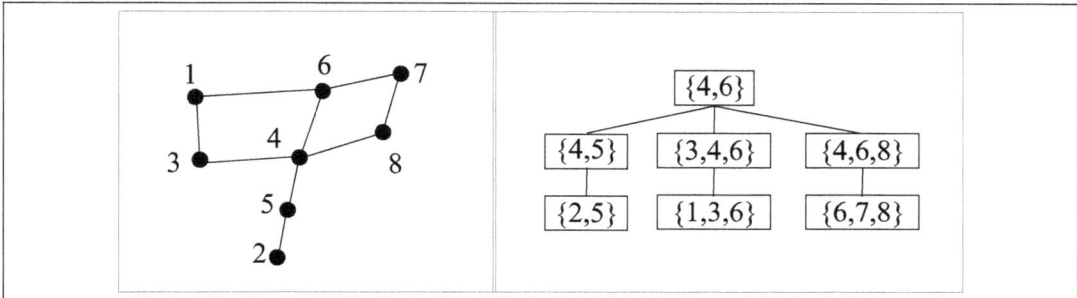

Figure 3: A graph and a tree decomposition for it.

In order to study the complexity of CSG-EXISTENCE problems over scenarios with associated tree-like interaction graphs, known results about structurally tractable *constraint satisfaction problems* (CSPs) are used as a technical tool.

Thus, in the remainder, the main notions arising in the context of CSPs are first discussed, by then focusing on showing how the reasoning problems of interest can be encoded in that formalism. Eventually, tractability results are derived from known tractability results for CSPs.

Preliminaries CSPs. Let us recall some preliminaries on *constraint satisfaction*. The reader interested in expanding on this formalism is referred to [11].

A constraint satisfaction problem instance is a triple $\mathcal{J} = \langle Var, U, \mathbf{C} \rangle$, where Var is a finite set of variables, U is a domain of values, and $\mathbf{C} = \{C_1, C_2, \ldots, C_q\}$ is a finite set of constraints. Each constraint C_v, for $1 \leq v \leq q$, is a pair (S_v, r_v), where $S_v \subseteq Var$ is a set of variables called the *constraint scope*, and r_v is a set of substitutions from variables in S_v to values in U indicating the allowed combinations of simultaneous values for the variables in S_v. A substitution from a set of variables $V \subseteq Var$ to U is extensively denoted as the set of pairs of the form X/u, where $u \in U$ is the value to which $X \in V$ is mapped. A substitution θ *satisfies* a constraint C_v if its restriction to S_v, i.e., the set of all pairs $X/u \in \theta$ such that $X \in S_v$, occurs as a tuple in r_v. A *solution* to $\mathcal{J}$ is a substitution $\theta : Var \mapsto U$ for which q tuples $t_1 \in r_1, \ldots, t_q \in r_q$ exist such that $\theta = t_1 \cup \ldots \cup t_q$. Thus, a solution satisfies all the constraints in $\mathcal{J}$. A *weighted* CSP (short: WCSP) instance consists of a tuple $\langle \mathcal{J}, w_{r_1}, \ldots, w_{r_q} \rangle$, where $\mathcal{J} = \langle Var, U, \mathbf{C} \rangle$ with $\mathbf{C} = \{C_1, C_2, \ldots, C_q\}$ is a CSP instance, and where, for each tuple $t_v \in r_v$, $w_{r_v}(t_v) \in \Re$ denotes the weight associated with t_v. For a solution $\theta = t_1 \cup \ldots \cup t_q$ to $\mathcal{J}$, its associated weight is defined as $w(\theta) = \sum_{v=1}^{q} w_{r_v}(t_v)$. An *optimal solution* to $\langle \mathcal{J}, w_{r_1}, \ldots, w_{r_q} \rangle$ is a solution[2] θ to

---
[2] Note that one can dually interpret weights as costs and look at minimization problems, rather than maximization ones.

$\mathcal{J}$ such that $w(\theta) \geq w(\theta')$, for each solution θ' to $\mathcal{J}$.

From Social Environment to CSPs. Let us now define an encoding mechanism that associates any given h-bounded decision-making scenario $(\Gamma, \mathcal{A})$ with a CSP instance $CSP(\Gamma, \mathcal{A})$, in a way that "preserves" its underlying semantics. Formally, $CSP(\Gamma, \mathcal{A}) = \langle Var, U, \mathbf{C} \rangle$ is defined as follows:

- For each agent $A_i \in \mathcal{A}$ and for each variable $X \in \text{dom}(A_i)$, Var contains the variable X_i. Moreover, for each agent $A_i \in \mathcal{A}$, Var contains the variable π_i. Finally, for each variable $X \in \text{dom}(\Gamma)$, Var contains the variable X_Γ.

- The domain U consists of the Boolean values $\top$ and $\bot$, plus the natural numbers in the set $\{1, ..., n\}$. No further constant is in U.

- The set $\mathbf{C}$ consists of the following two kinds of constraints.

  (C1) For each agent A_i, $\mathbf{C}$ contains the constraint $C_i = (S_i, r_i)$ such that $S_i = \{X_i \mid X \in \text{dom}(A_i) \cup \text{dom}(\Gamma)\} \cup \{\pi_i\}$. Moreover, r_i contains all possible substitutions θ that can be defined over S_i such that: $\theta(\pi_i) \in \{1, ..., n\}$; $\theta(X_i) \in \{\top, \bot\}$, for each $X \in \text{dom}(A_i) \cup \text{dom}(\Gamma)$; and $\theta' \models \Gamma$, where $\theta'(X) = \theta(X_i)$, for each $X \in \text{dom}(\Gamma)$.

  (C2) For each pair of agents A_i and A_j such that $A_i \in \text{neigh}(A_j, \mathcal{A})$, i.e., such that they are neighbors in the scenario, $\mathbf{C}$ contains the constraint $C_{i,j} = (S_{i,j}, r_{i,j})$ where $S_{i,j} = \{X_i, X_j \mid X_i \in S_i \wedge X_j \in S_j\} \cup \{\pi_i, \pi_j\}$. In particular, $r_{i,j}$ is the set of all substitutions θ such that: $\theta(\pi_i) \in \{1, ..., n\}$; $\theta(\pi_j) \in \{1, ..., n\}$; $\theta(X_\ell) \in \{\top, \bot\}$, for each $X_\ell \in S_{i,j}$, with $\ell \in \{i, j\}$; and, for each $\{X_i, X_j\} \subseteq S_{i,j}$, we have that $\theta(\pi_i) = \theta(\pi_j)$ implies $\theta(X_i) = \theta(X_j)$.

Note that since the scenario is h-bounded, then the construction is clearly feasible in polynomial time, for any fixed natural number $h > 0$. In particular, note that the size of the largest constraint scope is $h + 2$ at most and that the number of substitutions in each relation is (trivially) bounded by $|U|^{h+2}$, hence polynomially bounded when h is fixed.

Concerning the correctness of the encoding, hereinafter, for any substitution θ and for each agent A_i, let $C_{\theta,i}$ denote the maximal legal coalition including A_i as well as all agents A_j such that $\theta(\pi_i) = \theta(\pi_j)$. Moreover, let Π_θ denote the set $\{C_{\theta,i} \mid A_i \in \mathcal{A}\}$. The crucial result is a one-to-one correspondence between solutions to the CSP instance and coalition structures.

Lemma 16. *The following properties hold:*

(1) For each solution θ to $CSP(\Gamma, \mathcal{A})$, Π_θ is a coalition structure.

(2) For each coalition structure Π, there is a solution θ to $CSP(\Gamma, \mathcal{A})$ such that $\Pi = \Pi_\theta$.

Proof. *(1)* Let θ be a solution. Note that, for each agent A_i, the coalition $C_{\theta,i}$ is in Π_θ by definition. Therefore, to prove the result, it must be shown that for each pair of distinct coalitions $C_{\theta,i}$ and $C_{\theta,j}$ in Π_θ, it holds that $C_{\theta,i} \cap C_{\theta,j} = \emptyset$. By contradiction, assume that A_k is an agent in $C_{\theta,i} \cap C_{\theta,j}$. This means that $\theta(\pi_k) = \theta(\pi_i) = \theta(\pi_j)$. Moreover, the subgraph of $\text{IG}(\Gamma, \mathcal{A})$ induced over $C_{\theta,i}$ (resp., $C_{\theta,j}$) is connected. Therefore, the subgraph induced over $C_{\theta,i} \cup C_{\theta,j}$ is connected, too. W.l.o.g., assume that $C_{\theta,j}$ is not a subset of $C_{\theta,i}$. So, $C_{\theta,i} \cup C_{\theta,j}$ is a legal coalition with $C_{\theta,i} \cup C_{\theta,j} \supset C_{\theta,i}$. Moreover, all agents A_w in the coalition are such that $\theta(\pi_w) = \theta(\pi_i)$. This contradicts the maximality of $C_{\theta,i}$.

(2) Assume that $\Pi = \{C_1, .., C_\ell\}$ is a coalition structure. Consider the substitution θ such that for each agent $A_j \in \mathcal{A}$, if C_i is the coalition where A_j occurs, then $\theta(\pi_j) = i$. Moreover, recall that Γ is satisfiable and let σ be an interpretation in $\mathcal{I}(\text{dom}(\mathcal{A}) \cup \text{dom}(\Gamma))$ such that $\sigma \models \Gamma$. Then, $\theta(X_i) = \top$ (resp., $\theta(X_i) = \bot$) if, and only if, $X \in \text{dom}(A_i) \cup \text{dom}(\Gamma)$ and $\sigma(X) = \top$ (resp., $\sigma(X) = \bot$). It is immediate to check that θ satisfies all constraints and that $\Pi = \Pi_\theta$. □

In order to continue with the analysis, one has to further specialize the above lemma. Consider a coalition structure Π_θ associated with some solution θ, and let $C_{\theta,i}$ be the coalition associated with A_i. Consider two variables X_j and X_k such that $X \in (\text{dom}(A_j) \cup \text{dom}(\Gamma)) \cap (\text{dom}(A_k) \cup \text{dom}(\Gamma))$ for two agents $A_j, A_k \in C_{\theta,i}$. Because of the constraints of kind (C2), it holds that $\theta(X_j) = \theta(X_k)$, since $\theta(\pi_j) = \theta(\pi_k) = \theta(\pi_i)$ holds. In the light of this observation, it is meaningful to associate with θ and A_i an interpretation $\sigma_{\theta,i}$ in $\mathcal{I}(\text{dom}(C_{\theta,i}) \cup \text{dom}(\Gamma))$ such that $\sigma(X) = \top$ if, and only if, there is an agent $A_j \in C_{\theta,i}$ such that $\theta(X_j) = \top$.

Lemma 17. *The following properties hold:*

(1) *For each solution θ to $CSP(\Gamma, \mathcal{A})$ and for each agent $A_i \in \mathcal{A}$, $\sigma_{\theta,i} \models \Gamma$;*

(2) *Let Π be a coalition structure and, for each coalition $C \in \Pi$, let σ_C be an interpretation in $\mathcal{I}(\text{dom}(C) \cup \text{dom}(\Gamma))$ such that $\sigma_C \models \Gamma$. Then, there is a solution θ to $CSP(\Gamma, \mathcal{A})$ such that $\{\sigma_C \mid C \in \Pi\} = \{\sigma_{\theta,i} \mid A_i \in \mathcal{A}\}$.*

Proof. Concerning *(1)*, consider a solution θ and agent $A_i \in \mathcal{A}$. We know that $\sigma_{\theta,i}$ is an interpretation in $\mathcal{I}(\text{dom}(C_{\theta,i}) \cup \text{dom}(\Gamma))$. Moreover, by the properties observed above for $\sigma_{\theta,i}$ and given the constraints of kind (C1), it is immediate to check that

$\sigma_{\theta,i} \models \Gamma$. Concerning (2), assume that $\Pi = \{C_1, .., C_\ell\}$ is a coalition structure. Consider the interpretation θ built as follows. For each agent $A_j \in \mathcal{A}$, if C_i is the coalition where A_j occurs, then $\theta(\pi_j) = i$. Consider a variable X in $\text{dom}(A_j) \cup \text{dom}(\Gamma)$, for some agent A_j. Let C_ℓ be the coalition in Π where A_j occurs and note that, even though θ is not yet completely specified, one is nonetheless guaranteed about the existence of an agent, say $A_i \in \mathcal{A}$, such that $C_\ell = C_{\theta,i}$. Then, we complete the definition of θ so that $\theta(X_j) = \sigma_{C_\ell}(X)$. The result follows since it is immediate to check that $\sigma_{\theta,i} = \sigma_{C_\ell}$. $\square$

In order to complete the analysis, let $WCSP(\Gamma, \mathcal{A})$ be the weighted CSP instance whose underlying CSP instance is $CSP(\Gamma, \mathcal{A})$ and where each constraint relation r_i is equipped with the function w_{r_i} defined as follows. For each substitution $\theta \in r_i$, $w_{r_i}(\theta) = (1-\alpha) \times A_i(\sigma_{\theta,i})$. Moreover, $w_{r_{i,j}}(\theta) = \alpha \times \frac{1}{|\texttt{neigh}(A_i, \mathcal{A})|}$ if $\theta(\pi_i) = \theta(\pi_j)$. In all the other cases, w is the constant function mapping all interpretations to 0. By definition, it is immediate that the following holds.

Lemma 18. *Let θ be a solution to $CSP(\Gamma, \mathcal{A})$. Then, $w(\theta) = \sum_i \text{UT}^\alpha(\sigma_{\theta,i}, C_{\theta,i})$.*

Putting it All Together. All ingredients are now at hand to derive the main result of the section. First, by combing all the above lemmas, one immediately gets the following—note that $WCSP(\Gamma, \mathcal{A})$ always admits a solution.

Corollary 19. *Let θ be an optimal solution to $WCSP(\Gamma, \mathcal{A})$. Then, Π_θ is a coalition structure and $w(\theta) = \sum_{C \in \Pi_\theta} \text{UT}^\alpha\text{-}\text{OPT}(\Gamma, C)$.*

This means that UT^α-CSG-EXISTENCE has been reduced to solve a WCSP instance, which can be built in polynomial time. In fact, finding a solution is **NP**-hard in general, but is known to be feasible in polynomial time over instances having bounded treewidth. In particular, the structure of an instance $\mathcal{J}$ is defined as the graph $\texttt{G}(\mathcal{J})$ whose nodes are the variables and where an edge occurs between two variables if they occur together in the scope of some constraint (see, e.g., [21]).

It is easy to see that the treewidth $\texttt{G}(CSP(\Gamma, \mathcal{A}))$ is strongly related to the treewidth of $\texttt{IG}(\Gamma, \mathcal{A})$.

Lemma 20. *Assume that $(\Gamma, \mathcal{A})$ is h-bounded and such that $tw(\texttt{IG}(\Gamma, \mathcal{A})) \leq k$. Then, $tw(\texttt{G}(CSP(\Gamma, \mathcal{A}))) \leq (2h+2) \times k$.*

Proof. Let $\langle T, \chi \rangle$ be a tree decomposition of $(\texttt{IG}(\Gamma, \mathcal{A}))$, and consider the labeling function χ' defined as follows. For each vertex p of T, if $A_i \in \chi(p)$, then $\chi'(p)$ includes all variables in S_i (hence, $h+1$ variables at most); if $\{A_i, A_j\} \in \chi(p)$, then $\chi'(p)$ includes all variables in $S_{i,j}$ (hence, $2h+2$ variables at most).

It can be checked that $\langle T, \chi' \rangle$ is a tree decomposition of $G(CSP(\Gamma, \mathcal{A}))$. Its width is bounded by $(2h+2) \times k$, where k is the width of $\langle T, \chi \rangle$. □

Therefore, the following is obtained by the tractability results of weighted CSPs over structures having bounded treewidth [21]. Note that the tractability of the corresponding problem for the egalitarian social welfare is not established here, and constitutes an interesting avenue for further research.

Corollary 21. *Let h and k be two fixed natural numbers. Then, problem UT^α-CSG-EXISTENCE can be solved in polynomial time on h-bounded scenarios $(\Gamma, \mathcal{A})$ such that $tw(\text{IG}(\Gamma, \mathcal{A})) \leq k$.*

7 Conclusion and Discussion

The paper has described and studied a general framework for decision-making, where utility functions are encoded via weighted propositional formulas. In particular, the paper has considered the application of the framework to social environments, where the utilities of the agents are affected by their social relationships, by thoroughly analyzing the complexity of issues related to the formation of coalition structures.

Similar frameworks have been already considered in the literature to analyze classical decision-making scenarios, i.e., without dealing with social environments [24], or to focus on problems related to the formation of a single coalition, i.e., without dealing with coalition structures [25]. Moreover, orthogonally to the perspective of these works, game-theoretic issues arising with logic-based agents formalized via weighted propositional logic have been studied in earlier literature, too [17].

The most natural avenue of further research is to implement the proposed framework, in order to apply it for reasoning on real social environments. Actually, when the goal is to optimize the utilitarian social welfare, it is not hard to envisage that the problem can be recast into a standard (weighted) MaxSAT problem [7]. It is, therefore, natural to consider algorithms for MaxSAT, and try to extend them to the general case of multiple weight functions under the egalitarian social welfare.

A promising approach is to use pseudo-Boolean constraints to enforce improvements in the minimum level of satisfaction [13, 9, 1, 27]. Based on them, linear, binary, and progression search algorithms can be formalized, by taking advantage of *incremental* SAT solvers [12].

References

[1] Carlos Ansótegui, Maria Luisa Bonet, Joel Gabàs, and Jordi Levy. Improving WPM2 for (weighted) partial MaxSAT. In Proc. of *CP'13*, pages 117–132, 2013.

[2] N. Bansal and M. Sviridenko. The Santa Claus problem. In Proc. of *STOC'06*, pages 31–40, 2006.

[3] Ivona Bezáková and Varsha Dani. Allocating indivisible goods. *SIGecom Exchanges*, 5(3):11–18, 2005.

[4] Craig Boutilier and Holger H. Hoos. Bidding languages for combinatorial auctions. In Proc. of *IJCAI'01*, pages 1211–1217, 2001.

[5] Sylvain Bouveret and Jérôme Lang. Efficiency and envy-freeness in fair division of indivisible goods: Logical representation and complexity. *Journal of Artificial Intelligence Research*, 32:525–564, 2008.

[6] Vincent Conitzer, Felix Brandt, and Ulle Endriss. *Multiagent Systems*, chapter Computational Social Choices. MIT Press, 2012.

[7] Byungki Cha, Kazuo Iwama, Yahiko Kambayashi, and Shuichi Miyazaki. Local search algorithms for partial MAXSAT. In Proc. of *AAAI'97*, pages 263–268, 1997.

[8] Georgios Chalkiadakis, Gianluigi Greco, and Evangelos Markakis. Characteristic function games with restricted agent interactions: Core-stability and coalition structures. *Artificial Intelligence*, 232:76–113, 2016.

[9] Alessandro Cimatti, Anders Franzén, Alberto Griggio, Roberto Sebastiani, and Cristian Stenico. Satisfiability modulo the theory of costs: Foundations and applications. In Proc. of *TACAS'10*, pages 99–113, 2010.

[10] Sylvie Coste-Marquis, Jérôme Lang, Paolo Liberatore, and Pierre Marquis. Expressive power and succinctness of propositional languages for preference representation. In Proc. of *KR'04*, pages 203–212, 2004.

[11] Rina Dechter. *Constraint Processing*. Morgan Kaufmann Publishers Inc., San Francisco, CA, USA, 2003.

[12] Niklas Eén and Niklas Sörensson. An extensible sat-solver. In Proc. of *SAT'03*, pages 502–518, 2003.

[13] Niklas Eén and Niklas Sörensson. Translating pseudo-boolean constraints into SAT. *Journal on Satisfiability, Boolean Modeling and Computation*, 2(1-4):1–26, 2006.

[14] Edith Elkind, Leslie Goldberg, Paul Goldberg, and Michael Wooldridge. A tractable and expressive class of marginal contribution nets and its applications. *Mathematical Logic Quarterly*, 55(4):362–376, 2009.

[15] Edith Elkind and Michael Wooldridge. Hedonic coalition nets. In Proc. of *AAMAS'09*, pages 417–424, 2009.

[16] Edith Elkind, Talal Rahwan, and Nicholas R. Jennings. Computational coalition formation. *Multiagent Systems (2nd Edition), MIT Press*, pages 329–380, 2013.

[17] Erman Acar, Gianluigi Greco, and Marco Manna. Group Decision Making in Social Environments. In Proc. of *AAMAS'17*, 2017.

[18] Bruno Escoffier, Laurent Gourvès, and Jérôme Monnot. Fair solutions for some multi-agent optimization problems. *Autonomous Agents and Multi-Agent Systems*, 26(2):184–201, 2013.

[19] Christophe Gonzales, Patrice Perny, and Sergio Queiroz. Preference aggregation with

graphical utility models. In Proc. of *AAAI'08*, pages 1037–1042, 2008.

[20] Georg Gottlob and Gianluigi Greco. Decomposing combinatorial auctions and set packing problems. *Journal of the ACM*, 60(4), 2013.

[21] Georg Gottlob, Gianluigi Greco, and Francesco Scarcello. Tractable optimization problems through hypergraph-based structural restrictions. In *Proc. of ICALP'09*, pages 16–30, 2009.

[22] Georg Gottlob, Gianluigi Greco, and Francesco Scarcello. Treewidth and Hypertree Width. In Lucas Bordeaux, Youssef Hamadi, and Pushmeet Kohli, editors, *Tractability: Practical Approaches to Hard Problems*, 2012.

[23] Georg Gottlob, Nicola Leone, and Francesco Scarcello. A comparison of structural CSP decomposition methods. *Artificial Intelligence*, 124(2):243–282, 2000.

[24] Gianluigi Greco and Jérôme Lang. Group decision making via weighted propositional logic: Complexity and islands of tractability. In *Proc. of IJCAI'15*, pages 3008–3014, 2015.

[25] Gianluigi Greco and Antonella Guzzo. Coalition Formation with Logic-Based Agents. In *Proc. of AT'16*, 2016.

[26] Samuel Ieong and Yoav Shoham. Marginal contribution nets: A compact representation scheme for coalitional games. In *Proc. of PEC'05*, pages 193–202, 2005.

[27] Alexey Ignatiev, António Morgado, Vasco M. Manquinho, Inês Lynce, and João Marques-Silva. Progression in maximum satisfiability. In *Proc. of ECAI'14*, pages 453–458, 2014.

[28] Celine Lafage and Jérôme Lang. Logical representation of preferences for group decision making. In *Proc. of KR'00*, pages 457–468, 2000.

[29] Jérôme Lang and Lirong Xia. Voting in combinatorial domains. *Handbook of Computational Social Choice*, pages 1193–1195, 2014.

[30] Noam Nisan. Bidding languages for combinatorial auctions. In Peter Cramton, Yoav Shoham, and Richard Steinberg, editors, *Combinatorial auctions*, chapter 9. MIT Press, 2006.

[31] Christos H. Papadimitriou. *Computational Complexity*. Addison Wesley, Reading, MA, USA, 1994.

[32] Neil Robertson and P.D. Seymour. Graph minors. II. Algorithmic aspects of tree-width. *Journal of Algorithms*, 7(3):309–322, 1986.

[33] Tuomas Sandholm, Kate Larson, Martin Andersson, Onn Shehory, and Fernando Tohmé. Coalition structure generation with worst case guarantees. *Artificial Intelligence*, 111(1-2):209–238, 1999.

[34] Joel Uckelman, Yann Chevaleyre, Ulle Endriss, and Jérôme Lang. Representing utility functions via weighted goals. *Mathematical Logic Quarterly*, 55(4):341–361, 2009.

[35] Joel Uckelman and Ulle Endriss. Compactly representing utility functions using weighted goals and the max aggregator. *Artificial Intelligence*, 174(15):1222–1246, 2010.

www.ingramcontent.com/pod-product-compliance
Lightning Source LLC
Chambersburg PA
CBHW081124170426
43197CB00017B/2740